Ultrastructure Techniques for Microorganisms

Ultrastructure Techniques for Microorganisms

Edited by

Henry C. Aldrich

Department of Microbiology and Cell Science
University of Florida
Gainesville, Florida

and

William J. Todd

Department of Veterinary Science
Louisiana Agricultural Experiment Station
Louisiana State University Agricultural Center
and Department of Veterinary Microbiology and Parasitology
School of Veterinary Medicine
Louisiana State University
Baton Rouge, Louisiana

Plenum Press • New York and London

Library of Congress Cataloging in Publication Data

Ultrastructure techniques for microorganisms.

Includes bibliographies and index.
1. Electron microscope — Technique. 2. Micro-organisms — Morphology. 3. Ultra-structure (Biology) 4. Microscope and microscopy — Technique. I. Aldrich, Henry C. II. Todd, William J.
QR68.5.E45U48   1986   576'.028   86-12191
ISBN 0-306-42251-4

Front cover: Freeze-fracture electron micrograph of the plasma membrane of the bacterium *Escherichia coli,* showing lipid phase separation due to cold temperature, ×45,000. Photo by Dr. Gregory W. Erdos, Department of Microbiology and Cell Science, University of Florida, Gainesville, Florida.

Back cover: Scanning electron micrograph of the spores of the aquatic fungus *Clavatospora stellaculata,* ×10,400. Photo by Dr. Garry T. Cole, Department of Botany, University of Texas, Austin, Texas.

© 1986 Plenum Press, New York
A Division of Plenum Publishing Corporation
233 Spring Street, New York, N.Y. 10013

Printed in the United States of America

Contributors

H. C. Aldrich Department of Microbiology and Cell Science, University of Florida, Gainesville, Florida 32611

Barbara L. Armbruster Division of Infectious Diseases, Washington University Medical School, St. Louis, Missouri 63110. *Present address*: Monsanto Company, St. Louis, Missouri 63167

David L. Balkwill Department of Biological Science, Florida State University, Tallahassee, Florida 32306

Alan Beckett Department of Botany, University of Bristol, Bristol BS8 1UG, England

Edwin S. Boatman Department of Environmental Health, School of Public Health and Community Medicine, University of Washington, Seattle, Washington 98195

Russell L. Chapman Department of Botany, Louisiana State University, Baton Rouge, Louisiana 70803

Garry T. Cole Department of Botany, University of Texas, Austin, Texas 78713

Sylvia E. Coleman Research Service, Veterans Administration Medical Center, Gainesville, Florida 32602

Gregory W. Erdos Department of Microbiology and Cell Science, University of Florida, Gainesville, Florida 32611

Claude F. Garon Laboratory of Pathobiology, Rocky Mountain Laboratories, National Institute of Allergy and Infectious Diseases, Hamilton, Montana 59840

Thomas H. Giddings, Jr., Department of Molecular, Cellular, and Developmental Biology, University of Colorado, Boulder, Colorado 80309

Harvey C. Hoch Plant Pathology Department, New York State Agricultural Experiment Station, Cornell University, Geneva, New York 14456

Edward Kellenberger Abteilung Mikrobiologie, Biozentrum, Universität Basel, CH-4056 Basel, Switzerland

H. H. Mollenhauer Veterinary Toxicology and Entomology Research Laboratory, United States Department of Agriculture; and Department of Pathology and Laboratory Medicine, College of Medicine, Texas A&M University, College Station, Texas 77843

Sandra A. Nierzwicki-Bauer Department of Biology, Rensselaer Polytechnic Institute, Troy, New York 12181

Martha J. Powell Department of Botany, Miami University, Oxford, Ohio 45056

Nick D. Read Department of Botany, University of Bristol, Bristol BS8 1UG, England. *Present address:* Department of Botany, University of Edinburgh, Edinburgh EH9 3JH, Scotland

John Smit Naval Biosciences Laboratory, School of Public Health, University of California, Berkeley, California 94720

L. Andrew Staehelin Department of Molecular, Cellular, and Developmental Biology, University of Colorado, Boulder, Colorado 80309

Murray Stewart Medical Research Council, Laboratory of Molecular Biology, Cambridge CB2 2QH, England

William J. Todd Department of Veterinary Science, Louisiana Agricultural Experiment Station, Louisiana State University Agricultural Center; and Department of Veterinary Microbiology and Parasitology, School of Veterinary Medicine, Louisiana State University, Baton Rouge, Louisiana 70803

George P. Wray Department of Molecular, Cellular, and Developmental Biology, University of Colorado, Boulder, Colorado 80309

Preface

The modern microbiologist is often a real specialist who has difficulty understanding and applying many of the techniques beyond those in his or her own immediate field. On the other hand, most benefits to modern microbiology are obtained when a broad spectrum of scientific approaches can be focused on a problem. In early studies, electron microscopy was pivotal in understanding bacterial and viral morphology, and we still feel that we will understand a disease better if we have seen an electron micrograph of the causative agent. Today, because there is an increased awareness of the need to understand the relationships between microbial structure and function, the electron microscope is still one of the most important tools microbiologists can use for detailed analysis of microorganisms.

Often, however, the aforementioned modern microbiologist still thinks of ultrastructure as involving negative staining or ultrathin sectioning in order to get a look at the shape of a "bug." Many of the newer ultrastructure techniques, such as gold-labeled antibody localization, freeze-fracture, X-ray microanalysis, enzyme localization, and even scanning electron microscopy, are poorly understood by, and therefore forbidding to, the average microbiologist. Even many cell biologists admit to having difficulty staying in touch with current developments in the fast-moving field of electron microscopy techniques.

In the hope of collecting information on the most important and useful current ultrastructure techniques, we asked authors, who we consider experts in their fields, to write a series of reviews. Each author was asked to stress applications of ultrastructure techniques to microbes, wherever literature or unpublished data on the subject could be located. If microbial data were unavailable, then applications to other organisms were presented with suggestions as to how they might be profitably used on microorganisms. Techniques were evaluated and recipes given where appropriate. We have endeavored to process and publish the manuscripts as promptly as possible after completion, so that the current information is presented. We have avoided historical review unless it is relevant to the current techniques being evaluated.

The resulting volume is intended as a resource for the busy professional microbiologist. In practice, in our own laboratories, we have found the manuscripts of great interest to cell biologists working with many kinds of cells. Hence, we hope that this volume will find an audience of cell biologists and electron microscopists beyond the microbiological community for whom it was originally intended.

We wish to acknowledge with thanks the help and advice freely given by our colleagues at home and around the world during the preparation of this volume. The words of encouragement and support from these colleagues, as well as the universal cooperation of the authors of the individual chapters, made our tasks as editors bearable and even enjoyable!

Henry C. Aldrich
William J. Todd

Gainesville, Florida, and Baton Rouge, Louisiana

Contents

Chapter 4
Secrets of Successful Embedding, Sectioning, and Imaging
H. C. Aldrich and H. H. Mollenhauer

Chapter 5
Computer-Aided Reconstruction of Serial Sections
David L. Balkwill

Chapter 6
Electron Microscopy of Nucleic Acids
Claude F. Garon

Chapter 7
Freeze-Substitution of Fungi
Harvey C. Hoch

Chapter 8
Freeze-Fracture (-Etch) Electron Microscopy
Russell L. Chapman and L. Andrew Staehelin

Chapter 13
Digitizing and Quantitation

Edwin S. Boatman

Chapter 14
Localization of Carbohydrate-Containing Molecules

Gregory W. Erdos

Chapter 15
**Cytochemical Techniques for the Subcellular Localization
of Enzymes in Microorganisms**

Martha J. Powell

Chapter 1

Preparation of Microfungi for Scanning Electron Microscopy

Garry T. Cole

Department of Botany
University of Texas
Austin, Texas 78713

1. INTRODUCTION

The scanning electron microscope (SEM) has broad applications to fungal research but has been particularly valuable for taxonomic and developmental studies of microfungi (Moore and Grand, 1970; Cole and Behnke, 1975; Cole, 1976, 1979, 1981a; O'Donnell, 1979). As in other biological disciplines, SEM images at the outset generated widespread aesthetic appeal, as well as skepticism among investigators "who saw little application in a device the main function of which was to produce snapshots of surfaces" (Heslop-Harrison, 1979). During the early 1970s, however, such skepticism was largely put to rest with the development of high-resolution SEMs and scanning–transmission electron microscopes (STEMs) (Crewe, 1971, 1979) capable of providing structural data that previously were obtained only from thin sections and surface replicas of samples examined in the transmission electron microscope. Recognition that interaction between the incident electron beam and biological substrate yields many different signals (e.g., X-rays, backscattered electrons, cathodoluminescence) led to the development of the analytical STEM (Venables, 1976; Reimer, 1976) and a plethora of new information was made available to the researcher. The technological boom in ultrastructural instrumentation witnessed in the 1970s, however, confronted many workers with an "embarrassment of hardware" in that the "general corpus of knowledge of the time was inadequate for their full exploitation in application . . ." to investigations of the biological material of interest (Heslop-Harrison, 1979). To a large extent, this situation reflects the current status of mycological research. Applications of the analytical STEM to studies of fungi are relatively rare and by far the majority of mycological investigators employ the SEM for conventional, comparative morphology. The rapid technological advance also outpaced the development of improved methods of specimen handling. This is not surprising when one considers the complexity of structural features of biological samples, even within the fungal kingdom. Unlike the broad spectrum of applications characteristic of many newly developed SEM accessories, a particular method of specimen preparation is not equally suited even for members of the same group of organisms because of structural differences as well as the variety of questions posed by investigators. The approach used in this chapter, therefore, is an examination of both the kinds of information sought from the SEM and the preparatory techniques employed. A contention that I have maintained in my investigations of the development and ultrastructure of microfungi is that scanning electron microscopy is best used in concert with light microscopic and other electron microscopic techniques. Correlation of structural data provides for a more comprehensive study as well as a reasonable check on potential artifacts that may arise from a particular preparatory procedure. Limitations in the application of the preparatory techniques described in this chapter are also discussed. Each investigator must be cognizant of artifacts generated by sample preparation procedures, since none are artifact-free, and weigh these against the nature of the information required from the SEM.

2. MICROFUNGI IN PURE CULTURE

2.1. Agar Plate Cultures

Many species of microfungi sporulate well on the surface of solid, synthetic media contained in petri plates. A simple procedure that has been used for preparing these samples for the SEM requires the availability of only the most basic facilities in an electron microscope laboratory. Plate cultures were incubated until good sporulation was established and then quickly flooded with a cold (4°C) solution prepared by mixing equal volumes of 6% glutaraldehyde and 1% osmium tetroxide (OsO_4), each in 0.1 M cacodylate buffer (pH 7.1–7.4). The chemical fixative was carefully poured over the mycelium so as to minimize disruption of the structure and arrangement of the delicate fruiting bodies. For certain fungal preparations, one drop of a detergent (e.g., 0.5–1.0% Teepol; Cole and Aldrich, 1971) was added to each 10 ml of fixative to enhance wetting of hydrophobic cells. With the aid of a dissecting microscope placed inside a chemical hood, small blocks of agar (approx. 5 cm²) that had supported growth of sporulating mycelia were excised from different regions of the flooded plate culture. The blocks were cut below the level of fixative solution using an acetone-cleaned razor blade. Excess agar was removed, leaving only a 2- to 4-mm-thick layer with the fertile mycelia on its surface. The layer of agar functions as a support for later handling and mounting of the dried samples prior to their examination in the SEM. The excised and trimmed blocks were immediately transferred to a vial that contained fresh, cold fixative. Alternatively, very delicate specimens were transferred to shallow, Teflon cups (20 × 10 mm) filled with fresh fixative and placed on ice. Subsequent removal and addition of solutions was performed with care to minimize structural damage. The samples were left in the dark for 2 hr (4°C), washed in cold buffer (10×), and then placed in 1% OsO_4 buffered as above for an additional 2 hr at 4°C. The specimens were subsequently washed in buffer and then dehydrated in a graded ethanol series (15 min in each solution), followed by washes in a dilution series of amyl acetate in absolute ethanol, and finally in absolute amyl acetate (analytical reagent grade; Mallinckrodt, St. Louis, MO). The blocks were then carefully transferred to modified BEEM capsules (Nemanic, 1972; Cole, 1981b; Figure 1) that were three-fourths submerged in absolute amyl acetate contained in the base of a 50-mm-diameter clean, glass petri dish. Up to five capsules, each containing three to five sample blocks, were quickly transferred to the wire mesh basket contained within the pressure bomb of a Denton DCP-1 critical point-drying (CPD) apparatus (Denton Vacuum Inc., Cherry Hill, NJ). The bomb, which had been precooled in an ice bath, was sealed and then slowly pressurized by opening valves V_1 and V_2 (Figure 1). The liquid CO_2 in the cylinder was transported via a siphon and copper line to the pressure bomb. The fully charged cylinder was pressurized at approximately 900 psi. The level of liquid CO_2 in the cylinder was monitored on the basis of change in weight. Precooling the bomb increased the rate at which liquid CO_2 accumulated and mixed with the specimens. The amyl

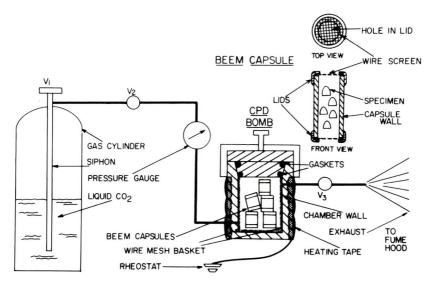

FIGURE 1. Critical point-drying apparatus and modified BEEM capsules. After Cole (1981b).

acetate was gradually replaced with liquid CO_2 during a series of flushes (Cole, 1981b). Valve V_3, which had been closed as the bomb was pressurized, was then opened slightly, allowing the liquid CO_2 and displaced amyl acetate to be exhausted into the fume hood. For particularly fragile samples, a low flow rate was maintained to reduce agitation of the specimens inside the bomb. This flushing process was continued for 3 min and then V_3 and V_2 were closed, in that order. The samples remained suspended in liquid CO_2 inside the isolated and pressurized bomb for 2 min. The level of liquid CO_2 was then lowered by slowly opening V_3 and reducing the line pressure to approximately 200 psi. The pressure was subsequently equalized to that of the cylinder by closing V_3 and slowing opening V_2. This same flushing process was repeated three to five times, or until the odor of amyl acetate could no longer be detected on a piece of filter paper that was held over the exhaust port for several seconds. Although this process is time-consuming and requires the operator's constant attention, the principal advantage is virtually total displacement of dehydrant with very little detectable damage to the specimen. Amyl acetate was chosen for this procedure rather than other dehydrants, such as acetone or ethanol, because of its relatively low volatility and distinctive odor. The former characteristic reduced the chance of accidental evaporative drying as the specimen was transferred to the CPD bomb (Hayat, 1978). The latter permitted detection of very low emission levels at the exhaust port using absorbant filter paper. Because of the toxicity of amyl acetate, vapors should be vented into a fume hood. With the bomb fully pressurized and both V_3 and V_2 closed, the wall of the chamber was heated to approximately 60°C (surface temperature) using insulated, flexible heating tape (Briscoe Manufacturing Co., Columbus, Ohio) attached to a rheostat (Figure 1). Since the internal temperature of the bomb of this apparatus cannot be monitored, change in chamber pressure was used as an indication of when the critical temperature of CO_2 was reached. The pressure was permitted to rise to 1500 psi, at which point the critical temperature was surpassed (depends on volume of apparatus occupied by CO_2) but not enough to cause heat damage to the specimen. The rheostat was turned off and V_3 was opened slightly to allow the chamber to gradually return to atmospheric pressure. If the chamber was vented too quickly, the interior of the bomb was cooled and condensation water accumulated on its surface and contents when exposed to atmospheric conditions. The critical-point-dried specimens were immediately transferred to a storage desiccator, or prepared for conductive metal coating. Murphy (1982) has presented an exhaustive review of the materials available and procedures used for specimen mounting prior to SEM examination. Conductive silver paint (Fullam, Schenectady, NY; PELCO, Tustin, Calif.) has been used as the standard adhesive in my laboratory for mounting dried specimens on clean, pin-type aluminum stubs. No recognized beam damage to this uncoated or metal-coated adhesive has been observed in the SEM (Muir and Rampley, 1969), although the specimen charging that occurs occasionally may be attributed to the mounting compound (Murphy, 1982).

Aluminum paint has also been used successfully as an adhesive for agar culture preparations. Much more care must be taken in choosing a mounting compound if X-ray analysis is to be performed, a technique that is discussed in a subsequent chapter of this volume. The mounted specimens are finally placed on the omnirotary table of a Denton DV-502 high-vacuum evaporator equipped with a DC diode sputter device. The latter deposits thin films of conductive metal [e.g., gold–palladium (60 : 40)] onto the sample surface in an argon atmosphere (Hayat, 1978). The thickness of the metal coat is reproducible by monitoring the distance between the surface of the specimen and the Au/Pd source, vacuum, DC current, and coating time (Cole, 1976). However, thickness monitors, now available at reasonable cost, permit more accurate control of the amount of metal deposited. The advantage of this high-vacuum apparatus over conventional "low-vacuum" sputter modules is that preliminary evacuation of the sample to 10^{-6} Torr ensures that it has outgassed before being transferred to the SEM, thus reducing the possibility of instrument contamination. The sputter module also permits deposition of a conductive metal coating with no appreciable substrate heating and specimen damage. Use of the omnirotary stage improves uniform distribution of the conductive metal coat over a highly irregular sample surface. The fixed, dried, mounted, and coated fungal specimens are now ready for examination in the SEM.

The micrographs of the coelomycetous fungus, *Sporonema* sp., in Figures 2–5 illustrate the range of developmental stages that may be observed with the SEM using this simple preparatory procedure. The agar surface was inoculated with a suspension of conidia, the asexual, deciduous propagules produced within hemispherical fruiting bodies (conidiomata) of this fungus. Formation of conidiomata in this case was not synchronized (Figure 2). Upon germination, conidia transform into budding yeast cells that adhere to the agar surface and can easily be prepared for the SEM (Figure 3). In Figure 4, small, raised colonies of yeast are seen to have formed and cells on the upper surface of the mound have given rise to short, filamentous outgrowths, or hyphae (arrows in Figure 4). During subsequent development, the hyphae appear to grow down over the colony toward the agar surface (Figure 5), become bound together with a mucilaginous deposit, and eventually form the outer peridium of the conidiomata (C, Figure 2). This unusual process of conidiomatal ontogeny was first revealed by the SEM (Cole, 1981c).

The degree of preservation of cellular form, structure, and arrangement using this same preparatory technique is illustrated in Figures 6–9. The fine, brittle spines of the sporangiospores of *Cunninghamella echinulata* are particularly susceptible to damage during preparation for SEM examination. On the other hand, very few spines in Figure 6 are fractured due to the careful manipulation of the specimen during fixation, dehydration, and CPD. The fertile vesicle of *Mycotypha poitrasii* is thin-walled and approximately 30 μm in diameter (Cole and Samson, 1979). The preparatory procedure has retained the spheroidal

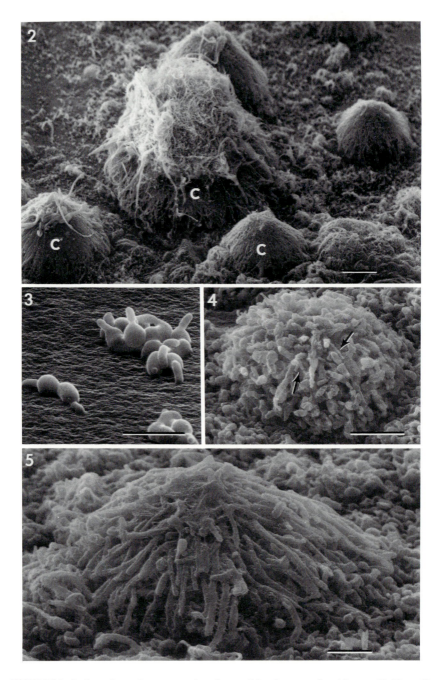

FIGURES 2–5. Agar plate culture preparation. Stages of development of conidiomata (C, Figure 2) of *Sporonema* sp. showing clusters of yeast (Figure 3), conidiomatal initial (Figure 4), and young conidioma with ''umbrella'' arrangement of hyphae on its surface (Figure 5). Arrows in Figure 4 locate filamentous outgrowths of yeast cells. Bars = 100, 10, 20, and 20 μm, respectively.

FIGURES 6, 7. Agar plate culture preparations. Echinulate sporangiospores (monosporous sporangiola) of *Cunninghamella echinulata* (Figure 6) and fertile vesicle of *Mycotypha poitrasii* fixed at an early stage of synchronized development of its sporangiola (Figure 7). Bars = 10 μm.

shape of this large, fragile sporangiophore (Figure 7). The short evaginations of the vesicle wall are sporangiolar initials that were fixed at a stage of synchronized development. The dichotomously branched sporangiophores of *Piptocephalis lepidula* revealed by the SEM in Figure 8 are seen in the same arrangement as that visible by light microscopy on the surface of plate cultures. The bulbous shape and natural tetraradiate arrangement of appendages of the aquatic conidia of *Orbimyces spectabilis* are also well preserved in Figure 9. This technique of preparation of agar plate cultures for SEM examination preserves gross morphology, shape, and arrangement of cells and has thereby contributed to interpretations of development of microfungi sporulating in pure culture.

2.2. Coverslip and Filter Disk Cultures

Glass slides and coverslips coated with growth media have long been used for culturing microorganisms and subsequent examination in the SEM (Biberfeld and Biberfeld, 1970; Galun, 1971). In most cases, however, the cellular preservation of such specimens has been relatively poor. The simple procedure described below yields good-quality preparations of microfungi.

A thin, circular (12-mm-diameter) glass coverslip was held with a pair of forceps while dipped into molten, nutrient agar. Both coverslip and forceps had previously been sterilized. The lower surface of the coverslip was quickly cleaned free of agar with filter paper and then placed on a glass slide and enclosed in a petri dish. The slide was supported above the surface of the petri dish on glass rods and the bottom of the dish was covered with two pieces of filter paper (Figure 10A). The latter were soaked with 10% glycerol to maintain a high relative humidity within the chamber. The surface of the agar-coated coverslip was inoculated with a yeast, mycelial, or conidial suspension in either liquid growth media or sterile distilled water. The suspension was obtained from well-sporulating stock cultures and applied to the surface of the coverslip in very low concentration. Care was taken to avoid penetration of the thin agar film during inoculation. The closed petri dish was then incubated for 12–72 hr. Because of the problem of rapid desiccation of the agar film, microfungi examined were restricted to those that sporulated within 2–3 days under these growth conditions. Fungal growth could be monitored at intervals by examining the coverslip with a stereomicroscope without removing the lid of the petri dish. For examination under the compound light microscope, a thin, flat Teflon ring was placed on the glass slide so that it surrounded the coverslip. The thickness of the ring was slightly greater than that of the coverslip culture. Both upper and lower surfaces of the ring were coated with a thin film of Vaseline. A second, larger coverslip (15 × 15 mm) was carefully placed over the coated coverslip and ring, making a thin, sealed sandwich (Figure 10A). The cells on the coverslip culture were undisturbed and their growth could be recorded by time-lapse photomicrography (Cole, 1975). In preparation for the SEM, the coated coverslip was

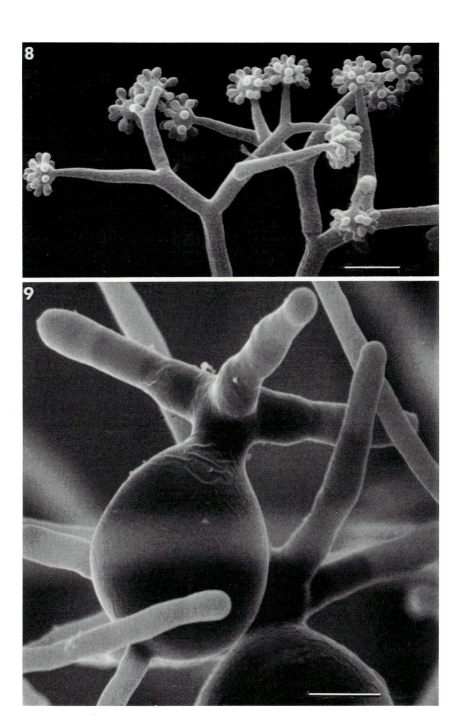

FIGURES 8, 9. Agar plate culture preparations. Dichotomously branched sporangiophores of *Piptocephalis lepidula* terminated at their apices by clusters of merosporangial initials (Figure 8), and conidium of *Orbimyces spectabilis* with tetraradiate appendages (Figure 9). Bars = 5, and 10 μm, respectively.

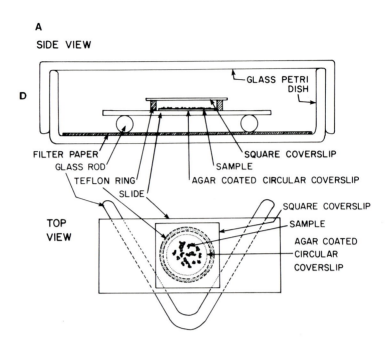

A

SIDE VIEW

GLASS PETRI DISH

D

FILTER PAPER
GLASS ROD
TEFLON RING
SLIDE

SQUARE COVERSLIP
SAMPLE
AGAR COATED CIRCULAR COVERSLIP

TOP VIEW

SQUARE COVERSLIP
SAMPLE
AGAR COATED CIRCULAR COVERSLIP

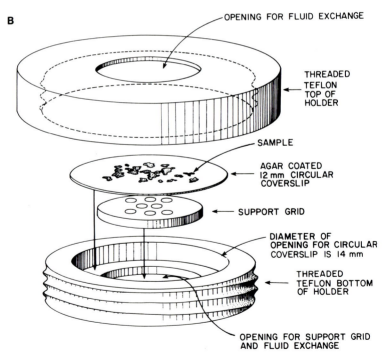

B

OPENING FOR FLUID EXCHANGE

THREADED TEFLON TOP OF HOLDER

SAMPLE

AGAR COATED 12 mm CIRCULAR COVERSLIP

SUPPORT GRID

DIAMETER OF OPENING FOR CIRCULAR COVERSLIP IS 14 mm

THREADED TEFLON BOTTOM OF HOLDER

OPENING FOR SUPPORT GRID AND FLUID EXCHANGE

FIGURE 10. (A) Apparatus used for inoculation and incubation of microfungi grown on agar-coated coverslips. (B) Apparatus used for SEM preparation of coverslip cultures.

exposed and flooded with a few drops of cold fixative solution. The same fixation procedure was used as outlined above (Section 2.1). The coverslip was placed at the bottom of a shallow Teflon cup (20-mm diameter) and solutions were exchanged carefully by aspiration. Once in absolute amyl acetate, the coverslip was transferred to the specially designed, Teflon holder illustrated in Figure 10B (Denton Vacuum Inc.). This step was performed while the coverslip and holder were maintained below the level of the amyl acetate. The entire apparatus was then placed in the wire mesh basket of the CPD device (Figure 1). The dried coverslip culture was mounted on an aluminum stub using conductive silver paint, coated, and examined in the SEM.

Preparations of coverslip cultures typically reveal homogeneous surfaces in the SEM (Figures 11 and 12). Precautions should be taken to ensure that solutions used for fixation and dehydration are clean (i.e., filtered if necessary), and that the liquid CO_2 used during CPD is oil-free. *Trichosporon beigelii,* shown in Figure 11, is an imperfect fungus that can give rise to conidia by fragmentation and disarticulation of its filamentous hyphae, or by formation of acropetal chains of budding cells. The latter arise as proliferations of swollen cells formed by hyphal fragmentation. Conidia formed by these two developmental processes are demonstrated by the SEM in Figure 11. *Candida albicans* is considered a dimorphic fungus because of its ability to alternate between mycelial and yeast phases under different growth conditions (Rippon, 1982). When growth on coverslips coated with cornmeal–Tween 80 agar (Beneke and Rogers, 1970) and incubated at 37°C for 2 hr, yeast, hyphae, and pseudohyphae are produced (Figure 12; Cole and Nozawa, 1981). Hyphae (H) arise as cylindrical outgrowths of parental yeast cells (PYC) while pseudohyphae (PH) arise as elongated, ellipsoidal proliferations of parental cells. Pseudohyphae may give rise to new daughter yeast cells by apical or lateral budding. The location of the hyphal septum (S) is evident in Figure 12. Thin sections have shown that the constriction at the junction of the pseudohypha and parental yeast cell (arrow) locates the septum between these two cell types. SEM examination of such coverslip culture preparations further reinforces the concept that greater morphogenetic variation exists in the cell cycle of *C. albicans* than simple yeast–hypha conversion (Cole and Nozawa, 1981).

A recently developed technique that may have wide application for microbiological preparations was used for examining gemma and gemmifer formation in *Mycena citricolor,* a plant pathogen causing leafspot of coffee. Gemmae are the asexual reproductive bodies and infectious agents of *M. citricolor* that are produced at the apex of stipitate fruiting structures called gemmifers (Figures 13–16; Buller, 1934; Wellman, 1961). Gemmae were used to inoculate the surface of Durapore filters (47-mm diameter, 0.2-μm pore size; Millipore Corp., Bedford, MA) that covered the surface of a defined glucose–salts–agar medium contained in 50-mm-diameter petri plates (Lanier, 1984). The filters were not digested by the fungus but provided adequate diffusion of nutrients to support

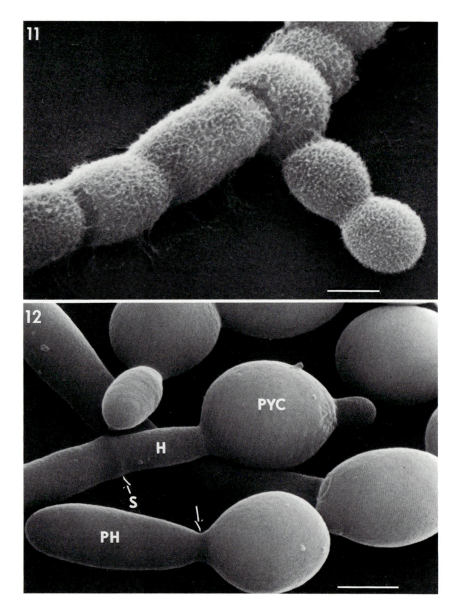

FIGURES 11, 12. Coverslip culture preparations. Figure 11 reveals two chains of conidia of *Trichosporon beigelii* that had formed by different mechanisms of development. Different cell types of *Candida albicans* are shown in Figure 12. H, hypha; PH, pseudohypha; PYC, parental yeast cell; S, septum. Arrow locates constriction (and septum) between parental yeast cell and pseudohypha. Bars = 2 μm.

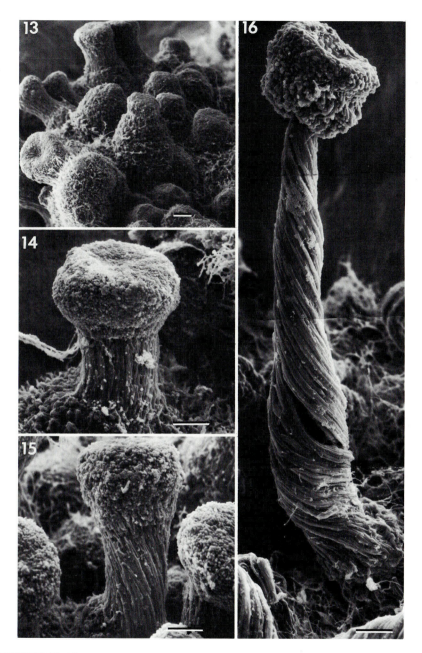

FIGURES 13–16. Filter disk culture preparation. Stages of gemmifer development in *Mycena citricolor* are revealed. Gemma inoculated directly on Durapore filters overlaying nutrient agar germinate (Figure 13) to form new gemmifers consisting of an apical gemma and tapered stipe. The latter is composed of a hollow cylinder of elongating hyphae that assume a helical arrangement (Figures 14–16). Figure 16 is a montage composed of four micrographs. Bars = 100, 50, 50, and 100 μm, respectively.

growth. The gemmae swelled and developed a mycelial mat by 48 hr postinoculation (Figure 13), and by 96 hr several new gemmifers had arisen from each gemma (Figures 14–16). Samples were prepared for the SEM by flooding the filter disk with fixative solution and then transferring individual microcolonies with a wire loop to vials containing fresh fixative. Since the filter was not penetrated by hyphae, the colonies could be easily dislodged and remained intact during subsequent processing. Orientation of the dried samples before mounting on the specimen stub was not a problem. All stages of development were easily recognized during SEM examination of Durapore filter cultures. Newly formed gemmae are attached to the apex of elongating stipes composed of parallel hyphal elements (Figure 14). The latter begin to twist soon after intercalary growth is initiated (Figure 15). The mature gemma is supported by a tapered, hollow stipe 1–4 mm long composed of a cylinder of hyphae that have assumed a helical arrangement (Figure 16). This simple culturing and preparatory procedure for SEM examination is applicable to other mycelial fungi.

2.3. Cell Suspensions

Preparation of suspensions of microorganisms for SEM examination poses certain difficulties. The cells must be concentrated for fixation and dehydration but handled in a manner that does not damage their delicate surface structures. Centrifugation of fungal spore suspensions usually results in fracture or distortion of cell appendages and should, therefore, be avoided. The chemically fixed and dried cells must be uniformly dispersed at low concentration on a suitable substrate for ultrastructural examination in the SEM rather than concentrated in a pellet. Several procedures have been described for processing suspensions of minute biological specimens for the SEM (Marchant, 1973; Ruffolo, 1974; Postek et al., 1974; Newell and Roth, 1975; Rostgaard and Christensen, 1975; Hayat, 1978). The following is a previously described procedure used for preparation of bacteria (Watson et al., 1980) and conidial fungi (Cole, 1981b). The apparatus employed (Figure 17) consists of a 10-ml interchangeable glass hypodermic syringe, a Millipore Swinny disk filter holder, and a Nucleopore filter (Nucleopore Corp., Pleasanton, Calif.). The Nucleopore filter was preferred over other filters tested because of its resistance to structural alterations during chemical processing, and the uniformity of its pore size (Kurtzman et al., 1974). A particular pore diameter was chosen (0.4–4.0 μm) on the basis of the sample under investigation. Approximately 1–2 ml of the solution containing the suspended conidia was drawn into the syringe with the filter holder detached. The syringe plunger was then inserted, and the Swinny holder attached and the suspension slowly and gently expelled. The sample was deposited on the filter but kept moist by not forcing all the liquid out of the Swinny apparatus. The

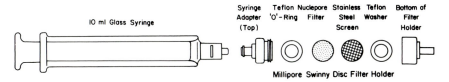

FIGURE 17. Apparatus used for preparing fungal cell suspensions for the SEM. After Watson *et al.* (1980) and Cole (1981b).

latter was removed from the syringe and quickly attached to a new syringe containing 3 ml of fixative solution (6% glutaraldehyde + 1% OsO_4 mixed 1 : 1). The sample was fixed by passing 1.5 ml of this solution through the filter and then allowing the syringe to sit at 4°C for 2 hr in the dark. After expulsion of the remaining fixative, the filter holder was transferred to a new syringe containing 10 ml of buffer (4°C) that was then used to wash the sample. This same procedure was repeated for subsequent treatment with 1% OsO_4, buffer, and dehydrating agents as outlined in Section 2.1. With a small volume of absolute amyl acetate remaining in the detached Swinny apparatus, it was dismantled and using a pair of fine-tipped forceps the Nucleopore filter was quickly transferred to the Teflon coverslip/filter disk holder described earlier (Figure 10B). The latter was immediately placed in the wire basket of the precooled pressure bomb of the CPD apparatus. Up to four Teflon filter holders can be stacked in the cylindrical basket. The critical-point-dried filters were mounted on aluminum stubs by attaching only the outer perimeter of the filter to the surface of the stub with conductive silver paint. Care is needed to avoid wetting the entire filter with paint. The samples were metal-coated and examined in the SEM.

Tripospermum camelopardus (Figures 18) is an aquatic fungus that produces four divergent, septate arms in a characteristic tetraradiate arrangement (Ingold, 1975; Ingold *et al.,* 1968). The conidial appendages cannot withstand the destructive forces of centrifugation but are well preserved after processing with the Swinny apparatus. Infectious conidia of the human respiratory pathogen, *Coccidioides immitis* (Figure 19), were collected from glucose–yeast extract–agar culture plates in a biological isolation hood using a vacuum harvesting technique described elsewhere (Cole *et al.,* 1983; Cole and Sun, 1985). High yields of the dry, airborne conidia were obtained by this collection procedure and were virtually free of mycelial elements. The conidia were suspended in fixative solution and prepared for the SEM in the Millipore Swinny apparatus using a 1.0-μm-pore-diameter Nucleopore filter. The fractured ends and outer sleeve of wall material visible in Figure 19 are characteristic of this kind of conidium (Sun *et al.,* 1979; Huppert *et al.,* 1982).

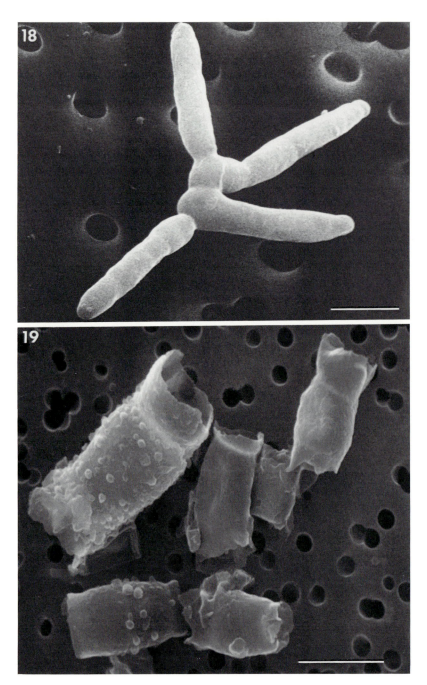

FIGURES 18, 19. Cell suspension preparations. Tetraradiate-appendaged, aquatic conidia of *Tripospermum camelopardus* (Figure 18) and arthroconidia of the human pathogen, *Coccidioides immitis* (Figure 19), were collected on Nucleopore filters (2.0- and 0.5-μm pore diameter, respectively). Bars = 5 and 2 μm, respectively.

3. FUNGUS–NATURAL SUBSTRATE ASSOCIATIONS

3.1. Whole Mount Preparations

Certain microfungi are obligate parasites or sporulate only when grown on natural substrates. The pycnidial fungi, many of which are plant pathogens or saprobes, exemplify this latter group (Sutton, 1980), although a significant number have now been shown to sporulate in pure culture (Cole, 1981a,c; B. C. Sutton and G. T. Cole, unpublished data). It is recognized, however, that some morphogenetic variations occur when such fungi are grown on artificial media (Nag Raj, 1981). Investigators interested in developmental aspects of pycnidial fungi, therefore, need to compare features of sporulation on both artificial and natural substrates (DiCosmo and Cole, 1980; Sutton and Cole, 1983). A similar situation confronts those concerned with morphogenesis of many fungal pathogens of animals. Unfortunately, comparative investigations of *in vitro* and *in vivo* fungal development are infrequently performed in medical mycology (Huppert *et al.*, 1983).

Obligate plant pathogens, such as the powdery mildew fungus, *Erysiphe graminis* (Figures 20 and 21), were obtained from infected plants grown in environmental chambers that supported good sporulation of the fungus (Cole, 1983). Infected plant parts (e.g., leaves) were excised and immediately placed in cold fixative solution. Because gases are usually trapped within the mesophyll tissue of leaves, it is necessary to subject the samples to a mild vacuum for adequate fixation of the plant material. This was accomplished by enclosing the sample in a small vacuum chamber to which a water aspirator was attached. The specimens were subsequently dehydrated, transferred to modified BEEM capsules (Figure 1), critical point-dried and metal-coated as previously described (Section 2.1). The leaf segments of St. Augustine grass (approx. 5 mm^2) shown in Figures 20 and 21 were processed in open 20 × 10-mm Teflon cups. The trichomes and fragile chains of conidia were well preserved. Day and Scott (1973) examined fresh segments of barley leaves infected with powdery mildew without specimen pretreatment. The preservation was remarkably good but the authors noted some distortion at 20 kV, probably due to vacuum desiccation, and severe charging after the specimen had been in the SEM for 10–15 min. These problems did not arise during prolonged examination of the material illustrated in Figures 20 and 21.

The parasitic cycle of the human respiratory pathogen, *Coccidioides immitis,* has been examined both *in vitro* and *in vivo* (Huppert *et al.*, 1983; Cole and Sun, 1985). In the latter case, mice (BALB/c, males, 23–27 g) were infected intranasally with viable conidia (Figure 19) and then sacrificed by sodium pentobarbital injection (0.05 ml of a solution of 50 mg/ml) at various times after challenge. Fungal cells were obtained from lung tissue by pulmonary lavage and prepared for the SEM by the following procedure (Sun *et al.*, 1985). The trachea

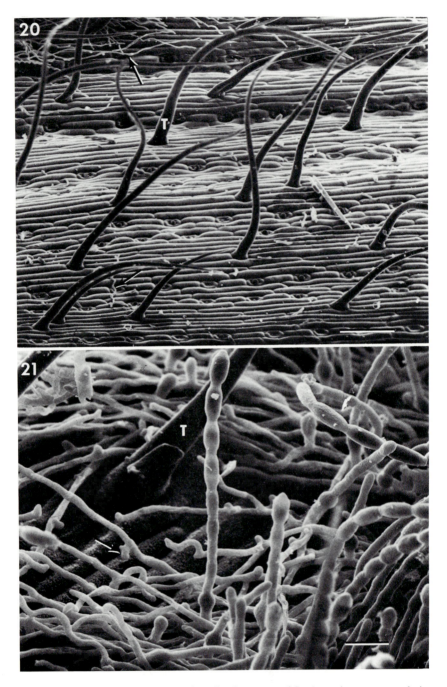

FIGURES 20, 21. Whole mount preparations. Leaf segments of St. Augustine grass sparsely infected (Figure 20) and heavily infected (Figure 21) with the powdery mildew fungus, *Erysiphe graminis*. Arrows in Figure 20 locate fungal hyphae. Arrow in Figure 21 locates infection peg that has penetrated the leaf surface. T, trichome. Bars 100 and 20 μm, respectively.

was exposed and a small incision made through the tracheal wall just large enough to accept the blunted end of a 22-gauge syringe needle tipped with polyethylene tubing, 20 mm in length. The needle was attached to a 1-ml syringe that contained 0.4 ml of 0.1 M sodium cacodylate buffer (pH 7.4). The buffer was gently forced into the lung, fluid was aspirated, and the lavage was then gently forced through a Millipore Swinny apparatus (Figure 17) containing a Nuclepore filter (0.4-μm pore diameter). The remaining preparatory steps for fixation, dehydration, and drying of the retentate are the same as those outlined in Section 2.3. Figure 22 shows a young spherule of *C. immitis* that developed within the lung of an infected mouse. Several polymorphonuclear neutrophils (PMNs) are attached to the cell surface, which may participate in some degree of breakdown of the outer wall material *in vivo* (Drutz and Huppert, 1983). The shape and size of the parasitic cells *in vivo* and *in vitro* were comparable.

3.2. Cryofracture Preparations

Further SEM investigations of the morphology of *C. immitis in vivo* were performed on cryofractured lung tissue (Figure 23; Sun *et al.*, 1985). The lungs

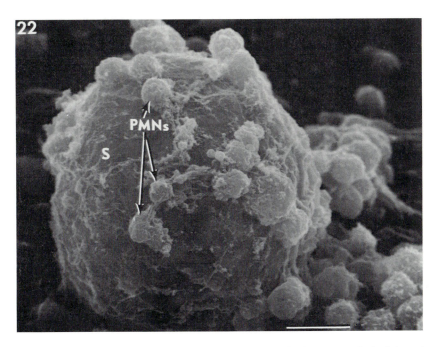

FIGURE 22. Whole mount preparation. Spherule (S) of *Coccidioides immitis* obtained from the lung of an intranasally infected mouse by a pulmonary lavage procedure and prepared for the SEM on a Nuclepore filter. PMNs, polymorphonuclear neutrophils. Bar 10 μm.

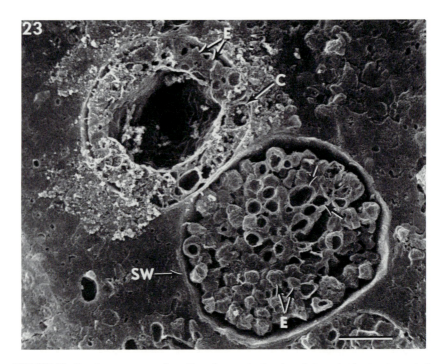

FIGURE 23. Cryofracture preparation. Cross-fractured spherules of *Coccidioides immitis* embedded in lung tissue of an infected mouse. C, compartment of young spherule within which endospores (E) differentiate; SW, spherule wall. Arrows locate empty shells from which endospores have been released. Bar = 10 μm.

of sacrificed, experimental animals were removed after perfusion with fixative solution (2.5% glutaraldehyde in 0.1 M sodium cacodylate, pH 7.4). Tiny lesions on the inner lining of the lungs identified under the dissecting microscope were excised as well as surrounding host tissue, and transferred to fresh fixative. After completion of OsO_4 treatment and buffer washes, the fragments of tissue were dehydrated in an ethanol series. The specimens were removed from absolute ethanol using a pair of fine-tipped forceps and quickly plunged into liquid nitrogen. The latter was contained in a small (50-mm-diameter) aluminum dish. The frozen tissue fragments were fractured into small pieces using an acetone-cleaned, liquid nitrogen-prechilled razor blade. The material was returned to fresh 100% ethanol and then transferred to modified BEEM capsules (Figure 1) while still immersed in absolute ethanol. The capsules were quickly placed in the CPD bomb and dried using liquid CO_2. The dried, mounted, and coated tissue fragments were examined in the SEM. The cross-fractured structures of *C. immitis* embedded in lung tissue shown in Figure 23 are endosporulating spherules (Huppert *et al.*, 1982). The wall (SW) of the lower spherule surrounds a mass of endospores (E) that represent the disseminating phase of the parasitic cycle of this pathogen. The empty shells (arrows) are the remains of envelopes that surround individual and multiple endospores. The fractured upper spherule is at an earlier stage of development and reveals a large, central cavity surrounded by immature endospores. Thin sections of this same developmental stage have shown that this central region is occupied by degenerated cytoplasmic material (Cole and Sun, 1985). During subsequent development the cavity becomes filled with endospores after their release from the peripheral compartments (C). The considerable cell damage and distortion visible in Figure 23 result from this preparatory technique. Some of the causes of this damage are discussed below. Nevertheless, the combined application of pulmonary lavage for whole mount preparations, and cryofracturing has provided valuable information on the ultrastructure of *C. immitis in vivo*.

Our SEM studies of the pathogenic yeast, *Candida albicans,* have largely concentrated on the ability of this microorganism to colonize the gastrointestinal (GI) tract of 5- to 6-day-old infant mice after oral–intragastric challenge (Pope *et al.,* 1979; Field *et al.,* 1981; Pope and Cole, 1981, 1982; Guentzel *et al.,* 1985). It was essential for these investigations that we attempt to preserve the natural microenvironment of the gut. Microbial colonization in the small intestine may be inhibited or retarded as a result of mechanical removal of microorganisms by peristaltic action aided by mucus secretion (Dixon, 1960). However, this barrier may be overcome by penetration of the mucin layer and adherence to villi (Guentzel *et al.,* 1977). Davis (1976) has reported poor preservation of mucin and loosely adherent microbial cells in germfree rat ilea after glutaraldehyde fixation. On the other hand, the author demonstrated that ileal segments that were initially frozen in liquid nitrogen prior to the same chemical fixation protocol, were characterized by well-preserved mucin and microbial associations with host tissue. In our studies of infant mice, inoculated and uninoculated control animals were sacrificed by decapitation, quickly dissected, and the entire GI tract placed in chilled saline. The GI tract was cut into the following segments: stomach, upper intestine, mid-intestine, ileum, cecum, and large bowel. The segments were quickly frozen in a dry petri dish (50-mm diameter) that had been sitting in a liquid nitrogen-flooded, larger petri dish (100-mm diameter) for at least 5 min. The level of liquid nitrogen was maintained just below the lip of the small petri dish during the following procedure. The frozen tissue was cleaved using a liquid nitrogen-prechilled razor blade. The cryofractured material was immediately placed in buffered fixative solution (2.5% glutaraldehyde) and left overnight at 4°C. After washing in buffer, the samples were transferred to 1% OsO_4 and left in the dark at 4°C for 2 hr. Subsequent washes, dehydration, CPD, and coating were performed as outlined in Section 2.1. Figure 24 shows a cryofractured, ileal segment of a control animal. The mucin is visible as strands and sheets associated with well-preserved villi. Figure 25 reveals the inner stomach wall of an infant mouse 3 hr after oral–intragastric challenge with stationary

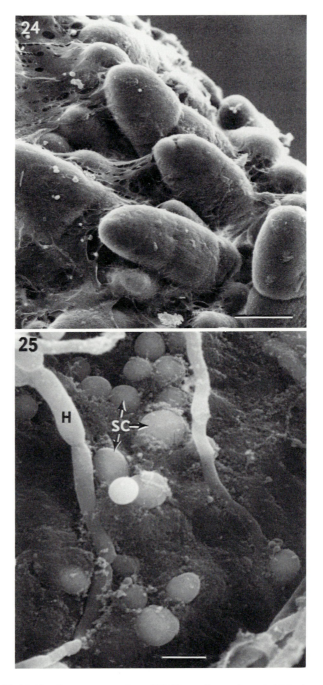

FIGURES 24, 25. Cryofracture preparations. (24) Ileum of normal, control infant mouse showing mucin and villi. (25) Inner stomach wall of infant mouse infected with *Candida albicans*. H, hypha; SC, stationary phase cells. Bars = 100 and 10 μm, respectively.

phase cells of *C. albicans*. The spherical, nonbudding cells (SC) are partially embedded in mucin while some have given rise to hyphae (H) that extend into the stomach lumen. This procedure also preserves the indigenous bacterial flora of the stomach (Pope and Cole, 1981), thus providing the opportunity to visualize the relative success of competitive microbes to colonize the same surface of host tissue (Guentzel *et al.*, 1985). The SEM has been useful in these investigations for demonstrating the pattern of colonization of *C. albicans* within the GI tract, the nature of pathogen interaction with the surface of host tissue, and the predominant morphogenetic phase (yeast or mycelial) within different colonized regions of the GI tract. The distribution data obtained from the SEM correlated well with counts (colony-forming units) of *C. albicans* derived from plating homogenates of each segment of the GI tract of infected animals on Sabouraud's dextrose agar (Pope and Cole, 1982).

Several variations of the above cryofracture techniques have been published (Nemanic, 1972; Brooks and Haggis, 1973; Humphreys *et al.*, 1974; Tokunaga *et al.*, 1974; Fujita *et al.*, 1982). In each case, however, the results demonstrated many of the same problems of preservation as revealed in Figure 23. Major improvements are now possible in the quality of such preparations by use of several alternative procedures. Of primary importance is the method of cryofixation (Pinto da Silva and Kachar, 1980; Chandler and Heuser, 1980; Phillips and Boyne, 1984). For rapid freezing with minimal damage due to ice crystal formation, it is important that the sample size be reduced to a minimum, and that heat extraction from the sample be as rapid as possible by exposure to a low-temperature environment with good heat-conducting properties (Frederik, 1982). This has been well achieved by plunging the samples into subcooled liquid nitrogen (in a liquid state but not boiling; Read, 1983), liquid helium (Heuser *et al.*, 1979), or liquid propane cooled with liquid nitrogen (Costello, 1980). In the last procedure, a real hazard is that oxygen in the air may be liquidized and mixed with liquid propane below −190°C to make a potentially explosive mixture. This problem was avoided by Inoue *et al.* (1982) in their freezing apparatus, which separated the air from liquid propane by a constant stream of nitrogen gas. Heath (1984) has described a simple liquid helium-cooled slam-freezing device that has been used for fungal preparations. The frozen samples may subsequently be transferred to liquid nitrogen and, as above, cryofractured using a liquid nitrogen-preincubated razor blade. Because of the bulk of certain fungal preparations, cryofixation alone is not suitable and samples may first be subjected to glutaraldehyde fixation followed by exposure to a cryoprotectant. An examination of the advantages and disadvantages of chemical fixation and/or use of cryoprotectants prior to sample freezing is beyond the scope of this discussion and the reader is referred to a review by Sleytr and Robards (1982). Pretreatment of samples with cryoprotectants that do not penetrate into cells (Skaer, 1982) will prevent extracellular ice crystal formation and may delay intracellular nucleation, resulting in a general improvement of cryopreservation (Frederik, 1982).

However, both chemical fixation and cryoprotection of biological samples result in extraction of certain cellular components (Echlin *et al.*, 1977; Barnard, 1980, 1982). On the other hand, selective extraction of undesirable compounds may be used to advantage in preparation of cryofractures for SEM examination (Tanaka, 1980, 1981; Tanaka and Naguro, 1981; Haggis, 1982; Tanaka and Mitsushima, 1984).

Cryofractures of fresh material may be prepared by rapid freezing without prior chemical fixation or alcohol dehydration. The frozen segments can then be freeze-dried (MacKenzie, 1976), sputter-coated, and observed in the SEM. This procedure was employed to examine the nature of fungal endophyte associations with different tissues of the forage grass *Lolium perenne*. Squash mounts of the mesocotyl region of young, infected plants were initially stained with aniline blue and examined with the light microscope (Figure 26). Septate hyphae (arrow) are revealed in the ground parenchyma (GP) adjacent and parallel to the procambium (Pc). No necrosis of plant tissues was evident and it appeared that the hyphae were intercellular. For SEM examination, the mesocotyl was excised from live plants of the same age, immediately plunged into subcooled nitrogen, fractured under liquid nitrogen, and transferred to a copper container while still immersed in liquid nitrogen. The container was placed in a Balzers BA 360M high-vacuum evaporator. Caution is required to prevent the liquid nitrogen from explosively boiling during the initial evacuation, which can result in discharge of the specimens from the copper container. The liquid nitrogen quickly converts to a sludge and sublimation is performed over 24 hr. The dried fragments were then properly oriented and mounted on stubs, sputter-coated, and examined in the SEM. Cross fractures of plant tissue revealed intercellular hyphae in the ground parenchyma tissue (Figures 27 and 28). The apparent fusion observed between the hyphal and host cell walls in Figure 28 supports observations of this same fungus–plant association derived from thin-section examination. Although the results of this cryofracture procedure indicate an improvement in preservation compared to Figure 23, cell shrinkage during the freeze-drying step is a major source of distortion (Boyde, 1978). Read (1983) used a procedure in which the frozen material was freeze-dried at $-65°C$ for 2 days in the presence of a molecular sieve or P_2O_5. A comparison of the quality of results using these two procedures was not performed during our studies.

The availability of cryo attachments for the SEM has made it possible for frozen samples to be cryofractured, sputter-coated, and transferred to the cold stage of the electron microscope while maintained under vacuum and free of contamination from condensing vapors (Echlin and Burgess, 1977; Pawley and Norton, 1978; Robards and Crosby, 1978; Pawley *et al.*, 1980; Beckett, 1982; Beckett and Porter, 1982; Becket *et al.*, 1982, 1984; Read and Beckett, 1983; Read *et al.*, 1983). Using biological material comparable to that demonstrated in Figures 26–28, Beckett (1982) mounted fresh leaf segments on a copper stub and then quickly froze the sample in subcooled nitrogen under dry argon gas. The

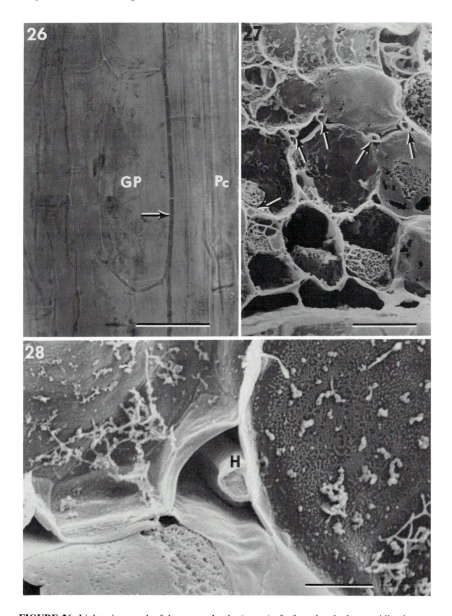

FIGURE 26. Light micrograph of the septate hypha (arrow) of a fungal endophyte residing between cells of the ground parenchyma (GP) and procambium (Pc) tissues in the mesocotyl of *Lolium perenne*. Tissue squash preparation stained with aniline blue. Bar = 40 μm.

FIGURES 27, 28. Cryofractured/freeze-dried preparations. Fractures through the mesocotyl tissue of *Lolium perenne* reveal intercellular, endophytic hyphae (arrows in Figure 27). The wall of the hypha (H, Figure 28) is typically juxtaposed to the wall of one of the host cells within the intercellular space. Bars = 10 and 4 μm, respectively.

stub was placed in an evacuation chamber and transferred under vacuum to a fracturing/coating chamber where the sample was cryofractured and sputter-coated. Finally, the specimen stub was transferred, still under vacuum, to the cold stage of the SEM for examination. The material remained in a frozen-hydrated state and it has been suggested that such biological specimen preparations are "the closest we can get to viewing . . . structure under the SEM in a state of *suspended animation*" (Read *et al.*, 1983). A major limitation of this sputter–cryo procedure is still the efficiency of the rapid-freezing step. The high cost of commercially available sputter–cryo attachments is an unfortunate re-stricting factor. On the other hand, this technique appears to provide excellent preservation, and allows fresh specimens to be prepared and ready for SEM examination in a remarkably brief period (from living organism to SEM image in approximately 13 min; Beckett, 1982). A detailed discussion of this procedure is presented by Beckett and Read in Chapter 2.

3.3. Post-CPD Sections

Certain biological specimens are suitable for sectioning at room temperature after they have been critical point-dried. The value of this technique was demon-strated in a morphological study of the interrelationship of a parasitic fungus, *Graphiola phoenicis,* and its host plant, the canary palm (Cole, 1983). Segments of infected leaf tissue were prepared as outlined in Section 3.1. However, after the CPD step the dried material was placed on a clean glass slide and infected regions of the leaf were identified under a dissecting microscope. Using an acetone-cleaned razor blade, transverse sections through these regions were ob-tained. A single, vertical cut was made through a site of infection with a smooth, continuous pass of the blade. The quality of the sections was adequate to easily identify different host tissues (Figure 29). It was possible to obtain information from such preparations on the degree of invasion of leaf tissue by the pathogen. As demonstrated in Figure 29, necrosis of host tissue appears to be limited to the chlorenchyma directly below the fructification of the parasite. The vascular tissue and costal bands seem to be unaffected by the presence of the biotrophic parasite. The latter host structures are composed of abundant, tough fibrous tissue and are visible both at the base of the fructification as well as at adjacent, uninfected regions. Microtome sections of plastic-embedded leaf tissue were also prepared for light microscopic examination. Good correlation of morphological data was obtained using these two microscopic techniques (Cole, 1983). A similar preparatory procedure has been employed for examining certain rust fungi parasitizing their host plants (Kakishima *et al.,* 1984). As above, good correlation was obtained between SEM images of post-CPD sections and light micrographs of hand sections of fresh material. Although a major drawback of this procedure is that some tissue distortion is inevitable when the dried material is sectioned, it is still useful for examinations of many fungus–plant associations

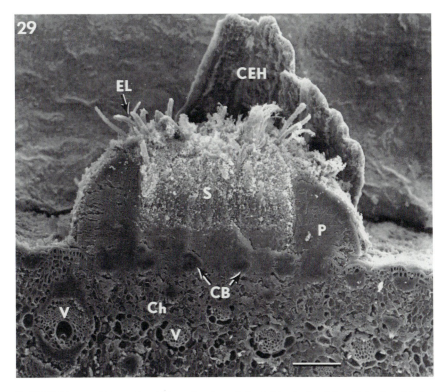

FIGURE 29. Post-CPD section. This section through the fruiting body of *Graphiola phoenicis* that had formed on the surface of a palm leaf reveals structure of both fungal tissues (e.g., P, peridium of fruiting body; S, spore mass; EL, elaters associated with spore discharge) and host tissues (e.g., CEH, cuticular–epidermal–hypodermal layer that ruptured during egress of the fungal fructification; CB, undigested costal bands of palm leaf; V, vascular tissue; Ch, chlorenchyma). Bar = 100 μm. From Cole (1983).

in which the host is particularly difficult to embed in plastic and section for light and transmission electron microscopy (Martens and Uhl, 1980).

4. CORRELATION OF MORPHOLOGICAL DATA

4.1. Developmental Studies

Correlative light and electron microscopy have been used effectively in ontogenetic studies of cultured cells (Price, 1974). The SEM has played an important role in developmental studies of microfungi, particularly the Zygomycetes (O'Donnell, 1979), Ascomycetes (Seale, 1973; Cole *et al.*, 1974; Harris *et al.*, 1975; Hock *et al.*, 1978; Read, 1983), and Deuteromycetes, or conidial fungi (Cole and Samson, 1979, 1983; Cole, 1981c). The ultrastructural images generated by the SEM have contributed to the formulation of developmental concepts within these fungal groups. In the case of the conidial fungi, scanning electron microscopy has been used in combination with time-lapse photomicrography to investigate different methods of conidium and conidiogenous (conidium-producing) cell formation. The original concepts of modes of conidiogenesis were derived from light microscopic examinations of both fixed material (Hughes, 1953), and live cells examined in specially designed growth chambers (Cole and Kendrick, 1968; Cole *et al.*, 1969). The role of the SEM has been to illustrate morphological details at selected stages of cell ontogeny that were beyond the resolving power of the light microscope, and thereby confirm or alter earlier developmental concepts. In fact, a pool of morphological data were obtained from the SEM as well as TEM examinations of thin sections and freeze-etch replicas of fungal cells, and these data were correlated for each species representing a particular mechanism of conidium and conidiogenous cell formation (e.g., Cole and Kendrick, 1969; Cole and Aldrich, 1971). An example of such correlation of images is presented in Figures 30–33. The primary conidium of *Dictyoarthrinium sacchari* is formed as a terminal outgrowth of the short lateral branch of a vegetative hyphae (Figure 30). This stage is followed by intercalary growth of the lateral branch just below the base of the first-formed conidium. The inelastic, outer wall layer of the former ruptures in the region of extension growth, leaving circumscissile tears that are visible with the SEM (arrows in Figure 31). The newly exposed inner wall continues to elongate and gives rise to a basipetal succession of laterally proliferated conidia (Figures 32 and 33). The fertile hypha is the conidiogenous cell, characterized by basauxic development, and the persistent, cup-shaped structure at its base below the zone of intercalary growth is the conidiophore mother-cell (Figures 31–33). These morphological features, revealed by combined SEM and light microscopic examinations, are diagnostic of basauxic conidiogenous cell development in *Dictyoarthrinium, Arthrinium, Cordella,* and *Pteroconium.*

4.2. Cell Wall Preparations

The SEM has been used to monitor the efficiency of removal of the outer conidial wall layer of several different fungi using a Ribi cell fractionator (Sorvall Model RF-1, refrigerated; Cole *et al.*, 1979). This apparatus, which is a modified French press, permitted conidia to be sheared but not ruptured at well-controlled pressure (approx. 12,000 psi) and temperature (5–10°C; cf. Figures 34 and 35). The stripped conidia (Figure 35) demonstrated viability comparable to that of untreated cells. The suspension of stripped conidia and wall fragments obtained from the cell fractionator were chemically fixed and prepared for the SEM on Nucleopore filters using the procedure outlined in Section 2.3 (Figure

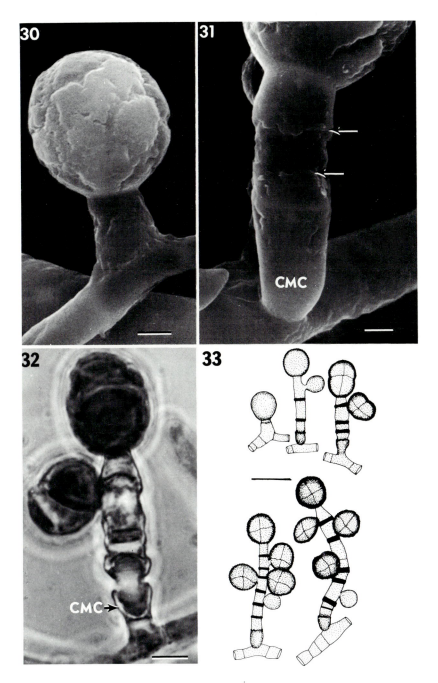

FIGURES 30–33. Morphological comparison of reproductive structures of *Dictyoarthrinium sacchari* examined with the SEM (Figures 30 and 31) and light microscope (Figure 32). Developmental concept is shown in Figure 33. Arrows in Figure 31 locate the zone of intercalary (basauxic) growth of the fertile cell. CMC, conidiophore mother cell. Bars = 2, 1, 5, and 20 μm, respectively.

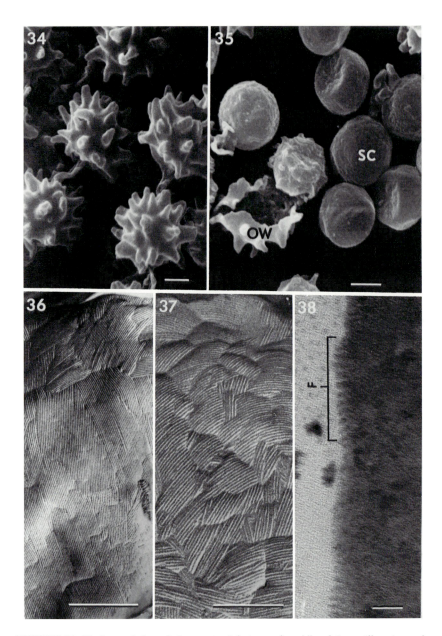

FIGURES 34–38. A correlation of ultrastructural features of conidia of *Aspergillus niger* using different electron microscopic techniques. (34) Chemically fixed, dehydrated, and critical-point-dried conidia; (35) wall preparation after fractionation of conidia. Wall fractions fixed as above but transferred to absolute acetone and air-dried (Cole *et al.*, 1979). (36, 37) Platinum/carbon shadow replicas of the outermost wall surface of untreated conidium (Figure 36) and Nonidet-P40-treated conidium (Figure 37) showing alteration in the morphology of rodlet fascicles; (38) thin section of isolated outer wall (OW, Figure 35) showing rodlet fascicle (F) on outermost surface. Preparations in Figures 36–38 examined with TEM. SC, sheared conidium after passage through the Ribi cell fractionator. Bars = 1, 1, 0.5, 0.5 μm, and 20 nm, respectively.

17). The aim of these investigations was to examine the ultrastructure, chemical composition, and immunogenicity of conidial surface wall components of selected members of the Fungi Imperfecti. The primary focus of these studies has been the characterization of the conidial surface of respiratory pathogens and allergens of man, including *Aspergillus fumigatus* (Cole *et al.*, 1982; Cole and Samson, 1984) and *Coccidioides immitis* (Cole *et al.*, 1983, 1985a,b). The rationale for this approach, at least for the pathogenic forms, is that conidia are the airborne infectious propagules, are the first to interact with host tissue, and are responsible for initiation of the fungal disease. Knowledge of the nature of the microbial surface and its role in host–pathogen interactions is crucial for our comprehension of fungal pathogenicity (Smith, 1977). The application of a mechanical procedure for fractionation of the cell envelope provides clean, reproducible preparations free of contaminants that are otherwise present after conventional chemical extraction techniques.

The high degree of reproducibility of sheared conidial fractions was confirmed by SEM examinations. The light microscope could not resolve the subtle alterations in conidial surface morphology after the shearing procedure, especially in the case of tiny propagules of *A. fumigatus* (diameter 2–3.5 μm) and *C. immitis* (L × W, 3–6 × 2–4 μm). SEM analyses of *A. fumigatus* conidial preparations indicated that approximately 90% of the cells were stripped of their surface wall layer but had not ruptured (Cole *et al.*, 1982). Viability tests (Cole *et al.*, 1983) confirmed that the sheared conidia were still intact. All conidia examined in these studies were extrmely hydrophobic. However, once the outer wall layer was removed by Ribi cell fractionation. the sheared conidia readily suspended in distilled water and the residual cell envelope swelled slightly as a result of hydration. Exposure of the inner wall layer of these cells to the aqueous medium also resulted in release of solubilized components, presumably of cell wall origin. The chemical and antigenic composition of both the outer particulate wall fraction and the soluble fraction in *C. immitis* have been characterized (Cole *et al.*, 1983), and the latter has been shown to be highly immunogenic (Cole *et al.*, 1986). The hydrophobic components removed with the outer wall layer (Figure 35) are largely present on the outermost surface of the cell as interdigitated fascicles of fibrous elements known as "rodlets" (Figure 36; Hess *et al.*, 1968; Beever and Dempsey, 1978). These structures are common wall components of air-dispersed conidia. Although the rodlet layer seems impervious to water, its hydrophobicity breaks down in the presence of mild detergents as well as L-α-phosphatidylcholine dipalmitoyl (PCD) or commercial surfactant (Cole and Pope, 1981; Cole, 1981d). Conidia of *A. niger* were incubated in various concentrations of nonionic detergents (e.g., Nonidet-P40) and PCD over a 12-hr period at 25°C and the incubation media were examined for presence of protein released from the conidia. Alterations in the surface ultrastructure of conidia after incubation were also investigated. In each case, significant amounts of protein were released from the conidia, which were still viable after incubation. The significance of these observations is that removal of the hydrophobic outer layer of the infectious propagule may not be necessary to release the underlying immunogenic, soluble wall components *in vivo*. It is possible in the case of *C. immitis,* for example, that interaction between surfactants and the conidial surface within the respiratory tract could release some of the same antigens as detected in the soluble outer wall fraction (Cole *et al.,* 1983). It has been important, therefore, to examine both the ultrastructure and the chemical composition of the hydrophobic surface layer of conidia. Our studies of the rodlet component of *A. niger* illustrate the value of correlating morphological data obtained by different electron microscopic methods (Cole and Pope, 1981). The SEM was capable of resolving individual fascicles of rodlets on the conidial surface (Cole and Ramirez-Mitchell, 1974). The morphology of rodlet fascicles of chemically fixed/critical-point-dried, freeze-dried, and air-dried conidia were comparable. It was not possible, however, to resolve individual rodlets with the SEM. For this it was necessary to examine either platinum/carbon shadow replicas (Figures 36 and 37; Cole, 1981b) or thin sections of intact conidia and isolated wall fractions (Figure 38). The arrangement of rodlets on the surface of untreated conidia and mechanically isolated wall fractions are shown in Figures 36 and 38, respectively. After treatment with Nonidet-P40, the SEM revealed that the fascicles had become distinctly swollen. Surface replicas of these same cells (Figure 37) showed that slight separation of adjacent rodlets had occurred, which may account for release of macromolecules from the underlying wall. Current investigations of *C. immitis* involve correlation of the results of these *in vitro* studies with surface ultrastructural examinations of conidia obtained from the respiratory tract of infected mice by pulmonary lavage (Section 2.3). Attempts are also under way to locate specific, wall-associated antigens using techniques of immunoelectron microscopy (Horisberger and Rosset, 1977; Kent *et al.,* 1978). A discussion of this latter procedure is presented elsewhere in this volume.

5. LIMITATIONS OF PREPARATORY PROCEDURES

5.1. Chemical Fixation and Dehydration

A common problem encountered in preparing microfungi, such as *Penicillium* and *Aspergillus,* for the SEM is the disruption of delicate chains of hydrophobic conidia during initial fixation using glutaraldehyde and/or OsO_4 solutions. Addition of a wetting agent to the fixative solution (see Section 2.1) does not eliminate the problem, although conidial chains of certain fungi have been successfully preserved using this procedure (Cole and Samson, 1979). However, Moore and Grand (1970) claimed that fixation of dry basidiospores using a glutaraldehyde solution containing a wetting agent introduced "artifactual depositions that rendered the spores unfit for scanning." Instead, the authors simply collected spores on glass coverslips or double sticky tape, dried and coated them in an evaporator (Grand and Moore, 1971), or spores on coverslips were first immersed in 80% ethanol, evaporated to dryness, and then coated (Grand and Van Dyke, 1976). Vapor fixation by exposure of sporulating cultures to glutaraldehyde (Williams and Veldkamp, 1974), or an OsO_4 atmosphere (Quattlebaum and Carner, 1980; Rykard et al., 1984) over 1–2 days have also been employed. The samples were then air-dried and coated to minimize mechanical disruption prior to SEM examination. Results of these procedures are generally poor due to severe cellular distortion, although the quality of vapor-fixed material is slightly better than that demonstrated by unfixed, air-dried specimens (Stolk and Samson, 1983) or alcohol/acetone-dried material (Kozakiewicz, 1978). These latter procedures should not be ruled out, however, especially for samples in which hydrophobic conidia are produced on a dry substrate, such as an insect cuticle (Samson and Evans, 1982), and have a low volume/surface area ratio. Cryofixation generally provides superior preservation of cells and tissues but the quenching process is also disruptive to delicate cellular arrangements, such as conidial chains (Williams and Veldkamp, 1974). No single fixation procedure is ideal for preparation of the variety of fungal specimens illustrated in this chapter. The choice ultimately depends on the precise nature of the problem under investigation.

Glutaraldehyde effectively cross-links proteins of biomembranes, but apparently has little effect in altering the selective permeability of the membrane (Wangensteen et al., 1981). On the other hand, OsO_4 reacts with unsaturated hydrocarbon portions of membrane lipids, causing a polymerization of hydrocarbon chains and loss of semipermeability properties of the membrane. These differences in features of the two chemicals commonly used for SEM preparations have significant effects on the response of cells and tissues during the fixation process. Glutaraldehyde alone is not considered to be osmotically active toward cells, at least at concentrations of 2% or less (Bullock, 1984), but instead it is the "fixative vehicle" (e.g., cacodylate buffer) that has osmotic effects and these may result in cell distortion (Arborgh et al., 1976; Boyde and Maconnachie, 1979). For example, Hoch and Howard (1981) have demonstrated that swelling of the septal apparatus of a basidiomycetous fungus was most noticeable after conventional fixation with 2.5% glutaraldehyde buffered with 75 mM phosphate (Na^+) buffer (pH 6.8, 20°C, 1 hr) followed by buffer rinses and postfixation with unbuffered 1% OsO_4. It is important for the osmolality of the fixative vehicle to approximate that of the extracellular fluid of plant or animal tissues, or the media in/on which cells are grown. In the case of intact tissues, however, changes in fixative osmolality are unavoidable due to limitations of fixative penetration and dilution by tissue fluid. This problem was encountered during preparation of infected lung tissue discussed in Section 3.2. Glutaraldehyde fixation logically results in some rigidification of cells and tissues because of the cross-linking of cytoskeletal and membrane-bound proteins. This may, in turn, provide the cells with some resistance to osmotic volume changes (Wangensteen *et al.*, 1981). However, glutaraldehyde penetrates into tissues rather slowly, compared with formaldehyde, and it is advantageous in some preparations to accelerate such stabilization by use of a combination of aldehydes (e.g., glutaraldehyde plus formaldehyde; Karnovsky, 1965; Hayat, 1978). It is important to note that aldehyde-fixed specimens remain osmotically active unless the duration of fixation is prolonged, or the samples are postfixed with OsO_4. In the case of dry, aerial cells produced by microfungi, it is desirable to both stabilize cells and disrupt selective permeability of their membranes as rapidly as possible. Application of simultaneous glutaraldehyde–OsO_4 fixation with postosmication (Franke *et al.*, 1969; Heintz and Pramer, 1972) appears successful in this regard. Prolonged OsO_4 exposure has the added advantage of imparting some conductivity to the tissue (Murphy, 1978), as well as hardening the tissue and making cells more resistant to osmotic volume changes.

Evidence is available that OsO_4 cross-links and cleaves proteins (Emerman and Behrman, 1982), and the latter can be prevented by the presence of ligands, such as pyridine or thiocarbohydrazide (Aoki and Tavassoli, 1981). When used alone, OsO_4 is a poor fixative and generally produces more artifacts than glutaraldehyde (Sleytr and Robards, 1982). However, in combination with glutaraldehyde, it may be effective in stabilizing protein cross-links already formed (Bullock, 1984).

Swelling of cells during dehydration steps below 50% ethanol after conventional glutaraldehyde–OsO_4 fixation has also been noted in many samples, particularly animal tissues (Boyde, 1978). It has been demonstrated that such swelling can be prevented if monovalent or divalent cations are added to ethanol washes below 70% EtOH. Conversely, specimens are most susceptible to shrinkage during the transition from 70 to 100% ethanol (Hayat, 1978). Both effects can be minimized in fungal tissue by passing samples through a multiple-step dehydration gradient (i.e., 10%, 20%, 30%, 50%, 70%, 90%, 95%, absolute EtOH; Section 2.1). This has the disadvantage, however, of being a time-consuming process, potentially destructive due to repeated specimen agitation and a source of artifact resulting from removal of nonvolatile extracellular material by the solvents employed. Although this last point (Hayat, 1970) can be cited as a disadvantage for use of any fixation and dehydration solution in preparation of fungal samples, selective removal of extracellular material may be beneficial for SEM examination of underlying structure. Samson *et al.* (1979) described a preparatory procedure for microfungi in which specimens were fixed in unbuffered, aqueous 2% OsO_4 or 6% glutaraldehyde, rinsed twice in water, and dehydrated in 2,2-dimethoxypropane (DMP) or methoxyethanol (MOE) followed by two washes in 100% acetone. Some collapse of fungal cells has been attributed to both DMP and MOE dehydration (Read, 1983), but good results have been obtained from this unconventional fixation and abbreviated dehydration procedure.

5.2. Drying

Boyde (1978, 1980) has suggested that a sense of complacency exists among many investigators who use CPD and freeze-drying techniques, and although dimensional changes associated with these procedures are ubiquitous, they are often unrecognized. Beckett *et al.* (1984) have cautioned that apparently good morphological preservation has led many workers to assume that CPD and associated preparatory steps cause little to no shrinkage. In fact, both CPD and freeze-drying cause some degree of shrinkage of most biological material (Boyde and Maconnachie, 1981; Boyde and Franc, 1981), and generally such dimensional changes appear to be greatest for tissues subjected to CPD. The amount of volume change varies with different specimens and has been shown to be severe in many animal tissues following CPD (Boyde and Boyde, 1980). Additional freeze-drying artifacts arise from cell rupture due to ice crystal formation and thermal stress cracking, as well as shrinkage resulting from treatment with cryoprotectants. These and other problems associated with freeze-drying are discussed in a subsequent chapter of this volume. An evaluation of CPD must take into account the preparatory events that precede the actual drying process. Thus, the degree of conformational change and extraction of cellular materials during fixation, dehydration, and CPD together influence the quality of the final product. A critical sequence of the final drying process, however, is exposure of the specimen to an appropriate intermediate fluid (e.g., amyl acetate) that is miscible with both the dehydration fluid and the transitional fluid (e.g., ethanol and liquid CO_2, respectively). Although absolute ethanol and acetone are miscible with liquid CO_2, their relatively high volatility can easily lead to specimen damage due to evaporative drying as the specimen is transferred to the CPD bomb. Gradual replacement of absolute ethanol by anhydrous amyl acetate, as outlined in Section 2.1, ensures good dehydration of the sample. Lower volatility and a distinctive odor are advantages for use of amyl acetate as an intermediate fluid (Hayat, 1978). However, exposure of the specimen to yet another solvent gradient may result in artifacts generated by further extraction of cellular material. Exchange of the intermediate fluid with liquid CO_2 in the CPD bomb also necessitates a slow purge, which prolongs exposure of the sample to the solvent action of liquid CO_2. On the other hand, specimen shrinkage after completion of the CPD process is a common problem arising from the presence of residual intermediate solvent (or water). It is important, therefore, that good miscibility exists between the dehydration, intermediate, and transitional fluids, and that complete substitution of intermediate by transitional fluid has occurred before the latter is taken through the critical point. Hoagland *et al.* (1980) have drawn attention to structural artifacts arising from contaminants in transitional fluid supply cylinders. They have suggested that inverted cylinders or cylinders with siphons be avoided, or at least tested for the presence of bottom solid and liquid contaminants. Use of on-line filters consisting of activated alumina for removing moisture from liquid CO_2, and activated bone charcoal for removal of any oily residue may be necessary.

5.3. Conductive Coating

The sputter-coating procedure outlined in Section 2.1 has the advantage of good deposition of a thin, conductive metal layer over a highly irregular surface, which results in very little to no charging in the SEM. The disadvantages are that any amount of coating will limit resolution of surface detail, in addition to the possibility of some heat damage to the specimen during the evaporation process (Holland, 1976; Ingram *et al.*, 1976). The alternative is to examine uncoated specimens (Murphy, 1978) that have been treated with solutions containing conductive heavy metals (e.g., Os, U, Pb) and mordants (e.g., 3-thiocarbazide, galloylglucoses, tannins). Impregnation of the sample with heavy metal also permits it to be dissected under vacuum in the SEM using a micromanipulator without loss of conductivity. A procedure that has been commonly employed involves treating OsO_4-fixed material with 3-thiocarbazide (TCH) to enhance conductivity either by ligand-mediated osmium binding (Seligman *et al.*, 1966; Kelley *et al.*, 1973) or by TCH functioning as a mordant (Murphy, 1978). Repetition of successive osmium and TCH treatments (Postek and Tucker, 1977) has permitted specimens to be examined at high accelerating voltages without loss of electron contrast (Melick and Wilson, 1975). In certain cases, metal impregnation of uncoated specimens has resulted in better secondary electron resolution than for sputter-coated samples (Murphy, 1978).

5.4. Specimen Examination

Several potential problems must be taken into account after the specimen is placed in the SEM and subjected to the electron beam. Interactions between the sample and incident beam may heat the specimen, alter its chemical bonding, and contaminate its surface (Joy and Maruszewski, 1978). A rise in temperature is especially critical during examination of samples in the frozen-hydrated state but can be minimized by use of a low beam current and good thermal conductivity between specimen and cold stage. Recent improvements in the performance of the low-voltage SEM (i.e., accelerating voltage of approx. 1 kV) also hold promise for application to such investigations (Pawley, 1984). Although some degree of alteration and destruction of chemical bonding of the specimen is unavoidable, surface contamination arising from reactions between the beam, residual gases in the vacuum, and specimen can be effectively reduced in a clean vacuum system. However, a conductive specimen that is examined using different accelerating voltages (e.g., 5–30 kV) may demonstrate different secondary electron images (Stein and Gay, 1974). At 30 kV, the thin, conductive metal coating is usually penetrated by the primary electron beam, which may result in

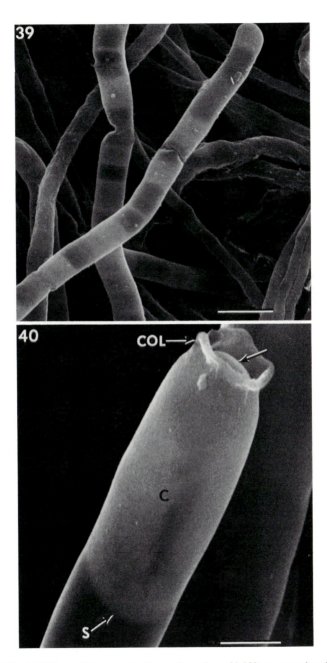

FIGURES 39, 40. Effects of high accelerating voltage (i.e., 30 kV) on secondary image. (39) Endogenously formed conidia of *Malbranchea sulphurea* (bright compartments) are distinguished from adjacent autolytic cells (dark compartments) within the fertile hyphae. (40) A conidium is visible within the sleeve, or collarette (COL), of the fertile cell (phialide) of *Chalara paradoxa*. The basal septum (S) is also visible as a result of the deep penetration of the electron beam and secondary ionization. Arrow locates the exposed apex of the conidium. Bars = 5 and 2 μm, respectively.

secondary ionizations from subsurface regions of the specimen. Under these circumstances, the surface detail is obscured and valuable information may be lost. On the other hand, generation of subsurface images using high acceleration voltages has been beneficial for examination of certain microfungi. *Malbranchea sulphurea* gives rise to conidia that are differentiated within fertile hyphae (Figure 39). Certain hyphal compartments are converted into conidia while others undergo autolysis. The wall of the degenerative cells eventually breaks down and the endogenously formed conidia are released. The thick-walled compartments are outlined as bright regions in Figure 39 as a reult of the additive effects of reflected electrons and secondary electrons from ionizations within the specimen. Many more electrons are collected from these regions than the adjacent, thin-walled, partially autolysed compartments, which appear darker on the micrograph (Cole and Samson, 1979). A similar phenomenon is illustrated in Figure 40. In this case, however, the reproductive structure is a phialide and the conidium (C) resides within the open end, or collarette (COL), of the fertile cell (Cole and Samson, 1979). The basal septum (S) of the conidium is visible inside the phialide wall and the conidial apex is exposed at the tip of the collarette (arrow). At this accelerating voltage (30 kV), however, surface detail of the phialide wall was lost. It is advantageous, therefore, to examine specimens using a range of accelerating voltages to obtain maximum information from SEM images.

6. REFERENCES

Aoki, M., and Tavassoli, M., 1981, OTO method for preservation of actin filaments in electron microscopy, *J. Histochem. Cytochem.* **29**:682–683.

Arborgh, B., Bell, P., Brunk, B., and Collins, V. P., 1976, The osmotic effect of glutaraldehyde during fixation: A transmission electron microscopy, scanning electron microscopy and cytochemical study, *J. Ultrastruct, Res.* **56**:339–350.

Barnard, T., 1980, Ultrastructural effects of the high molecular weight cryoprotectants dextran and polyvinyl pyrrolidone on liver and brown adipose tissue in vitro, *(Oxford) J. Microsc.* **120**:93–104.

Barnard, T., 1982, Thin frozen-dried cryosections and biological X-ray microanalysis, *J. Microsc. (Oxford)* **126**:317–332.

Beckett, A., 1982, Low temperature scanning electron microscopy of the bean rust fungus *Uromyces viciae-fabae*, *Philip's Electron Optics Bull.* No. 117, pp. 6–8.

Beckett, A., and Porter, R., 1982, *Uromyces viciae-fabae* on *Vicia faba:* Scanning electron microscopy of frozen-hydrated material, *Protoplasma* **111**:28–37.

Beckett, A., Porter, R., and Read, N. D., 1982, Low-temperature scanning electron microscopy of fungal material, *J. Microsc. (Oxford)* **125**:193–199.

Beckett, A., Read, N. D., and Porter, R., 1984, Variations in fungal spore dimensions in relation to preparatory techniques for light microscopy and scanning electron microscopy, *J. Microsc. (Oxford)* **136**:87–95.

Beever, R. E., and Dempsey, G. P., 1978, Function of rodlets on the surface of fungal spores, *Nature (London)* **272**:608–610.

Beneke, E. S., and Rogers, A. L., 1970, *Medical Mycology Manual,* Burgess, Minneapolis, Minnesota.

Biberfeld, G., and Biberfeld, P., 1970, Ultrastructural features of *Mycoplasma pneumoniae, J. Bacteriol.* **102**:855–861.

Boyde, A., 1978, Pros and cons of critical point drying and freeze drying for SEM, *Scanning Electron Microsc.* **II**:303–314.

Boyde, A., 1980, A review of basic preparation techniques for biological scanning electron microscopy, *Electron Microsc.* **2**:768–777.

Boyde, A., and Boyde, S., 1980, Further studies of specimen volume changes during processing for SEM: Including some plant tissue, *Scanning Electron Microsc.* **II**:117–124.

Boyde, A., and Franc, F., 1981, Freeze-drying shrinkage of glutaraldehyde fixed liver, *J. Miscrosc. (Oxford)* **122**:75–86.

Boyde, A., and Maconnachie, E., 1979, Volume changes during preparation of mouse embryonic tissue for SEM, *Scanning* **2**:149–163.

Boyde, A., and Maconnachie, E. 1981, Morphological correlations with dimensional change during SEM specimen preparation, *Scanning Electron Microsc.* **IV**:27–34.

Brooks, S. E. H., and Haggis, G. H., 1973, Scanning electron microscopy of rat's liver, *Lab. Invest.* **29**:60–64.

Buller, A. H. R., 1934, *Researches on Fungi,* Vol. 6, Longmans, Green, New York.

Bullock, G. R., 1984, The current status of fixation for electron microscopy: A review, *J. Microsc. (Oxford)* **133**:1–15.

Chandler, D. E., and Heuser, J., 1980, The vitelline layer of the sea urchin egg and its modification during fertilization, *J. Cell Biol.* **84**:618–632.

Cole, G. T., 1975, A preparatory technique for examination of imperfect fungi with the SEM, *Cytobios* **12**:115–121.

Cole, G. T., 1976, Application of scanning electron microscopy to developmental and taxonomic studies of the Fungi Imperfecti, *Scanning Electron Microsc.* **VII**:527–534.

Cole, G. T., 1979, Contributions of electron microscopy to fungal classification, *Am. Zool.* **19**:589–608.

Cole, G. T., 1981a, Conidiogenesis and conidiomatal ontogeny, in: *Biology of Conidial Fungi,* Vol. 2 (G. T. Cole and B. Kendrick, eds.), pp. 271–327, Academic Press, New York.

Cole, G. T., 1981b, Techniques for examining ultrastructural aspects of conidial fungi, in: *Biology of Conidial Fungi,* Vol. 2 (G. T. Cole and B. Kendrick, eds.), pp. 577–634, Academic Press, New York.

Cole, G. T., 1981c, Application of scanning electron microscopy to studies of conidiomatal development in the Fungi Imperfecti, *Scanning Electron Microsc.* **III**:305–312.

Cole, G. T., 1981d, Architecture and chemistry of the cell walls of higher fungi, in: *Microbiology— 1981* (D. Schlessinger, ed.), pp. 227–231, American Society for Microbiology, Washington, D.C.

Cole, G. T., 1983, *Graphiola phoenicis:* A taxonomic enigma, *Mycologia* **75**:93–116.

Cole, G. T., and Aldrich, H. C., 1971, Ultrastructure of conidiogenesis in *Scopulariopsis brevicaulis, Can. J. Bot.* **49**:745–755.

Cole, G. T., and Behnke, H.-D., 1975, Electron microscopy and plant systematics, *Taxon* **24**:3–15.

Cole, G. T., and Kendrick, W. B., 1968, A thin culture chamber for time-lapse photomicrography of fungi at high magnifications, *Mycologia* **60**:340–344.

Cole, G. T., and Kendrick, W. B., 1969, Conidium ontogeny in hyphomycetes: The annellophores of *Scopulariopsis brevicaulis, Can. J. Bot.* **47**:925–929.

Cole, G. T., and Nozawa, Y., 1981, Dimorphism, in: *Biology of Conidial Fungi,* Vol. 1 (G. T. Cole and B. Kendrick, eds.), pp. 97–133, Academic Press, New York.

Cole, G. T., and Pope, L. M., 1981, Surface wall components of *Aspergillus niger* conidia, in: *The Fungal Spore: Morphogenatic Controls* (G. Turian and H. Hohl, eds.), pp. 195–215, Academic Press, New York.

Cole, G. T., and Ramirez-Mitchell, R., 1974, Comparative scanning electron microscopy of *Penicillium* conidia subjected to critical point drying, freeze-drying and freeze-etching, *Scanning Electron Microsc.* **II**:367–374.

Cole, G. T., and Samson, R. A., 1979, *Patterns of Development in Conidial Fungi,* Pitman, London.

Cole, G. T., and Samson, R. A., 1983, Conidium and sporangiospore formation in pathogenic microfungi, in: *Fungi Pathogenic for Humans and Animals,* Part A (D. H. Howard, ed.), pp. 437–524, Marcel Dekker, New York.

Cole, G. T., and Samson, R. A., 1984, The conidia, in: *Mould Allergy* (Y. Al-Doory and J. F. Domson, eds.), pp. 66–103, Lea & Febiger, Philadelphia.

Cole, G. T., and Sun, S. H., 1985, Arthroconidium–spherule–endospore transformation in *Coccidioides immitis,* in: *Fungal Dimorphism: With Emphasis on Fungi Pathogenic for Humans* (P. J. Szaniszlo, ed.), pp. 281–333, Plenum Press, New York.

Cole, G. T., Nag Raj, T. R., and Kendrick, W. B., 1969, A simple technique for time-lapse photomicrography of microfungi in plate culture, *Mycologia* **61**:726–730.

Cole, G. T., Hardcastle, R. V., and Szaniszlo, P. J., 1974, *Subbaromyces splendens:* Development and ultrastructure, *Can. J. Bot.* **52**:2453–2457.

Cole, G. T., Sekiya, T., Kasai, R., Yokoyama, T., and Nozawa, Y., 1979, Surface ultrastructure and chemical composition of the cell walls of conidial fungi, *Exp. Mycol.* **3**:132–156.

Cole, G. T., Starr, M. E., Sun, S. H., and Kirkland, T. N., 1986, Antigen identification in *Coccidioides immitis,* in: *Microbiology-1986* (L. Leive, ed.), pp. 1–6, American Society for Microbiology, Washington, D.C.

Cole, G. T., Sun, S. H., and Huppert, M., 1982, Isolation and ultrastructural examination of conidial wall components of *Coccidioides* and *Aspergillus, Scanning Electron Microsc.* **IV**:1677–1685.

Cole, G. T., Pope, L. M., Huppert, M., Sun, S. H., and Starr, P., 1983, Ultrastructure and composition of conidial wall fractions of *Coccidioides immitis, Exp. Mycol.* **7**:297–318.

Cole, G. T., Chinn, J., Pope, L. M., and Starr, P., 1985a, Characterization and distribution of 3-*O*-methylmannose in *Coccidioides immitis,* in: *Coccidioidomycosis: Proc. 4th Int. Conf. Coccidioidomycosis* (H. E. Einstein and A. Catanzaro, eds.), pp. 130–145, National Foundation for Infectious Diseases, Washington, D.C.

Cole, G. T., Pope, L. M., Sun, S. H., Huppert, M., and Starr, P., 1985b, Wall composition in different cell types of *Coccidioides immitis,* in: *Coccidioidomycosis Proc. 4th Int. Conf. Coccidioidomycosis* (H. E. Einstein and A. Catanzaro, eds), pp. 112–129, National Foundation for Infectious Diseases, Washington, D.C.

Costello, M. J., 1980, Ultra-rapid freezing of thin biological samples, *Scanning Electron Microsc.* **11**:361–370.

Crewe, A. V., 1971, A high resolution scanning electron microscope, *Sci. Am.* **224**(4): 26–35.

Crewe, A. V., 1979, Development of high resolution STEM and its future, *J. Electron Microsc.* **28**(Suppl.):S9–S16.

Davis, C. P., 1976, Preservation of gastrointestinal bacteria and their microenvironmental associations in rats by freezing, *Appl. Environ. Microbiol.* **31**:304–312.

Day, P. R., and Scott, K. J., 1973, Scanning electron microscopy of fresh material of *Erysiphe graminis* f. sp. *hordei, Physiol. Pathol.* **3**:433–435.

DiCosmo, F., and Cole, G. T., 1980, Conidiomatal development in *Chaetomella acutiseta* (Coelomycetes), *Can. J. Bot.* **58**:1127–1137.

Dixon, J. M. S., 1960, The fate of bacteria in the small intestine, *J. Pathol. Bacteriol.* **79**:131–140.

Drutz, D. J., and Huppert, M., 1983, Coccidioidomycosis: Factors affecting the host–parasite interaction, *J. Infect. Dis.* **147**:372–390.

Echlin, P., and Burgess, A., 1977, Cryofracturing and low temperature scanning electron microscopy of plant material, *Scanning Electron Microsc.* **I**:491–500.

Echlin, P., Skaer, H. le B., Gardiner, B. O. C., Franks, F., and Asquith, M. H., 1977, Polymeric cryoprotectants in the preservation of biological ultrastructure. II. Physiological effects, *J. Microsc. (Oxford)* **110**:239–255.

Emerman, M., and Behrman, E. J., 1982, Clearage and cross-linking of proteins with osmium (viii) reagents, *J. Histochem. Cytochem.* **30**:395–397.

Field, L. H., Pope, L. M., Cole, G. T., Guentzel, M. N., and Berry, L. J., 1981, Persistence and spread of *Candida albicans* following intragastric challenge in infant mice, *Infect. Immun.* **31**:783–791.

Franke, W. W., Krien, S., and Brown, R. M., 1969, Simultaneous glutaraldehyde–osmium tetroxide fixation with postosmication, *Histochemie* **19**:162–164.

Frederik, P. M., 1982, Cryoultramicrotomy—Recognition of artifacts, *Scanning Electron Microsc.* **II**:709–721.

Fujita, T., Kashimura, M., and Adachi, K., 1982, Scanning electron microscopy (SEM) studies of the spleen—Normal and pathological, *Scanning Electron Microsc.* **I**:435–444.

Galun, E., 1971, SEM of intact *Trichoderma* colonies, *J. Bacteriol.* **108**:938–940.

Grand, L. F., and Moore, R. T., 1971, Scanning electron microscopy of basidiospores of species of Strobilomycetaceae, *Can. J. Bot.* **49**:1259–1261.

Grand, L. F., and Van Dyke, C. G. 1976, Scanning electron microscopy of basidiospores of species of *Hydnellum, Hydnum, Phellodon,* and *Bankera* (Hydnaceae), *J. Elisha Mitchell Sci. Soc.* **92**:114–123.

Guentzel, M. N., Field, L., Cole, G. T., and Berry, L. J., 1977, The localization of *Vibrio cholerae* in the ileium of infant mice, *Scanning Electron Microsc.* **II**:275–282.

Guentzel, M. N., Cole, G. T., and Pope, L. M., 1985, Animal models for candidiasis in: Current topics in Medical Mycology (M. R. McGinnis, ed.), pp. 57–116, Springer Verlag, New York.

Haggis, G. H., 1982, Contribution of scanning electron microscopy to viewing internal cell structure, *Scanning Electron Microsc.* **II**:751–763.

Harris, J. L., Howe, H. B., and Roth, J. L., 1975, Scanning electron microscopy of surface and internal features of developing perithecia of *Neurospora crassa, J. Bacteriol.* **122**:1239–1246.

Hayat, M. A., 1970, *Principles and Techniques of Electron Microscopy: Biological Applications,* Vol. I, Van Nostrand–Reinhold, Princeton, New Jersey.

Hayat, M. A., 1978, *Introduction to Biological Scanning Electron Microscopy,* University Park Press, Baltimore.

Heath, I. B., 1984, A simple and inexpensive liquid helium cooled 'slam freezing' device, *J. Microsc. (Paris)* **35**:75–82.

Heintz, C. E., and Pramer, D., 1972, Ultrastructure of nematode-trapping fungi, *J. Bacteriol.* **110**:1163–1170.

Heslop-Harrison, J., 1979, Botanical microscopy 1979—Retrospect and prospect, *Proc. R. Microsc. Soc.* **14**:256–266.

Hess, W. M., Sassen, M. M. A., and Remsen, C. C., 1968, Surface characteristics of *Penicillium* conidia, *Mycologia* **60**:290–303.

Heuser, J. E., Reese, T. S., Dennis, M. J., Jan, Y., Jan, L., and Evans, L., 1979, Synaptic vesicle exocytosis captured by quick freezing and correlated with quantal transmitter release, *J. Cell Biol.* **81**:275–300.

Hoagland, K. D., Rosowski, J. R., and Cohen, A. L., 1980, Critical point drying: Contamination in transitional fluid supply cylinders, *Scanning Electron Microsc.* **IV**:133–138.

Hoch, H. C., and Howard, R. J., 1981, Conventional chemical fixations induce artifactual swelling of dolipore septa, *Exp. Mycol.* **5**:167–172.

Hock, B., Bahr, M., Walk, R.-A., and Nitschke, U., 1978, The control of fruiting body formation in the ascomycete *Sordaria macrospora* Auersw. by regulation of hyphal development: An analysis based on scanning electron and light microscopic observation, *Planta* **141**:93–103.

Holland, V. F., 1976, Some artifacts associated with sputter-coated samples observed at high magnification in the SEM, *Scanning Electron Microsc.* **I**:71–74.

Horisberger, M., and Rosset, J., 1977, Colloidal gold, a useful marker for transmission and scanning electron microscopy, *J. Histochem. Cytochem.* **25:**295–305.

Hughes, S. J., 1953, Conidiophores, conidia and classification, *Can. J. Bot.* **31:**577–659.

Humphreys, W. J., Spurlock, B. O., and Johnson, J. S., 1974, Critical point drying of ethanol-infiltrated, cryofractured biological specimens for scanning electron microscopy, *Scanning Electron Microsc.* **I:**275–282.

Huppert, M., Sun, S. H., and Harrison, J. L., 1982, Morphogenesis throughout saprobic and parasitic cycles of *Coccidioides immitis, Mycopathologia* **78:**107–122.

Huppert, M., Cole, G. T., Sun, S. H., Drutz, D. J., Starr, P., Frey, C. L., and Harrison, J. L., 1983, The propagule as an infectious agent in coccidiodomycosis, in: *Microbiology—1983* (D. Schlessinger, ed.), pp. 262–267 American Society for Microbiology, Washington, D.C.

Ingold, C. T., 1975, Hooker lecture 1974. Convergent evolution in aquatic fungi: The tetraradiate spore, *Biol. J. Linn. Soc.* **7:**1–25.

Ingold, C. T., Dann, V., and McDougall, P. J., 1968, *Tripospermum camelopardus* sp. nov., *Trans. Br. Mycol. Soc.* **51:**51–56.

Ingram, P., Morosoff, N., Pope, L., Allen, F., and Tisher, C., 1976, Some comparisons of the techniques of sputter (coating) and evaporative coating from SEM, *Scanning Electron Microsc.* **I:**75–81.

Inoue, K., Kurosumi, K., and Deng, Z.-P., 1982, An improvement of the device for rapid freezing by use of liquid propane and the application of immunochemistry to the resin section of rapid-frozen, substitution-fixed anterior pituitary gland, *J. Electron Microsc.* **31:**93–97.

Joy, D. C., and Maruszewski, C. M., 1978, The physics of the SEM for biologists, *Scanning Electron Microsc.* **II:**379–390.

Kakishima, M., Sato, T., and Sato, S., 1984, *Ceraceospora,* a new genus of Uredinales from Japan, *Mycologia* **76:**969–974.

Karnovsky, M. J., 1965, A formaldehyde–glutaraldehyde fixative of high osmolarity for use in electron microscopy, *J. Cell Biol.* **27:**137a.

Kelley, R. D., Dekker, R. A. F., and Bluemink, J. G., 1973, Ligand-mediated osmium binding: Its application in coating biological specimens for scanning electron microscopy, *J. Ultrastruct. Res.* **48:**254–258.

Kent, S. P., Caldwell, T. C., and Siegel, A. L., 1978, Gold particles coated with anti-A versus ferritin of [125]iodine-labeled anti-A reactions with cell surface antigen, *Ala. J. Med. Sci.* **16:**22–25.

Kozakiewicz, Z., 1978, Phialide and conidium development in the aspergilli, *Trans. Br. Mycol. Soc.* **70:**175–186.

Kurtzman, C. P., Baker, F. L., and Smiley, M. J., 1974, Specimen holder to critical point dry microorganisms for scanning electron microscopy, *Appl. Microbiol.* **28:**708–712.

Lanier, W., 1984, Light synchronously induces development and bioluminescence in *Mycena citricolor,* in: *Abstr. Annu. Meet. Am. Soc. Microbiol.,* St. Louis (March 1984), p. 131.

MacKenzie, A. P., 1976, Principles of freeze-drying, *Transplant. Proc.* **8**(Suppl. 1):181–188.

Mallick, L. E., and Wilson, R. B., 1975, Modified thiocarbohydrazide procedure for scanning electron microscopy: Routine use for normal, pathological, or experimental tissues, *Stain Technol.* **50:**265–269.

Marchant, H. J., 1973, Processing small delicate biological specimens for scanning electron microscopy, *J. Microsc. (Oxford)* **97:**369–371.

Martens, J., and Uhl, N. W., 1980, Methods for the study of leaf anatomy in palms, *Stain Technol.* **55:**241–246.

Moore, R. T., and Grand, L. F., 1970, Application of scanning electron microscopy to basidiomycete taxonomy, *Scanning Electron Microsc.* **I:**138–144.

Muir, M. D., and Rampley, D. N., 1969, The effect of the electron beam on various mounting and coating media in SEM, *J. Microsc. (Oxford)* **90:**145–150.

Murphy, J. A., 1978, Non-coating techniques to render biological specimens conductive, *Scanning Electron Microsc.* **II**:175–194.

Murphy, J. A., 1982, Considerations, materials, and procedures for specimen mounting prior to scanning electron microscopic examination, *Scanning Electron Microsc.* **II**:657–696.

Nag Raj, T. R., 1981, Coelomycete systematics, in: *Biology of Conidial Fungi,* Vol. 1 (G. T. Cole and B. Kendrick, eds.), pp. 43–84, Academic Press, New York.

Nemanic, M. K., 1972, Critical point drying, cryofracture, and serial sectioning, *Scanning Electron Microsc.* **V**:297–304.

Newell, D. G., and Roth, S., 1975, A container for processing small volumes of cell suspensions for critical point drying, *J. Microsc. (Oxford)* **104**:321–323.

O'Donnell, K. L., 1979, *Zygomycetes in Culture,* Department of Botany, University of Georgia, Athens.

Pawley, J., 1984, Low voltage scanning electron microscopy, *J. Microsc. (Oxford)* **136**:45–68.

Pawley, J. B., and Norton, J. T., 1978, A chamber attached to the SEM for fracturing and coating frozen biological samples, *J. Microsc. (Oxford)* **112**:169–182.

Pawley, J. B., Hook, G., Hayes, T. L., and Lai, C., 1980, Direct scanning electron microscopy of frozen-hydrated yeast, *Scanning* **3**:219–226.

Phillips, T. E., and Boyne, A. F., 1984, Liquid nitrogen-based quick freezing: Experiences with bounce-free delivery of cholinergic nerve terminals to a metal surface, *J. Electron Microsc. Tech.* **1**:9–29.

Pinto da Silva, P., and Kachar, B., 1980, Quick freezing vs. chemical fixation: Capture and identification of membrane fusion intermediates, *Cell Biol. Int. Rep.* **4**:625–640.

Pope, L. M., and Cole, G. T., 1981, SEM studies of adherence of *Candida albicans* to the gastrointestinal tract of infant mice, *Scanning Electron Microsc.* **III**:73–80.

Pope, L. M., and Cole, G. T., 1982, Comparative studies of gastrointestinal colonization and systemic spread by *Candida albicans* and nonlethal yeast in the infant mouse, *Scanning Electron Microsc.* **IV**:1667–1676.

Pope, L. M., Cole, G. T., Guentzel, M. N., and Berry, L. J., 1979, Systemic and gastrointestinal candidiasis of infant mice after intragastric challenge, *Infect. Immun.* **25**:702–707.

Postek, M. J., and Tucker, S. C., 1977, Thiocarbohydrazide binding for botanical specimens for scanning electron microscopy: A modification, *J. Miscrosc. (Oxford)* **110**:71–74.

Postek, M. J., Kirk, W. L., and Cox, E. R., 1974, A container for the processing of delicate organisms for scanning or transmission electron microscopy, *Trans. Am. Microsc. Soc.* **93**:265–267.

Price, Z. H., 1974, Correlative light and electron microscopy of single cultured cells, in: *Principles and Techniques of Electron Microscopy: Biological Applications,* Vol. 4 (M. A. Hayat, ed.), pp. 45–63, Van Nostrand–Reinhold, Princeton, New Jersey.

Quattlebaum, E. C., and Carner, G. R., 1980, A technique for preparing *Beauveria* spp. for scanning electron microscopy, *Can. J. Bot.* **58**:1700–1703.

Read, N. D., 1983, A scanning electron microscopic study of the external features of perithecium development in *Sordaria humana, Can. J. Bot.* **61**:3217–3229.

Read, N. D., and Beckett, A., 1983, Effects of hydration on the surface morphology of urediospores, *J. Microsc. (Oxford)* **132**:179–184.

Read, N. D., Porter, R., and Beckett, A., 1983, A comparison of preparative techniques for the examination of the external morphology of fungal material with the scanning electron microscope, *Can. J. Bot.* **61**:2059–2078.

Reimer, L., 1976, Electron–specimen interactions and applications in SEM and STEM, *Scanning Electron Microsc.* **I**:1–8.

Rippon, J. W., 1982, *Medical Mycology: The Pathogenic Fungi and the Pathogenic Actinomycetes,* Saunders, Philadelphia.

Robards, A. W., and Crosby, P., 1978, A transfer system for low temperature scanning electron microscopy, *Scanning Electron Microsc.* **II**:927–936.

Rostgaard, J., and Christensen, P., 1975, A multipurpose specimen-carrier for handling small biological objects through critical point drying, *J. Microsc. (Oxford)* **105**:107–113.

Ruffolo, J. J., 1974, Critical point drying of protozoan cells and other biological specimens for scanning electron microscopy: Apparatus and methods of specimen preparations, *Trans. Am. Microsc. Soc.* **93**:124–131.

Rykard, D. M., Luttrell, E. S., and Bacon, C. W., 1984, Conidiogenesis and conidiomata in the Clavicipitoideae, *Mycologia* **76**:1095–1103.

Samson, R. A., and Evans, H. C., 1982, *Clathroconium*, a new helicosporous hyphomycete genus from spiders, *Can. J. Bot.* **60**:1577–1580.

Samson, R. A., Stalpers, J. A., and Verkerke, W., 1979, A simplified technique to prepare fungal specimens for scanning electron microscopy, *Cytobios* **24**:7–11.

Seale, T., 1973, Life cycle of *Neurospora crassa* view by scanning electron microscopy, *J. Bacteriol.* **113**:1015–1025.

Seligman, A. M., Wasserkrug, H. L., and Hanker, T. J., 1966, A new staining method (OTO) for enhancing contrast of lipid-containing membranes and droplets in OsO_4 fixed tissue with osmophilic TCH, *J. Cell Biol.* **30**:424–432.

Skaer, H., 1982, Chemical cryoprotection for structural studies, *J. Microsc. (Oxford)* **125**:137–147.

Sleytr, U. B., and Robards, A. W., 1982, Understanding the artefact problem in freeze fracture replication: A review, *J. Microsc. (Oxford)* **126**:101–122.

Smith, H., 1977, Microbial surfaces in relation to pathogenicity, *Bacteriol. Rev.* **41**:475–500.

Stein, O. L., and Gay, J. L., 1974, Effect of accelerating voltage in scanning electron microscopy, *Trans. Br. Mycol. Soc.* **63**:404–408.

Stolk, A. C., and Samson, R. A., 1983, The ascomycete genus *Eupenicillium* and related *Penicillium* anamorphs, *Studies in Mycology*, Centraalbureau Voor Schimmelcultures Publ., Baarn, Netherlands.

Sun, S. H., Sekhon, S. S., and Huppert, M., 1979, Electron microscopic studies of saprobic and parasitic forms of *Coccidioides immitis*, *Sabouraudia* **17**:265–273.

Sun, S. H., Cole, G. T., and Harrison, J. L., 1985, In vivo ontogeny of the *Coccidioides immitis* parasitic cycle, in: *Abstr. Annu. Meet. Am. Soc. Microbiol.*, Las Vegas (March 1985), p. 368.

Sutton, B. C., 1980, *The Coelomycetes: Fungi Imperfecti with Pycnidia, Acervuli and Stromata*, Commonw. Mycol. Inst., Kew, Surrey, England.

Sutton, B. C., and Cole, G. T., 1983, *Thozetella* (Hyphomycetes): An exercise in diversity, *Trans. Br. Mycol. Soc.* **81**:97–107.

Tanaka, K., 1980, Scanning electron microscopy of intracellular structures, *Int. Rev. Cytol.* **68**:97–126.

Tanaka, K., 1981, Demonstration of intracellular structures by high resolution scanning electron microscopy, *Scanning Electron Microsc.* **II**:1–8.

Tanaka, K., and Mitsushima, A., 1984, A preparation method for observing intracellular structures by scanning electron microscopy, *J. Microsc. (Oxford)* **133**:213–222.

Tanaka, K., and Naguro, T., 1981, High resolution scanning electron microscopy of cell organelles by a new specimen preparation method, *Biomed. Res.* **2**(Suppl.)63–70.

Tokunaga, J., Edanaga, M., Fujita, T., and Adachi, K., 1974, Freeze cracking of scanning electron microscope specimens: A study of kidney and spleen, *Arch. Histol. Jpn.* **37**:165–182.

Venables, J. A., 1976, *Developments in Electron Microscopy and Analysis*, Academic Press, New York.

Wangensteen, D., Bachofen, H., and Weibel, R., 1981, Effects of glutaraldehyde or osmium tetroxide fixation on the properties of lung cells, *J. Microsc. (Oxford)* **124**:189–196.

Watson, L. P., McKee, A. E., and Merrell, B. R., 1980, Preparation of microbiological specimens for scanning electron microscopy, *Scanning Electron Microsc.* **II**:45–56.

Wellman, F. L., 1961, *Coffee: Botany, Cultivation and Utilization*, Interscience, New York.

Williams, S. T., and Veldkamp, C. J., 1974, Preparation of fungi for scanning electron microscopy, *Trans. Br. Mycol. Soc.* **63**:408–412.

Chapter 2

Low-Temperature Scanning Electron Microscopy

Alan Beckett and Nick D. Read

Department of Botany
University of Bristol
Bristol BS8 1UG, England

1. INTRODUCTION

Low-temperature scanning electron microscopy (LTSEM) involves three main operational phases:

1. The specimen is rapidly frozen ("quench-frozen") after which it is maintained either under vacuum or in a dry, argon atmosphere at a temperature below −130°C (143K). This is generally considered to be the point above which the recrystallization of pure water will occur (e.g., see Talmon, 1982a, and Read *et al.,* 1983, and references therein). Quench-freezing rapidly transforms freezable cellular and extracellular water into its solid state (ice) and the specimen is considered to be fully frozen-hydrated (FFH).
2. The FFH specimen may be fractured, dissected, or retained intact and, if required, it may be heated (etched) and/or coated. If the sample is etched, variable amounts of water are removed by sublimation and the specimen may then be considered partially freeze-dried (PFD).
3. The sample is observed at low temperature [approx. −175°C (98K)] on a temperature-controlled stage in the scanning electron microscope.

Uncoated specimens may be repeatedly dissected or etched if necessary during the second phase of operation and they may also be etched while under observation in the microscope using a conductive heater built into the base of the cold stage.

Present address for NDR: Department of Botany, University of Edinburgh, Edinburgh EH9 3JH, Scotland.

Why was the concept of LTSEM developed and why should the method be used for biological samples? In answer to these questions, the following points must be considered.

Before a specimen is examined at ambient temperature in an electron microscope, the major component of it, namely water, is normally removed. This is necessary because the high-vacuum environment ($< 10^{-5}$ Torr) within the microscope causes water to rapidly evaporate from a fresh sample, resulting in catastrophic alterations in specimen morphology. For ambient-temperature scanning electron microscopy (ATSEM), water removal is conventionally achieved by either (1) dehydration with organic solvents and critical point-drying or (2) freeze-drying, commonly at approximately $-65°C$ (208K) in the presence of a drying agent (e.g., see Read et al., 1983; see also Chapter 1). The removal of specimen water by dehydration and drying even under controlled conditions is a potentially damaging process and must significantly affect the structure and dimensions of a sample. It follows therefore that any process that immobilizes the water and retains it stabilized in situ within a specimen offers a considerable advantage over conventional procedures.

For practical purposes we are unable to study living organisms in the scanning electron microscope and so the dynamic activities of the individual cell or groups of cells must be immobilized and stabilized as quickly as possible. This is done by the process of fixation for which there are two major aims:

1. We aim to halt and preserve the momentary distribution or location of components within a system and in so doing provide a temporal or dynamic resolution of that system.
2. To preserve the structural relationships between the various components of the system and so obtain a spatial or morphological resolution of that system.

Conventional fixation procedures require that specimens be immersed in chemical fluids or subjected to chemical vapors. Chemical immersion fixation is a relatively slow process and chemical vapor fixation may be a very slow process (Colotelo, 1978; see also Read et al., 1983). For example, Mersey and McCully (1978) found that with a variety of chemical fluid fixatives, the average time needed to completely immobilize the cytoplasm in the petiolar hairs of tomato plants was 6–30 min. The observed penetration rate also varied for different fixative and buffer mixtures, but for glutaraldehyde in cacodylate buffer an average rate of 140 $\mu m/min$ was obtained (Mersey and McCully, 1978). The rate of penetration of a chemical fixative into and throughout a specimen will depend, among other things, on specimen size, structure, and composition. The highly heterogeneous nature of many biological structures can significantly influence the mode in which these structures become fixed. For example, it is unlikely that the main route for the entry of chemical fixatives into a mature perithecium of the fungus Sordaria humana (see Read and Beckett, 1985) is through the thick, pigmented walls of the peridial cells. If we assume that entry is via the ostiole and, to some extent, through the mycelial hyphae subtending the perithecium, then using the results obtained by Mersey and McCully (1978), an estimate for time taken to immobilize all of the cells of such a perithecium would be in excess of 5 min. Osmium tetroxide vapor treatment of similar "bulk" samples at ambient temperatures takes considerably longer to immobilize the cells and with some specimens can result in secondary structural changes. Two examples of the latter are: (1) asci and excessive mucilage are extruded from vapor-fixed perithecia (Read et al., 1983), and (2) gills on segments from the pileus of Coprinus cinereus mature and deliquesce during the period of exposure to OsO_4 vapor (e.g., 48–96 hr). This suggests that fixation may not have reached completion during this time.

In contrast to chemical fixation, fixation for LTSEM is a physical process; it is achieved by rapid cooling (cryofixation) and as such not only avoids the use of chemicals but, more importantly, is a much quicker process. If we again use the mature perithecium of S. humana as an example, then the time taken to freeze all of the cells in the perithecium may be estimated with the aid of an equation derived by Bald (1975, his equation 20). If we assume that the mature perithecium is a cylindrical specimen of high water content and has a radius of 150 μm, then the time taken for the freezing front (0°C) to penetrate to the center when the perithecium is plunged into subcooled nitrogen (using a heat transfer coefficient of 7230 $Wm^2 K^{-1}$; Bald, 1984) is of the order of 50 msec. In practice, the efficiency of this cooling method will be reduced at atmospheric pressure because of the mediating effect of the temperature difference between the specimen and the cryogenic fluid during plunging. Furthermore, specimen freezing will occur at temperatures below 0°C because of the effects of undercooling (see Franks, 1981). However, it is clear that the immobilization time obtained by rapid cooling is several orders of magnitude shorter than that achieved by chemical fixation.

In a recent theoretical analysis of the times required to freeze biological tissues at depths of 10 μm, Jones (1984) calculated that cooling rates in excess of $4 \times 10^4 Ks^{-1}$ were associated with freezing times of less than 0.5 msec. These times are much shorter than the time estimated to freeze perithecia for LTSEM because (1) we have to freeze tissues at depths in excess of 10 μm, and (2) plunge freezing in subcooled nitrogen provides relatively slow cooling rates (1.1 $\times 10^2 Ks^{-1}$ Robards and Crosby, 1978). Quench-freezing for LTSEM is not as rapid as that employed for freeze-substitution (see Chapter 7) or for low-temperature transmission electron microscopy (LTTEM) (Dubochet et al., 1983; Lepault et al., 1983; Adrian et al., 1984) of microorganisms. Nevertheless, we have found it to be rapid enough for most morphological studies, although examples have been found where the freezing rate seems to be too slow (see Section 4.3).

Quench-freezing of bulk specimens is the technique of choice for the preservation and immobilization of migratory components such as electrolytes within cells and tissues. Moreover, in FFH material the pattern of elemental distribution, particularly within extracellular, fluid-filled spaces, is not destroyed by the effects of dehydration and/or drying. With FFH bulk specimens, an analytical spatial resolution on the order of 1 μm may be achieved (Hall and Gupta, 1984). Although the shape of the analyzed volume in these specimens is uncertain (Hall and Gupta, 1984), recent experimental work by Oates and Potts (1985) has provided data on electron beam penetration and X-ray excitation depth in ice. These data are in accord with the theoretical estimates of Marshall (1982). These and earlier studies illustrate the ability to perform spatial analysis of immobilized ions in untreated tissues and together have been a major stimulus for the development of apparatus by which these specimens can be prepared.

Finally, it should be noted that although LTSEM cannot at present compete in terms of resolution with the well-established technique of freeze-etching for transmission electron microscopy (TEM), it does enable us to observe specimen surfaces directly over large areas. Repeated fractures or dissections of the specimen can also be made and furthermore the intact or fractured specimen can, if necessary, be retrieved either from the cryoapparatus work chamber or from the microscope after viewing (Pawley et al., 1980). With freeze-etching the specimen is of course destroyed to release the replica.

2. HISTORICAL REVIEW

Numerous papers have been published describing apparatus for preparing and maintaining samples at low temperature for SEM. The following account will not be a comprehensive review of these reports but rather a guide to some of the major developments in the field during the last 15 years. Further details of various aspects of LTSEM are given in several review-type papers (Nei, 1974; Robards, 1974; Koch, 1975; Echlin, 1978; Robards and Crosby, 1979) and in the book edited by Echlin et al. (1978).

Instruments for cryopreparation may be conveniently classified as "nondedicated" or "dedicated" (Robards and Crosby, 1979). Nondedicated systems are not attached directly to the microscope and frozen specimens, once prepared, are transferred to it within a transfer device that can be made to interface with any make or model of microscope. Dedicated systems are attached directly and more or less permanently to the microscope and as such are usually only compatible with that particular model.

2.1. Nondedicated Instruments

The feasibility of LTSEM of biological specimens was first demonstrated with a series of publications by Echlin and colleagues working in conjunction with Cambridge Scientific Instruments Ltd. (Echlin et al., 1970; Echlin, 1971;

Echlin and Moreton, 1973, 1976). Their procedures for freezing, coating, transfer onto the microscope cold stage, and etching the frozen specimen have been described in detail (Echlin and Moreton, 1973, 1976). Their results included the first micrographs by LTSEM of microorganisms (algae and protozoans) to be published (Echlin et al., 1970; Echlin, 1971). Frozen specimens were coated with carbon and aluminum or gold in a Balzers BAE 120 vacuum evaporator equipped with a rotating cold stage. Transfer to the microscope was at atmospheric pressure with the specimen stub protected by an aluminum cap. The fact that superficial ice, formed either from natural surface water or from condensed atmospheric water, could be removed by etching was also demonstrated (Echlin et al., 1970; Echlin, 1971). The specimen stage in this system was a modified Cambridge 200-SEM hot–cold stage and, as with many subsequent stages, was cooled by heat extraction through copper braid attached to a liquid nitrogen reservoir. By this method, under high vacuum and normal viewing conditions, the temperature of the cold stage was maintained at −175°C (98K).

The need for adequate protection during the transfer of an FFH specimen in the nondedicated apparatus from one part of the system to another or to the microscope, stimulated the design and development of a purpose-built transfer device (Robards and Crosby, 1978). This was potentially capable of interfacing with any microscope and with the more comprehensive preparatory system subsequently developed (Robards and Crosby, 1979). Their detailed development studies (see Robards and Crosby, 1979) resulted in the production of the first commercially available nondedicated system in the United Kingdom. This was manufactured by EMscope Laboratories Ltd., as their Sputter Cryo System. The design criteria to be embodied in this instrument were considered in depth by Robards and Crosby (1979) and briefly included the following:

1. The acceptable maximum and minimum temperature range limits during the whole preparatory operation
2. The specimen freezing method
3. Whether the system should be dedicated or nondedicated; in the latter case, what type of transfer device would be needed
4. The method of fracturing and manipulating the frozen specimen and how it might be etched and coated
5. The type of cold stage and its construction and the provision of an anticontaminator.

All of the requirements imposed by these criteria were satisfied by the EMscope Sputter Cryo System produced in 1978 and much of the early application work done in the senior author's laboratory was accomplished with this instrument interfaced with a Philips SEM 501B (Beckett, 1982; Beckett and Porter, 1982; Beckett et al., 1982, 1984; Campbell and Porter, 1982; Campbell, 1983; Read and Beckett, 1983, 1985; Read et al., 1983; Williams et al., 1985). In 1980, EMscope Laboratories produced a redesigned instrument that was marketed as the SP2000 Sputter-Cryo Cryogenic-Preparation System. This model is now routinely used in the senior author's laboratory and several of the micrographs presented in this chapter are of specimens prepared with this apparatus. The SP2000 will be described in detail in Section 3.

2.2. Dedicated Instruments

The first dedicated system was that developed by JEOL Ltd. and used, with modifications, by several workers in Japan (Nei *et al.*, 1971, 1973; Hasegawa and Yotsumoto, 1972; Tokunaga and Tokunaga, 1973; Harada and Okuzumi, 1973; Hasegawa *et al.*, 1974). This system comprised a preevacuation chamber permanently bolted to the side of the microscope specimen chamber that housed a cold stage, a precooled knife, an evaporative coating head, and a heater (see: Hasegawa *et al.*, 1974; Nei and Fujikawa, 1977). Subsequent modifications allowed for gold to be sputter-coated onto the specimen at low temperature, a technique that was first developed and demonstrated by Ken Oates at Lancaster University (see below).

The most detailed report of a dedicated system is that by Pawley and Norton (1978) on the Biochamber attached to the AMR-1000A scanning electron microscope. This is a high-vacuum, preevacuated device with gate valve interlocks between the evaporating head and the microscope specimen chamber. A "shuttle" enables the specimen to be moved from one position to another within the preevacuation chamber so that it may be fractured, etched, and coated as required. The microscope cold stage accommodates a Joule–Thompson refrigerator and allows for considerable X, Y, tilt, and rotation of the specimen.

During the period 1978–1981, the prototype of a similar dedicated system was developed at Lancaster University (Lancaster, United Kingdom) by Ken Oates. This was based on the original JEOL cryo system but incorporated a temperature-controlled microscope stage, an improved anticontaminator, a transfer mechanism, and a preparation chamber with facilities for sputter-coating, carbon and metallic evaporation at high vacuum. In 1980, Hexland Ltd. (Oxfordshire, United Kingdom) produced a commercial version of this system, and in recent years this has been substantially modified and improved; the most recent models are the CT1000 Cryotrans System with modular electronic controls and the CP2000 Cryo Preparation Chamber adapted for use with samples for both SEM and TEM.

A recent addition to the Balzers range of equipment for SEM is the SCU 020 Cryo Preparation Chamber. This is also a dedicated system.

Significant advances have been made in the design of instrumentation often with the intention of performing X-ray microanalysis on frozen-hydrated samples. This important aspect of LTSEM will not be dealt with in this chapter because very little microanalytical work has been done on frozen-hydrated microorganisms. X-ray microanalysis of frozen-hydrated specimens both in bulk form and as sections has, however, been reviewed in several papers and the reader is referred to these for further details (Fuchs and Linderman, 1975; Fuchs *et al.*, 1978; Fuchs and Fuchs, 1980; Marshall, 1980; Echlin *et al.*, 1982; Zierold, 1983; Hall and Gupta, 1984).

The suitability of quench-freezing to immobilize soluble ions in fungal hyphae was demonstrated by Galpin *et al.* (1978). These workers did not provide details of the cryo apparatus used but it was a version of the JEOL system, progressively modified by Ken Oates, already described above.

3. THE APPARATUS AND METHODS USED

The operational and experimental procedures for LTSEM obviously vary in detail depending on the apparatus used and the particular specimen(s) under investigation. However, because relatively little work has been published on LTSEM of microorganisms, we will confine our detailed discussion in this section to the instrumentation used and the specimens studied in our laboratory. Reference to work done elsewhere will be made where relevant.

3.1. The SP2000 Sputter-Cryo Cryogenic Preparation System

The apparatus (Figure 1) consists of the following components.

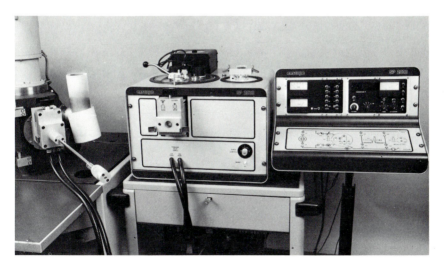

FIGURE 1. The EMscope SP2000 Sputter-Cryo Cryogenic-Preparation System interfaced with a Philips scanning electron microscope 501B.

3.1.1. The Control Module

This houses: switches for the main electricity supply, rotary pump, vacuum and argon valves; the temperature preset control; thermocouple selection switch; and the various coating and heating head switches. An illuminated mimic display of the vacuum and argon circuits enables the sequence of operations to be readily followed. A temperature readout displays the temperature at any one of three thermocouple sites: two in the work chamber, one on the microscope cold stage.

3.1.2. The Preparation Unit

This consists of a freezing chamber and a preparation (or work) chamber. Both chambers may be evacuated by a rotary pump or purged with argon.

The freezing chamber contains a polystyrene cup within which liquid nitrogen is held prior to slushing. A small pressure valve enables argon to be released from the chamber when under positive argon pressure.

A T-shaped, shrouded, cold stage is located in the base of the preparation chamber and is connected by a copper rod to a liquid nitrogen Dewar (Figures 2 and 3). A port on the front of the chamber allows access for specimens and is sealed by a gate valve. The transparent cover of the chamber is fitted with: a manipulator/probe and fracturing device; a carbon evaporation head; a radiant-heating filament; and a sputter-coating head (Figure 3). One of the specimen-locating bays on the cold stage is fitted with a conductive heater.

3.1.3. The Transfer Device

This comprises a transfer chamber, sealed by a gate valve and a transfer rod (Figure 2). The chamber may be evacuated or purged with argon. The rod has a threaded, nylon end piece for attachment to the specimen stub (Figure 4).

3.1.4. The Specimen Stub

The copper stub has a stainless steel shroud that may be opened or closed by the rotation of a molded nylon sleeve at the end of the transfer rod (Figures 4 and 5). Depending on the type of specimen to be mounted, the surface of the stub may be flat or grooved and/or drilled with holes.

3.1.5. The Microscope Cold Stage

This is custom-built for the make and model of microscope. It is cooled by heat extraction through a flexible copper rope that is connected to a liquid nitrogen Dewar. A copper anticontaminator is also connected to the Dewar and projects into the microscope specimen chamber beneath the final polepiece. A custom-built, gate-valved, interface port provides access to the cold stage.

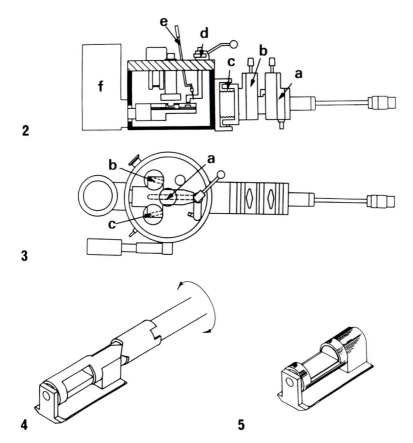

FIGURE 2. Side elevation of the SP2000 preparation chamber and transfer device. a, transfer device; b, access valve; c, pivot assembly; d, macro fracture device; e, macro manipulator; f, liquid nitrogen Dewar.

FIGURE 3. Plan elevation of the SP2000 preparation chamber and transfer device. a, sputter-coating head/stage; b, metal evaporation–etching head/stage; c, carbon evaporation head/stage.

FIGURE 4. Specimen stub on the end of the transfer rod. Note the molded, nylon sleeve and the half-closed stub shroud.

FIGURE 5. Specimen stub with the shroud opened.

3.2. The Operational Sequence

3.2.1. Cooling the Apparatus

When not in use, the preparation chamber is normally continuously pumped to vacuum (10^{-2} Torr) at ambient temperature. With the addition of liquid nitrogen to the Dewar, the temperature falls and the vacuum improves to approximately 10^{-3} Torr at $-100°C$ (173K). After 30 min, the temperature of the cold stage within the preparation chamber reaches −180°C (93K). It is necessary to continually "top up" the nitrogen Dewar during the cooling process and approximately 3–4 liters of liquid nitrogen are needed to lower the cold stage temperature to −180°C. During this time the cold stage in the microscope, which is already under vacuum, can be cooled by adding liquid nitrogen to the Dewar mounted on the side of the microscope specimen chamber. The time taken for this cold stage to cool from ambient temperature to −178°C (98K) is between 10 and 15 min. The vacuum at this temperature, as measured with a penning gauge situated immediately above the diffusion pump, is 10^{-6} Torr.

3.2.2. Mounting and Freezing the Specimen

Two processes are involved here: (1) the mounting of a living specimen onto the surface of a copper stub, and (2) the rapid freezing of the mounted specimen in subcooled nitrogen. It is important to avoid unnecessary delay between these two steps because secondary changes to the specimen can result from premature dehydration or other environmental stress. For this reason the freezing chamber is filled with liquid nitrogen and pumping is initiated before the specimen is prepared for mounting. While the chamber is pumping a specimen stub is secured onto the end of the transfer rod. After about 45 sec, the nitrogen begins to solidify and at this point delicate specimens may be quickly excised (if necessary), removed from their culture vessel, and placed onto the stub surface. Dry samples can be attached, prior to freezing, with a thin film of 2.0% aqueous methyl cellulose. Depending on their nature and whether they are to be fractured or not, dry specimens may be placed flat on the stub surface or inserted in grooves or holes cut into its surface. Wet samples may be placed as a drop on the stub surface, in which case they are usually held in place by surface tension. Alternatively, excess water can be removed by mounting the drop on a pad of Millipore filter (0.3-μm pore size) overlying a similar pad of No. 1 Whatman filter paper that is attached by a thin film of methyl cellulose to the stub surface. One or two drops of cell suspension is added to the filter, taking care not to allow the drops to overflow the sides of the pad. The Millipore filter provides a suitably permeable surface to allow a slow leakage of excess water into the filter paper pad while maintaining a water film around the cells.

Once mounted, the specimen must be quench-frozen as quickly as possible. The stub and mounted sample is withdrawn into the transfer chamber. The vacuum line to the freezing chamber is now closed and argon is released into the chamber. The nitrogen melts and as the pressure becomes positive the chamber lid may be removed and the transfer device can be fitted over the top of the freezing chamber. The gate valve is opened and the specimen stub and sample are rapidly plunged into the subcooled nitrogen. The stub reaches equilibrium with the temperature of the nitrogen after about 40 sec. With practice, the interval between interfering with the specimen and plunging it into the nitrogen can be as short as 25 sec but this will obviously depend on the type of specimen, how it is grown (cultured) or otherwise maintained, and therefore how much manipulation may be necessary prior to attaching it to the stub.

The stub shroud may now be closed over the specimen. The transfer chamber is quickly purged with argon gas and the stub is withdrawn into it. The chamber is closed by the gate valve, argon is switched off, and the vacuum line is opened. The protected specimen can now be transferred either directly to the microscope as an FFH, uncoated intact specimen (Figure 6), or to the preparation chamber for further manipulation (Figure 6).

3.2.3. Fracturing, Etching Blind, and Coating the Specimen

When in the work chamber, the shroud is opened and the specimen can be checked by eye through the Perspex cover to ensure that it has been adequately frozen without obvious damage. This simple procedure can save much time later because it avoids time spent on possibly etching and coating an inferior specimen that may prove unstable in the microscope. For this reason, in our laboratory the specimen is routinely transferred to the work chamber for checking even if it is to be subsequently viewed intact and uncoated.

If the specimen is suitable, several options are available (Figure 6, images 2–13). These include fracturing or dissecting the specimen (e.g., plant leaves infected with fungal pathogens; soil particles and roots; rotting leaves), etching the surface by radiant or conductive heat, and sputter- or evaporative-coating. Specimens in liquid suspension can often be prepared successfully by subliming off the excess water by subjecting the specimen surface to radiant heat from a coiled tungsten wire situated 6–9 mm above it. Extra cold trapping to catch subliming water molecules can be provided by placing a precooled knife blade close to the specimen. Etching done in the preparation chamber, whether by radiant or conductive heat, must be performed empirically, i.e., it is done "blind" and the extent of the etching is unknown until the specimen is inspected on the cold stage of the microscope. If insufficient etching has been achieved, the process may be repeated, either "blind" in the work chamber or, alternatively, the specimen may be retained on the microscope cold stage and, using the conductive heater in this stage, it can be etched while observed in the microscope. This combination of "blind" and visual etching has been successfully performed on some algal suspensions in the senior author's laboratory but because it is not a routine operational procedure, it is not included in the flow chart (Figure 6).

Once the temperature of the specimen has been raised sufficiently to cause water loss by sublimation, the specimen may be considered to be PFD. Normally, four types of PFD, "blind" preparations may be recorded (Figure 6, images 4, 5, 9, 10). These will depend on whether the specimen is coated or not and whether it is fractured/dissected or not. Manipulative procedures may be

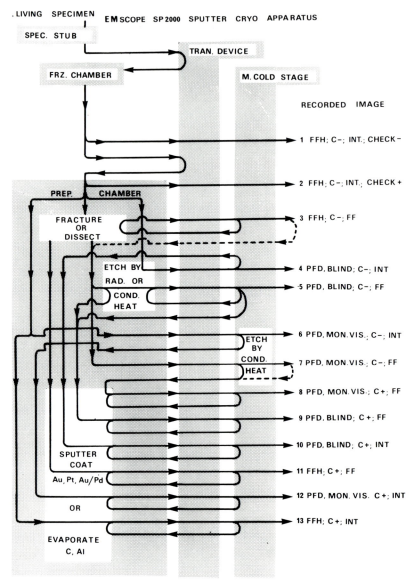

FIGURE 6. Flow diagram illustrating the possible operational routes and different recorded images using the EMscope SP2000 Sputter-Cryo Apparatus. Any of the uncoated, fractured specimens (recorded images 3, 5, 7) may be coated by evaporation and subjected to X-ray microanalysis. In addition, most specimens may be repeatedly fractured or dissected and recoated. Key: Apparatus.— COND., conductive; FRZ., freezing; M., microscope; PREP., preparation; RAD., radiant; SPEC., specimen; TRAN., transfer. Recorded images—C−, uncoated; C+, coated; CHECK−, unchecked; CHECK+, checked; FF, freeze-fracture; FFH, fully frozen-hydrated; INT., intact; MON, VIS., monitored visually; PFD, partially freeze-dried.

repeated if necessary on the same specimen (Figure 6) and in this way, successive layers, of for example soil and mucilage, can be removed to expose microorganisms *in situ* in the rhizosphere region of the root.

Samples may be sputter-coated, normally with gold, but platinum and gold/palladium targets may be used instead. The amount of gold coating should ideally be as little as is sufficient to provide an acceptable image, free of charging and structural artifacts. This amount may, however, vary significantly with specimen type, surface topography, and the resolution and magnification required for recording. It is necessary therefore that each worker adopts his/her own routine for coating his/her particular specimens. For most fungal specimens studied in our laboratory, coating for 2 min with a new gold target at 10^{-1} Torr and with a current of 25 mA provides an acceptable image for relatively low-magnification ($< \times 3700$) work. However, for higher magnifications, for example of the surfaces of uredospores of *Uromyces viciae-fabae* (Figure 16), coating for only 30 sec is required because more than this results in visible coating artifacts. The flexibility of the nondedicated cyro apparatus such as the SP2000 is such that specimens can if necessary be manipulated and coated repeatedly. These possible repetitive loops are illustrated in Figure 6.

3.2.4. Transfer to the Microscope Cold Stage and Etching under Observation

When all operations within the work chamber have been completed, the stub and attached specimen can be withdrawn into the transfer chamber and both this and the work chamber are sealed by closing the relevant gate valves (Figure 2). For this operation the specimen may be protected by closing the stainless steel shroud over the stub. However, we have not found this to be essential and our studies on FFH fracture faces of both plant and fungal material, transferred with an open shroud, have produced no visible evidence of surface contamination. The transfer chamber is immediately evacuated while the space, termed the "dead space," between the valves of the transfer device and the preparation chamber interface (Figure 2) is pressurized with argon gas. This facilitates the removal of the transfer device, which is then quickly fitted to the interface on the microscope specimen chamber. When in place, the argon flow to the "dead space" is switched off and the vacuum line is switched on. When sufficient vacuum has been attained in both "dead space" and transfer chamber ($< 5 \times 10^{-2}$ Torr) the gate valves are opened and the stub is positioned onto the prealigned cold stage by means of the transfer rod. The rod is unscrewed, withdrawn from the microscope, and both gate valves are again closed. The specimen is now ready for viewing.

Etching in the microscope is not practicable for specimens actually suspened in water. This is because the time needed to sublime off relatively large amounts of water is excessively long and the extent of contamination in the microscope will soon reach unacceptable levels. Such specimens must either be first etched "blind" (Section 3.2.3) to remove excess water or they can be mounted on pads of Millipore filter and filter paper (see Section 3.2.2). We have found cells prepared in the latter way to be suitable for etching while visually monitored in the microscope. Suspension of *Micrasterias,* various diatoms, zoo-sporic fungi, and dental plaque bacteria have been PFD by this method (e.g., Figures 26 and 27; see also Section 4.10).

Prior to etching, the stage temperature may be preset with a selector switch on the control module. Depending on the specimens, the stage temperature selected for etching will vary. Most of our preliminary experiments were carried out at a stage temperature of $-65°C$ (208K) (the temperature at which we routinely "fully" freeze-dry specimens to be subsequently observe at ambient temperature; e.g., Read *et al.,* 1983). More recently, however, we have opted for a temperature of $-90°C$ (183K). If biological material contained only pure water, then in theory the rate of sublimation at $-65°C$ will be in excess of 60 nm S^{-1} while at $-90°C$ it will be 10 nm S^{-1} (Robards and Crosby, 1979). However, neither of these rates will actually apply to biological specimens because:

1. The two temperatures cited are measured by thermocouples in the base of the cold stage and could therefore be several degrees higher than the actual specimen surface from which water is sublimed.
2. Much, if not all, cellular water differs in its physical properties from pure water (Clegg, 1982). In multicellular tissues, different cells and extracellular materials contain different amounts of water. This water will be unevenly distributed and show varying properties depending on its particular ionic or molecular environment. Sublimation rates will therefore vary spatially over tissues and their components (see also Section 4.10).

With these points in mind, we would maintain that for this process of visually monitored etching, there is no substitute for experimentation mediated by experience. Thus, with our apparatus, $-90°C$ provides adequate control over etching rates and has, for most microorganisms, been a suitable temperature from which to proceed with trial runs. The efficiency of the stage heater is such that it takes only 1.5 min for the stage to reach the preset temperature ($-90°C$). A typical etching period would be a further 8–10 min. During this time, frozen water droplets are often observed clumping together and then gradually decreasing in size until they disappear from the surface. These changes vary in nature over different regions of a specimen.

When it is deemed that sufficient sublimation has occurred, the stage heater is switched off and the specimen may be either transferred immediately to the preparation chamber, or, as is more usual in our laboratory, retained on the cold stage until the temperature has again fallen to below $-130°C$ (143K). At this point it is transferred to the preparation chamber for coating. It is of course possible, and indeed it is good practice, to record samples at low accelerating voltage (e.g., 5 kV) in the uncoated, FFH state; then again in the uncoated, PFD state; and finally in the coated, PFD state (Figure 6, e.g. images 3, 7, 8). For the latter recording, we now routinely use accelerating voltages of 8–10 kV. A series of micrographs produced in this way provides useful information on the nature and location of sublimable fluids (mainly water) in any one part of the specimen (see also Section 4.10).

4. A COMPARATIVE EVALUATION OF THE TECHNIQUE

4.1. The Need to Compare LTSEM with Conventional Procedures

To obtain a reliable understanding by microscopy of the structure–function relationship within a given biological system, it is essential to employ as many methods of preparation and observation as are available. This is particularly true with regard to preparatory procedures for SEM. No one technique surpasses all others and each has advantages and disadvantages depending on the type of material under investigation and the information required (Read *et al.*, 1983). Accurate interpretation of results often depends on correlated information from several methods. We wish to emphasize therefore that LTSEM does not represent a panacea for preparatory artifacts. In view of this, throughout this section we will compare results from LTSEM with those from the more conventional procedures such as critical point-drying and freeze-drying, further details on which are provided in Chapter 1. It is our aim, however, to demonstrate that, if the facilities are available, frozen-hydrated (used here to include FFH and/or PFD) material should be used as the baseline against which other results are judged.

4.2. Criteria Considered and Organisms Studied for Evaluation

We have established a number of criteria by which an evaluation can be made of results from a variety of organisms. These criteria are listed in Table I and the organisms with which our evaluation will be illustrated are presented in Table II.

4.3. Specimen Immobilization and Stabilization

The comparative attributes of rapid cooling versus various conventional methods for immobilizing and stabilizing biological specimens have already been discussed in detail (see Section 1). Although freezing bulk samples with subcooled nitrogen (as described in Section 3.2.2) provides rapid immobilization and stabilization, in some cases it still seems too slow to fix certain biological processes. One example that we have found relates to the appearance of the perithecial ostiolar pore region of *S. humana*. The ostiolar pore is the opening at the distal end of the perithecium neck within which individual asci are gripped and from which asci forcibly discharge their spores. The location of an ascus within an ostiolar pore prior to discharge is very short-lived. In a detailed study of the morphology of the ostiolar region prepared in different ways (Read *et al.,* 1983), it was found that the ostiolar pores of chemically fixed perithecia were always constricted (Figure 19) while those that had been cryofixed exhibited a range of diameters (Figures 20 and 21). Undischarged asci were never observed within the ostiolar pores of chemically fixed perithecia. On only two occasions were undischarged asci observed within the ostiolar pores of cryofixed perithecia (Read, unpublished observations). These asci were probably abnormal, however, because considerable superficial water had condensed on their surfaces (see Figure 35 in Read *et al.,* 1983), indicating that they may have occupied the pore for much longer than normal. Many hundreds of frozen-hydrated ostiolar pores were examined in the aforementioned study, and because asci discharge their spores regularly from individual perithecia, it is surprising that normal, undischarged asci were not observed more frequently. For example, Ingold (1928) found that during the day, asci of *Podospora curvula* were discharged at rates up to at least one every 6 min from a single perithecium. We now consider it likely that the large open appearance of ostiolar pores in cryofixed preparations is very transitory and represents the stretched condition just after an ascus has discharged its spores and retracted out of sight down the ostiolar canal (see Read and Beckett, 1985). Whether faster freezing with different cryogenic fluids (e.g., propane) can fix the undischarged ascus gripped within the ostiolar pore still

Table I
Criteria Considered for Evaluating the Technique
of LTSEM

	Criterion	Section
a.	Specimen immobilization and stabilization	4.3
b.	General preservation of external morphology	4.4
c.	General preservation of internal morphology	4.5
d.	Exposure to solvents	4.6
e.	Overall dimensional changes	4.7
f.	Cell surface texture	4.8
g.	Differential conformational changes	4.9
h.	Etching frozen-hydrated material	4.10
i.	Beam damage	4.11
j.	Specimen resolution	4.12
k.	Specimen life	4.13

Table II
Systems Studied and Specific Criteria Illustrated

Organism	Structures	Criteria illustrated	Figures
Bacteria	Dental plaque	h	26, 27
Coprinus cinereus	Basidia, basidiospores, spore droplets	a, g	38–41
Erysiphe graminis f. sp. *hordei*	Conidia, haustoria, surface mycelium	a–d, g	7–9
Flammulina velutipes	Gills	c, d, h	10–12
Gyrosigma sp.	Whole cells	h	28–30
Puccinia striiformis	Germinating uredospores	h	31, 32
Sordaria humana	Perithecia, ascospores	a, d, g, i	19–25, 36, 37
Uromyces viciae-fabae	Uredospores	e, f	13–18

needs to be determined. Another possibility to consider is that initial contact with the coolant prior to freezing might even have stimulated premature ascus dehiscence.

4.4. General Preservation of External Morphology

It is logical to assume that most cells and tissues of living, growing organisms, unless in hypertonic media or in naturally desiccating environments, should appear turgid. This is not the case, however, with most dried cells illustrated in scanning electron micrographs in the literature. Obviously, some specimens are more susceptible to changes in external morphology, as a consequence of preparatory procedures, than others (e.g., see Beckett *et al.*, 1982; Read, 1983; Read *et al.*, 1983; Read and Beckett, 1985). An extreme example is illustrated by the conidia and mycelial hyphae of *Erysiphe graminis* f. sp. *hordei* on barley leaves (Figures 7 and 8). FFH hyphae and conidia appear turgid with no evidence of distortion or collapse (Figure 7) and provide a marked contrast to the severely collapsed and distorted cells, which have been chemical immersion-fixed, dehydrated, and critical point-dried (Figure 8). Various modifications to the standard critical point-drying procedure can be employed to minimize damage, but this may be time-consuming and is often tedious. It is also important to note that in this particular example, the preservation of wax platelets on the epidermal cells of the host leaves is consistently superior in FFH preparations (Figure 7) than in critical point-dried ones (Figure 8). This feature may be linked with the use of solvents during critical point-drying (see Section 4.6).

Cells often collapse naturally as a result of desiccation or autolysis (Read *et al.*, 1983; Beckett *et al.*, 1984; see also Section 5.3). LTSEM represents the most reliable method involving SEM for identifying these states.

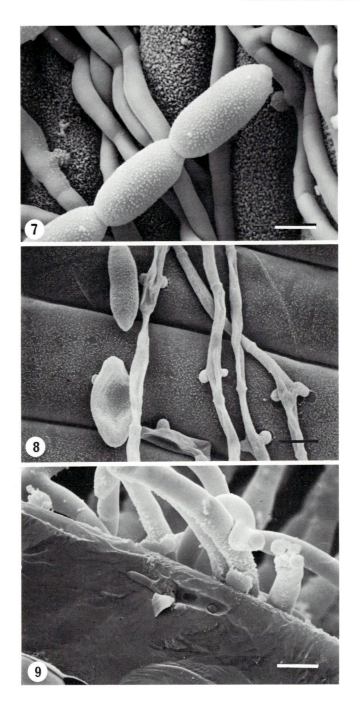

4.5. General Preservation of Internal Morphology

Frozen specimens are rendered amenable to fracturing and this exposes internal surfaces and structures. It is therefore relatively easy to break open a frozen leaf or similar multicellular structure, thus revealing internal surfaces and cell–cell structural relationships. This technique is particularly useful for observing intercellular and intracellular hyphae of fungal pathogens within plant hosts. For example, Figure 9 shows the digitate haustorium of *E. graminis* in a cross-fractured epidermal cell. The intercellular hyphae and haustoria of *U. viciae-fabae* have also been studied using LTSEM (Beckett *et al.*, 1982; Beckett and Porter, 1982).

Freeze-fracturing can of course be employed prior to freeze-drying or critical point-drying but the internal details revealed will differ significantly depending on which technique is used (see Section 4.6).

4.6. Exposure to Solvents

Many algal, fungal, and bacterial cultures produce extracellular secretions and cells may often be surrounded by these materials. In addition, cells may be covered by surface water droplets and/or films. Conventional preparatory procedures, for the most part, dissolve and wash away these labile components during fluid fixative treatments and, particularly, during dehydration with organic solvents. Mucilage remains may be preserved in drastically altered form in dried samples (Read *et al.*, 1983; Figure 19), and some extracellular matrix materials may be almost totally destroyed. For example, cross-fractured, FFH gills of the toadstool fungus *Flammulina velutipes* (Figure 10) reveal tramal and subhymenial cells embedded in a conspicuous extracellular matrix. Similar gills following freeze-drying (Figure 11) or critical point-drying (Figure 12) show a progressive loss of this matrix material. Only a few strands remain traversing apparent spaces in the critical-point-dried sample (Figure 12).

4.7. Overall Dimensional Changes

Biological specimens normally have high water contents and should be expected to shrink when this water is removed during dehydration and/or drying by conventional methods. When prepared in this way, they may appear mor-

←

FIGURE 7. *Erysiphe graminis* f. sp. *hordei*. FFH hyphae and conidia. Note turgid appearance of the fungus and the abundance of wax on the leaf surface. Sputter-coated (Au). Bar = 10 μm.

FIGURE 8. *E. graminis*. CPD hyphae and conidia. Both hyphae and conidia are distorted and collapsed and much of the wax has been removed from the leaf surface. Sputter-coated (Au). Bar = 10 μm.

FIGURE 9. *E. graminis*. FFH, freeze fracture through part of a barley leaf showing a digitate haustorium within an epidermal cell. Surface mycelium and conidiophores are also visible. Sputter-coated (Au). Bar = 10 μm.

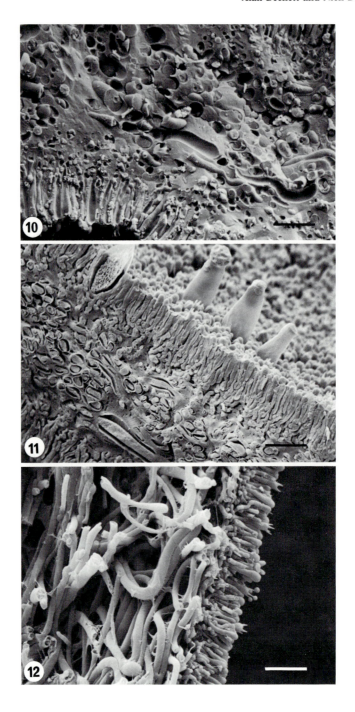

phologically intact and undamaged and probably because of this there has been little information published relating to the dimensional changes that must have occurred with respect to the original, living turgid sample. Water is retained in FFH specimens and LTSEM provides the opportunity to demonstrate that, in some cases, water loss during conventional preparatory treatments can lead to severe shrinkage (Beckett *et al.*, 1984; Figures 13–15). The technique also enables us to observe and monitor dimensional changes that are related to different water contents. This is particularly well illustrated by the uredospores of *U. viciae-fabae*, the water content of which varies according to the conditions under which they are formed (Beckett *et al.*, 1984; Beckett, unpublished results). Once formed, these spores are able to imbibe fluid water and, as a result, they swell. If swollen, "wet" uredospores are critical point-dried, they shrink to approximately 34% of their original volume. Ascospores of *S. humana* show similar but less drastic shrinkage.

Two types of water are traditionally considered to be associated with biological cells. They are (1) unfreezable (hydration or "bound") water and (2) ordinary ("bulk" or "free") water, which is freezable (Cooke and Kuntz, 1974; Finney, 1979; Franks, 1981; Clegg, 1982). However, this is probably a much simplified view of cellular water (Clegg, 1982) and so theoretical considerations of the possible expansion of the freezable water when frozen, are at least complex and possibly misleading. Thus, although pure water expands by approximately 9% when frozen at 0°C (273K), in living cells it is (1) not pure; (2) nonuniformly distributed, even within a single cell; and (3) is in different amounts in different cells according to their environmental conditions (see Section 3.2.4). It is likely then that whatever the expansion is that results from ice formation, it is almost certainly insignificant when compared to the differences in dimensions between "wet" and dried spores. Nevertheless, certain structures with high water content (e.g., mucilage, agar) may sometimes exhibit cracking or rupturing as a result of "freezing damage" involving slight specimen expansion (e.g., see Figure 25; and Section 4.9).

4.8. Cell Surface Texture

The surface textures of some fungal and plant cells that we have studied are seen at suitable magnification ($> \times 5000$) to vary considerably depending on

---

FIGURE 10. *Flammulina velutipes.* FFH, freeze fracture across a gill. The tramal, subhymenial, and hymenial cells are embedded in a conspicuous extracellular matrix. Sputter-coated (Au). Bar = 20 μm.
FIGURE 11. *F. velutipes.* FFD, freeze fracture across a gill. The extracellular matrix shows signs of rupture. Sputter-coated (Au). Bar = 20 μm.
FIGURE 12. *F. velutipes.* CPD, freeze fracture across a gill. The extracellular matrix is almost completely absent. A few remaining strands are visible between the tramal cells. Sputter-coated (Au). Bar = 20 μm.

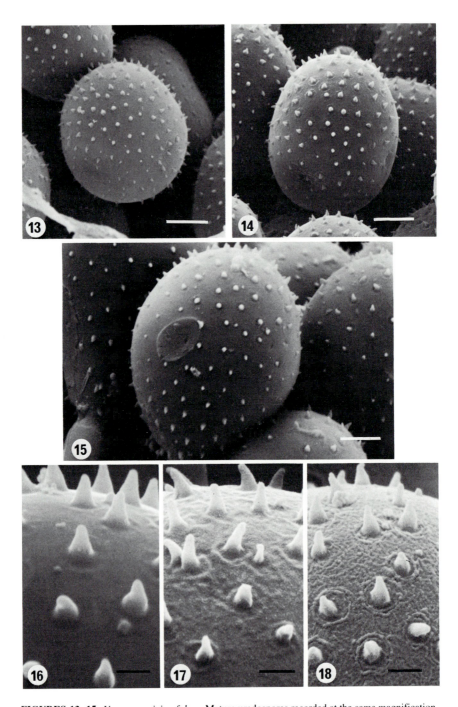

FIGURES 13–15. *Uromyces viciae-fabae*. Mature uredospores recorded at the same magnification. Sputter-coated (Au). Bars = 5 μm. Figure 13, CPD; Figure 14, FFD; Figure 15, FFH.

FIGURES 16–18. *U. viciae-fabae*. Surface texture of mature uredospores recorded at the same magnification. Sputter-coated (Au). Bar = 1 μm. Figure 16, FFH; Figure 17, FFD; Figure 18, CPD.

whether they are FFH, freeze-dried, or critical point-dried (Figures 16–18). It is possible that exposure to chemical fixative and solvents (Section 4.6) might influence this texture and certainly the shrinkage known to occur in critical-point-dried preparations (Section 4.7) is likely to cause surface wrinkling. Coating, if excessive, will also impart an artifactual surface texture, although this, it is claimed, would only normally be resolved within the magnification range of ×10,000–×25,000 (Bråten, 1978; Echlin and Kaye, 1979).

Recent work with uredospores of *U. viciae-fabae* (Read and Beckett, 1983) suggests that another factor that may influence surface texture is hydration water. In hydrated specimens, water interacts strongly with the polar groups of macromolecules to form hydration shells around them. The formation of these hydration shells is essential for maintaining the structural stability of many of the polymers that are important for providing the specimen surface with its characteristic morphology (see Read and Beckett, 1983, and references therein). If unfreezable water (see also Section 4.7) is removed from the specimen by drying, then the individual structures and associations of its constituent polymers may be significantly altered (e.g., Finney, 1977) and, thus, we have suggested (Read and Beckett, 1983) that the gross texture of the cell surface might also be affected. The important point is that dehydration and critical point-drying seem to eliminate all unfreezable water (Boyde *et al.*, 1981) and "complete" freeze-drying in the presence of a desiccant removes most, or possibly all, of it (Boyde and Franc, 1981). However, in FFH specimens, all unfreezable water is retained.

4.9. Differential Conformational Changes

Dimensional and structural changes may occur to different extents in different parts of a specimen. These artifacts, which we have termed *differential conformational changes* (Read and Beckett, 1982), may be manifested at the tissue, cellular, subcellular, and extracellular levels (Read *et al.*, 1983).

At the tissue level, these changes are exemplified by the greater shrinkage of tissues within the ostiolar canal than shrinkage exhibited by the surrounding peridium after complete freeze-drying (Figure 22). Differential shrinkage of tissues is absent from FFH perithecia (Read and Beckett, 1985).

Conformational changes at the cellular level are also most obvious in dried samples. Collapse and distortion of critical point-dried conidia (Figure 8) and fully freeze-dried (FFD) uredospores (Beckett *et al.*, 1982) are examples of this type of change. In contrast, FFH cells are always turgid except when they have dehydrated naturally as is the case for released spores of many fungi (Beckett *et al.*, 1984).

Disruption of cellular contents in frozen-fractured cells and tissues after complete and partial freeze-drying results in characteristic reticulate patterns at the subcellular level. These patterns (Figures 23 and 24) result from ice crystal damage (Read and Beckett, 1985; see also Section 4.10).

Finally, conformational changes may be found at the extracellular level.

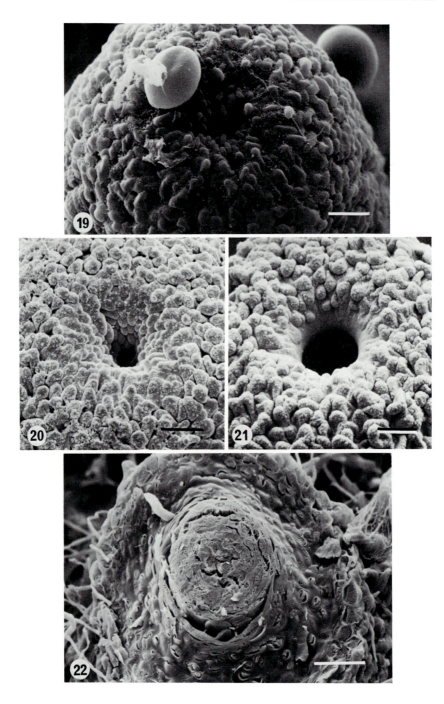

Following critical point-drying, soluble components of the perithecial mucilage in *S. humana* are lost and excessive rupturing or loss of the mucilage in the ostiolar pore region occurs (Figure 19). In contrast, mucilage is retained in FFD, PFD, and FFH preparations (Read *et al.*, 1983; Read and Beckett, 1985; Figure 25). In the former, tearing commonly occurs to varying extents, possibly as a result of the mucilage shrinking, during drying, more than the underlying cells. Surface cracking of FFH mucilage may result from differential thermal, dimensional changes in ice and ice-encapsulated water at various temperatures through the bulk of the perithecium. The strains developed in this way could be released at the surface in the form of cracks (Boyde, 1978; Read *et al.*, 1983; Figure 25). After complete and partial freeze-drying, rupturing can also be seen in the extracellular matrix within the gills of *F. velutipes* (Figure 11), and in the membranous pellicle that covers the hymenium of *Coprinus cinereus* (McLaughlin *et al.*, 1985; Figures 39 and 40). In the latter, tearing of the pellicle was evident in gills that had been PFD by conductive heating on the temperature-controlled stage in the microscope but was absent in gills subjected to radiant heating. The conductive heater is located in the base of the stage and would set up a thermal gradient across the stub to the specimen surface. Differential shrinkage of underlying cells at a higher temperature than the pellicle could have caused rupturing at the surface.

4.10. Etching Frozen-Hydrated Material

The ability to remove water under controlled conditions from frozen-hydrated specimens is a valuable option embodied within the routine operational procedures for LTSEM (see Section 3.2.4). Water is removed during vacuum sublimation by raising the temperature of the specimen so that the saturation vapor pressure of ice within it exceeds the partial pressure of water in the vacuum environment immediately surrounding the specimen. The provision of a liquid nitrogen-cooled cold trap just above the specimen ensures that water molecules are rapidly removed from the vacuum, thus preventing condensation and subsequent freezing of water droplets on the specimen surface.

FIGURE 19. *Sordaria humana.* CPD ostiolar pore region. Note constricted pore and the granular, membranous, and filamentous remains of the perithecial mucilage. Sputter-coated (Au). Bar = 10 μm.
FIGURES 20, 21. *S. humana.* FFH ostiolar pore regions. Note difference in diameter of the pores and the presence of mucilage between the cells of the periphyses lining the pores. Sputter-coated (Au). Bars = 10 μm.
FIGURE 22. *S. humana.* OsO$_4$ vapor-fixed, FFD, freeze fracture through the base of the perithecium neck illustrating differential conformational changes at the tissue level. Note greater shrinkage of tissues within the ostiolar canal than of the surrounding peridium. Sputter-coated (Au). Bar = 25 μm.

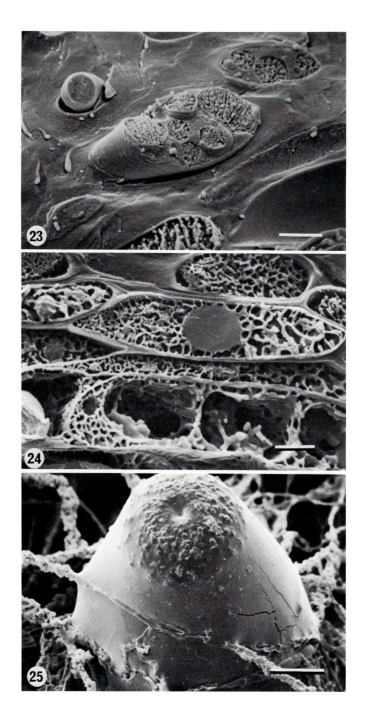

Reasons for etching are as follows:

1. It will remove superficial water droplets and films derived from environmental sources either naturally or as contaminants. Thus, features of the specimen surface can be exposed that would otherwise be obscured.
2. It provides information on the nature, location, and source of extracellular secretions.
3. It accentuates the differences between fluid-filled and gas-filled compartments.
4. It enhances mass density differences between aqueous and organic regions of fracture faces, which enables greater distinctions to be made between cell types or layers within a tissue.
5. It can give us qualitative information on the amount and distribution of water, concentration of ice nucleation sites, and possibly the rate of freezing within cells and tissues. This information is based on the spacing of segregation zones caused by ice crystal damage.

As we have seen (Sections 3.2.3 and 3.2.4) with the SP2000 instrument, etching may be performed (1) "blind" within the preparation chamber by either conductive or radiant heating, or (2) it may be visually monitored on the cold stage within the microscope using conductive heating (see Figure 6 for operational routes). Provision of both radiant and conductive heaters is itself a useful feature because the two methods of heating satisfy somewhat different requirements. Radiant heating provides a very rapid, localized increase in temperature at the specimen surface. It is a means of removing relatively large amounts of water from, for example, cell suspensions (see Section 3.2.3). However, it lacks the fine control of the conductive heating method and although it has been used successfully by other workers (Hasegawa *et al.*, 1974; Echlin and Burgess, 1977; Nei and Fujikawa, 1977; Pawley and Norton, 1978; Echlin *et al.*, 1979a,b, 1980a,b, 1982; Lewis and Pawley, 1981) as well as by us (Read *et al.*, 1983; and unpublished work in the senior author's laboratory), few experimental studies have been done to quantify this type of etching (however, see: Talmon, 1980; Umrath, 1983).

In contrast, conductive heating, using the SP2000 system, is accomplished with a thin film heating element sandwiched within the base of the microscope stage or with a cartridge heater enclosed in the base of the cooling block in the

←————————————————————————————

FIGURES 23, 24. *S. humana.* Freeze fractures through the peridial cells of the perithecium illustrating differential conformational changes at the subcellular level. Note variations in the spacing and configuration of the segregation zones. Sputter-coated (Au). Figure 23, PFD; bar = 5 μm. Figure 24, FFD; bar = 2.5 μm.

FIGURE 25. *S. humana.* FFH, perithecial neck and ostiolar pore illustrating differential conformational changes at the extracellular level. Note surface cracks in the perithecial mucilage. Sputter-coated (Au). Bar = 50 μm.

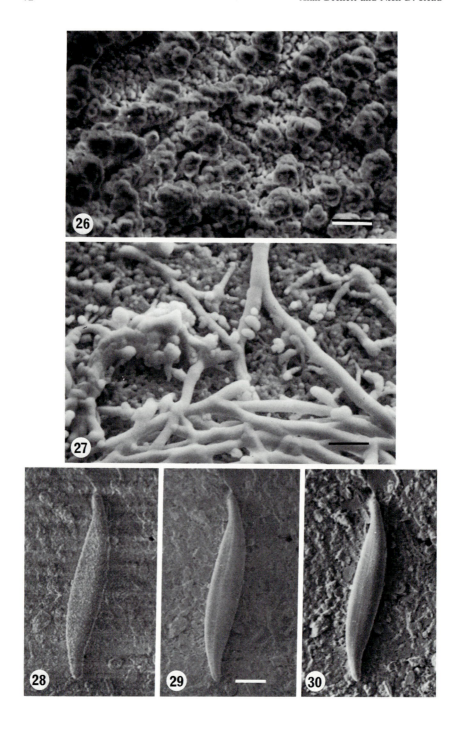

preparation chamber. These types of heater raise the temperature of the whole specimen by conduction through the copper stub. The temperature, as measured by a thermocouple embedded in the microscope stage, can be preset and the heater, once activated, rapidly reaches this temperature and then holds it until switched off. A temperature gradient will occur across the stage, stub, and specimen but in practice it is relatively easy to etch with a high degree of control and because the process may be monitored on the viewing screen, the operation can be performed with some degree of precision. In addition, visual etching has the important advantage that we can directly observe the rates and extent of etching within different parts of tissues and cells. Furthermore, the identification of the remains, if any, of etched water droplets can be determined, and these can be distinguished from "droplets" that will not etch. In essence then, visually monitored, controlled etching gives us an insight into some of the structural roles of water within biological material. A full discussion of this topic must await further, detailed experimentation but the following points serve to illustrate some of these features.

4.10.1. Removal of Superficial Water from Intact Specimens

Many microorganisms normally exist within an aquatic environment. These organisms cannot be usefully observed by LTSEM without etching. Visual etching of such bulk water is not practicable (see Section 3.2.4) unless the excess water is first removed either "blind," by radiant heating, or by the Millipore filter/filter paper wick mounting method (Section 3.2.2). However, once this is achieved, cells can be progressively exposed from within a "watery" environment by extending the etching times, and dental plaque bacteria (Figures 26 and 27) may be prepared in this way. Diatoms within estuarine mud samples can be successfully "cleaned" of superficial water by etching (Figures 28–30) following freezing in the field, within tinfoil containers.

The mycelial hyphae and germ tubes of fungi and spores cultured or grown in humid environments, are frequently coated with water droplets or water films. When FFH, this water may conceal surface features of the specimen (Read *et al.,* 1983). This is illustrated by the FFH, uncoated germ tubes from a uredospore of *Puccinia striiformis* (Figure 31). When etched and coated, the nature of the germ tube surface is revealed (Figure 32).

←——

FIGURES 26, 27. Dental plaque bacteria. Different regions of the same specimen recorded before and after etching by conductive heating in the microscope. Figure 26, FFH, uncoated, recorded at 5 kV; Figure 27, PFD, sputter-coated (Au), recorded at 9 kV. Bars = 5 μm.
FIGURES 28–30. *Gyrosigma* sp. Micrographs illustrating the removal of superficial water by etching in the microscope. Figure 28, FFH, uncoated, recorded at 4 kV; Figure 29, PFD, uncoated, recorded at 4 KV; Figure 30, PFD, sputter-coated (Au). recorded at 5 kV. Bar = 20 μm.

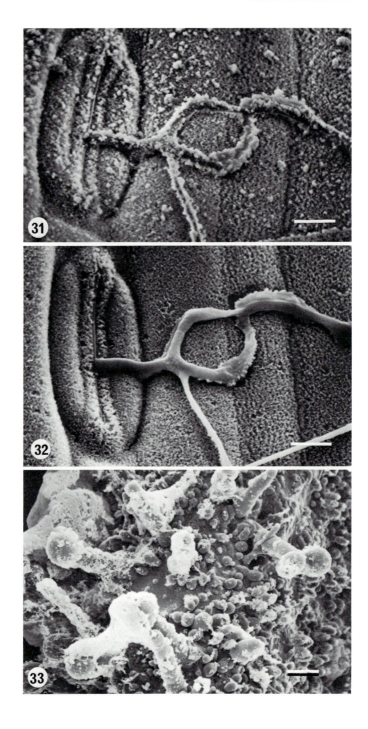

4.10.2. Nature and Location of Extracellular Exudates

Fungal cultures may produce many extracellular exudates that often are only partially preserved or even destroyed by conventional preparatory techniques (Colotelo, 1978). An example of this are the extracellular droplets produced, at least in part, at the tips of cystidia on the developing fruit body and gills of the toadstool of *F. velutipes* (Williams *et al.,* 1985). In FFH, coated preparations (Figure 33), it is very difficult to distinguish between superficial frozen water droplets (from the humid environment) and secreted droplets. If, however, an FFH, uncoated sample (Figure 34) is etched under observation in the micro-scope, it soon becomes apparent that the water droplets can be removed by sublimation but that the secreted droplets at the tips of cystidia cannot (Figure 35). From this we may conclude that they contain something less labile than water (e.g., mucilage). The presence of smooth endoplasmic reticulum, seen by TEM of thin sections through cystidial apices (Williams *et al.,* 1985), supports the suggestion of a secretory role for these cystidia.

4.10.3. Formation of Segregation Zones

Ice crystallization and recrystallization can cause intracellular disruption as a result of the physical separation of two phases within the frozen specimen. These phases are (1) the freezable water that forms ice crystals and (2) the so-called biological matrix. The biological matrix will contain supramolecular structures, macromolecules, solutes, and hydration water. The remains of the biological matrix following ice sublimation have been termed *eutectic margins* (Lewis and Pawley, 1981). Because the biological matrix is not a pure eutectic, either in the frozen-hydrated or dried state, we have proposed the terms *protoplasmic segregation zone(s)* and *ice segregation zone(s)* to describe the phases (1) within FFH specimens that are not vitrified, and (2) within freeze-dried samples respectively (Read and Beckett, 1985). In such samples, protoplasmic segregation zones tend to be most widely spaced in regions or cells that contain the fewest ice nucleation sites and the highest water content. Thus, the conformation of these zones provides us with qualitative information on the distribution of water within cells and tissues (see Figures 23 and 24). On this basis, it was concluded that paraphyses of *S. humana*, which contained few or no protoplasmic segregation zones when freeze-dried, had a high water content in the

←———————————————————————————————————————

FIGURES 31, 32. *Puccinia striiformis.* Germ tubes on the surface of a wheat leaf recorded before and after etching in the microscope. Figure 31, FFH, uncoated, recorded at 5 kV; Figure 32, PFD, sputter-coated (Au), recorded at 10 kV. Bars = 10 μm.
FIGURE 33. *F. velutipes.* Part of the edge of an intact gill. FFH, sputter-coated (Au). Note ice, giving a strong electron signal (white areas), obscuring parts of the surface. Bar = 10 μm.

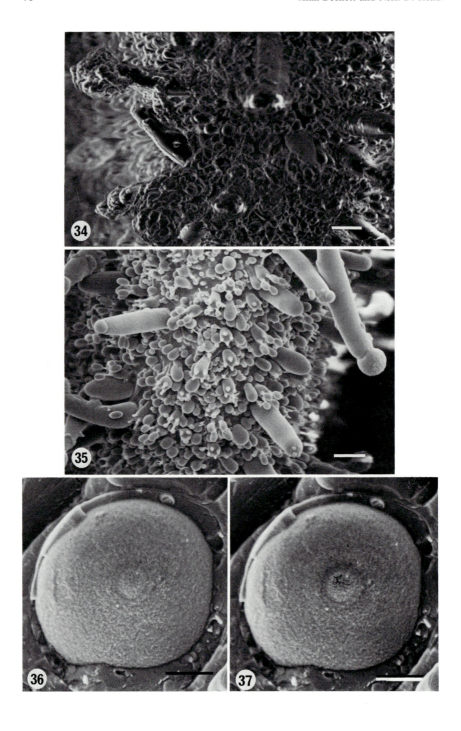

living, hydrated state. Correlative light microscopy of living paraphyses shows them to be highly vacuolated (Read and Beckett, 1985).

The spacing of segregation zones can also be affected by the freezing rate: the slower the freezing rate, the more time available for large ice crystals to grow at the expense of smaller ones, resulting in greater spacing of the protoplasmic segregation zones. We might expect slower freezing rates at greater depths within a sample because ice formed toward the outside of a specimen progressively insulates the inner regions of it. In practice, however, we found no obvious correlation between segregation zone patterns and their positions within perithecia (Read and Beckett, 1985).

4.11. Beam Damage

Damage due to electron beam–specimen interactions are of two types: (1) damaged caused by the generation of heat within the specimen ("beam heating"), and (2) damage resulting from chemical and physical changes in the sample ("radiation damage") (Talmon, 1982b). It has been argued by Talmon and co-workers (Talmon and Thomas, 1977; Talmon et al., 1979; Talmon, 1982b) that under normal operating conditions, thermal changes due to beam heating are relatively insignificant in thick or bulk frozen-hydrated specimens mounted on a conductive stub, as compared with radiation damage. A loss of mass from the specimen occurs as a result of the ionization of water molecules by the electron beam (Talmon et al., 1979; Talmon, 1982b, 1984). This radiolysis of ice, by the beam, produces free radicals (Taub and Eiben, 1968) that diffuse through the ice lattice and probably etch away organic components in contact with it (Talmon et al., 1979; Heide, 1982; Talmon, 1982b, 1984). Although these experiments were carried out with distilled water, it has been shown that the inclusion of high concentrations of organic solutes found in biological material (e.g., sucrose) considerably increase the sensitivity of the sample to radiation (Talmon, 1982b).

Bulk, FFH samples therefore tend to be particularly sensitive to radiation damage. Little information is available on the manifestations of this damage in biological specimens, but fine cracking of specimen surfaces following exposure to the beam has been observed in fungal material (Beckett et al., 1982; Read et al., 1983). That such cracks are induced by beam irradiation can be demonstrated on the same material before (Figure 36) and after (Figure 37) it has occurred (see also Read et al., 1983).

FIGURES 34, 35. *F. velutipes.* The same gill edge recorded before and after etching in the microscope. Figure 34, FFH, uncoated, recorded at 5 kV; Figure 25, PFD, sputter-coated (Au), recorded at 10 kV. Note removal of ice and exposure of droplets on cystidia. Bars = 10 μm.
FIGURES 36, 37. *S. humana.* FFH ascospore germ pore recorded before and after beam damage. Sputter-coated (Au). Figure 36, undamaged; Figure 37, note cracks induced by beam. Bars = 5 μm.

4.12. Specimen Resolution

LTSEM cannot be regarded as a very-high-resolution technique (see Section 1) and will not provide detailed images to rival those from freeze-etched replicas viewed with the transmission electron microscope (see also Robards and Crosby, 1979). Specified resolution figures guaranteed by manufacturers for a particular make and model of scanning electron microscope are largely irrelevant in the context of LTSEM of bulk biological samples. Even a modern, dedicated scanning electron microscope (as opposed to a SEM/STEM system) with a specified resolution of 2–5 nm will hardly approach this optimum with such samples (see also Pawley *et al.*, 1980). One reason for this is that the bulk of the cold stage and specimen stub on it, usually dictates that long working distances must be used. In the senior author's laboratory, a modified interface port allows the insertion of the stub onto the stage in the upper eucentric position of a Philips 501B microscope. Depending on the bulk of the specimen, this enables us to operate at the top of the Z range, at a working distance of 6 mm. The achievable resolution has not been quantified but with optimum optical settings for the fungal specimens that we have studied, the maximum useful magnification on the microscope is on the order of ×15,000. The recent availability of field emission guns on some microscopes will presumably improve the resolution attained (Pawley *et al.*, 1980). However, for most biologists, the value of LTSEM as a technique will best be judged by the quality of the micrographs (results) produced of his/her particular specimen, and of the information contained therein, rather than by the acquisition of a numerical level of resolution. Furthermore, the use of high accelerating voltages, in an attempt to improve resolution, will normally introduce interference of the signal from FFH bulk specimens. The increase in backscattered electrons, penetration and charging of the specimen as well as the possible effects of beam damage, all contribute toward a loss of useful information. Thus, although, for example, intramembranous particles in yeast plasma membrane could not be resolved by LTSEM (Pawley *et al.*, 1980), information was obtained that was not available using other techniques.

4.13. Specimen Life

Unless special facilities are developed, one of the disadvantages of LTSEM is that once a living specimen is frozen in subcooled nitrogen, and viewed in the microscope (whether or not it is fractured and coated), it must be kept at liquid nitrogen temperatures and protected from atmosphere. Normally, therefore, unless a frozen-hydrated sample is kept at these temperatures, it has a finite life. It cannot be stored on its stub in a desiccator and repeatedly viewed in the microscope weeks or even months later as can a critical point-dried or FFD specimen. It is possible to freeze several samples on one stub and to "park" two stubs in the T-shaped cooling block of the SP2000 preparation chamber (Figure 3) while maneuvering a third for fracturing, dissection, or coating. This, however, provides a "storage" facility only for as long as the preparation unit is cooled with liquid nitrogen.

Recent work on the maturation and discharge of ballistospores and the formation of their associated spore droplets in *C. cinereus* (McLaughlin *et al.*, 1985) has demonstrated a need to be able to freeze multiple specimens over a short sampling period, and to store them FFH for subsequent microscopy. With this in mind, a combined storage and stub-loading device that can be operated within a reservoir of liquid nitrogen has been developed by EMscope Laboratories for the SP2000. Such devices on this and other cryo instruments will significantly extend the versatility of LTSEM.

5. THE VALUE AND IMPORTANCE OF THE TECHNIQUE

5.1. Speed and Convenience of Preparation

A very considerable advantage of LTSEM is the extremely short specimen preparation time. Depending on the complexity of the manipulations (Figure 6), the type of specimen and of information required from it, it may take only 13 min to produce an image of an FFH fractured and coated specimen (Beckett *et al.*, 1982; Figure 6, image 11). If a single specimen is observed and photographed in the (1) uncoated, unetched; (2) uncoated, etched; and (3) coated, etched states (Figure 6, images 3, 7, 8), then the complete sequence of events can be achieved in approximately 70 min. This also includes the time required for recooling the stage following etching and prior to recording the uncoated, etched specimen. These times compare extremely favorably with those needed to prepare similar specimens by conventional means (usually 1–2 days; see Chapter 1). Using LTSEM, it is now practicable to study dynamic processes (e.g., infection of a leaf by a fungus; fungal development in culture) with repeated sampling at short intervals from the same material.

5.2. Rapid Immobilization of Dynamic Processes

Dynamic biological processes cannot be reliably immobilized and preserved by the relatively slow methods of conventional, chemical fixation (see Section 1). For example, three stages were recognized in the development of the droplet associated with the maturing ballistospores of *C. cinereus* (McLaughlin *et al.*, 1985). The "stage 3," fully expanded droplet was seen only briefly in living material with the light microscope (Figure 38). This stage of droplet enlargement could only be preserved for SEM (and reliably identified) in visually monitored,

PFD preparations (Figure 39). Early droplet enlargement (stage 2), however, was seen in both PFD (Figure 40) and FFD material (Figure 41). In the latter, a shrinkage in volume was detected.

5.3. Avoidance of Dehydration and Prolonged Drying

Using LTSEM, cells and tissues, with their constituent water immobilized *in situ,* may be observed directly at liquid nitrogen temperature and in a high vacuum. This facility removes the need to dehydrate and completely dry specimens and therefore provides us with a relatively high-resolution, microscopical method for monitoring changes in form and dimensions that are directly related to specimen water content (Beckett *et al.,* 1984; see also Section 4.7). Similarly, when used in conjunction with information on the environmental/growth conditions and on the physiological state of the specimen, LTSEM can help us determine whether, for example, cell collapse is due to natural dehydration or whether it is an artifact of preparatory technique (Beckett *et al.,* 1984). The retention of water, as ice, within delicate biological samples also provides a means of accurately preserving such organisms in a ''lifelike'' state. Because chemical fixatives and organic solvents for dehydration are not required, their deleterious effects on biological samples are avoided. Thus, the extraction or spatial redistribution of compounds from or within a multicomponent sample is not a problem with LTSEM.

Besides the numerous applications of LTSEM to microbiology that have already been given, two further examples are worthy of mention. First, studies on the behavior of liquid solutions such as herbicides and pesticides following their application to plant surfaces are uniquely suited for LTSEM (Hart, 1979). Such applied droplets cannot be preserved by conventional procedures involving drying and this represents just one aspect of crop protection and plant pathology for which LTSEM may be very useful. Second, the technique has recently been used to study the structure and condition of biofilms used in wastewater treatment plants (Richards and Wilson, 1983; Richards and Turner, 1984). It was found that LTSEM preserved the integrity of the biofilm, and particularly bacterial slime, while conventional methods disrupted and/or destroyed various components of it.

5.4. Versatility

One of the most important aspects of LTSEM stems from the large number of options available with regard to manipulation of the specimen once it has been frozen. Numerous different types of preparation can be produced according to the information required from the one original sample. Results obtained, when correlated with those from ATSEM and with those from TEM of freeze-etched

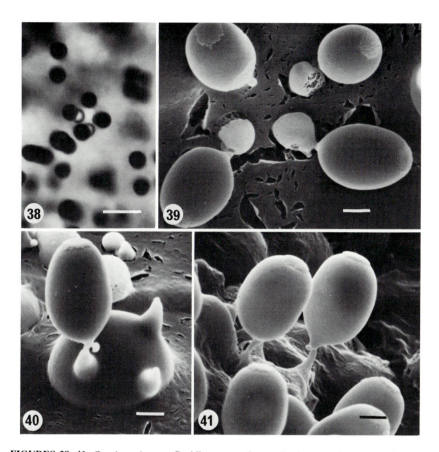

FIGURES 38–41. *Coprinus cinereus.* Basidiospores and spore droplets at various stages of maturation. Figure 38, living culture. Note "fully expanded" droplets on two of the spores of the basidium in the center. Bar = 14.5 μm. Figure 39, PFD basidium showing spores and "fully expanded" droplets. Also note ruptures in the membranous covering of the hymenium. Sputter-coated (Au). Bar = 2.0 μm. Figure 40, PFD basidium and one spore with a droplet at the "early enlargement" stage. Sputter-coated (Au). Bar = 2.0 μm. Figure 41, FFD basidiospores. Spore droplet at the "early enlargement" stage. Note shrinkage compared with that in Figure 40. Sputer-coated (Au). Bar = 2.0 μm.

replicas, provide a valuable supplement to cytological and biochemical data from other sources.

The range of options for LTSEM is suitably illustrated by Figure 6. Here, with specific reference to the EMscope SP2000 Cryo Preparation unit, the number of routes or flow lines is directly proportional to the versatility of the technique. As explained in the legend, further procedures are also available for microanalysis and multiple fracturing.

6. CONCLUSIONS

These may be conveniently tabulated in the form of pros and cons as follows:

Pros	Cons

1. Excellent preservation
  a. Specimen water retained
  b. Rapid immobilization
  c. No exposure to chemicals, fixatives, or solvents
  d. No deformation due to prolonged drying
  e. Relatively insignificant dimensional changes unless PFD
2. Retention of water enables controlled etching of specimen to reveal both external and internal features not otherwise observable
3. Speed and convenience of preparation
	1. Small samples required
	2. Slight expansion on freezing may cause surface rupture
	3. Susceptible to beam damage
	4. Resolution impaired
	5. Normally finite specimen life

ACKNOWLEDGMENTS. We are indebted to: Roy Sutcliffe of EMscope Laboratories Ltd., for information on the development of instrumentation and for providing the original drawings of Figures 2–5; to Richard E. W. Banfield of Hexland, Electron Microscope Technology, for details of the Hexland range of instruments; to Arnold Cleaver, JEOL (U.K.) Ltd., for assistance in tracing certain references; to Ken Oates, Lancaster University, for unpublished information and for supplying us with a preprint of his paper on beam penetration and X-ray excitation depth; to Dr. Simon Richards, AFRC Food Research Institute, Norwich, for providing reprints and information on his recent work. Figures 28–30 were kindly supplied by Dr. David Patterson, Department of Botany, Bristol University; Figures 31 and 32 were prepared in conjunction with and from material supplied by Dr. Oisín C. MacNamara, Department of Agriculture, Newcastle University; and Figures 38–41 were supplied by Professor David J. McLaughlin, University of Minnesota. Bob Porter provided invaluable technical expertise in the modification, adaptation, and general maintenance of equipment, without which the work would not have been possible. We thank the SERC and the AFRC for research grants during the tenure of which many of the results reported here were obtained. Finally, we thank Mrs. Jean Summers for her skill and patience in typing the manuscript.

7. REFERENCES

Adrian, M., Dubochet, J., Lepault, J., and McDowall, A. W., 1984, Cryo-electron microscopy of viruses, *Nature (London)* **308:**32–36.

Bald, W. B., 1975, A proposed method for specifying the temperature history of cells during the rapid cool-down of plant specimens, *J. Exp. Bot.* **26:**103–119.

Bald, W. B., 1984, The relative efficiency of cryogenic fluids used in the rapid quench cooling of biological samples, *J. Microsc. (Oxford)* **134:**261–270.

Beckett, A., 1982, Low temperature scanning electron microscopy of the bean rust fungus *Uromyces viciae-fabae*, *Philips Electron Optics Bull.* No. 117, pp. 6–8.

Beckett, A., and Porter, R., 1982, *Uromyces viciae-fabae* on *Vicia faba:* Scanning electron microscopy of frozen-hydrated material, *Protoplasma* **111:**28–37.

Beckett, A., Porter, R., and Read, N. D., 1982, Low temperature scanning electron microscopy of fungal material, *J. Microsc. (Oxford)* **125:**193–199.

Beckett, A., Read, N. D., and Porter, R., 1984, Variations in fungal spore dimensions in relation to preparatory techniques for light microscopy and scanning electron microscopy, *J. Microsc. (Oxford)* **136:**87–95.

Boyde, A., 1978, Pros and cons of critical point-drying and freeze-drying for SEM, *Scanning Electron Microsc.* II:303–314.

Boyde, A., and Franc, F., 1981, Freeze-drying shrinkage of glutaraldehyde fixed liver, *J. Microsc. (Oxford)* **132:**75–86.

Boyde, A., Franc, F., and Maconnachie, E., 1981, Measurements of critical point shrinkage of glutaraldehyde fixed liver, *Scanning* **4:**69–82.

Bråten, T., 1978, High resolution scanning electron microscopy in biology: Artifacts caused by the nature and mode of application of the coating material, *J. Microsc. (Oxford)* **113:**53–59.

Campbell, R., 1983, Ultrastructural studies of *Gaeumannomyces graminis* in the water films on wheat roots and the effect of clay on the interaction between this fungus and antagonistic bacteria, *Can. J. Microbiol.* **29:**39–45.

Campbell, R., and Porter, R., 1982, Low temperature scanning electron microscopy of microorganisms in soil, *Soil Biol. Biochem.* **14:**241–245.

Clegg, J. S., 1982, Alternative views on the role of water in cell function, in: *Biophysics of Water* (F. Franks and S. F. Mathias, eds.), pp. 365–383, Wiley, New York.

Colotelo, N., 1978, Fungal exudates, *Can. J. Microbiol.* **24:**1173–1181.

Cooke, R., and Kuntz, I. D., 1974, The properties of water in biological systems, *Annu. Rev. Biophys. Bioeng.* **3:**95–126.

Dubochet, J., McDowall, A. W., Menge, B., Schmid, E. N., and Lickfield, K. G., 1983, Electron microscopy of frozen-hydrated bacteria, *J. Bacteriol.* **155:**381–390.

Echlin, P., 1971, The examination of biological material at low temperatures, *Scanning Electron Microsc.* I:225–232.

Echlin, P., 1978, Low temperature scanning electron microscopy: A review, *J. Microsc. (Oxford)* **112:**47–61.

Echlin, P., and Burgess, A., 1977, Cryofracturing and low temperature scanning electron microscopy of plant material, *Scanning Electron Microsc.* I:491–500.

Echlin, P., and Kaye, G., 1979, Thin films for high resolution conventional electron microscopy, *Scanning Electron Microsc.* II:21–30.

Echlin, P., and Moreton, R., 1973, The preparation, coating and examination of frozen biological materials in the SEM, *Scanning Electron Microsc.* III:325–332.

Echlin, P., and Moreton, R., 1976, Low temperature techniques for scanning electron microscopy, *Scanning Electron Microsc.* I:753–762.

Echlin, P., Paden, R., Dronzek, B., and Wayte, R., 1970, Scanning electron microscopy of labile biological material maintained under controlled conditions, *Scanning Electron Microsc.* **1970:**49–56.

Echlin, P., Ralph, B., and Weibel, E. R. (eds.), 1978, *Low Temperature Biological Microscopy and Microanalysis,* Royal Microscopical Society, Blackwell, Oxford.

Echlin, P., Hayes, T. L., and Pawley, J. B., 1979a, Freeze-fracture SEM of differentiating phloem parenchyma of *Lemna minor* L. (Duckweed), *Beitr. Elektronenmikroskop. Direktabb. Oberfl.* **12:**95–104.

Echlin, P., Pawley, J. B., and Hayes, T. L., 1979b, Freeze-fracture scanning electron microscopy of *Lemna minor* L. (Duckweed), *Scanning Electron Microsc.* **III:**69–76.

Echlin, P., Lai, C., Hayes, T., and Saubermann, A., 1980a, Cryofixation of *Lemna minor* roots for morphological and analytical studies, *Cryo-Letters* **1:**289–298.

Echlin, P., Lai, C. E., Hayes, T. L., and Hook, G., 1980b, Elemental analysis of frozen-hydrated differentiating phloem parenchyma in roots of *Lemna minor* L., *Scanning Electron Microsc.* **II:**383–394.

Echlin, P., Lai, C. E., and Hayes, T. L., 1982, Low-temperature X-ray microanalysis of the differentiating vascular tissue in root tips of *Lemna minor* L., *J. Microsc. (Oxford)* **126:**285–306.

Finney, J. L., 1977, The organization and function of water in protein crystals, *Philos. Trans. R. Soc. London Ser. B* **278:**3–32.

Finney, J. L., 1979, The organization and function of water in protein crystals, in: *Water: A Comprehensive Treatise,* Vol. 6 (F. Franks, ed.), pp. 47–122, Plenum Press, New York.

Franks, F., 1981, Biophysics and biochemistry of low temperatures and freezing, in: *Effects of Low Temperature on Biological Membranes* (G. J. Morris and A. Clarke, eds.), pp. 3–19, Academic Press, New York.

Fuchs, W., and Fuchs, H., 1980, The use of frozen-hydrated bulk specimens for X-ray microanalysis, *Scanning Electron Microsc.* **II:**371–382.

Fuchs, W., and Lindemann, B., 1975, Electron beam X-ray microanalysis of frozen biological bulk specimens below 130K. I. Instrumentation and specimen preparation, *J. Microsc. Biol. Cell* **22:**227–232.

Fuchs, W., Lindemann, B., Brombach, J. D., and Trösch, W., 1978, Instrumentation and specimen preparation for electron beam X-ray microanalysis of frozen hydrated bulk specimens, *J. Microsc. (Oxford)* **112:**75–87.

Galpin, M. F., Jennings, D. H., Oates, K., and Hobot, J. A., 1978, Localization by X-ray microanalysis of soluble ions, particularly potassium and sodium, in fungal hyphae, *Exp. Mycol.* **2:**258–269.

Hall, T. A., and Gupta, B. L., 1984, The application of EDXS to the biological sciences, *J. Microsc. (Oxford)* **136:**193–208.

Harada, H., and Okuzumi, H., 1973, Some applications of the freeze method with cryo-unit in scanning electron microscopy, *JEOL News* **11e**(2):11–15.

Hart, C. A., 1979, Use of scanning electron microscope and cathodoluminescence in studying the application of pesticides in plants, *Pestic. Sci.* **10:**341–357.

Hasegawa, Y., and Yotsumoto, H., 1972, Direct observation by the freeze method, *JEOL News* **10e**(3):22–23.

Hasegawa, Y., Hasegawa, M., Suzuki, T., and Yotsumoto, H., 1974, Soft tissue observation by cryoscan fitted with vacuum evaporating device, *JEOL News* **12e**(2):26–27.

Heide, H. G., 1982, On the irradiation of organic samples in the vicinity of ice, *Ultramicroscopy* **7:**301–302.

Ingold, C. T., 1928, Spore discharge in *Podospora curvula, Ann. Bot.* **42:**567–570.

Jones, G. J., 1984, On estimating freezing times during tissue rapid freezing, *J. Microsc. (Oxford)* **136:**349–360.

Koch, G. R., 1975, Preparation and examination of specimens at low temperatures, in: *Principles and Techniques of Scanning Electron Microscopy,* Vol. 4 (M. A. Hayat, ed.), pp. 1–33, Van Nostrand–Reinhold, Princeton, New Jersey.

Lepault, J., Booy, F. P., and Dubochet, J., 1983, Electron microscopy of frozen biological specimens, *J. Microsc. (Oxford)* **129:**89–102.

Lewis, E. R., and Pawley, J. B., 1981, Direct SEM of frozen inner ear, *Scanning* **4:**131–140.

McLaughlin, D. J., Beckett, A., and Yoon, K. S., 1985, Ultrastructure and evolution of ballistosporic basidiospores, *Bot. J. Linn. Soc.* **91:**253–271.

Marshall, A. T., 1980, Frozen-hydrated bulk specimens, in: *X-Ray Microanalysis in Biology* (M. A. Hayat, ed.), pp. 167–196, University Park Press, Baltimore.

Marshall, A. T., 1982, Application of $\phi(pz)$ curves and a windowless detector to the quantitative X-ray microanalysis of frozen-hydrated bulk biological specimens, *Scanning Electron Microsc.* **I:**243–260.

Mersey, B., and McCully, M. E., 1978, Monitoring of the course of fixation of plant cells, *J. Microsc. (Oxford)* **114:**49–76.

Nei, T., 1974, Cryotechniques, in: *Principles and Techniques of Scanning Electron Microscopy,* Vol. 1 (M. A. Hayat, ed.), pp. 113–124, Van Nostrand–Reinhold, Princeton, New Jersey.

Nei, T., and Fujikawa, S., 1977, Freeze-drying process of biological specimens observed with a scanning electron microscope, *J. Microsc. (Oxford)* **111:**137–142.

Nei, T., Yotsumoto, H., Hasegawa, Y., and Nagasawa, Y., 1971, Direct observation of frozen specimens with a scanning electron microscope, *J. Electron Microsc.* **20:**202–203.

Nei, T., Yotsumoto, H., Hasegawa, Y., and Nagasawa, Y., 1973, Direct observation of frozen specimens with a scanning electron microscope, *J. Electron Microsc.* **22:**185–190.

Oates, K., and Potts, W. T. W., 1985, Electron beam penetration and X-ray excitation depth in ice, *Micron Microsc. Acta* **16:**1–4.

Pawley, J. B., and Norton, J. T., 1978, A chamber attached to the SEM for fracturing and coating frozen biological samples, *J. Microsc. (Oxford)* **112:**169–182.

Pawley, J. B., Hook, G., Hayes, T. L., and Lai, C., 1980, Direct scanning electron microscopy of frozen-hydrated yeast, *Scanning* **3:**219–226.

Read, N. D., 1983, A scanning electron microscopic study of the external features of perithecium development in *Sordaria humana, Can. J. Bot.* **61:**3217–3229.

Read, N. D., and Beckett, A., 1982, Fact or artifact under the SEM?, *Proc. R. Microsc. Soc.* **17:**S35.

Read, N. D., and Beckett, A., 1983, Effects of hydration on the surface morphology of urediospores, *J. Microsc. (Oxford)* **132:**179–184.

Read, N. D., and Beckett, A., 1985, The anatomy of the mature perithecium in *Sordaria humana* and its significance for fungal multicellular development, *Can. J. Bot.* **63:**281–296.

Read, N. D., Porter, R., and Beckett, A., 1983, A comparison of preparative techniques for the examination of the external morphology of fungal material with the scanning electron microscope, *Can. J. Bot.* **61:**2059–2078.

Richards, S. R., and Turner, R. J., 1984, A comparative study of the techniques for the examination of biofilms by scanning electron microscopy, *Water Res.* **18:**767–773.

Richards, S. R., and Wilson, A. J., 1983, Rapid scanning electron microscope techniques to investigate colonization of biomass support particles in the captor process, *Environ. Technol. Lett.* **4:**183–188.

Robards, A. W., 1974, Ultrastructural methods for looking at frozen cells, *Sci. Prog. (Oxford)* **61:**1–40.

Robards, A. W., and Crosby, P., 1978, A transfer system for low temperature scanning electron microscopy, *Scanning Electron Microsc.* **II:**927–936.

Robards, A. W., and Crosby, P., 1979, A comprehensive freezing, fracturing and coating system for low temperature scanning electron microscopy, *Scanning Electron Microsc.* **II:**325–343.

Talmon, Y., 1980, Rate of sublimation of ice by radiative heating in freeze-etching, in: *Proceedings*

Electron Microscopy Society of America, 38th (G. W. Bailey, ed.), pp. 618–619, Claitor, Baton Rouge.

Talmon, Y., 1982a, Frozen hydrated specimens, *Electron Microscopy 1982* **1:**25–32.

Talmon, Y., 1982b, Thermal and radiation damage to frozen hydrated specimens, *J. Microsc. (Oxford)* **125:**227–237.

Talmon, Y., 1984, Radiation damage to organic inclusions in ice, *Ultramicroscopy* **14:**305–316.

Talmon, Y., and Thomas, E. L., 1977, Beam heating of a moderately thick cold stage specimen in the SEM/STEM, *J. Microsc. (Oxford)* **111:**151–164.

Talmon, Y., Davis, H. T., Scriven, L. E., and Thomas, E. L., 1979, Mass loss and etching of frozen hydrated specimens, *J. Microsc. (Oxford)* **117:**321–332.

Taub, I. R., and Eiben, K., 1968, Transient solvated electron, hydroxyl, and hydroperoxy radicals in pulse-irradiated crystalline ice, *J. Chem. Phys.* **49:**2499–2513.

Tokunaga, J., and Tokunaga, M., 1973, Cryo-scanning microscopy of conidiospore formations in *Aspergillus niger, JEOL News* **11e**(1)**:**3–7.

Umrath, W., 1983, Calculation of the freeze-drying time for electron-microscopical preparations, *Mikroskopie* **40:**9–34.

Williams, M. A. J., Beckett, A., and Read, N. D., 1985, Ultrastructural aspects of fruit body differentiation in *Flammulina velutipes,* in: *Developmental Biology of Higher Fungi,* (D. Moore, L. A. Casselton, D. A. Wood, and J. C. Frankland, eds.), pp. 429–450, Cambridge University Press, Cambridge.

Zierold, K., 1983, X-ray microanalysis of frozen-hydrated specimens, *Scanning Electron Microsc.* **II:**809–826.

Chapter 3

Effects of Specimen Preparation on the Apparent Ultrastructure of Microorganisms

William J. Todd

Department of Veterinary Science
Louisiana Agricultural Experiment Station
Louisiana State University Agricultural Center
and
Department of Veterinary Microbiology and Parasitology
School of Veterinary Medicine
Louisiana State University
Baton Route, Louisiana 70803

As microbiologists, we have available to us a variety of both established and newly developed electron microscopy (EM) techniques useful for detailed ultrastructural analysis of microorganisms. Although applications of the new methods will undoubtedly lead to a better understanding of microbial ultrastructure and function, the traditional approach of investigations using fixed, embedded, and sectioned specimens will remain as a central EM method for the foreseeable future. Among the many useful protocols for traditional processing of specimens, the procedures recommended by Ito and Rikihisa (1981) are consistently effective for preserving the ultrastructure of both microorganisms and host cells. These procedures have evolved through the input of many investigators and are summarized here.

1. SPECIMEN PREPARATION FOR EXAMINATION OF THIN SECTIONS

1.1. Glutaraldehyde

Primary fixation is done for 1 hr at room temperature in 0.05 M to 0.1 M cacodylate buffer at pH 7.4 containing 2.5% glutaraldehyde, 1.25% paraformaldehyde, 0.03% $CaCl_2$, and 0.03% of either trinitrocresol or picric acid.

1.2. Osmium Tetroxide

After primary fixation the specimens are rinsed in several changes of 0.1 M cacodylate buffer (pH 7.4) and postfixed for 1 hr at room temperature in 0.1 M cacodylate buffer (pH 7.4) containing 1% osmium tetroxide.

1.3. Uranyl Acetate

After unreacted osmium tetroxide is removed by several rinses in cacodylate buffer, the specimens are rinsed in 0.1 M maleate buffer (pH 5.2) and then incubated for 1 hr at room temperature in 0.1 M maleate buffer (pH 5.2) containing 1% uranyl acetate.

1.4. Dehydration

Following removal of unbound uranyl acetate by three rinses in maleate buffer over several hours, the specimens are rinsed in water and then dehydrated in a graded series of cold ethanol (30, 50, 75, 95%). This dehydration requires only a few minutes at each step. The specimens are then warmed to room temperature and the dehydration completed using several changes of bonded absolute alcohol for a total time of 30 min.

1.5. Embedding

Embedding media such as Epon, Araldite, Spurr's, or 1, 2, 7, 8-diepoxyoctane (DEO) are gradually infiltrated into the specimen usually with propylene oxide as a transition solvent. DEO as described by Luft (1973) provides the most effective penetration of specimens that are difficult to infiltrate. DEO medium is prepared by mixing 18 parts DEO with 32 parts nonenyl succinic anhydride. Catalyst [tri(dimethylaminoethyl)-phenol (DMP-30), 2–3%] is added. The hardness of the block is controlled by altering the ratio of DEO to nonenyl succinic anhydride. To infiltrate, the dehydrated specimens are rinsed in propylene oxide followed sequentially by propylene oxide/DEO medium at ratios of 2 : 1, 1 : 1, 1 : 2 and two to three changes of DEO medium over several hours or overnight. The specimens are sealed and the resin cured by incubation for 18 to 24 hr at 60°C. Silver to gold sections of DEO-embedded specimens can easily be cut and effectively stained with heavy metals such as uranyl acetate and lead citrate.

1.6. Discussion

This procedure for preparing specimens for thin section analysis in transmission electron microscopy (TEM) evolved in part from efforts to preserve the ultrastructure of both host cells and pathogenic microorganisms that propagate within them. With this in mind, the following notes and rationale are pertinent to development of the procedure.

The inclusion of formaldehyde (freshly prepared from paraformaldehyde by heating and dropwise addition of sodium hydroxide) along with the almost universally used glutaraldehyde, functions primarily to increase the rate of infiltration of the fixative. To preserve the internal features of our intracellular gram-negative microorganisms, the fixative must first penetrate the host cell plasmalemma, cytoplasmic matrix, and both inner and outer membranes of the bacterium. A small molecule, formaldehyde penetrates rapidly and provides some stabilization prior to the more effective fixation by glutaraldehyde. The glutaraldehyde that provides the most reliable results is purchased as purified 25% glutaraldehyde in sealed ampoules and mixed with a few drops of 30% hydrogen peroxide just prior to use (Peracchia et al., 1970). Including small amounts of trinitro compounds also seems to improve preservation and fixation consistency. However, it should be noted that these trinitro compounds are potentially explosive especially when dry. Including larger amounts as is common in light microscopy may have a negative effect on ultrastructure preservation.

Although osmium tetroxide is known to hydrolyze at least some proteins (Emerman and Behrman, 1982; Maysin-Szamier and Pollard, 1978) and when used alone can induce vesiculation of membranes (Aldrich, 1974), the reagent is essential to prevent major extraction of lipids that otherwise occurs during embedding. Osmium tetroxide is also an excellent electron-dense stain, resulting in improved contrast throughout the specimen. When used after primary fixation in concentrations of 1.0–1.5%, the negative effects of osmium are usually tolerable.

Including uranium treatment of fixed specimens, prior to dehydration, accounts for a surprising amount of difference in the preservation of the internal membrane of intracellular bacteria, limits clumping of the microbial DNA, and provides improved contrast throughout the specimen. Although the maleate buffer is excellent for maintaining the solubility of uranyl salts, the salts must be extensively washed from the specimen prior to dehydration to prevent precipitation.

In choosing an embedding medium, DEO has the advantage of excellent penetration in comparison to most other media, and when used with gram-negative intracellular bacteria, seems to prevent wrinkling of the outer membrane, a common shrinkage artifact. On the negative side, DEO is expensive and likely to be more carcinogenic than other embedding media.

For many investigations this procedure provides adequate preservation to permit resolution of most ultrastructural constituents of both microorganisms and host cells, and is a reasonable way to begin ultrastructural studies. An example of an obligate intracellular pathogen (*Rickettsia tsutsugamushi*) prepared as described is given by Ito and Rikihisa (1981). The good preservation shown by these authors is important before functional interpretations of electron micrographs can be attempted. For example, inadequate preservation of membranes, and section angles oblique to the plane of the membrane, can lead to the illusion that the cytoplasm of *R. tsutsugamushi* is at times actually continuous with the cytoplasm of the host cell (Hase, 1983a,b). Based on this observation, Hase concluded that this bacterium multiplies by *de novo* assembly within the cytoplasm of the host cell in a manner analogous to viruses. However, when membranes are well preserved, images reminiscent of bacteria dividing by binary fission are obtained (Rikihisa and Ito, 1980). The proposal of a new method of bacterial multiplication for *R. tsutsugamushi* may thus not be warranted. Adequate preservation of morphology is essential before realistic interpretations can begin, and even then interpretations can be difficult.

2. FIXATION ADDITIVES TO IMPROVE DETECTION OF MICROBIAL STRUCTURES

Many different compounds have been included during different phases of fixation procedures in attempts to improve detection and resolution of structures. As mentioned earlier, in embedded specimens only the staining patterns of the heavy metals provide enough contrast for adequate resolution. Structures that do not bind metals usually remain invisible within the section, i.e., the inherent contrast is not enough to distinguish the structure from the embedding medium. The rationale for use of additives is either to coat these structures with a mordant to provide more efficient binding of heavy metals or to directly stain with heavy metal compounds in addition to including the usual osmium, lead and uranyl salts. Two such commonly used additives are tannic acid and ruthenium red.

2.1. Tannic Acid

Tannic acid functions as a mordant; it will both coat biological structures and react extensively with osmium tetroxide and other metals. The results are sometimes significant. For example, using tannic acid, Matsumoto *et al.* (1976) detected surface projections on elementary bodies of chlamydiae. In these studies the tannic acid appeared to function as a negative stain to highlight the projections. Without the tannic acid, the limited contrast of the projections was simply lost in the electron density of the embedding medium. The presence of these projections was confirmed in specimens processed by the critical point method and viewed by high-resolution scanning electron microscopy (SEM) (Gregory *et al.*, 1979) or high-voltage EM (Stokes, 1978). They are also revealed in replicas of frozen, fractured, and etched elementary bodies (Louis *et al.*, 1980). The elementary bodies in tannic acid-treated specimens also appear to be surrounded by an electron-dense coat. This coat is most likely just stained aggregates of tannic acid and are of no biological significance; however, this coat could easily be misinterpreted as a bacterial glycocalyx. The phenomenon of adding electron-dense structures where none actually exist is an important potential artifact in the use of mordants or other electron-dense additives.

2.2. Ruthenium Red

Ruthenium red is commonly used to detect glycocalyx-like coats found on the surfaces of many bacteria (Costerton and Irvin, 1981), including intracellular bacteria such as rickettsiae (Silverman *et al.*, 1978). These coats are difficult to preserve and detect in thin sections unless first reacted with either antibodies or stains such as ruthenium red. Although ruthenium red is often described as a stain specific for carbohydrates, its specificity is somewhat in doubt. Ruthenium red does react nicely with carbohydrates, but, as pointed out by Luft (1971a,b), it also binds to acidic amino acids, DNA, RNA, and some lipids, and cannot be considered specific for the presence of carbohydrates. Ruthenium red will also cross-bridge with osmium tetroxide (Feria-Valasco and Arauz-Contreras, 1981) and will enhance the contrast of structures with attached molecules of osmium tetroxide, whether or not those structures are rich in carbohydrates. In general, the heavy metal stains cannot be considered specific for reactions with any one class of molecules. Although ruthenium red can be very effective in detecting surface features of some microorganisms that frequently are not stained well by osmium, lead or uranyl salts, it is difficult to conclude as to what chemical contituent the ruthenium red is actually bound.

2.3. Diamines

Perhaps the most exciting addition to the well-established fixation regimen is the inclusion of diamines during the early stages of glutaraldehyde fixation (Boyles, 1984). Direct evidence that free amines enhance the action of glutaraldehyde at the molecular level was provided by Jost *et al.* (1984). Using electron spin resonance spectroscopy to detect an androstan spin label in red cell ghost membranes, these investigators documented a decrease in lipid motion when amines were added to glutaraldehyde fixation. Reduction in movement did not occur with glutaraldehyde alone, or with the combination of formaldehyde and amines. A drop in the pH of the buffer was also noted when the amines were included along with glutaraldehyde. This drop in pH is most likely the result of hydrogen ions generated when the positive charge on the parent amine is neutralized by the amine–aldehyde reaction. Stronger buffers may be required if amines are included along with glutaraldehyde as a fixative.

At physiological pH the diamines are positively charged and would tend to bind to negatively charged sites on cells and tissues such as the glycocalyx, improving the preservation and detection of such structures. As pointed out by Boyles (1984), linear diamine–glutaraldehyde polymers of almost any length can also be formed. Cross-bridges of various lengths could provide for more effective cross-linking and stability of tissue constituents than with glutaraldehyde alone. The amines will also bind osmium tetroxide. Although at high resolution some of the electron-dense structures observed could be merely the osmium-stained diamine–glutaraldehyde polymers themselves, this possibility for creating artifacts seems less likely with diamines than with other higher-molecular-weight additives such as tannic acid or ruthenium red. At this writing, however, the application of diamines to the analysis of microbial ultrastructure has not been reported in the literature.

3. MESOSOMES AND FIXATION

Considering that living cells are so abused by methods of specimen preparation required for ultrastructural analysis by TEM, our progress in gaining insight into microbial ultrastructure and its relationship to function is remarkable. However, as methods improve and newly developed techniques and equipment become available, it is necessary to reconsider our previously held interpretations. Our understanding of ultrastructure is more of an evolving process than any unequivocal description of microbial morphology. An example of such evolving interpretations is that for the microbial structure called the mesosome.

Mesosomes were defined ultrastructurally as invaginations of the cytoplasmic membrane that evidently formed compartments within the bacterial cell. Throughout the 1960s and 1970s the mesosome was extensively studied and documented in the literature. These structures have reportedly been isolated as mesosome-rich fractions and extensively investigated to determine unique biochemical functions, albeit without definitive results. A good summary of this work is the review article "Mesosomes: Membranous Bacterial Organelles" by Greenawalt and Whiteside (1975).

More recently, with the application of rapid freezing and freeze-substitution techniques, the extensive membrane invaginations can be shown to be induced by traditional methods of chemical fixation (Dubochet et al., 1983). Based on these results, Dubochet et al. (1983) and others now interpret mesosomes to be fixation artifacts generated during the early stages of fixation. The extensive invaginations of the cytoplasmic membrane, and apparent compartmentations of the cytoplasm are indeed almost certainly artifacts of fixation. However, careful and quantitative studies by Higgins and associates (Higgins and Daneo-Moore, 1974; Higgins et al., 1976, 1981) indicate a clear correlation between the number of mesosomes present per cell and the physiological status of the cell. Their data also show that mesosome formation is site specific within the bacteria and dependent on the phase of the cell cycle at the time fixation occurs. By this reasoning, mesosomes represent unique physiologically significant portions of the cytoplasmic membrane that are, when exposed to fixatives such as osmium tetroxide, distorted to form the large membrane invaginations. The mesosomes are likely to be DNA attachment points on the cytoplasmic membrane. Condensation of the DNA during the early stages of fixation may pull the still-fluid cytoplasmic membrane into the nuclear region, along with the DNA, thus forming the observed membrane invaginations and apparent cytoplasmic compartments.

Are the fixation-induced invaginations of the cytoplasmic membranes isolated examples of artifacts in microbial electron microscopy, or are our interpretations of other microbial structures also as greatly influenced by methods of specimen preparation? In an attempt to gain some answers to this question, effects of specimen preparation were studied using bacterial pili as a relatively simple model. Piliation in one colony type of a strain of *Neisseria gonorrhoeae* was studied using a variety of EM techniques. It was found that even the simplest of microbial structures, such as pili or fimbriae, can appear quite differently depending on the method of specimen processing.

4. IMAGES OF GONOCOCCAL PILI

Pili are threadlike structures that radiate from the surfaces of many different types of bacteria. The common or somatic pili are believed to be important in attachment of bacteria to mucosal or other surfaces. These long thin polymers are generally constructed from identical subunits of the protein pilin, or glycoprotein. There are a wide variety of pili that can often be expressed at different times by the same organism. Some types of pili are specialized for unique functions such as conjugation. In *N. gonorrhoeae,* the expression of pili is associated with the establishment of urogenital infections (Punsalang and Sawyer, 1973). When cultured on solid media, the expression of pili will influence the colony morpholgy. Piliation morphology designated p^{2+} (Swanson, 1973, Swanson *et al.,* 1971) was used in this study. Comparison of colonies of *N. gonorrhoea* that express one colony type of pili with those that do not, provides a useful model to study the effects of specimen preparation on the morphology of a relatively simple structure.

4.1. Negative Staining

When piliated gonococci are removed from colonies and then negatively stained on coated grids with phosphotungstic acid, air-dried, and examined by TEM, the pili are seen primarily as long thin threads scattered across the surface of the grid (Figure 1). The images are similar whether or not the specimens are fixed in glutaraldehyde and osmium tetroxide prior to negative staining and drying in air.

4.2. Embedding in Epoxy

When similar colonies are fixed as described earlier in this chapter, embedded in epoxy, and thin sections cut, stained with heavy metals, and examined by TEM, the pili are also seen primarily as individual strands. Because the sections are thin, the long pili can usually be traced only over short distances before they wander out of the plane of the section. By using thick sections of 1 μm or greater, and analysis by high-voltage EM, the pili can be followed for longer distances within the three-dimensional area of the section; nonetheless, they still appear

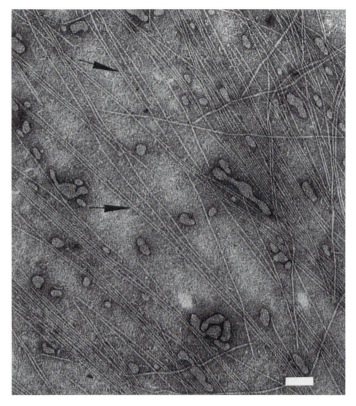

FIGURE 1. Negatively stained and air-dried pili of *N. gonorrhoeae* appear as linear arrays of unbranched threads (arrows). Bar = 0.1 μm.

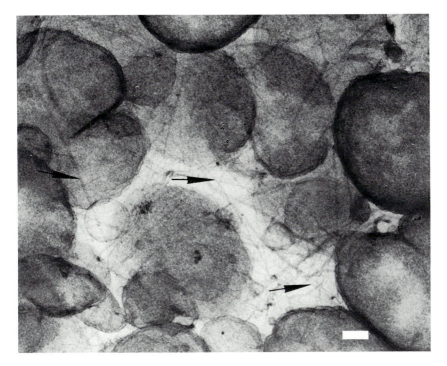

FIGURE 2. In 1-μm thick sections of epoxy-embedded colonies, the pili of *N. gonorrhoeae* appear as thin threads (arrows), both singly and in aggregates, when studied by high-voltage EM. Bar = 0.1 μm.

primarily as thin individual threads scattered throughout the embedding medium (Figure 2).

Our interpretations of the EM images of pili are similar whether the specimens are prepared by negative staining and drying in air, or studied after embedding in epoxy. These results are consistent with the traditional view of piliation in *N. gonorrhoeae*. In contrast, specimen processing by critical point-drying (CPD) provides us with apparently different information about the nature of piliation in the colonies.

4.3. Critical Point-Drying

Three methods of specimen preparation that incorporate a CPD step were used to prepare specimens of piliated gonococci. For each method, the specimens were fixed in glutaraldehyde and osmium tetroxide as described in this chapter. For SEM, the fixed colonies were dehydrated in a graded series of ethanol, dried by the CPD method using carbon dioxide, and sputter-coated with

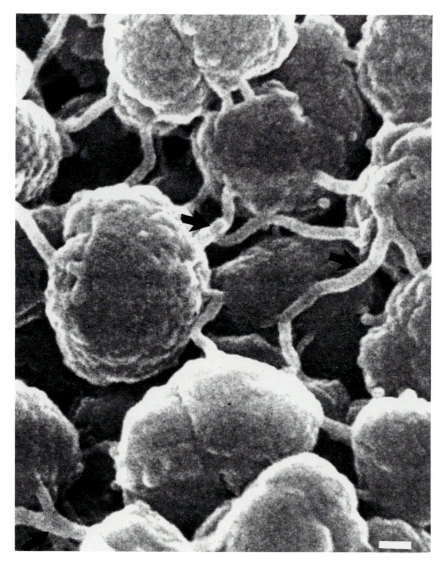

FIGURE 3. By scanning electron microscopy, piliated colonies of *N. gonorrhoeae*, processed by the critical point method, reveal thick branched structures (arrows) that appear to interconnect the colony members. Bar = 0.1 μm.

gold. SEM examination seemed to reveal large branched structures that interconnect the colony members (Figure 3). These structure were not observed in nonpiliated colonies of gonococci processed in the same way. To view the interior portions of the colonies, fixed colonies were embedded in polyethylene glycol (PEG) and sectioned. The sections were placed on Formvar-coated grids and the

PEG removed with water. The specimens were dehydrated in a graded series of ethanol and dried by CPD using carbon dioxide. Because of low specimen contrast, the specimens were rotary shadowed with platinum, and carbon replicas made. The pili were again detected as thick structures that seemed to branch and rejoin (Todd *et al.*, 1984). Specimens of piliated gonococci fixed, alcohol dehydrated, and processed by CPD were carbon-coated for stability and were examined directly by high-voltage EM, with similar results (see Figure 13 of Chapter 11). To determine whether the apparent increase in the thickness of pili was caused by coating the specimens with carbon, platinum, or gold, fixed specimens on Formvar-coated grids were prepared by CPD and examined directly by TEM. Thick structures that appear to branch and rejoin were also detected without the metal or carbon coating (Todd *et al.*, 1984).

Clearly, the morphology of gonococcal pili in specimens processed by CPD is different from that in specimens either air-dried and negatively stained, or embedded and studied in sections of epoxy. One interpretation of this apparent dichotomy in results is that processing by CPD induces aggregation to form the artifact of a latticelike network. This explanation assumes that pili radiate from the gonococcal surfaces as individual threads prior to the specimen processing required for negative staining or epoxy embedding. These interpretations can be tested by rapid freezing methods.

4.4. Rapid Freezing

4.4.1. Freeze-Drying

Piliated gonococci were fixed and then rapidly frozen by sudden immersion in propane cooled to the temperature of liquid nitrogen. The frozen specimens were dried by sublimation in a vacuum of 7.8×10^{-8} m Hg at $-100°C$ for 24 hr. The images of pili prepared by this freeze-drying method (Figure 5 of Chapter 9) are similar to images of pili prepared by CPD. Because of this result, it is unlikely that the larger structures of pili are formed as a consequence of specimen preparation through CPD. To try to understand how the different images are related, the techniques of freeze fracture were applied to piliated and, as controls, to nonpiliated colonies.

4.4.2. Freeze Fracture

Freeze fracture techniques have the advantages that specimen morphology is retained by sudden freezing, and that high-resolution images of the specimen can be generated through shadowing with high-contrast heavy metals. When piliated colonies are examined by freeze fracture (Figure 4), the individual threadlike pili are detected only across the surfaces of the gonococci. As the pili leave the gonococcal surfaces, they are seen to form thick bundles that branch and rejoin to create a latticelike network interconnecting the colony members.

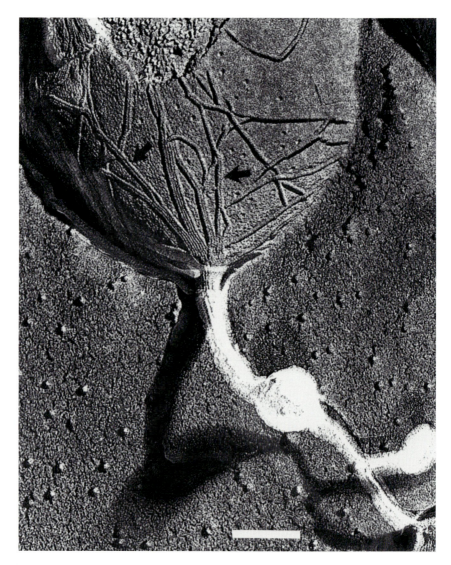

FIGURE 4. Carbon and platinum replicas made by rapidly freezing unfixed colonies of *N. gonor-rhoeae*, followed by fracture and etching, reveal both the individual linear pili across the bacterial surface (arrows) and the large aggregates that are formed when pili leave the gonococcal surface. These aggregates of pili branch and interconnect the colony members. Individual pili were not detected away from the gonococcal surface. This micrograph is reproduced as a negative reversal from Todd *et al.* (1984). Bar = 0.1 μm.

These structures were similar whether or not the colonies were fixed in glutaraldehyde and osmium tetroxide prior to rapid freezing. Apparently, these large structures are disassociated by the surface tension of air-drying and by heating at 60°C in epoxy-embedding media, to yield the individual threadlike arrangement that appears to radiate from the gonococcal surfaces in both epoxy-embedded and air-dried specimens.

Clearly, the apparent morphology of even the simplest of biological structures such as pili can vary considerably depending on the procedure chosen for specimen processing. How extensive then are alterations of the more complex microbial structures during similar procedures of specimen processing? How confident can we be of our results and interpretations? Unfortunately, there are no simple or absolute answers to these questions of specimen processing and interpretations, no single technique appropriate to all specimens and objectives. However, by understanding the advantages and limitations of the procedures applicable to microorganisms, and by examining the same subject from a variety of perspectives, accurate information concerning the structure and function of microorganisms can be ascertained. The chapters in this book provide a basis for understanding the advantages and limitations of each method and provide insight into appropriate interpretations.

ACKNOWLEDGMENTS. Use of the JEM-1000 high-voltage electron microscope was made possible by the Division of Research Resources, NIH, and the Department of Molecular, Cellular and Developmental Biology, University of Colorado, Boulder. The assistance and cooperation of the Electron Microscopy Section and Laboratory of Microbial Structure and Function, Rocky Mountain Laboratories, NIAID, NIH, Hamilton, Montana, are also appreciated.

5. REFERENCES

Aldrich, H. C., 1974, Spore cleavage and the development of wall ornamentation in two myxomycetes, *Proc. Iowa Acad. Sci.* **81**:28–35.

Boyles, J., 1984, The use of primary amines to improve glutaraldehyde fixation, in: *The Science of Biological Specimen Preparation for Microscopy and Microanalysis* (J.-P. Revel, T. Barnard, G. H. Haggis, and S. A. Bhatt, eds.), pp. 7–21, Scanning Electron Microscopy, Inc., O'Hare, Illinois.

Costerton, J. W., and Irvin, R. T., 1981, The bacterial glycocalyx in nature and disease, *Annu. Rev. Microbiol.* **35**:299–324.

Dubochet, J., McDowall, A. W., Menge, B., Schmid, E. N., and Lickfeld, K. T., 1983, Electron microscopy of frozen-hydrated bacteria, *J. Bacteriol.* **155**:381–390.

Emerman, M., and Behrman, E. J., 1982, Cleavage and cross-linking of proteins with osmium (VIII) reagents, *J. Histochem. Cytochem.* **30**:395–397.

Feria-Velasco, A., and Arauz-Contreras, J., 1981, Ruthenium red-mediated osmium binding for examining uncoated biological material under the scanning electron microscope, *Stain Technol.* **56**:71–78.

Greenawalt, J. W., and Whiteside, T. L., 1975, Mesosomes: Membranous bacterial organelles, *Bacteriol. Rev.* **39**:405–463.

Gregory, W. W., Gardner, M., Byrne, G. I., and Moulder, J. W., 1979, Arrays of hemispheric projections on *Chlamydia psittaci* and *Chlamydia trachomatis* observed by scanning electron microscopy, *J. Bacteriol.* **138**:241–244.

Hase, T., 1983a, Assembly of *Rickettsia tsutsugamushi* progeny in irradiated L cells, *J. Bacteriol.* **154**:976–979.

Hase, T., 1983b, Growth pattern of *Rickettsia tsutsugamushi* in irradiated L cells, *J. Bacteriol.* **154**:879–892.

Higgins, M. L., and Daneo-Moore, L., 1974, Factors influencing the frequency of mesosomes observed in fixed and unfixed cells of *Streptococcus faecalis*, *J. Cell Biol.* **61**:288–300.

Higgins, M. L., Tsien, H. C., and Daneo-Moore, L., 1976, Organization of mesosomes in fixed and unfixed cells, *J. Bacteriol.* **127**:1519–1523.

Higgins, M. L., Parks, L. C., and Daneo-Moore, L., 1981, The mesosome, in: *Organization of Prokaryotic Cell Membranes* (B. Ghosh, ed.), pp. 75–94, CRC Press, Cleveland.

Ito, S., and Rikihisa, Y., 1981, Techniques for electron microscopy of rickettsiae, in: *Rickettsiae and Rickettsial Diseases* (W. Burgdorfer and R. L. Anacker, eds.), pp. 213–227, Academic Press, New York.

Jost, P. C., McMillen, D. A., and Griffith, O. H., 1984, Effect of fixation on molecular dynamics in membranes, in: *The Science of Biological Specimen Preparation for Microscopy and Microanalysis* (J.-P. Revel, T. Barnard, G. H. Haggis, and S. A. Bhatt, eds.), pp. 23–30, Scanning Electron Microscopy, Inc., O'Hare, Illinois.

Louis, C., Nicolas, G., Eb, F., Lefebvre, J. F., and Orfila, J., 1980, Modifications of the envelope of *Chlamydia psittaci* during its developmental cycle: Freeze-fracture study of complementary replicas, *J. Bacteriol.* **141**:868–875.

Luft, J. H., 1973, Embedding media—Old and new, in: *Advanced Techniques in Biological Electron Microscopy* (J. K. Koehler, ed.), pp. 1–66, Springer-Verlag, Berlin.

Luft, J. H., 1971a, Ruthenium red and violet. I. Chemistry, purification, methods of use for electron microscopy and mechanism of action, *Anat. Rec.* **171**:347–368.

Luft, J. H., 1971b, Ruthenium red and violet. II. Fine structural localization in animal tissues, *Anat. Rec.* **171**:369–416.

Matsumoto, A., Fujiwara, E., and Higashi, N., 1976, Observations of the surface projections of infectious small cell of *Chlamydia psittaci* in thin sections, *J. Electron Microsc.* **25**:169–170.

Maysin-Szamier, P., and Pollard, T. D., 1978, Actin filament destruction by osmium tetroxide, *J. Cell Biol.* **77**:837–852.

Peracchia, C., Mittler, B. S., and Frenk, F., 1970, Improved fixation using hydroxyalkylperoxides, *J. Cell Biol.* **47**:156 (abstr.).

Punsalang, A. P., and Sawyer, W. D., 1973, Role of pili in the virulence of *Neisseria gonorrhoeae*, *Infect. Immun.* **8**:255–263.

Rikihisa, Y., and Ito, S., 1980, Localization of electron-dense tracers during entry of *Rickettsia tsutsugamushi* into polymorphonuclear leukocytes, *Infect. Immun.* **30**:231–234.

Silverman, D. J., Wisseman, C. L., Jr., Waddell, A. D., and Jones, M., 1978, External layers of *Rickettsia prowazekii* and *Rickettsia rickettsii*: Occurrence of a slime layer, *Infect. Immun.* **22**:233–246.

Stokes, G. V., 1978, Surface projections and internal structure of *Chlamydia psittaci*, *J. Bacteriol.* **133**:1514–1516.

Swanson, J., 1973, Studies on gonococcus infection. IV. Pili: their role in attachment of gonococci to tissue culture cells, *J. Exp. Med.* **137**:571–589.

Swanson, J., Kraus, S. J., and Gotshlich, E. C., 1971, Studies on gonococcus infection. I. Pili and zones of adhesion: their relation to gonococcal growth patterns, *J. Exp. Med.* **134**:886–905.

Todd, W. J., Wray, G. P., and Hitchcock, P. J., 1984, Arrangement of pili in colonies of *Neisseria gonorrhoeae*, *J. Bacteriol.* **159**:312–320.

Chapter 4

Secrets of Successful Embedding, Sectioning, and Imaging

H. C. Aldrich
Department of Microbiology and Cell Science
University of Florida
Gainesville, Florida 32611

and

H. H. Mollenhauer
Veterinary Toxicology and Entomology Research Laboratory
United States Department of Agriculture
College Station, Texas 77840
and
Department of Pathology and Laboratory Medicine
College of Medicine
Texas A & M University
College Station, Texas 77843

1. INTRODUCTION

For those of us who have worked with resistant structures such as spores, embedding media have always been central to success in ultrathin sectioning. Development of an effective embedding protocol often meant success or failure of a research project. However, as microbiologists have gained more experience with a variety of embedding resins, it has become clear that we must address criteria beyond success or failure of penetration. Choice of embedding resin can affect image contrast, stability under the electron beam, image granularity, and size of cells and organelles. Furthermore, when postsectioning manipulations such as immunocytochemistry with colloidal gold are used, we have to consider such characteristics as hydrophobicity and charge on the sections when we select embedding protocols.

In this chapter, we will summarize selected observations on embedding from our own laboratories and from the recent literature, concentrating on those methodologies of most utility for morphological and cytochemical studies on microorganisms. We will then relate some observations concerning how to optimize the sectioning process, how to obtain clean poststaining, and how to deal with image defects caused by various factors.

2. HISTORICAL ASPECTS

An excellent historical survey of embedding media has been published by one of the pioneers in the field (Luft, 1973); readers seeking a comprehensive treatment should consult this source.

Polymerized methacrylates became the first widely used embedding resins, and image quality sometimes approached what we expect today. However, several problems have been identified with these resins; e.g., instability in the electron beam, tissue swelling, and tissue distortion during polymerization (Luft, 1973). Therefore, only water-miscible methacrylates are now regularly used, mainly to improve penetration of stains for light microscopy (Feder and O'Brien, 1968), to enhance access of enzymes for extraction and identification of components within sections (Leduc and Bernhard, 1961), and for immunocytochemistry (Bendayan, 1984). An improved protocol for water-miscible methacrylate embedding was recently developed (Spaur and Moriarty, 1977).

To avoid the aforementioned problems, some workers turned to another class of plastics, the epoxy resins. Glauert *et al.* (1956) successfully embedded tissues in Araldite M, and Luft (1961) described the use of Epon 812, which is still frequently employed. However, Shell Chemical Corp. no longer manufactures the original formula, although several EM supply firms offer acceptable substitutes. A series of Epon–Araldite mixtures was detailed by Mollenhauer (1964), and these also remain in frequent use today. All of these epoxies were rather viscous after mixing, and consequently penetration of hard and resistant structures such as spores (Aldrich, 1967) and pollen (Skvarla and Larson, 1966) was slow and difficult to achieve. The low-viscosity mixtures of Spurr (1969) exhibited greatly improved penetration properties, with only slight sacrifice in cutting qualities, image contrast, and stability in the beam as compared to the higher-viscosity resins. They are probably the resins in widest use today, although other nonepoxy mixtures of very low viscosity, such as LR White (Causton, 1984), have recently been introduced. Only a few epoxy resins (e.g., Durcupan) are miscible with water.

Techniques for embedding at low temperatures have recently been published, offering the advantages of reduced lipid extraction and better preservation of protein structures for subsequent immunocytochemical labeling (Armbruster *et al.*, 1982; and Chapter 10). Lowicryl K4M has been used in gold-labeled antibody studies because the entire process of infiltration and polymerization can proceed at cold temperatures, thus avoiding denaturation of tissue antigens by heat polymerization. LR White has also proven useful for immunogold labeling (Causton, 1984).

3. FIXATION

Fixation of microbial cells for electron microscopy has been dealt with in detail in Chapter 3. We wish to point out here only a few fixation-caused artifacts that may be commonly encountered.

Two cellular components are unusually susceptible to fixative alteration or damage: nucleoids and glycogen. Prokaryotic nucleoids were identified very early as difficult to maintain in their native state during processing for electron microscopy. The Ryter–Kellenberger fixative (1958) was developed to meet this need and is still sometimes employed. Recently, *en bloc* staining methods employing aqueous uranyl acetate following osmium tetroxide or uranyl in 75% ethanol during dehydration have become popular. However, this method clumps the DNA after several hours of exposure. Consequently, we recommend that uranyl acetate *en bloc* be applied for no more than 1 hr or eliminated.

Glycogen is also extracted by uranyl *en bloc,* and may be leached by osmium or rendered unstainable by overfixation in glutaraldehyde (Hayat, 1981). Glycogen can be preserved by potassium ferricyanide, usually added to the osmium tetroxide fixative (De Bruijn, 1973). Ferricyanide enhances membrane contrast markedly, although image granularity may result, and some structures such as ribosomes may show less contrast (Mollenhauer and Droleskey, 1980).

Dimensional changes occur during the various steps of fixation, as well as during the dehydration and plastic infiltration steps to be discussed below. In general, it appears that cells shrink slightly in glutaraldehyde and then swell to somewhat more than their original volume in osmium. These processes are accentuated after longer fixation times, so one should aim at fixing in both these steps only long enough to achieve satisfactory results. Complete recent reviews of this problem may be found in Hayat (1981) and Lee (1984).

Sometimes it is not possible to design fixation and embedding protocols to eliminate all undesirable artifacts, but if the diverse factors interacting on the sample during EM preparation are understood, informed choices can be made. For example, we have recently been examining the meristematic region of corn roots. It has proved difficult to obtain adequate contrast of intracellular membranes such as nuclear envelope and endoplasmic reticulum with 1–2 hr of glutaraldehyde and 1–2 hr of osmication. Even though shrinkage problems were expected, we deliberately processed roots though a protocol that included 2 hr fixation in paraformaldehyde–glutaraldehyde, osmium tetroxide overnight at room temperature, and 1 hr in aqueous 0.5% uranyl acetate. Contrast improved markedly (Figure 1), but signs of tissue shrinkage were evident. The nuclear envelope profile was wavy, as were those of the plasmalemma, the mitochondria, and the endoplasmic reticulum. Such wavy profiles are never seen in well-frozen freeze-fractured root tissue, hence we conclude that they result from shrinkage. We view this distortion as an acceptable sacrifice balanced against the contrast improvement, but interpretation must be adjusted to take the shrinkage into account.

Other evidence of alterations in the transmitted electron image by the fixation process was provided by Dubochet *et al.* (1983), who compared thin sections of several bacteria with fixed and unfixed frozen-hydrated cells of the same species. In frozen-hydrated material, they found a constant 33-nm distance between inner and outer membranes, with no wrinkling of the outer membrane as one finds in plastic-embedded material (Figure 2). Both membranes of unfixed gram-negative cells exhibited only a single density profile, while a trilaminar structure was characteristic of unfixed gram-positive plasma membranes. Osmication caused trilaminar profiles of all three membrane types, condensation of nucleoid regions, and induction of mesosomes. Hobot *et al.* (1984) reached similar conclusions concerning the regular spacing of the two parallel limiting membranes of gram-negative cells and referred to this region as the "periplasmic gel" layer.

These conclusions agree with our own observations concerning freeze-fractured versus fixed, embedded, and sectioned cells. The propane jet freezing technique (Müller *et al.*, 1980) is a method by which unfixed cells in suspension are snap frozen without cryoprotectants and then fractured and replicated as in other freeze-fracture procedures. Plasma membranes and organelle membranes in eukaryotes and plasma and outer wall membranes in prokaryotes always appear smooth after jet freezing and replication (Figure 3), as if the cells were turgid. After fixation, embedding, and sectioning, however, the same cells have ruffled membranes (Figure 2). We believe that the jet frozen material represents images as close to the *in vivo* state as possible and that the undulating membranes in fixed material represent shrinkage at one or more steps in the preparation. This agrees with the conclusions reviewed in the previous paragraph based on observation of frozen-hydrated sections.

4. SHRINKAGE

As discussed above, artifacts arising from shrinkage have become so familiar as to be acceptable to most electron microscopists. Only when examining certain kinds of material or comparing nonembedding preparation methods do the problems become evident. Lee (1984) has recently provided quantitative data on the considerable extent of this problem. Shrinkage phenomena can arise at all stages

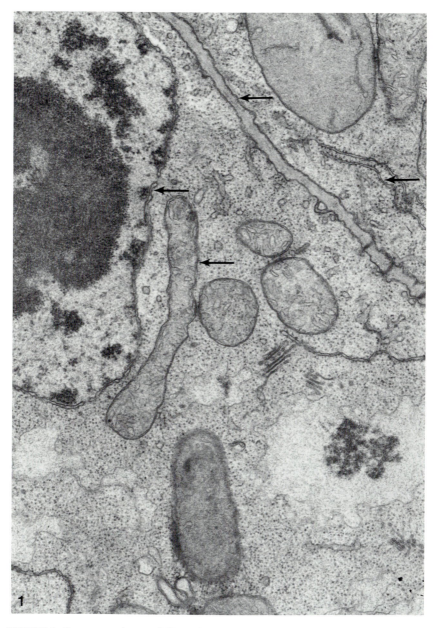

FIGURE 1. Corn root meristem cell illustrating shrinkage phenomena. Root was fixed for 2 hr in glutaraldehyde–paraformaldehyde, postfixed in 2% OsO_4 at room temperature overnight, and then stained in 1% aqueous uranyl acetate for 1 hr at room temperature. Note the undulating membrane profiles (arrows) of plasmalemma, nuclear envelope, mitochondria, and endoplasmic reticulum. Unfixed freeze-fractured cells show no undulations. ×25,3000.

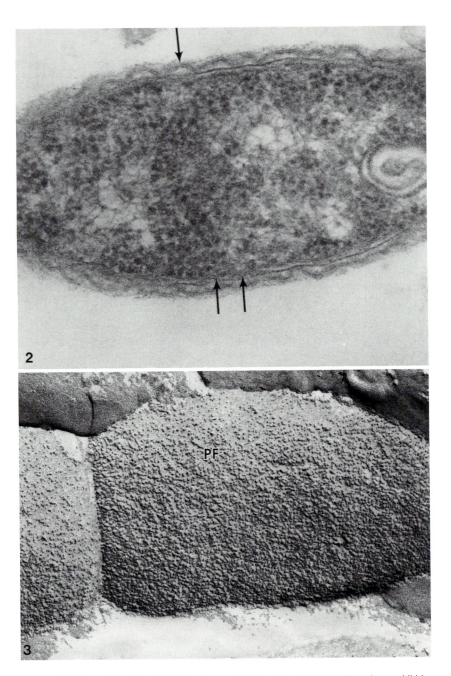

FIGURE 2. Thin section of *E. coli* cell fixed sequentially in glutaraldehyde and osmium, exhibiting wavy outer membrane and plasma membrane (arrows) typical of conventionally processed cells. The wavy membranes represent shrinkage artifacts. Compare with Figure 3. × 123,000.

FIGURE 3. Protoplasmic (PF) face of plasma membrane of unfixed *E. coli* cell, propane jet frozen and freeze-fractured. Membrane profile is smooth, unlike that of fixed cell in Figure 2. × 76,000. Micrograph courtesy of G. W. Erdos.

of preparation. For example, several years ago in our laboratory, workers sectioning clover roots found the root hairs consistently flattened as if collapsed (Napoli and Hubbell, 1975). After monitoring their appearance at each step, it was found that these delicate cells appeared normal until stepwise dehydration was begun in 25% ethanol. Collapse was solved by reducing the increment to 10% at each step and holding tissue 15–30 min in each solution. Neophytes are well advised to examine their cells with the light microscope at each step when fixing and embedding unfamiliar cells, to monitor for deleterious changes such as this.

Signs of possible shrinkage we have learned to look for include: (1) spacing between cells and walls, with no intervening stainable material; (2) matching irregularities in adjacent cells separated by several micrometers of resin, reminding one of the famous case wherein the correspondence between the coastlines of South America and Africa strengthened the case in favor of continental drift; (3) ruffled surface and intracellular membranes, such as nuclear envelopes and endoplasmic reticulum. These membranes are almost never wavy in freeze-fractured preparations.

In general, shrinkage can be minimized by spending the least amount of time possible in the osmium and dehydration steps of a protocol. We find that 1 hr in osmium tetroxide at room temperature is usually enough, while 5–10 min in each 25% ethanol increment will normally serve. In our experience, choice of acetone, propylene oxide, or ethanol followed by acetone in dehydration steps during epoxy embedding makes little difference in the final image. It does seem to be important to process the material from aldehyde into 100% plastic in one continuous series if shrinkage is to be minimized. Leaving cells overnight in 75% ethanol or dilute plastic, even in the cold, can lead to shrinkage. An alternative approach is to treat the fixed tissue with 2,2-dimethoxypropane, which reacts with specimen water to form acetone and methanol (Muller and Jacks, 1975). We have no information on whether this method affects specimen size.

The choice of embedding resin can also be critical in minimizing or eliminating shrinkage. In our experience, methacrylates such as Lowicryl K4M or glycol methacrylate cause shrinkage and morphological distortion. LR White causes minimal distortion and shrinkage. Spurr's resin causes some shrinkage and little or no distortion. Epon 812 and Epon–Araldite mixtures (Mollenhauer, 1964) cause minimal to no distortion providing the above caveats are observed. The Ladd ultralow-viscosity formulation, conversely, causes slight cell swelling, as evidenced by broken plasma membranes. We have had some success with a hybrid resin between the Spurr and Ladd recipes; this mixture neither shrinks nor swells cells and also has acceptable contrast characteristics (Aldrich, 1982):

ERL 4206 (vinyl cyclohexene dioxide)	2.5 g
DER 736	1.0 g
NSA (nonenyl succinic anhydride)	3.9 g
HXSA (hexenyl succinic anhydride)	2.6 g
DMAE (dimethylaminoethanol)	0.1 g

In working with this formulation and with Spurr's resin, we find it important to oven polymerize the resin in covered containers. Otherwise, volatile plastic components can escape and result in sticky, unpolymerized surface layers and/or brittle bottom layers.

5. FLAT EMBEDMENT FOR LIGHT MICROSCOPE SELECTION OF CELLS

Reymond and Pickett-Heaps (1983), among others, have described methods of embedding filamentous or unicellular material in thin layers, examining them with phase contrast after polymerization, and selection of specific cell types or stages for subsequent ultrathin sectioning. Plastic molds are available from EM supply houses to use in this protocol. We have also performed flat embedments in tissue culture flasks (Roos, 1973) and on carbon- or chemical-coated slides. The utility of this procedure is hard to overemphasize, especially if one is seeking somewhat rare events such as mitosis or cell fusion events. Days of time can be saved if flat embedding is employed in preference to random searching of pelleted material.

6. OTHER EMBEDDING TRICKS

It hardly seems necessary to mention, but we save much trouble and minimize cell loss when carrying cell suspensions through dehydration into plastic by embedding cells in agar after the osmium tetroxide fixation step. Since the cells are somewhat brittle after osmium treatment, repeated centrifugations may cause cell breakage. We routinely wash twice in water and then resuspend the cells in warm 2% agar in conical centrifuge tubes. Before the agar solidifies, the cells are repelleted by centrifugation. After hardening, the excess agar is then trimmed away and the pellet cut into ca. 1-mm^3 fragments, which are then dehydrated and embedded like pieces of tissue.

Cells too delicate to centrifuge can be concentrated by pulling them down onto a Millipore (or similar) filter over a vacuum flask and then covering them with warm agar before the cells can dry out. The agar and filter can then be cut into convenient-sized pieces for subsequent handling. The agar will usually separate easily from the filter, bringing the cells with it. If not, the filter will dissolve in acetone during dehydration.

7. MICROTOMY

The actual process of microtomy is very personal, so we will not treat it in detail here. For readers interested in the subject, Reid's book, *Ultramicrotomy*

(1975), is recommended. In the following sections, we will offer some observations gained during our years of experience in dealing with various embedding media, knife types and brands, and microtomes. Finally, we will include hints on how to recognize and minimize poststaining and beam damage problems, once sections have been obtained.

8. STABILITY OF EPOXY RESINS

Polymerized epoxy resins are inherently unstable, and movement of tissue within them can occur. The degree of instability depends on (1) the character of the resin (e.g., kind of resin, brand name); (2) ratios of hardeners, diluents, and accelerators; and (3) tissue characteristics. Epoxy resins are seldom formulated to achieve maximum cross-linking and are, in reality, quite fluid. This fluidity is usually (but not always) of little consequence so long as the tissue remains in the block, since the dimensions of tissue movement are much smaller (in the block) than that induced by the fixation and dehydration procedures (e.g., Kushida, 1962). However, significant movement almost always occurs in the section unless suitable preventive measures are exercised. Both *in block* and *in section* movement are discussed briefly.

Movement of tissue within a section is due both to beam–section interaction and to fluidity of the embedding resin. Beam–section interaction may be quite vigorous. The section may move hundreds of angstroms in spots immediately under the beam. Movement is due primarily to the heat produced by the beam, but electrical charges that deposit in the section probably also cause movement of the section. Heat causes the embedding resins to undergo a partial breakdown under the beam, which results in both a gross linear movement of the section and a rapid vaporization of resin or resin constituents (Luft, 1973). Loss of resin by vaporization may be as high as 20% of total mass for commonly used epoxy embedding resins. Loss of resin is manifested as an increase in image contrast.

Electrical charges probably also play a role in section stability. Embedding resins (and therefore sections) are insulators and the electrons lost to the section do not readily move through the section to the grid bars where they can be dissipated to ground. Charges trapped in the section repel each other, thus developing forces that can move or (occasionally) tear the section. However, the insulating properties of the section presumably decrease as it is heated by the beam (insulators often become conductors at elevated temperatures) and as it becomes carbonized by the heat. Eventually, the section achieves a reasonable degree of both thermal and electrical stability. Surprisingly, these section instabilities do not often prevent the acquisition of useful data. Nonetheless, it must be recognized that tissue components move as a result of beam–section interaction. These effects can be reduced by use of the minimum-beam radiation devices on some microscopes, section stabilization by carbon, or careful (slow) aging of sections in the beam (i.e., exposing the sections to low beam intensity).

Due to resin fluidity, tissues may move slowly both within the section and within the resin block even at room temperature. In extreme cases, this can be verified by looking at the smooth surfaces of blocks sectioned several months previously. In these instances, outlines of tissue components are visible that verify movement of tissue or resin into or out of the original plane of the block face (Figure 4). The amount of movement depends primarily on the resin. Some resins are considerably less stable than others. The point here is that movements of several micrometers can be demonstrated in some resins, yet only a few nanometers' movement (which would not be directly visible) would be sufficient to degrade a micrograph. Sectioned tissues, because of their thinness, should be particularly susceptible to movement. Thus, we postulate that sections degrade rapidly in regard to tissue movements of 0.5–1.0 nm. In some instances, section life is probably less than 2–3 days. However, recognition of degradation is not straightforward since the major tissue components appear unchanged. What usually occurs is a feeling that the microscope is not quite "tuned" and the pictures are not quite as sharp as they should be. The obvious solution is to use sections as soon as possible after cutting. The relationship between storage conditions such as temperature or humidity and section life, is not known.

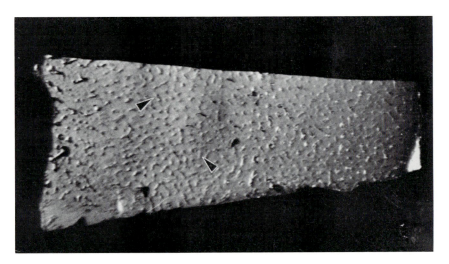

FIGURE 4. Photograph of the face of an epoxy resin block in which *Euglena* are embedded. Sections were cut from the block approximately 6 months before this picture was taken and, at that time, the face of the block was mirror smooth and no cell outlines could be discerned. With storage, however, the cells at the surface of the sectioned block became visible (arrowheads). We presume that this effect is caused primarily by movement of the tissues in the resin since this part of the block is less rigid than those parts of the block composed only of resin (i.e., the tissue dilutes, and thus weakens, the resin). The amount of movement depends on plastic and plastic formulation and, in extreme cases, may be as much as several micrometers per month.

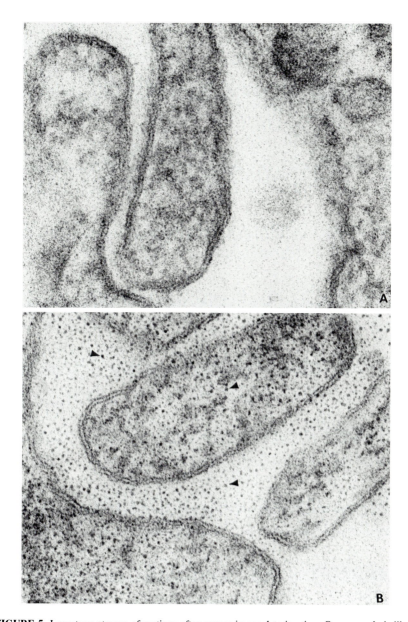

FIGURE 5. Long-term storage of sections often causes image deterioration. One example is illustrated here. Both sections from which these micrographs were made are from the same block of liver embedded in Spurr resin (Spurr, 1969) and poststained in uranyl acetate and lead citrate. The section in (A) was examined immediately after it was cut and stained, whereas the section in (B) was stored in the laboratory at room temperature for 1 month before being examined. In both instances, the beam was near crossover and relatively intense when it first struck the section. The pepper in (B) (arrowheads) formed during the first few seconds after the section was exposed to the beam. Slowly aging the section under low beam intensity would have reduced the amount of pepper substantially. Both ×250,000.

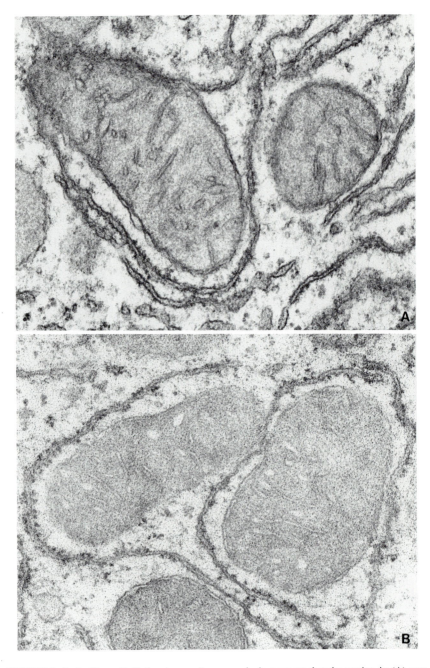

FIGURE 6. As in Figure 5. Both micrographs are equivalent except that the section in (A) was examined immediately after sectioning and that in (B) was stored for 1 month before examination. In addition to increased beam pepper, there was a marked change in the appearance and contrast of mitochondria in some sections. In both (A) and (B), section thickness, poststaining, and photographic procedures were comparable. Both ×60,000.

A side effect of section instability and storage occurs occasionally; namely there is a greater propensity to form "beam pepper" (Figure 5; see also Section 9.2). We postulate that unbound components of the resin mixture, or unstabilized components of the tissue, gradually separate into micropools within the section and these components are burned by the beam or are vaporized over the section. Additionally, there may be marked changes in image contrast (Figure 6).

9. TISSUE AND SECTION CONTAMINANTS

Contamination in the form of extraneous material on or in the section has been experienced by most, if not all, electron microscopists. Micrographs of sections with contamination are not readily accepted for publication even when the contamination does not affect interpretation of data. Methods for preventing contamination are numerous (e.g., Millonig, 1961; Feldman, 1962; Reynolds, 1963; Björkman and Hellström, 1965; Frasca and Parks, 1965; Avery and Ellis, 1978; Kuo, 1980) but the problem is difficult and, to our knowledge, no procedure is universally successful. The problem is compounded because there seem to be several kinds of contamination. We have identified four (Mollenhauer and Morré, 1978) and these are discussed briefly.

9.1. Surface Contamination

Surface contamination occurs as extraneous material deposited on the surface of a section. This kind of contamination usually occurs during sectioning or section staining and the "dirty-looking" particles characteristic of this form are often large and dense. The primary causes of section contamination are bacteria, lint, dirt, oil, and dried residues (or reaction products) from the poststaining solutions.

Bacteria often grow in distilled water supplies and end up in both the poststaining solutions and the bottles used to fill boats and wash sections. The presence of bacteria is almost always associated with a static water storage system, i.e., the water is kept in bottles that are continuously replenished with freshly distilled water. Bacteria are seldom present in water from purification systems in which the water is continuously circulated through filtration beds. If a water system clear of particulate matter is unavailable, then simple Millipore (or equivalent) filters attached to the syringe used to dispense the water, can be used to eliminate the bacteria.

Lint often comes from the filter paper on which grids may be placed before use or which are used to dry the section after staining. The lint is easily trapped between the grid and the forceps and pulled loose from the filter paper. Airborne lint (as well as other particulate matter) seldom falls onto a section even after

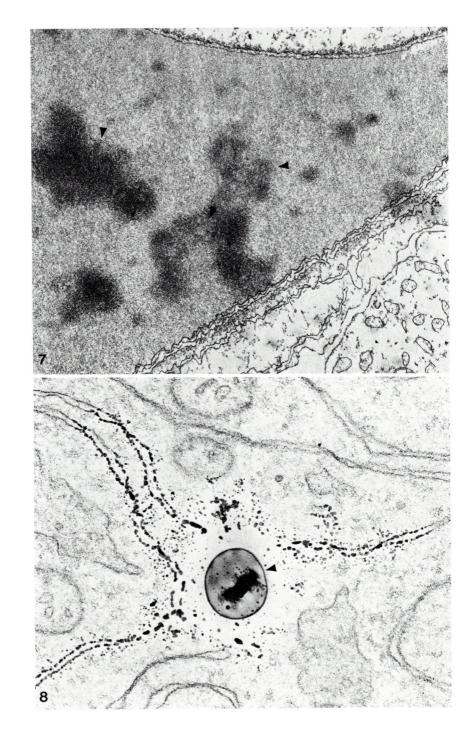

hours and sometimes days of exposure to the atmosphere and, therefore, is usually not a serious problem.

Particulate matter (dirt) most often comes from water in the boat, liquid trapped between the tongs of the forceps, or the poststaining solutions to which the section is exposed. Perhaps the most critical aspect here is the way that the section is handled during transfer through the staining and washing solutions and/or during drying. Two points need to be considered: (1) A dry section should never be put into or on a solution containing extraneous material. (2) Contaminated solutions should never be drained off a section or allowed to dry on a section. However, the fact that a solution contains suspended material is usually of little consequence since this does not, in itself, lead to dirty sections. The problem is almost always associated with the passage of the section into or out of the water and/or staining solutions. Simplified staining procedures that use these ideas have been described and work reasonably well (Mollenhauer, 1974, 1975).

Oil most often comes from hands, uncleaned razor blades used for block trimming, forceps (Figure 7). It is good practice to clean hands, razor blades, and forceps before starting work. Additionally, epoxy resins often contain unpolymerized components that leach out onto the water surface during sectioning. These contaminants may be carried to the dry (back) face of the knife and be transmitted from one block to the next.

Dried residues of stain, or reaction products between stain and air, are usually obvious because of their large mass or their crystalline nature. However, some also may be more subtle as indicated below (see Section 9.3).

9.2. Beam Pepper

This fine particulate granularity is caused by interaction between the beam and the section (see Section 8). Sections are rapidly altered by beam energy as indicated by movement of the EM image, clearing of the section to give increased image contrast, and the eventual carbonization and stabilization of the section. Plastic and tissue components, or dirt in or on the section, may be vaporized by the beam and redeposited on the surface of the section (Figures 5B and 8). Beam-induced recrystallization of metal shadowing films (Böhler, 1975) may be a similar phenomenon. Beam pepper probably cannot be eliminated under the usual operating conditions because even the lowest beam intensities

←⎯⎯

FIGURE 7. Section showing one form (oil) of surface contamination (arrowheads). In this instance, the oil came from a new and uncleaned forceps used during poststaining. ×34,000.

FIGURE 8. Beam pepper is the result of interaction between the beam and section. Material in or on the section is vaporized by the beam and redeposited on the section. Redeposition is sometimes random (see Figure 5B) and sometimes highly oriented as illustrated here. In this example, the beam impinged on the dense substance indicated by the arrowhead, which then vaporized outward to settle on discontinuities (endoplasmic reticulum profiles) within the section. ×50,000.

generate enough heat to change the section. However, beam pepper usually can be reduced to acceptable levels by aging sections at low beam intensity.

9.3. Embedding Pepper

Embedding paper can be recognized as small black specks that resemble black pepper (Figure 9A). Embedding pepper is usually associated with mitochondria, peroxisomes, and red blood cells and is formed during poststaining of the section. The origin of embedding pepper is not known but it appears to be caused by an improperly formulated plastic mixture or by poor penetration of resin or resin constituents into the tissue or into selected tissue components. In the latter case, the tissue components appear to act as barriers or columns separating out the components of the resin mixture and altering resin composition. Poststaining solutions then penetrate into, or through, these regions of improperly polymerized plastic and deposit in, or on, the sections. Precipitated lead is insoluble in water and, if it forms, remains as a contaminant on the section. Embedding pepper can be minimized during poststaining by submerging the section into the staining solutions (Figure 9B; Mollenhauer, 1975). Resins containing NSA as the sole hardening agent seem particularly susceptible to the formation of embedding pepper.

Further insight into the origin of embedding pepper is indicated in Figure 10A. Here the pepper is oriented parallel to the cutting stroke and in association with discontinuities in the knife edge. Presumably, the knife alters the surface of the section (Peters *et al.,* 1971), which then traps, or somehow precipitates, the staining solution. This kind of pepper is most conspicuous in organelles such as red blood cells and occurs most often when sections are floated on the poststaining solutions (Mollenhauer, 1974) and in tissues block-stained in uranyl acetate. It can be reduced by submerging the section into the staining solutions (Mollenhauer, 1975) and can be eliminated by submerging the section into HCl (Figure 10B) or EDTA (Figure 10C) before poststaining the section. HCl will lower image contrast and solubilize some cellular constituents depending on concentration and length of exposure.

9.4. Fixation Pepper

This problem appears much like embedding pepper except that it is usually present in or around Golgi apparatus, smooth endoplasmic reticulum, nuclei, chloroplasts, and bile canaliculi (Figure 11A). It is seldom found in mitochondria or red blood cells. The cause of fixation pepper is unknown and, therefore, no means for preventing its appearance can be suggested. However, several factors are possible causes and these will be discussed briefly. The general premise is that fixation pepper results from interaction between the glutaraldehyde and the osmium tetroxide. This is not easy to verify, however, because deliberate mixing of glutaraldehyde and osmium tetroxide does not necessarily yield pepper-laden

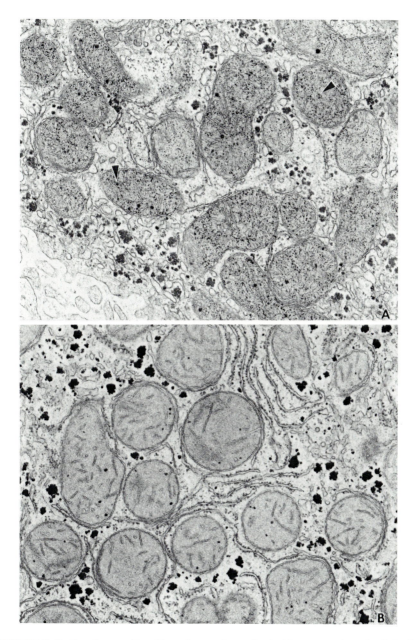

FIGURE 9. Sections from rat liver illustrating one form of embedding pepper (arrowheads in A). The section in (A) was poststained by floating it on the surface of a drop of lead citrate (Mollenhauer, 1974). The upper (dry) surface of the section was not washed. The section in (B) differed from that in (A) only in that it was submerged into the lead citrate during poststaining as per Mollenhauer (1975). Embedding pepper is often associated with red blood cells and mitochondria (see also Figure 10A) and seems to be related in part to lack of penetration of the embedding resin (or some components of the embedding resin) into the red blood cells and organelles. Uranyl acetate used as block stain also may contribute to formation of pepper. A, ×30,000; B, ×20,000.

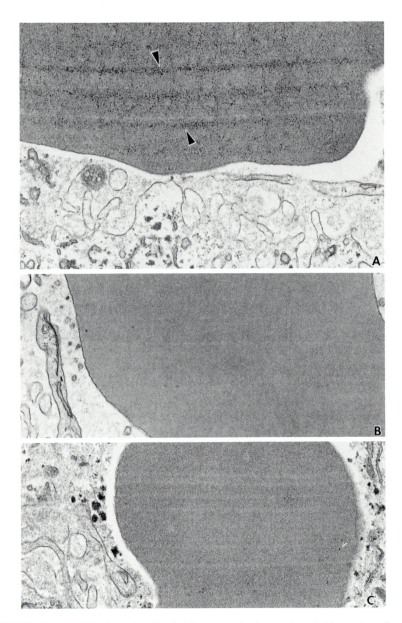

FIGURE 10. Another manifestation of embedding pepper is related to irregularities on the surface of the section. These irregularities appear to come from plastic or tissue constituents that are displaced by the knife and possibly spread over the surface of the section. These displaced constituents either stain or trap stain to form pepper (arrowheads). (A) Section through a red blood cell poststained with uranyl acetate and lead citrate. Pepper is oriented along the direction of cut as indicated by the knife marks. This pepper can be prevented or eliminated by submerging the grid and section into HCl (about 0.1–0.5 N; Figure 10B) or EDTA (about 1% aqueous; Figure 10C) before staining with uranyl acetate and/or lead citrate. The HCl may bleach some parts of the section depending on HCl concentration and exposure time. Bleached sections often appear exceptionally "clean." All ×31,000.

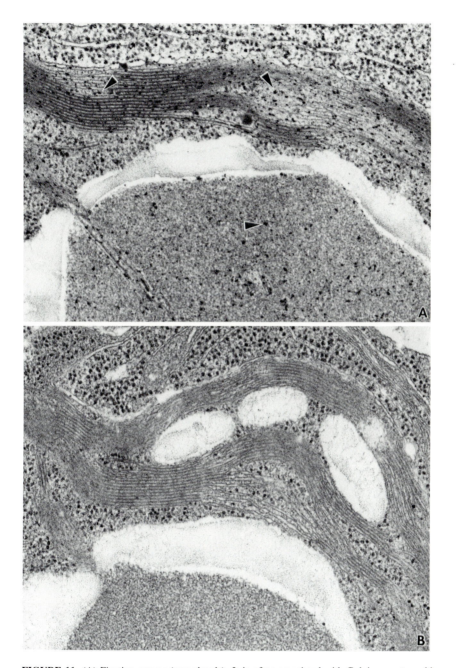

FIGURE 11. (A) Fixation pepper (arrowheads). It is often associated with Golgi apparatus, chloroplasts, bile canaliculi, smooth endoplasmic reticulum, and other membranous constituents of the cell. (B) A companion section to that in (A) but destained with periodic acid before poststaining with uranyl acetate and lead citrate. Almost all of the beam pepper was removed and the resulting section was relatively clean. From Mollenhauer and Morré (1978). Both ×50,000 approx.

tissues. Nonetheless, slowing the glutaraldehyde–osmium tetroxide reaction often helps eliminate fixation pepper. The most effective approach is to use ice-bath temperatures whenever glutaraldehyde and osmium tetroxide may be in contact. This would include the last buffer rinse before osmium tetroxide fixation and all subsequent fixation and block-staining procedures. Sucrose also is useful in slowing the glutaraldehyde–osmium tetroxide reaction. The amount of sucrose is not critical and about 0.05 M is usually sufficient for the exposure times involved. Too much sucrose will alter the fixation image and lower image contrast. Purity of glutaraldehyde also has been suggested as a cause of fixation pepper, but this has been difficult to verify. Nonetheless, use of a good grade and purity of glutaraldehyde is highly recommended, or a simple redistillation method as described by Smith and Farquhar (1966). Use of glutaraldehyde purified in this way has virtually eliminated fixation pepper in the senior author's laboratory. When pepper is present, it sometimes can be solubilized from the section by destaining the section in 1–2% periodic acid for 10–20 min and then restaining the section in the usual uranyl acetate and/or alkaline lead solution (Figure 11; Mollenhauer and Morré, 1978; Ellis and Anthony, 1979). Some contrast may be lost using this procedure but this is preferable to the pepper.

10. RESIN MODIFIERS FOR IMPROVING SECTIONING

Numerous substances have been added to resins in order to improve sectioning (Stäubli, 1960; Freeman and Spurlock, 1962; Spurlock et al., 1963; Mollenhauer, 1964; Langenberg, 1982; Wakefield, 1984). These often were oils or lubricants rather than active modifiers of the molecular structure of the resin. Perhaps the best known of these is dibutyl phthalate, which was used for years in various epoxy formulations (Stäubli, 1960; Freeman and Spurlock, 1962; Mollenhauer, 1964). Most of these additives, though useful, achieved only marginal success and none were of sufficient benefit to achieve universal use. Recently, silicone modifiers have again been recommended (Langenberg, 1982; Wakefield, 1984) but we have not evaluated those currently available, and they have not been available long enough to establish strong user support. However, there is some concern about introducing silicones into the microscope since contaminants of them are very difficult to remove from lenses and column.

An entirely different class of modifiers (surfactants) has been explored and these have imparted some possibly useful properties to the resins (Mollenhauer, 1984, 1986). The object of this study was to improve sectioning sufficiently to allow the routine use of glass knives. It was felt that two aspects had to be addressed, i.e., the forces necessary to cleave a section from the block as well as the coefficient of friction between the section and knife face had to be lowered. The first item would probably require an active diluent to weaken the molecular structure that holds the plastic together, whereas the second item would probably

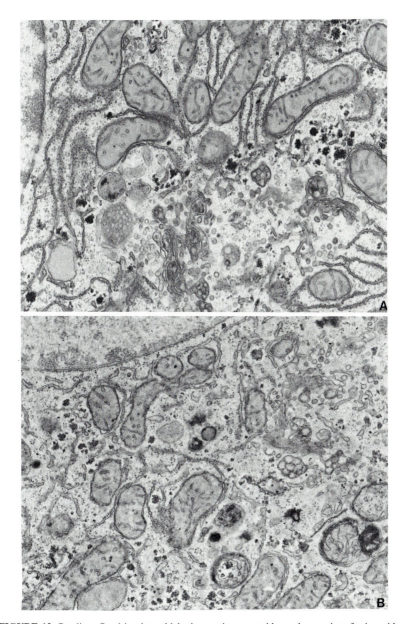

FIGURE 12. Rat liver fixed in glutaraldehyde, osmium tetroxide, and potassium ferricyanide and then embedded in Spurr resin containing 1% lecithin. The section in (A) was cut with a diamond knife and poststained with uranyl acetate and lead citrate. The section in (B) is similar except that it was cut with a glass knife. Glass knives do not cleave plastics cleanly and this characteristic imparts a "mottled" appearance to the tissue. (This characteristic is present with or without lecithin but is sometimes enhanced by lecithin and other surfactants.) In this instance, the mitochondria appear to be poorly embedded (see discussion in Section 9.3) and show more mottling than the other parts of the cell (also compare with the mitochondrial matrices in A). A, ×26,000; B, ×27,000.

require only a lubricant to lower the frictional forces between knife and section. Surfactants seemed promising because many are excellent lubricants (e.g., soaps) and many are compatible with both aqueous and nonaqueous media.

Numerous surfactants were tested but lecithin seemed to offer the most promise. The results with lecithin were spectacular though not all beneficial. Specifically, adding 4–5% lecithin (Sigma No. P3644) to the hard epoxy mixture of Spurr (1969) produced a plastic that could be sectioned easily with glass or diamond. Lecithin softened the polymerized block of epoxy resin and caused the block to trim as though it were waxy. Lecithin also rendered the epoxy resin partially soluble in acetone and similar solvents, thus indicating a weakening of intermolecular bonds in the resin and possibly a reduction of the forces necessary

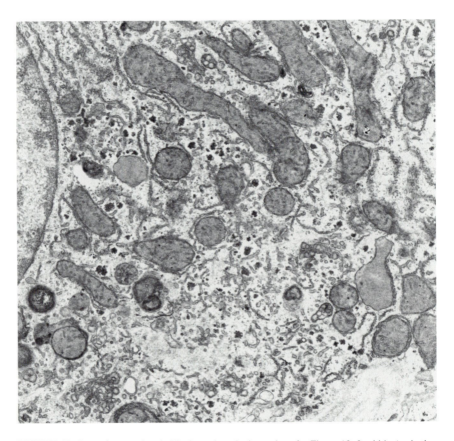

FIGURE 13. Same tissue and resin block used to obtain sections for Figure 12. Lecithin (and other surfactants) markedly improve the useful life of the cutting edge of glass knives. This micrograph was made from the 100th thin section (approximately) cut from the same place on the glass knife that was used also for facing the block using 1- to 2-μm-thick sections. ×19,000.

to cleave a section. With glass, it was possible to face a block and then cut several hundred thin sections at the same spot on the knife. Unfortunately, there were two undesirable characteristics: (1) the sections appeared rough or mottled and sometimes had chatter and (2) contrast was low. Nonetheless, results were sufficiently good to suggest trials with intermediate concentrations of lecithin. To this end we recommend $\frac{1}{2}$–2% lecithin (Sigma No. P3644) added to the firm or hard resin mixture of Spurr (1969). Lecithin may appear difficult to dissolve into the resin mixture. However, it is easy to put into solution by first dissolving it into the accelerator. A mixture containing equal parts (by weight) of lecithin and DMAE may be achieved within a day or so at room temperature or 2–4 hr at 60°C. Concentrations of 1% lecithin are sufficient to markedly improve sectioning (for diamond knives) without significantly altering any other aspect of the section/staining process or the appearance of the EM image (Figure 12A). However, a slight (Figure 12B) to moderate "mottled" appearance of the section may occur when sections are cut with a glass knife. Glass knives have been shown to cut rough sections as compared to diamond knives (Peters *et al.*, 1971) and this is apparently an inherent characteristic of the knife. However, mottling is variable depending on resin formulation and embedding procedure and can be minimized by careful adjustment of these parameters. Thus, lecithin (or some other surfactant) will allow the routine use of glass knives to produce sections of sufficient quality for much work. Because these resins are so easy to section, they should be particularly beneficial to students.

Surfactants markedly increase the longevity of the glass knife, often making it possible to cut over a hundred good sections from the same spot on the knife edge even after facing the block with 1- to 2-μm-thick sections (Figure 13). The same effect also may apply to diamond knives. Presumably, the increased ease of cutting sections will prolong the life of the diamond edge. The lecithin plastics cut well with the diamond knife and section quality is excellent.

11. IMAGE

Embedding resins influence the EM image of tissues and tissue components. Three image characteristics—general appearance, contrast, and dimensional variations—are considered. Though differences in image are easily documented, the reasons for them often remain obscure. For example, each brand or formulation of epoxy resin may impart specific image characteristics to the tissue (Mollenhauer and Droleskey, 1985; Figure 14). Differences in resin–tissue interaction (e.g., plastic constituents may react with the tissue; Luft, 1973), solvent action, as well as differences in the molecular structure of the cross-linked plastic could influence appearance.

Most plastics are of similar density after polymerization and should not, of themselves, cause gross differences in image contrast. Yet image contrast does

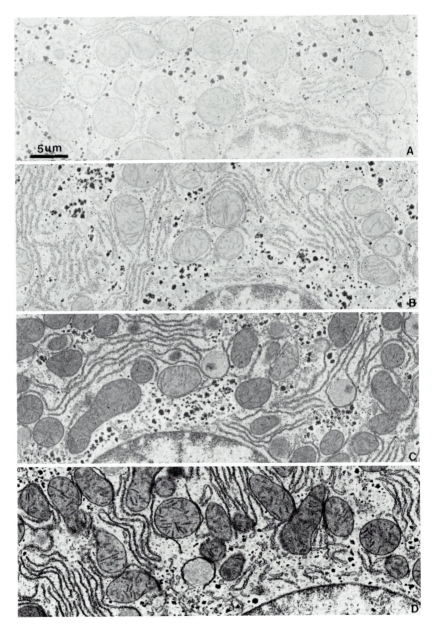

FIGURE 14. The general appearance of the tissue is dependent on the embedding resin. A series of sections from four epoxy resins illustrating differences in contrast and organelle size. (A) Ladd resin with RD-2 (Ladd Research Industries, Inc., Burlington, VT.). (B) Ladd resin with DER 736 instead of RD-2. (C) Spurr resin (Spurr, 1969). (D) Quetol resin (Kushida, 1974, 1975). All tissues were from the same animal, were processed through fixation and dehydration in the same vial, and were separated only at the embedding stage. All prints were made on contrast grade 1 paper so that the background densities were approximately equal. From Mollenhauer and Droleskey (1985).

vary markedly between resins. In epoxy resins, some of these differences apparently reflect loss of resin by heat-induced vaporization (Luft, 1973). This accounts for the "clearing" of a section immediately following the first exposure to the electron beam. Additionally, the resin may affect tissue staining either by hindering stain penetration into the section or by blocking reaction sites in the tissue. The amount of contrast variation between resins may be quite large, as illustrated in Figure 14, which compares four commonly used formulations.

Size differences of organelles as great as 15% in linear dimension also have been noted in epoxy-embedded tissues (Mollenhauer and Droleskey, 1985) but the reasons for this are unclear. Epoxy resins generally (and including those used for EM) exhibit very low shrinkage (Kushida, 1962) and this cannot account for the differences in organelle size. As postulates, however, three factors seem worthy of consideration: (1) lack of penetration of resin into the tissue or into some organelles, resulting in regions of poor preservation; (2) direct interaction of the resin or constituents of the resin with tissue components; (3) differential organelle swelling or shrinking during infiltration.

12. EVALUATION OF DIAMOND KNIVES

Judging the quality of a diamond knife is not easy unless the knife is either very good or very bad. Unfortunately, the vast majority of knives fall between good and bad. Evaluation is highly subjective and the decision is dependent on the use for which the knife is intended, the urgency to put the knife into service, the available resources of the user, and managerial requirements such as bids and contracts (common to federal agencies and many universities). In almost all cases, the decision to keep or return a knife is based on a weighted average of these factors. Published procedures for testing knives are available (e.g., Persson and Persson, 1976) as are manufacturer's recommendations (see the appendix, "Suggested Reading") and personal assistance from the manufacturer if necessary. In addition, we include the following generalized statements, the ideas of which bear on the sectioning process and which have been found helpful in evaluating a knife.

1. Diamonds vary markedly in physical characteristics depending on the temperature and pressure under which they were formed and impurities, which affect crystal structure (Fernández-Morán, 1985). Additionally, careful selection and orientation of the diamond crystal is necessary to obtain a good knife (Fernández-Morán, 1985). These physical characteristics translate directly into edge quality (or lack thereof) and knife life.

2. Perhaps the easiest way to test a diamond knife is to put it into service during the test period and slowly determine its character.

3. In spite of what most of us have been led to believe, the clearance angle between the back (dry) face of the diamond knife and the path of the tissue block

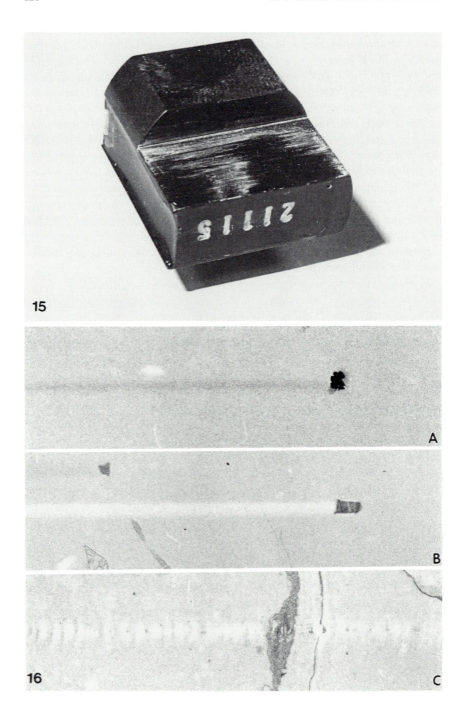

15

16

A

B

C

being sectioned is not critical *if the knife is good.* Conversely, *if the clearance angle is critical for good sectioning, then the knife is of less than optimum quality.* A good knife will cut sections of excellent quality at clearance angles from 1 to over 22° (Mollenhauer, 1981). (Note, however, that *large clearance angles should never be used* since they place extremely large stresses across the knife edge.)

4. Clearance angles of less than 0° (i.e., the block rubs the back face of the knife) are probably even more deleterious to the knife edge than large clearance angles. As the knife cuts the section, it is forced back by the block, putting large stresses across the knife edge.

5. Clearance angles of 5–8° are suitable for most knives.

6. Occasionally, the physical characteristics of the knife holder and microtome markedly influence sectioning, especially in regard to chatter. An example is illustrated by the following anecdote. A new knife being tested produced sections with chatter at all clearance angles. The problem was traced eventually to the boat in which the knife was mounted, i.e., the boat could not be properly clamped into the microtome because of a nonflat mounting surface (Figure 15). The problem was eliminated by a slight modification of the microtome clamping device so that matching surfaces between the knife boat and microtome clamping device were not critical. Problems relating to the microtome–knife interface have been observed by one of us on six occasions. Unfortunately, mechanical problems related to the microtome, or to the interaction between the knife and microtome, are difficult to diagnose. About the only way available to most microtomists is to test questionable knives in at least two microtomes.

7. The life of a diamond knife depends in part on its sharpness. An exceptionally sharp knife may well last 4–5 years cutting biological material, whereas a barely acceptable knife is often dull or scratched after only 2–5 months of use. Edge sharpness relates directly to the prorated cost of the knife and is a very important factor in knife evaluation.

←_____

FIGURE 15. A diamond knife mount and boat showing, especially, the lower part that clamps into the microtome. The white marks were formed by rubbing a chalk-coated glass microscope slide over the surface. The uneven distribution of chalk indicates a lack of flatness of the surface. This lack of flatness can be a problem if the area of contact between the knife and microtome is small and if a high spot happens to be superimposed over this contact area. Under these conditions, the knife cannot be firmly mounted into the microtome and this sometimes results in section chatter (see Mollenhauer, 1981). (Note: Other factors such as poor fit between tissue block and clamping chuck or loose mounting screws in the microtome, also may contribute to chatter but these problems are relatively obvious and are not discussed or illustrated.)

FIGURE 16. (A, B) Sections showing common types of striations formed by debris attached to the edge of a diamond knife. In both examples, the debris that formed the striations became dislodged at the points where the striations ended. Striations of this type can be either dark or light and are typically smooth in appearance. (C) Striations caused by chips in the knife edge are usually more sharply delineated than those formed by debris on the knife edge and they often appear jagged. A, ×22,000; B, ×40,000; C, ×12,000.

8. Knife quality, and particularly sharpness, often can be evaluated by observing how a section moves off the knife as it is being cut. Sections being cut by a good knife will "flow" onto the water surface almost as if by magic. If there is any visible reaction between the forming section and the knife, then the knife is not as good as it should be. Literally, a good knife will cut a section as easily as a "hot knife cuts butter."

9. The quality of a knife is usually consistent across the entire edge; i.e., if a knife is good or bad at one spot, it is likely to be good or bad, respectively, at other spots.

10. Cleaning a knife seldom improves its sharpness or its ability to cut sections although it may remove debris attached to the knife edge (e.g., see Figure 16A,B). Therefore, we do not recommend regular cleaning of a knife or wiping the edge with pith, plastic, or any other material. However, sections, as well as any nonvolatile constituents from the water in the boat, should not be allowed to dry on the edge or face of the knife. Always rinse the knife with clean distilled water from a squirt bottle before these substances have a chance to dry. If cleaning is necessary, follow the suggestions of the manufacturer. A suggestion of potential merit is to use the jet stream from a device such as a Water Pik to wash strongly bound debris from the knife edge (item 1 in the appendix).

13. SAFETY

All epoxy resins, and perhaps other embedding resins as well, are hazardous to health, and extreme care should be exercised in handling them. Ringo *et al.* (1982, 1984) have shown that 13 of the most commonly used epoxy resins are mutagenic and that VCD (vinyl cyclohexene dioxide—ERL 4206) is carcinogenic in animals. Additionally, these epoxies, as well as the other components of the resin mixtures, may be extremely allergenic to some individuals. Direct contact and absorption through the skin is the most obvious route into the body. However, some resin components are quite volatile even at room temperature and especially so at oven temperatures. Crude tests in our laboratories with open embedding dishes indicate that at 60°C, VCD evaporates at about $\frac{1}{8}$ inch every 4 hr and DMAE about $\frac{1}{4}$ inch every hour. One technician associated with us was so sensitive to epoxies that she could not remain in a room containing unpolymerized plastic (even in capped bottles). Another of our technicians can no longer work with DMAE outside of the hood (even for weighing). The problem is a real one and suitable handling procedures should be observed consistently.

Perhaps the best general outline for handling embedding resins is that of David Ringo (A. B. Chandler Medical Center, University of Kentucky, Lexington, Ky. 40536) and this is quoted in its entirety.

1. Only disposable labware should be used, so that anything contaminated by resin or resin components can be disposed of and need not be washed. Pasteur pipets are recommended for transfer from the bottle to other containers. One dram screw-cap glass vials, or scintillation counting vials, are convenient for infiltration of tissue samples. Disposable 50 ml plastic beakers with caps should be used for measuring the resin and for mixing by gentle swirling. Measurement of resin components by weight is preferred. Alternatively, measurement can be done by volume using calibrated disposable pipets and a pipetting bulb. Lab-mat should be used to cover the work surface where the resin is used.

2. No skin contact should be allowed with the epoxide or the mixed resin. Plastic gloves should be used as a safety precaution, but not as a barrier (i.e., resin should not be expected to contact the gloves under normal circumstances). In the event of contact, the skin should be rinsed immediately with water and washed with soap.

3. All measurement and use of the resin, including polymerization, should be done in a fume hood. A well-ventilated room should not be considered acceptable.

4. To insure that the outsides of bottles are not contaminated, all transfers should be made by pipet and not by pouring. Infiltrated tissue samples should be transferred from vials to embedding capsules by a wooden applicator stick or Pasteur pipet.

5. All resin, including waste, should be kept in closed containers. Capped vials should be used for infiltration, capped beakers for mixed resin, and closed capsules (e.g., BEEM capsules) for embedding.

6. Only as much resin should be prepared as will be needed for one day's use. (A guide for estimating this is 2 gm per 1 dram vial and 0.65 gm per BEEM capsule).

7. Disposal of labware contaminated with resin components should be handled as follows:

    (a) The anhydride hardener (DDSA, NMA, NSA, HXSA) by itself are relatively non-toxic and disposal requires no precautions.

    (b) Waste resin can be accumulated in the hood in a disposable beaker; all resin can be thrown away without precaution after it has hardened.

    (c) Pipets used to transfer or measure the epoxides should be accumulated in a special container in the hood for disposal (contact your local Health and Safety officer).

    (d) Pipets used for accelerator (DMAE, BDMA, DMP-30) should be handled as in (c). These substances are less toxic but are potent skin irritants.

8. Storage: Room temperature storage of resin components is perfectly adequate over periods of 6 months to a year; desiccator storage is probably desirable in humid climates. Some workers prefer to freeze the components, but this requires a long wait or heating before use; water can condense inside a bottle which is still cool when opened. It is recommended that resin components, with the possible exception of hardeners, be stored in the fume hood where they will be used.

14. APPENDIX: SUGGESTED READING

1. Diamond Knives for Ultra-microtomy. Micro Engineering, Inc., Et. 10 268 Summer Place St., Huntsville, TX 77340 (409-291-6891).
2. Diamond Knife Applications Brief, Nos. 1–6. DuPont Company, Biomedical Products Division, Wilmington, DE 19898.

3. Ultramicrotomy Newsletter, No. 2 (June 1972) and No. 4 (Nov. 1973), DuPont Company, Biomedical Products Division, Wilmington, DE 19898.
4. Diamond Knives—Their Myths and Myth-conceptions (by Leo Casio); also, there are several untitled general reports on use and care of diamond knives. Diatome—US, P.O. Box 125, Fort Washington, PA 19034 (215-646-1478).

15. REFERENCES

Aldrich, H. C., 1967, Ultrastructure of meiosis in three species of *Physarum, Mycologia* **59**:127–148.
Aldrich, H. C., 1982, Electron microscopy techniques, in: *Cell Biology of Physarum and Didymium* (H. C. Aldrich and J. W. Daniel, eds.), Vol. 2, pp. 255–260, Academic Press, New York.
Armbruster, B. L., Carlemalm, E., Chiovetti, R., Garavito, R. M., Hobot, J. A., Kellenberger, E., and Villiger, W., 1982, Specimen preparation for electron microscopy using low temperature embedding resins, *J. Microsc. (Oxford)* **126**:77–85.
Avery, S. W., and Ellis, E. A., 1978, Methods for removing uranyl acetate precipitate from ultrathin sections, *Stain Technol.* **53**:137–140.
Bendayan, M., 1984, Enzyme–gold electron microscopic histochemistry: A new approach for the ultrastructural localization of macromolecules, *J. Electron Microsc. Tech.* **1**:349–372.
Björkman, N., and Hellström, B., 1965, Lead–ammonium acetate; a staining medium for electron microscopy free of contamination by carbonate, *Stain Technol.* **40**:169–171.
Böhler, S., 1975, Artefacts and specimen preparation faults in freeze etch technology, Balzers AG, Liechtenstein.
Causton, B. E., 1984, The choice of resins for electron immunocytochemistry, in: *Immunolabelling for Electron Microscopy* (J. M. Polak and I. M. Varndel, eds.), pp. 29–36, Elsevier, Amsterdam.
De Bruijn, W. C., 1973, Glycogen, its chemistry and morphologic appearance in the electron microscope. I. A modified OsO_4 fixative which selectively contrasts glycogen, *J. Ultrastruct. Res.* **42**:29–50.
Dubochet, J., McDowall, A. W., Menge, B., Schmid, E. N., and Lickfeld, K. G., 1983, Electron microscopy of frozen-hydrated bacteria, *J. Bacteriol.* **155**:381–390.
Ellis, E. A., and Anthony, D. W., 1979, A method for removing precipitate from ultrathin sections resulting from glutaraldehyde–osmium tetroxide fixation, *Stain Technol.* **54**:282–285.
Feder, N., and O'Brien, T. P., 1968, Plant microtechnique: Some principles and new methods, *Am. J. Bot.* **55**:123–142.
Feldman, D. G., 1962, A method of staining thin sections with lead hydroxide for precipitate-free sections, *J. Cell Biol.* **15**:592–595.
Fernández-Morán, H., 1985, Cryo-electron microscopy and ultramicrotomy: Reminiscences and reflections, in: *Advances in Electronics and Electron Physics,* Suppl. 16 (P. W. Hawkes, ed.), pp. 167–223, Academic Press, New York.
Frasca, J. M., and Parks, V. R. 1965, A routine technique for double-staining ultrathin sections using uranyl lead salts, *J. Cell Biol.* **25**:157–169.
Freeman, J. A., and Spurlock, B. O., 1962, A new epoxy embedment for electron microscopy, *J. Cell Biol.* **13**:437–443.
Glauert, A. M., Rogers, G. E., and Glauert, R. H., 1956, A new embedding medium for electron microscopy, *Nature (London)* **178**:803.
Hayat, M. A., 1981, *Fixation for Electron Microscopy,* Academic Press, New York.
Hobot, J. A., Carlemalm, E., Villiger, W., and Kellenberger, E., 1984, Periplasmic gel: New

concept resulting from reinvestigation of bacterial cell envelope ultrastructure by new methods, *J. Bacteriol.* **160**:143–152.

Kuo, J., 1980, A simple method for removing stain precipitates from biological sections for transmission electron microscopy, *J. Microsc. (Oxford)* **120**:221–224.

Kushida, H., 1962, A study of cellular swelling and shrinkage during fixation, dehydration and embedding in various standard media, *J. Electron Microsc. (Japan)* **11**:135–138.

Kushida, H., 1974, A new method for embedding with a low viscosity epoxy resin "Quetol 651," *J. Electron Microsc. (Japan)* **23**:197.

Kushida, H., 1975, Hardness control of the Quetol 651 cured block, *J. Electron Microsc. (Japan)* **24**:299.

Langenberg, W. G., 1982, Silicone additive facilitates epoxy plastic sectioning, *Stain Technol.* **57**:79–82.

Leduc, E. H., and Bernhard, W., 1961, Ultrastructural cytochemistry: Enzyme and acid hydrolysis of nucleic acids and proteins, *J. Biophys. Biochem. Cytol.* **10**:437–455.

Lee, R. M. K. W., 1984, A critical appraisal of the effects of fixation, dehydration and embedding on cell volume, in: *The Science of Biological Specimen Preparation for Microscopy and Microanalysis* (J.-P. Revel, T. Barnard, and G. H. Haggis, eds.), pp. 61–70, SEM, Inc., Chicago.

Luft, J. H., 1961, Improvements in epoxy resin embedding methods, *J. Biophys. Biochem. Cytol.* **9**:409–414.

Luft, J. H., 1973, Embedding media—old and new, in: *Advanced Techniques in Biological Electron Microscopy* (J. K. Koehler, ed.), Vol. 1, pp. 1–34, Springer-Verlag, Berlin.

Millonig, G., 1961, A modified procedure for lead staining of thin sections, *J. Biophys. Biochem. Cytol.* **11**:736–739.

Mollenhauer, H. H., 1964, Plastic embedding mixtures for use in electron microscopy, *Stain Technol.* **39**:111–114.

Mollenhauer, H. H., 1974, Poststaining sections for electron microscpy, *Stain Technol.* **49**:305–308.

Mollenhauer, H. H., 1975, Poststaining sections for electron microscopy: An alternate procedure, *Stain Technol.* **50**:292.

Mollenhauer, H. H., 1981, Diamond knives and chatter: A case history of six knives, Sorval Applications Brief No. 4.

Mollenhauer, H. H., 1984, Surfactants improve sectioning of epoxy embedding resins, *Tex. Soc. Electron Microsc. J.* **15**:30.

Mollenhauer, H. H., 1986, Surfactants as resin modifiers and their effect on sectioning, *J. Electron Microsc. Tech.* **3**:217–222.

Mollenhauer, H. H., and Droleskey, R. E., 1980, Some specific staining reactions of potassium ferricyanide in cells of guinea pig testes, *J. Ultrastruct. Res.* **72**:385–391.

Mollenhauer, H. H., and Droleskey, R. E., 1985, Some characteristics of epoxy embedding resins, *J. Electron Microsc. Tech.* **2**(6):557–562.

Mollenhauer, H. H., and Morré, D. J., 1978, Contamination of thin sections, cause and elimination, in: *Electron Microscopy 1978*, Vol. II (J. M. Sturgess, C. K. Kalnins, F. P. Ottensmeyer, and G. T. Simon, eds.), pp. 78–79. Imperial Press, Ontario.

Muller, L. L., and Jacks, T. J., 1975, Rapid chemical dehydration of samples for electron microscopic examinations, *J. Histochem. Cytochem.* **23**:107–110.

Müller, M., Meister, N., and Moor, H., 1980, Freezing in a propane jet and its application in freeze-fracturing, *Mikroskopie* **36**:129–140.

Napoli, C. A., and Hubbell, D. H., 1975, Ultrastructure of *Rhizobium* induced infection threads in clover root hairs, *Appl. Microbiol.* **30**:1003–1009.

Persson, A., and Persson, K., 1976, Rational and complete testing of diamond knives, *J. Ultrastruct. Res.* **57**:213–214.

Peters, A., Hinds, P. L., and Vaughn, J. E., 1971, Extent of stain penetration in sections prepared for electron microscopy, *J. Ultrastruct. Res.* **36**:37–45.

Reid, N., 1975, *Ultramicrotomy,* North-Holland, Amsterdam.

Reymond, O., and Pickett-Heaps, J. D., 1983, A routine flat embedding method for electron microscopy of microorganisms allowing selection and precisely orientated sectioning of single cells by light microscopy, *J. Microsc. (Oxford)* **130**:79–84.

Reynolds, E. S., 1963, The use of lead citrate at high pH as an electron opaque stain in electron microscopy, *J. Cell Biol.* **17**:208–212.

Ringo, D. L., Brennan, E. F., and Cota-Robles, E. H., 1982, Epoxy resins are mutagenic: Implications for electron microscopists, *J. Ultrastruct. Res.* **80**:280–287.

Ringo, D. L., Read, D. B., and Cota-Robles, E. H., 1984, Glove materials for handling epoxy resins, *J. Electron Microsc. Tech.* **1**:417–418.

Roos, U.-P., 1973, Light and electron microscopy of rat kangaroo cells in mitosis. I. Formation and breakdown of the mitotic apparatus, *Chromosoma* **40**:43–82.

Ryter, A., and Kellenberger, E., 1958, Etude au microscope electronique de plasmas contenant de l'acide desoxyribonucleique, *Z. Naturforsch.* **13B**:597–605.

Skvarla, J. J., and Larson, D. A., 1966, Fine structural studies of *Zea mays* pollen. I. Cell membranes and exine ontogeny, *Am. J. Bot.* **53**:1112–1125.

Smith, R. E., and Farquhar, M. G. 1966, Lysosome function in the regulation of the secretory process in cells of the anterior pituitary gland, *J. Cell Biol.* **31**:319–348.

Spaur, R., and Moriarty, G., 1977, Improvements of glycol methacrylate. I. Its use as an embedding medium for electron microscope studies, *J. Histochem. Cytochem.* **25**:163–174.

Spurlock, B. O., Kattine, V. C., and Freeman, J. A., 1963, Technical modifications in Maraglas embedding, *J. Cell Biol.* **17**:203–204.

Spurr, A. R., 1969, A low viscosity epoxy resin embedding medium for electron microscopy, *J. Ultrastruct. Res.* **26**:31–43.

Stäubli, M. W., 1960, Nouvelle matiere d'inclusion hydrosoluble pour la cytologie electronique, *C. R. Acad. Sci.* **250**:1137.

Wakefield, J. St. L., 1984, Epon plus silicon fluid, *Electron Microsc. Soc. Am Bull.* **14**:93.

Chapter 5

Computer-Aided Reconstruction of Serial Sections

David L. Balkwill

Department of Biological Science
Florida State University
Tallahassee, Florida 32306

1. INTRODUCTION

Thin sectioning probably has been used more than any other method to prepare biological specimens for transmission electron microscopy. The enduring popularity of this technique is hardly surprising, considering its ability to reveal intracellular features at high resolution and its compatibility with a wide range of sophisticated staining procedures. Yet, an inherent drawback of thin sectioning is the fact that a single section represents only a very small portion (essentially a two-dimensional plane) of each cell through which it passes. The information provided by such a section can be deceptive or misleading if it is not interpreted within the context of the entire cell and in view of the direction in which the section traveled through that cell. This is especially true if (as in most thin sectioning studies) one views only a limited number of randomly cut sections. For example, the micrograph in Figure 1a shows a section that appears to pass through four independent cells. Subsequent serial sections (Figure 1b,c), on the other hand, demonstrate that two of these "cells" actually are portions of a single spiral-shaped cell that was transected twice. Similar, but often far more intricate, deceptions may be encountered when randomly cut sections are used to examine intracellular ultrastructure.

The misleading data sometimes provided by randomly cut thin sections may be of little consequence in studies where the cell system of interest has been well characterized or where they are not relevant to the specific information sought. However, such data can lead to significantly flawed conclusions in studies that involve poorly characterized systems or that require an accurate understanding of three-dimensional (3-D) relationships. The limitations and potential interpretive

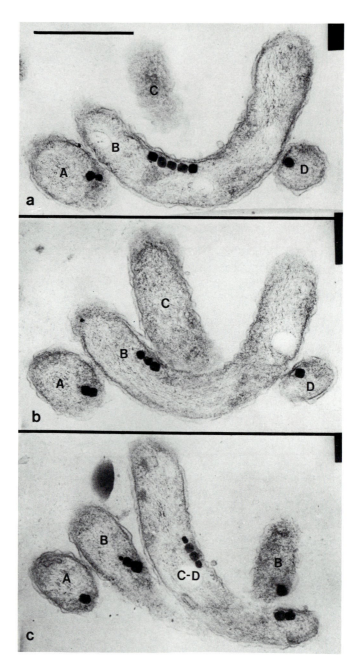

FIGURE 1. Consecutive serial thin sections through a small group of spiral-shaped bacterial cells. Bar = 0.05 μm. (a) First section of series appears to transect four independent cells (A, B, C, and D). (b) Second section appears to transect same four cells. (c) Third section shows that cells C and D in first two sections are actually just different portions of a single cell. Note that cell B in first two sections appears to be two independent cells in this section.

dangers of randomly cut sections can be overcome in such instances by preparation of serial sections through entire cells (or important portions thereof) and reconstruction of these sections in three dimensions. This approach eliminates the need to extrapolate conclusions regarding 3-D structure from two-dimensional (2-D) information, and it allows one to consider each internal substructure within the context of the entire cell. Moreover, the investigator gains a better understanding of relationships between intracellular structure and function as the organism's 3-D architecture is elucidated.

Three-dimensional reconstruction of serial sections was once a burdensome and time-consuming process because there were no efficient methods for reassembling the data from the many sections. During the past few years, however, the task has been facilitated and simplified greatly by the development of semi-automated computer systems for 3-D assembly and subsequent manipulation of structural information. This chapter briefly reviews the current technology for 3-D reconstruction of serial sections, explains how such reconstructions are carried out, and describes the potential benefits that may be derived when this approach is applied to microorganisms. A recently described reconstruction of the unicellular cyanobacterium *Agmenellum quadruplicatum* (Nierzwicki-Bauer *et al.*, 1983a,b) will be used as an example to illustrate the methodology and typical results of this technique.

2. LITERATURE REVIEW

2.1. Serial Sectioning

Serial thin sectioning became an established technique for studying 3-D ultrastructure after Sjöstrand (1958) showed that it could be used to examine biological tissue at the electron microscopy (EM) level. It soon became evident that serial sections provide structural information that is not readily obtained from individual, randomly cut sections. Consequently, serial sectioning has been used extensively in a wide variety of applications (for reviews, see Hayat, 1981; Kay, 1961; Knobler *et al.*, 1978; Pease, 1960; Stevens *et al.*, 1980; Ware and LoPresti, 1975).

2.2. Manual Methods for Reconstructing Serial Sections

Most manual (pre-computer-aided) methods for 3-D reconstruction of serial sections are adaptations of procedures developed by German anatomists in the late 1800s. These methods and their successful applications have been reviewed extensively by Ware and LoPresti (1975). Briefly, they involve one of three basic approaches: graphical reconstruction, model building, and serial section cinematography. In graphical reconstructions, the outlines of objects in micrographs of each section are parallel projected (by hand) onto straight lines separated by a distance corresponding to the magnified section thickness. The outlines are then connected, and the resulting image can be shaded to produce a 3-D view of the reconstructed object perpendicular to the direction of sectioning. Variations of the procedure permit the production of oblique views. Model building, a more cumbersome approach, involves the production of space-filling models to illustrate the 3-D architecture of a specimen. The investigator makes such models by tracing the outlines of the feature(s) to be reconstructed from micrographs of each section onto separate sheets of wax, cardboard, or other flat material of appropriate thickness. The outlined areas on each flat sheet are then cut out and stacked vertically to produce the space-filling model. Alternatively, one can trace the outlines of selected cell structures onto transparent sheets (e.g., Plexiglas) and then look through the transparent block formed when the sheets are stacked to get an idea of what the reconstructed object looks like in three dimensions. In serial section cinematography, the micrographs of the sections are recorded onto sequential frames of movie film. The film can be viewed in forward and reverse, so as to produce the effect of passing back and forth through the specimen. Repeated viewing of such a film yields an impression of the specimen's 3-D architecture.

Manual methods for reconstructing serial sections all have serious drawbacks. They are complicated, tedious, and time-consuming. Except for serial section cinematography, they are better suited for illustrating surface topography than internal architecture (i.e., the 3-D distribution of substructures throughout the cell). Worse, they produce more or less fixed images (or models) that generally cannot be manipulated or processed for the derivation of additional information. These drawbacks probably explain why manual methods have been used only infrequently, even though they have sometimes produced very impressive results (see Ware and LoPresti, 1975).

2.3. Computer-Aided Reconstruction Systems

The use of computers to facilitate 3-D reconstruction of serial sections was pioneered by Cyrus Levinthal and his co-workers at Columbia University. This group's efforts led to the development of CARTOS (for "computer-aided reconstruction by tracing of serial sections"), a computer-driven device originally designed to generate 3-D images of neurons (Levinthal et al., 1974; Macagno et al., 1979). The CARTOS system has been steadily upgraded and expanded in capability since it was introduced in the mid-1970s, and it now includes sophisticated features like the Loaner System (a compact version that can be shipped to other laboratories), color graphics display capabilities, and a shading program (see Section 6.6). Several alternative computer-aided reconstruction systems have been developed in recent years, among the more powerful of which are the modified CARTOS-type system used by Stevens et al. (1980) and the TROTS system devised by Veen and Peachey (1977). For researchers who wish to build their own reconstruction facility, Moens and Moens (1981) have explained how to assemble a workable system for under $10,000 and have provided simple BASIC-language reconstruction programs for operating it.

All computer-aided reconstruction systems operate on the same general principle (detailed procedures are described in Sections 5 and 6). The locations of structures in electron micrographs of each section in the series are converted to pairs of X and Y coordinates that are stored in computer's memory (or, more commonly, on some type of magnetic storage device). The computer may then be instructed to retrieve this digitized information, compile it in three dimensions, and display it graphically as a 3-D reconstruction image. The advantages of this approach over the manual techniques are considerable. It is faster and less subject to error, but more important, it enables one to manipulate the reconstruction data in many ways. Images can be rotated in 3-D space, selected cell features can be added or deleted, specific portions of cells can be viewed separately, and so on. Moreover, the stored data can be used to obtain quantitative information such as lengths, widths, surface areas, and volumes (Stevens et al., 1980). In short, the advent of computer-aided methods has made 3-D reconstruction of serial sections a practicable and highly versatile technique.

2.4. 3-D Reconstructions of Microorganisms

The great majority of studies involving reconstruction of serial sections have focused on the 3-D architecture of neurons (see Stevens et al., 1980; Ware and LoPresti, 1975). Three-dimensional reconstructions of microorganisms have been comparatively rare, and many of these have dealt with eukaryotic forms. Hoffmann and Avers (1973) used serial sections to elucidate the configuration of mitochondria in *Saccharomyces* cells, finding that each cell contains only a single branched mitochondrion. Schötz et al. (1972) reconstructed gametes of *Chlamydomonas reinhardii* in order to understand better the spatial relationships between subcellular organelles. Among other things, they found the *Chlamydomonas* cells to be highly asymmetric. Vivier and Petitprez (1972) reconstructed the protozoan red blood cell parasite *Anthemosoma garnhami* to study the process by which host hemoglobin is digested and concluded that the process is extracellular. More recently, Pellegrini (1980a,b) has published two rather elegant studies in which 3-D reconstructions were used to monitor qualitative and quantitative changes of *Euglena gracilis* organelles during bleaching of a culture in darkness.

Several prokaryotic microorganisms have been partially reconstructed from short series of sections [e.g., Remsen et al. (1968) reconstructed a portion of the photosynthetic membrane system in *Ectothiorhodospira mobilis*], but complete reconstructions have been infrequent. Van Iterson et al. (1975) used serial sectioning to study conformational changes in the *Bacillus subtilis* nucleoid during activation of stationary-phase cells and, later (Van Iterson and Aten, 1976), to study changes in both nucleoids and mesosomes during spore germination. Waters and Hunt (1980) reconstructed mycoplasma-like organisms residing in plants in order to elucidate their cell shapes. Reconstructions from serial sections were required to complete this seemingly simple task because the bacteria could not be isolated without alteration of their morphological characteristics. Most recently, Nierzwicki-Bauer *et al.* (1983a,b, 1984) published the first complete 3-D reconstruction of a cyanobacterium, an investigation that revealed several interesting and unexpected aspects of cyanobacterial intracellular architecture (see Section 6).

3. SERIAL SECTIONING

3.1. Fixation and Embedding

Any fixation that provides adequate structural preservation can be used to prepare specimens for serial sectioning. Therefore, the selection of a specific protocol should be based on the purpose of the reconstruction project. A general fixation that preserved a wide variety of cellular features was used in the 3-D reconstruction of *A. quadruplicatum* because it was important to know how all of those features were arranged within the cell. The 3-D arrangement of the organism's photosynthetic membrane system proved to be especially interesting because it divided the cell into several distinct compartments (Balkwill *et al.*, 1984; Nierzwicki-Bauer *et al.*, 1983a,b, 1984); thus, a likely future study might deal specifically with this membrane system and how its conformation responds to changes in growth conditions. A fixation that selectively stains membranes would be more appropriate than a general procedure for such a study and, moreover, might be amenable to automatic digitizing techniques (see Section 5.5). Other possibilities include the use of cytochemical or immunocytochemical (Roth, 1983) stains in conjunction with serial sectioning to determine the 3-D distribution of specific molecules throughout a cell, an approach that could provide a direct link between ultrastructural and molecular investigations.

It is helpful to concentrate the cells by centrifugation prior to dehydration and embedding when microorganisms are prepared for serial sectioning, regardless of the fixation used to preserve the specimen. The resulting cell pellet is then embedded in agar or agarose and cut into small blocks that can be treated much like pieces of tissue. The advantage of this method is twofold: (1) it eliminates the need for centrifugation between dehydration and infiltration steps, and (2) it produces distinct groups of densely packed cells that are helpful in the reconstruction process (see Section 4). Cell concentration should be done after the fixation; any structural changes that might be brought about by centrifugation are thus avoided.

The choice of an embedding medium is less critical than that of a fixation.

Epon epoxy and Spurr's low-viscosity epoxy (Spurr, 1969) resins have been used in many of the published reconstructions, and both have worked well in my own laboratory. The embedding medium with which the investigator is most familiar may be the best one to try initially. If necessary, minor adjustments in hardness can be made to facilitate the serial sectioning.

3.2. Cutting the Sections

Anyone who has tried it knows that producing long ribbons of usable serial sections can be an onerous and, often, frustrating task. Aside from recognizing that a diamond knife is required for success, there is little general agreement on how this task should be performed so as to minimize its difficulties (for reviews, see Hayat, 1981; Kay, 1961; Knobler *et al.*, 1978; Pease, 1960; Sjöstrand, 1967; Stevens *et al.*, 1980). Fortunately, Fahrenbach (1984) has recently published a thorough analysis of the serial sectioning process in which he offers detailed technical advice that should be useful even to an experienced microtomist. The salient features of his method for cutting the sections are well worth noting:

1. The block is trimmed so that the sides and leading edge of the trapezoid mesa are vertical (i.e., perpendicular to the face of the trapezoid); in this way the extent to which the trapezoid gets larger as long series of sections are cut is minimized. (The trailing edge of the trapezoid mesa is tapered in the conventional manner, to provide stability during cutting.)

2. A slow cutting speed (< 0.3 mm/sec) and a firm plastic consistency are used to avoid vibration.

3. Prior to cutting, the leading and trailing edges of the trapezoid mesa are coated with a thin layer of diluted contact cement. This procedure helps to ensure that the sections stay together in a cohesive ribbon, thereby making it possible to cut single ribbons of up to 1 cm or more in length.

3.3. Accurate Placement of the Serial Sections on Grids

Placement of the section ribbons on grids is the most difficult and disaster-prone step in serial sectioning. The placement must be done precisely enough that critical portions of the sections are not obscured by grid bars, which could hide important features. Grids with large, single slots are used most frequently because the large open area minimizes the likelihood of a fatal error. A multislot grid with large slots of four different widths (SIPI No. 3220C; SIPI Supplies, West Chester, PA) is an interesting alternative to the single-slot grid because it allows the investigator to match the width of the slot to the width of the ribbon. The amount of open (not covered with sections) coating film, which is very easily ruptured by physical stress or uneven heating in the electron beam, is thereby reduced. A coating film is necessary with either type of grid and, owing to its strength, Formvar is the coating of choice. The Formvar solutions should be prepared carefully (Fahrenbach, 1984; Stevens *et al.*, 1980), and the coated grids must be inspected thoroughly for defects.

Several methods are available for executing the actual placement of the sections. Fahrenbach (1984) uses a custom-built pedestal mounted in the boat of the diamond knife. This pedestal holds a grid in place just beneath the surface of the water in the boat. Sections are positioned directly above the grid, and the water level in the boat is carefully lowered to place them on the Formvar coating. Fahrenbach gives detailed directions for constructing the pedestal and manipulating the ribbons. Stevens *et al.* (1980) prefer to remove ribbons of appropriate length from the knife boat by picking them up in a droplet of water with a wire loop. This loop is mounted on a micromanipulator and carefully lowered onto a grid (which is held in place by a vertical chuck) until the water droplet contacts the Formvar coating. The sections are fastened to the coating by gentle aspiration of the water droplet with a micropipette. Although somewhat more complicated than either of the above methods, the ''sol-gel'' technique of Anderson and Brenner (1971) has worked effectively in my laboratory. The ribbons are transferred (with a wire loop) from the boat to the surface of a container of liquid gelatin, after which the gelatin is allowed to solidify. The solid gelatin serves to hold the ribbons in place so that grids can be placed on top of them (Formvar-coated side down) very precisely. The gelatin is then liquefied and removed from the grids by washing with dilute acetic acid and water.

3.4. Poststaining

Properly mounted serial sections can be subjected to any poststaining protocol, from the standard uranyl acetate/lead citrate routine to highly specialized immunocytochemical labeling procedures. The main pitfall to avoid during this step is accidental breakage of the delicate Formvar films that support the sections. The fairly conventional method of placing grids specimen-side-down on droplets of stain or rinse solutions can be used successfully if one is exceedingly careful. Because this approach is risky, however, several authors recommend using gentler alternatives. Fahrenbach (1984) poststains his grids in the wells of an LKB grid box (LKB-Produkter AB, Bromma, Sweden), thereby minimizing the kind of lateral motion that tends to disrupt the coating films. Several others have developed special devices for holding the grids during staining, a good example of these being the modified piece of plastic tubing described by Tyler (1981).

3.5. Care and Handling of the Finished Sections

It is necessary to handle grids containing serial sections with extreme caution, even after poststaining, because the Formvar films on large-slot grids are so easily disrupted. Standard copper grids are quite flexible and, as a result, can easily be twisted or bent enough to overstress the film. Stevens *et al.* (1980) avoid this problem by using specially fabricated, extrarigid grids (No. 1-EY00828-04; Chem Fab Co., Doylestown, PA). These grids are then mounted in a protective cassette that fits into the specimen holder of the TEM; thus, the grids are never handled directly after poststaining.

Great care must also be exercised when grids containing serial sections are examined in the electron microscope. A liquid nitrogen-cooled anticontamination device is virtually essential. The beam should be condensed very slowly because the Formvar film is so easily disrupted by uneven heating. It is also advisable to use the highest accelerating voltage available, because doing so will minimize sublimation and heating of the specimen. The resultant loss in contrast can be compensated for by slight underfocusing (Stevens *et al.,* 1980). If a grid is to be examined more than once (as is often the case), it must be removed from the column very carefully because the Formvar films are easily disrupted by the sudden rush of air into the specimen airlock. This problem can be avoided if the specimen holder is pulled out of the air lock very slowly, or if a needle valve is installed (Fahrenbach, 1984) that is used bring the air lock to atmospheric pressure very slowly.

4. OBTAINING THE ULTRASTRUCTURAL DATA FOR COMPUTER-AIDED RECONSTRUCTION

After acceptable serial sections have been obtained, it is necessary to extract from them the data actually used for 3-D reconstruction. Low-magnification survey micrographs should be produced first in the case of microbial specimens. Distinct clusters of cells that are easily recognized from one section to the next will be visible at low magnifications (Figure 2), if the cells were concentrated by centrifugation prior to dehydration and embedding (Section 3.1). Several such cell clusters are photographed throughout the entire series of sections at magnifications low enough to include whole clusters within the photographic frame. Enlarged prints of the resulting micrographs are then prepared and examined closely for identification of specific cells that are suitable for reconstruction. After marking the selected cells on the micrographs (see Figure 2), the investigator must return to the microscope, relocate each of the marked cells in the original sections, and rephotograph each one of them through all sections of the series at magnifications sufficient to resolve internal ultrastructural details. The resulting high-magnification micrographs represent the final data for 3-D reconstruction.

The above approach may seem tedious, but it can significantly reduce the amount of time required to obtain suitable micrographs for reconstruction. By first examining the cells in the low-magnification micrographs, one can avoid cells that have not been sectioned completely or that are not free of sectioning defects (such

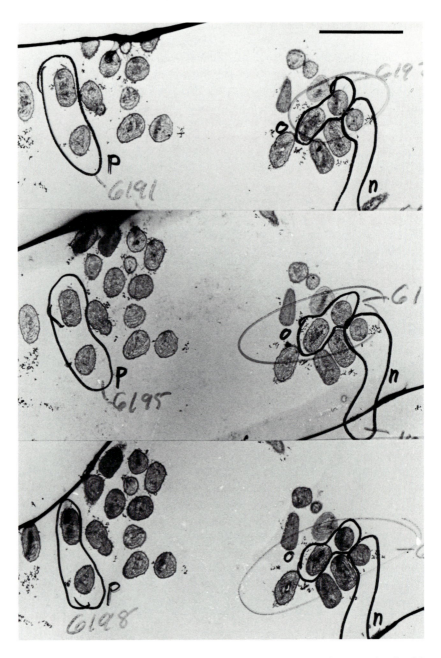

FIGURE 2. Low-magnification micrographs of consecutive serial thin sections through cells of the unicellular cyanobacterium *Agmenellum quadruplicatum*. Bar =5.0 μm. Micrographs like these are examined to identify individual cells suitable for reconstruction. Note that cells occur as distinct clusters because they were concentrated by centrifugation prior to dehydration and embedding. Note also that clusters are easily recognized from one section to the next. Markings designate cells to be rephotographed at high magnification, to produce final data for reconstruction.

as stain precipitates and wrinkles) throughout the entire series of sections. It is also possible to select deliberately cells that were sectioned in different directions (i.e., serially longitudinally sectioned, serially cross-sectioned, and so on). A varied sampling like this provides an internal quality control during computer-aided reconstructions because reconstructed cells should look similar no matter how they were originally sectioned (for an example, see Section 6.4).

5. ENTRY OF ULTRASTRUCTURAL DATA INTO THE COMPUTER SYSTEM

5.1. Introduction

The operation of the CARTOS Loaner System that was used in the 3-D reconstruction of *A. quadruplicatum* is described here as an example of how computer-aided reconstructions are performed. The basic manipulations of the CARTOS Loaner System are typical of those used with all reconstruction devices, although some of the specific procedures do vary from one system to another. The most significant of these variations are detailed below.

The CARTOS Loaner System was developed in Cyrus Levinthal's laboratory at Columbia University, as an extension of the original CARTOS device for reconstructing neurons (Levinthal *et al.*, 1974; Macagno *et al.*, 1979). The primary advantage of the Loaner System is that it can be shipped to and used in any qualified investigator's laboratory, thereby making it easy to coordinate electron microscopy with the reconstruction work. The Loaner System is made available to qualified users without charge (the user pays only the shipping costs), as an NIH-funded National Biotechnology Research Resource (for further information, see Research Resources Information Center, 1981, 1983).

5.2. Prealignment of Serial Section Images

Before a series of micrographs can be reconstructed in three dimensions, each "frame" of the series must be prealigned with respect to the X and Y axes in 3-D space. (The X and Y axes in 3-D reconstructions are defined as being parallel to the plane of each section and perpendicular to each other; the Z axis is perpendicular to the X and Y axes and parallel to the direction of knife advance during sectioning.) This is one of the more difficult steps in the reconstruction process and, if it is not done correctly, distorted 3-D images will be produced when the sections are "reassembled." Precise prealignment requires fairly sophisticated optical and photographic equipment, the theory and operation of which have been described in detail elsewhere (Levinthal and Ware, 1972; Levinthal *et al.*, 1974, 1975; Macagno *et al.*, 1979; Stevens *et al.*, 1980; Ware and LoPresti, 1975). Simpler methods like the external marker technique originally developed for manual reconstructions (see Ware and LoPresti, 1975) can be used instead, but are less likely to produce accurate results. The NIH-funded CARTOS service offers a very significant advantage in this regard, because prealignment is performed for the user by the Columbia University Computer Graphics and Image Processing Facility. The original high-magnification negatives from the electron microscope are used to produce a positive-image, 35-mm filmstrip (or "movie"), in which each frame depicts one section (correctly aligned with respect to the X and Y axes) in the series (Figure 3).

5.3. Selection of Cell Features for Entry into the Computer

Entering data into a computer-aided reconstruction system is usually a time-consuming process (see below). Consequently, the investigator should first decide which cell features are absolutely essential with respect to the project's goals and then include only those features in the subsequent reconstructions. This approach can reduce the length of a project's data-entry phase considerably.

5.4. Organization of Structural Data for Entry into Separate Files

All data entered into a reconstruction system's computer are stored (usually on magnetic disks) in defined memory locations called "files." Each file generally contains structural information from an entire series of sections so that, when this information is "displayed" on a video monitor, a reassembled (more or less 3-D) image of the entire specimen is produced. The more sophisticated reconstruction systems permit the simultaneous display of several different files in the form of a combined image. The investigator using such a system should consider entering each cell feature to be reconstructed in a separate file. In the reconstruction of *A. quadruplicatum,* for example, the outline of the cell, the outlines of each type of intracellular inclusion body, and the positions of each set of photosynthetic membranes were all entered into different files. This approach has two advantages: (1) it accelerates the tracing process described below, and (2) it permits files to be codisplayed later in any desired combination. The latter is important because it allows one to create reconstructions with different sets of cell features, a capability that facilitates the production of meaningful and easily interpreted images. However, it is necessary to decide how the separate memory files will be organized (i.e., what will be placed in each) before entry begins. Some experimentation with rapidly executed, rough reconstructions might be required to help devise a good organizational plan.

5.5. Entry of Data by "Tracing"

Structural data from the electron micrographs are entered into the computer by a process known as "tracing." With the CARTOS Loaner System, this

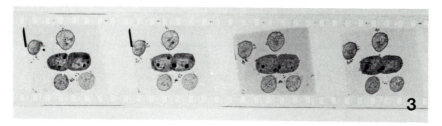

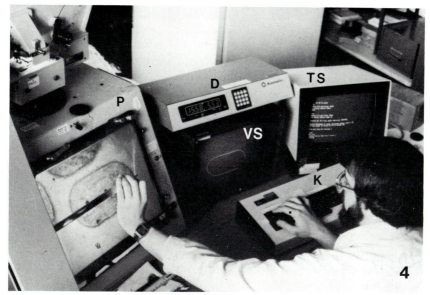

FIGURE 3. Segment of "movie" for computer-aided reconstruction of serial sections; i.e., a 35-mm filmstrip in which each frame depicts a single section in the series and in which each section image has been precisely prealigned with respect to the X and Y axes in 3-D space.

FIGURE 4. Operation of CARTOS Loaner System developed in Cyrus Levinthal's laboratory at Columbia University. System includes rear-screen projector (P), digitizing circuitry (D), video screen for display of reconstruction data (VS), terminal screen (TS), keyboard (K), and microcomputer (not shown). Operator is entering structural data into computer memory by "tracing"; i.e., by moving pointer on end of digitalizing arm (in left hand) around cell feature while tapping command key to enter pairs of $X–Y$ coordinates. Note that cell feature (in this case, outline of cell) appears on the video screen as it is being traced.

process is carried out as shown in Figure 4. (Other systems use different equipment, but the general principle is the same.) The 35-mm "movies" are loaded into a rear-projection enlarging system that produces a magnified image of each frame on a piece of frosted glass. A digitalizing arm is then used to transmit data from each movie frame to the computer. This arm automatically converts the location (on the projector screen) of a small pointer at the end of the arm to X and

Y coordinates that can be stored in digital form. To enter the position of a cell structure into the memory, the operator slowly moves ("traces") the digitalizing arm's pointer around the outline of the structure (or along the length of a linear feature) while periodically tapping a command key on the computer terminal's keyboard. A pair of $X-Y$ coordinates is entered every time the command key is tapped, and the computer assumes that the single points specified by these coordinate pairs should be connected with lines. The resolution of the graphics image is directly proportional to the number of separate points entered while tracing. Straight lines can be delineated accurately by only two points, but complex curves will look jagged and irregular unless many coordinates are used to define their shapes. On the other hand, excessive numbers of coordinates produce very large memory files that only the more powerful computer systems can process in a reasonable period of time. In practice, then, the investigator must strive for a reasonable compromise between high resolution and data manageability.

Each desired cell feature must be traced on each section of the series in order to accomplish a complete 3-D reconstruction. This is the most time-consuming portion of a reconstruction study and may require several months if the organism being examined is complex or if the investigator wants to reconstruct a large number of cells. The task is simplified to some extent with the CARTOS system because the computer, through an interface with the projector, automatically keeps track of which section in the series is being traced. The operator merely sets "frame 1" on the projector to match the first section after loading a new filmstrip, tells the computer the distance between sections (i.e., distance along the Z axis), and then starts tracing. This design allows one to trace a specific structure through the whole series of sections rapidly, thereby making it easier to "follow" the structure from one section to the next.

The tedious manual tracing process may someday be facilitated or, perhaps, eliminated entirely by the development of highly sophisticated automatic tracing systems. Briefly (for more detailed discussions, see Sobel *et al.,* 1980; Ware and LoPresti, 1975), these systems use scanning microdensitometers, laser scanners, or other image-processing devices to convert each micrograph of a series into digital information that can be examined by a computer. The computer then has to "recognize" the cell features being reconstructed so that it can store their locations in a memory file. Computers in automatic reconstruction systems that work from electron micrographs of thin sections do so by a process known as "gray-level gradient thresholding." The computer simply considers any sudden and large change in image density to represent the border of a cellular feature. Once these "borders" have been found, contour lines marking their locations are automatically created and stored for reconstruction. The final result is identical to that obtained when the same structures are traced manually. However, dependence on gray-level gradient thresholding means that automatic tracing systems currently work best on samples in which the features to be reconstructed can be strongly and selectively contrasted to ensure accurate "recognition" by the computer. An automated version of the CARTOS system, for example, can trace the cell boundaries of selectively stained neurons in tissue sections and then reconstruct the cells with very little manual input (Sobel *et al.,* 1980). Unfortunately, the automatic tracing approach is not yet sophisticated enough to work this effectively on samples that are not amenable to selective staining. The computer will simply trace all of the strongly contrasted cell features in such cases, thus making it difficult to create reconstructions that include only the features of interest. The unwanted features can be rejected or removed from the tracings, but only with comparatively tedious manual operations. Similarly, cell features that do not stand out sharply from their surroundings will not be traced automatically at all (the computer will not "recognize" them) and, therefore, must be entered into the memory files by hand. Several research groups are now working to overcome these and other limitations of automatic tracing. Their task is not an easy one because, essentially, the computers are being asked to develop the kind of judgement that an electron microscopist uses to interpret his/her micrographs. Nevertheless, it is not unreasonable in view of the rapid progress made thus far to expect significantly more powerful and versatile automatic tracing systems to emerge within the next few years.

6. PRODUCTION AND MANIPULATION OF COMPUTER-GENERATED RECONSTRUCTIONS

6.1. Introduction

It is relatively easy to produce 3-D reconstruction images after all of the structural data have been entered into the computer system. One generates images simply by telling the computer which memory file(s) should be reconstructed and the form in which the results should be displayed. All reconstruction devices have a set of straightforward commands for doing so, and most of them prompt the operator in plain English at each step to facilitate rapid operation of the system. Little or no knowledge of computers is needed to carry out a successful reconstruction project. The computer itself eventually produces the desired reconstruction automatically by "drawing" a video image on its monitor from the data in the designated file(s). The operator can record this image in permanent form either by photographing the monitor screen or by sending the image to a high-quality plotting device.

Different reconstruction systems vary considerably with respect to their capabilities to produce and manipulate images. The CARTOS Loaner System used in the reconstruction of *A. quadruplicatum* is somewhat limited in display capacity because it is run by a microcomputer with a comparatively small active memory. The Loaner System is very useful for collecting and storing the structural data and for determining in a general way how well the reconstructions are turning out. It can also be used to produce intermediate-resolution images of simple reconstructions that are suitable for publication (see Figure 5). Nevertheless, many of the more sophisticated image manipulations described below were performed at the Columbia University Computer Graphics and Image Processing Facility. This facility provides (as a part of the NIH-funded CARTOS service) sophisticated display and image analysis equipment that can be used to process Loaner System data very effectively. The data can be transferred to Columbia University via telephone link or in the form of magnetic storage disks. The investigator must then travel to New York to make full use of the processing equipment, but the system is so powerful that all of the needed reconstructions for even a major project can be produced in just a few working days.

6.2. Production and Utility of Stereo Pairs

Visual interpretation of any computer-generated 3-D reconstruction can be made easier if a stereo pair of it is produced. Stereo pairs consist of two images that depict the reconstructed object from two slightly different directions, the difference in perspective being referred to as the separation angle. Stereo pairs facilitate interpretation because they provide a true 3-D image in which it is easier to see where different structures are situated with respect to each other in 3-D space (see Figures 5–8).

Some reconstruction systems will automatically display stereo pairs with optimal separation angles on the monitor screen, thereby allowing one to record both images with a single photograph. With less sophisticated systems, the investigator must produce an initial reconstruction image, photograph it, instruct the computer to rotate that image about the vertical axis through an appropriate separation angle, and then photograph the rotated image. The latter approach is admittedly somewhat time-consuming, but the extra effort is usually justified by the increased understandability of the resulting images.

6.3. Selective Inclusion or Exclusion of Cell Features

If the structural data corresponding to different cell features are initially entered into separate memory files as suggested above, one can arbitrarily select the features to be included in each reconstruction image. In effect, the investigator can increase the visual force of the image by excluding cell features that are not relevant to the purpose of the reconstruction. It is also possible to obtain unobstructed views of features that would normally be surrounded and hidden from view by other components in a complete cell. This strategy was used in the reconstruction of A. *quadruplicatum* to determine and compare the 3-D distributions of several different intracellular inclusion bodies. Reconstructions that depicted only the inclusion bodies and a cell outline (Figure 5) showed effective-

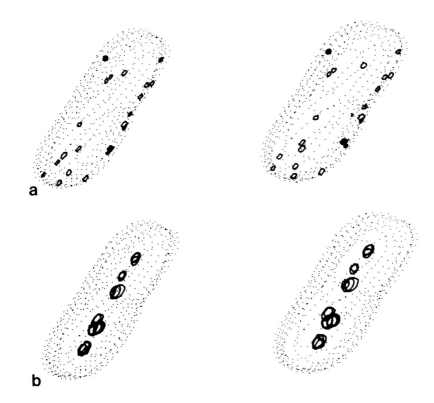

FIGURE 5. Stereo pairs of CARTOS reconstructions depicting intracellular distribution of specialized inclusion bodies (solid circles) in *A. quadruplicatum* (cell outline in dots).(a) Lipid bodies are distributed only about periphery of cell. (b) Carboxysomes are situated only along cell's central longitudinal axis.

ly that this organism's lipid bodies were always distributed about the periphery of the cell, while its carboxysomes and polyphosphate granules were always situated along the cell's central longitudinal axis.

An alternative method for limiting the amount of information in a reconstruction is to display data from only a selected portion of the entire series of sections. This approach can be used even if different cell features were not stored in separate memory files as the data were originally entered into the computer. However, its effect on the reconstruction image is not to delete specific features from the whole cell, but rather to delete a specific portion of the cell itself along with all of the features that happen to be situated therein. The portions of the cell that can be deleted in this way are determined solely by the original direction of cutting. In the case of a rod-shaped cell that was serially cross-sectioned, for example, it would not be possible to split the cell along its central longitudinal axis and then display one of the resulting halves; that could be done only with a longitudinally sectioned cell. On the other hand, one could easily delete the cross-sectioned cell's poles and display only the cylindrical portion of its body. This particular strategy was, in fact, used during the reconstruction of *A. quadruplicatum* to prevent polar structures from obscuring objects deep within the cell cylinder (see Figures 6–8).

6.4. Image Rotation

Most computer-aided systems permit the investigator to rotate their reconstruction images freely about the X, Y, and Z axes in 3-D space, so that the reconstructed object can be viewed and studied from any direction. Each rotation must be executed by a complete redrawing of the image from the original data in the appropriate memory file(s) if the active memory capacity of the system's computer is limited (e.g., as with a system run by a small microcomputer). In this case, the operator "calls up" the reconstruction image to be rotated (thereby causing it to appear on the video screen) and enters a series of commands to identify the axis about which rotation should occur and to specify the amount of rotation in degrees. The video screen will blank out in response to these commands and then display the rotated image after the computer completes the necessary calculations. Thus, the image is seen before and after rotation, but not while it is being rotated. The operator must simply guess the rotation parameters that will produce the correct result if a particular viewing angle or image effect is desired. Several such guesses are often required to achieve success although, with practice, one does develop some practical intuition for manipulating the images more efficiently. Rotation is more straightforward on reconstruction systems that have a large active memory capacity (e.g., the equipment at the Columbia University Computer Graphics and Image Processing Facility), because these systems can provide "real-time" image rotation. One accomplishes "real-time" rotation simply by turning (individually or simultaneously) three control dials that effect rotation about the X, Y, and Z axes, respectively. The reconstruction on the screen moves smoothly and continuously (without blanking out or flickering) while these dials are being turned, thereby permitting the operator to monitor the effect of rotation on the image constantly. Even an inexperienced operator can "maneuver" a reconstructed object to the desired position rapidly and accurately with this kind of system.

The ability to rotate reconstruction images by any method is helpful because, by examining a reconstructed object from many different directions, the investigator can more readily develop an intuitive understanding of how its structural components are related to one another in 3-D space. Figure 6 illustrates two views of a reconstructed *A. quadruplicatum* cell cylinder that, together, show clearly the peripheral versus central locations of this organism's intracellular inclusion bodies. Image rotation was especially helpful in elucidating the

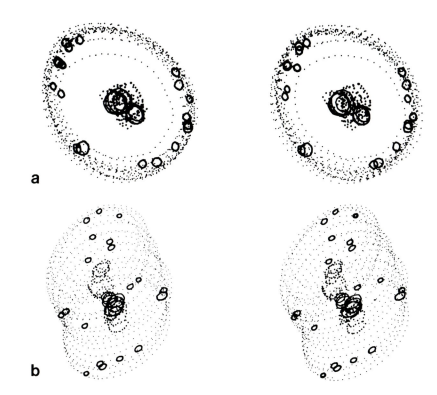

a

b

FIGURE 6. Stereo pairs of CARTOS reconstructions of the cylindrical portion of an *A. quadruplicatum* cell (outline in dots), including lipid bodies (small solid circles), carboxysomes (large dotted circles), and polyphosphate granules (large solid circles). Note central location of carboxysomes and polyphosphate granules versus peripheral location of lipid bodies. Ends of cell have been deleted from reconstruction to prevent polar features from obscuring structures within cell cylinder. (a) Cell viewed down its central longitudinal axis. (b) Same reconstruction, rotated so that cell is viewed from oblique angle.

3-D architecture of the photosynthetic (thylakoid) membrane system in *A. quadruplicatum*. It was eventually ascertained that the thylakoid system was an anastomosing series of roughly concentric shells, and that each of these shells was constructed somewhat like a hollow prism in the cylindrical portion of the cell body (Nierzwicki-Bauer *et al.*, 1983a,b). This arrangement was not readily understood, however, until separate reconstructions of the cell cylinder with a single thylakoid shell were rotated until the 3-D configuration of each membrane became evident (Figure 7).

Image rotation can be used to maximize the clarity and understandability of most reconstruction images. To do so, the operator rotates the image so as to minimize confusing superimposition of different cell features. Figure 8 depicts two views of a cell-cylinder reconstruction that includes several thylakoid shells instead of just one (compare with Figure 7). This reconstruction presents only a confused and uninterpretable jumble of overlapping lines when viewed from the side of the cell (Figure 8a), but, if the image is rotated so that the cell is viewed down its central longitudinal axis (Figure 8b), the concentric arrangement and shapes of the thylakoid shells are easier to visualize. Similar (though sometimes less extensive) improvements in image clarity can be realized by rotation of almost any type of reconstruction image.

Image rotation was also used in the reconstruction of *A. quadruplicatum* to exert an internal quality control on the reconstruction process. It was reasoned that reconstructed cells should appear similar, regardless of the direction in which they were originally sectioned. To check for such similarity, reconstructions from cells that had been cross-sectioned, longitudinally sectioned, and

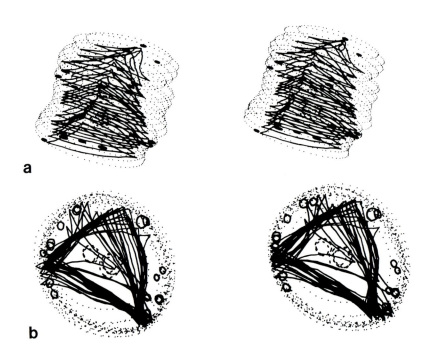

FIGURE 7. Stereo pairs of CARTOS reconstructions of the cylindrical portion of an *A. quadruplicatum* cell (outline in dots), including one thylakoid shell (solid lines), lipid bodies (small solid circles) and carboxysomes (large dotted circles). Note hollow-prism-like configuration of thylakoid shell. Thylakoid membrane surfaces are depicted as series of disconnected contour lines, rather than as planar sheets. Ends of cells have been deleted as in Figure 6. (a) Cell viewed from its side. (b) Same reconstruction, rotated so that cell is viewed down its central longitudinal axis.

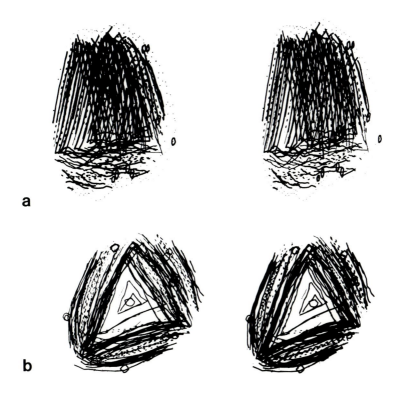

a

b

FIGURE 8. Stereo pairs of CARTOS reconstructions of the cylindrical portion of an *A. quad-ruplicatum* cell (outline in dots), including several thylakoid shells (solid lines and dotted lines) and lipid bodies (small solid circles). One end of the cell has been deleted to prevent polar features from obscuring structures within cell cylinder. (a) Cell viewed from side at an oblique angle. Image is very confusing because of excessive overlap of concentric structures. (b) Same reconstruction, rotated so that cell is viewed down its central longitudinal axis. Concentric arrangement of thylakoid shells is now clearer because overlap has been reduced. Note that thylakoids are tapering (closing) at back of cell.

obliquely sectioned were all rotated to show a view of the cell down its central longitudinal axis and then compared (Figure 9).

6.5. Display Modes and Enhancements

Most computer-aided systems provide display modes and image enhancements that can be used to clarify a 3-D reconstruction or to increase the number of cell structures that can be included without producing a confusing image. Display mode options typically include the ability to depict structures from different memory files as solid lines, dotted lines, dashes, dots, and so on. In the

FIGURE 9. CARTOS reconstructions of three *A. quadruplicatum* cells that were sectioned in different directions. All three have been rotated so that cells are viewed down their central longitudinal axes. Note similarity in appearance of thylakoid membrane system (solid lines).

reconstruction of *A. quadruplicatum*, the dot mode was often used to superimpose a somewhat ghostly (and, therefore, noninterfering) image of the cell outline over the internal features being reconstructed (Figures 5–8). Different types of lines were used to indicate different kinds of inclusion bodies (Figure 6) or to delineate alternating layers of concentric thylakoid shells (Figure 8). Many of the reconstruction images were eventually improved by testing of different groupings of display modes until a more or less optimal combination was found.

Depicting different cell features in different colors can enhance reconstruction images effectively. This can be done either with a color graphics display monitor or with multiple photographic exposures through color filters of the image on a high-resolution black-and-white monitor. The latter approach was used with *A. quadruplicatum* reconstructions, and it produced some of the most effective images in that study (Nierzwicki-Bauer *et al.*, 1983a; Balkwill *et al.*, 1984). In fact, these color reconstructions came the closest to depicting the complete 3-D architecture of *A. quadruplicatum* because they permitted the effective inclusion of the widest variety of cell features. The color reconstructions also showed most clearly that the thylakoid membrane system entirely surrounded the central cytoplasmic portion of the cell, thereby walling it off as an independent compartment.

6.6. Limitations and Possible Solutions

Computer-generated reconstruction images are not unlimited in their ability to portray 3-D architecture clearly, especially when one is working with a structurally complex cell system or a system that contains concentric features like the thylakoid shells of *A. quadruplicatum*. Confusing overlap of different cell features cannot be avoided entirely in such situations, even if the investigator makes

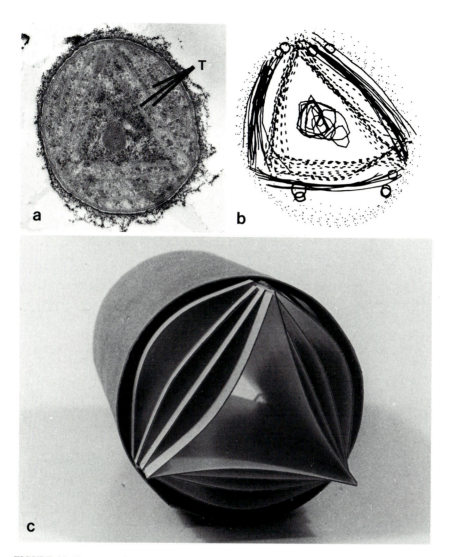

FIGURE 10. Summary of major phases in a typical computer-aided 3-D reconstruction project, as illustrated by stages in reconstruction of the thylakoid membrane system in cylindrical portion of an *A. quadruplicatum* cell. (a) Original high-magnification electron micrograph of cross section through cell cylinder, showing positions of several thylakoids (T). (b) CARTOS reconstruction of cell cylinder (ends have been deleted) made by tracing cell outline (dots), lipid bodies (small solid circles), carboxysomes (large solid circles), and two thylakoid shells (solid lines and dashed lines). (c) Simple 3-D model of complete thylakoid system, made after extensive examination of CARTOS reconstructions. Note that model depicts thylakoid system more effectively than computer-drawn reconstructions (compare with Figure 8b).

judicious use of display-mode and image-rotation options while producing the reconstructions. Thus, a simple 3-D model (Figure 10) depicted the complete thylakoid system of *A. quadruplicatum* more effectively than did any of the varied types of reconstructions that were created throughout that study. Similarly, it was easier to use an artist's drawing (Figure 11) than to produce a computer-generated image that could accurately summarize the overall architecture of the complete cell. Such limitations in no way detract from the importance of computer-generated images, however. After all, it is these images that make the production of realistic models and drawings possible.

Some complex reconstruction images can be clarified by processing with a "hidden-line" program. Hidden-line displays delete all of the reconstruction lines that are situated behind (from the vantage point of the viewer) other structures. As a result, the viewer sees the surfaces of the nearest objects, but not the structures situated within or behind those objects. Such images are easier to interpret, although their usefulness varies with the nature of the specimen being reconstructed. Veen and Peachey (1977) give some particularly effective examples to illustrate the considerable potential of this approach. Present reconstruction systems vary widely as to their hidden-line capabilities and the ease with which these capabilities can be used.

An obvious limitation of the computer-generated images presented here is that planar features like the thylakoid membrane shells are displayed as a series of disconnected lines, rather than as smooth, solid objects. These lines are

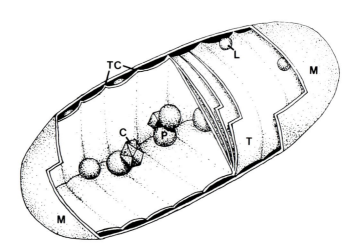

FIGURE 11. Artist's drawing of overall 3-D architecture of *A. quadruplicatum*. C, carboxysome; L, lipid body; M, cytoplasmic membrane; P, polyphosphate body; T, photosynthetic thylakoid membrane system; TC, contacts between thylakoids and cytoplasmic membrane. Thylakoids are depicted as solid sheets, each of which represents a pair of closely apposed unit membranes. Illustration does not include nuclear material, ribosomes, or cell wall.

similar to the contour lines on a topographic map, and as with such a map, the viewer must imagine them to be connected in actual 3-D space. "Shading" programs that display planar objects much as they would be drawn by an artist, instead of representing them with the contour lines, are currently being developed in hopes of overcoming this problem. Such a program has been implemented at the Columbia University Computer Graphics and Image Processing Facility and can already be applied effectively to certain types of CARTOS reconstructions. Increasingly sophisticated shading programs, perhaps in combination with advanced color graphics display capabilities, will enable computer-aided reconstruction systems to produce increasingly realistic and understandable images in coming years.

7. CONCLUSIONS

Computer-aided reconstruction of serial sections is already an effective and practical technique for elucidating the complete 3-D architecture of microbial cells at the ultrastructural level. As automatic tracing systems continue to develop in sophistication, however, the technique will become increasingly efficient and applicable to a wider variety of problems. At the same time, advances in imaging capabilities will make it easier to produce meaningful reconstruction images automatically. Such developments should increase the popularity of computer-aided reconstruction methods among microbial cytologists.

The potential scientific value of computer-aided 3-D reconstructions is demonstrated by the results that were produced when this approach was applied to the unicellular cyanobacterium *A. quadruplicatum.* Among other things (see Nierzwicki-Bauer *et al.,* 1983a,b), it was discovered that (1) different kinds of specialized inclusion bodies occupy specific locations within the cell, perhaps in keeping with their differing functions; (2) the photosynthetic (thylakoid) membrane system contacts the cytoplasmic membrane; (3) the thylakoid system completely surrounds the central part of the cytoplasm, thereby separating it (and the inclusion bodies, ribosomes, and nuclear material that are contained therein) from the remainder of the cell in a prokaryotic form of compartmentalization; and (4) the 3-D arrangement of all intracellular features is exceedingly consistent from one cell to another. These findings are all somewhat surprising because they contradict several established conceptions about cyanobacterial ultrastructure and because one would not expect a prokaryotic organism to produce such a complex cell.

The general message to be derived from the above example is that, by examining the 3-D architecture of entire microbial cells, one can sometimes obtain considerable amounts of novel or unexpected information about their internal organization. Such information can, in turn, lead to a better understanding of the relationships between structure and function within the cell and to a better understanding of the organism's basic cell biology. Therefore, computer-aided 3-D reconstruction of serial sections is likely to be highly productive as it is applied to an increasing variety of microbial structure/function problems in the future.

ACKNOWLEDGMENTS. The active involvement of Dr. Sandra A. Nierzwicki-Bauer and Dr. S. Edward Stevens, Jr. in the *A. quadruplicatum* reconstruction project is gratefully acknowledged. I am also grateful to Dr. Cyrus Levinthal, Amalia Grumfeld, and Noel Kropf for their valuable advice and assistance in using the CARTOS Loaner System and other instrumentation of the Columbia University Computer Graphics and Image Processing Facility. The *A. quadruplicatum* investigation was supported in part by United States Public Health Service Grant 5P41-RR00442 to establish the Columbia University Computer Graphics and Image Processing Facility as a National Institutes of Health Biotechnology Research Resource. Figures 5 and 11 are reproduced from *The Journal of Cell Biology* (1983; Vol. 97, pp. 713–722) by copyright permission of The Rockefeller University Press. Figures 6 and 9 are reproduced from the *Journal of Ultrastructure Research* (1983; Vol. 84, pp. 73–82) by copyright permission of Academic Press, Inc. Figures 4, 7, 8, and 10 are reproduced from *BioTechniques* (1984; Vol. 2, pp. 242–251) by copyright permission of Eaton Publishing Co.

8. REFERENCES

Anderson, R. G. W., and Brenner, R. M., 1971, Accurate placement of ultrathin sections on grids: Control by sol-gel phases of a gelatin flotation fluid, *Stain Technol.* **46**:1–6.
Balkwill, D. L., Stevens, S. E., Jr., and Nierzwicki-Bauer, S. A., 1984, Use of computer-aided reconstructions and high-voltage electron microscopy to examine microbial three-dimensional architecture, *Biotechniques* **2**:242–251.
Fahrenbach, W. H., 1984, Continuous serial thin sectioning for electron microscopy, *J. Electron Microsc. Tech.* **1**:387–398.
Hayat, M. A., 1981, *Principles and Techniques of Electron Microscopy: Biological Applications*, Vol. 1, 2nd ed., University Park Press, Baltimore.
Hoffmann, H.-P., and Avers, C. J., 1973, Mitochondrion of yeast: Ultrastructural evidence for one giant, branched organelle per cell, *Science* **181**:749–750.
Kay, D., 1961, *Techniques for Electron Microscopy*, Blackwell, Oxford.
Knobler, R. L., Stempak, J. G., and Laurencin, O., 1978, Preparation and analysis of serial section in electron microscopy, in *Principles and Techniques in Electron Microscopy*, Vol. 8 (M. A. Hayat, ed.), pp. 113–155, Van Nostrand–Reinhold, Princeton, New Jersey.
Levinthal, C., and Ware, R., 1972, Three dimensional reconstruction from serial sections, *Nature (London)* **236**:207–210.
Levinthal, C., Macagno, E. R., and Tountas, C., 1974, Computer-aided reconstruction from serial sections, *Fed. Proc.* **33**:2336–2340.
Levinthal, F., Macagno, E. R., and Levinthal, C., 1975, Anatomy and development of identified cells in isogenetic organisms, *Cold Spring Harbor Symp. Quant. Biol.* **40**:321–331.

Macagno, E. R., Levinthal, C., and Sobel, I., 1979. Three-dimensional computer reconstruction of neuronal assemblies, *Annu. Rev. Biophys. Bioeng.* **8**:323–351.

Moens, P. B., and Moens, T., 1981, Computer measurements and graphics of three-dimensional cellular ultrastructure, *J. Ultrastruct. Res.* **73**:131–141.

Nierzwicki-Bauer, S. A., Balkwill, D. L., and Stevens, S. E., Jr., 1983a, Three-dimensional ultrastructure of a unicellular cyanobacterium, *J. Cell Biol.* **97**:713–722.

Nierzwicki-Bauer, S. A., Balkwill, D. L., and Stevens, S. E., Jr., 1983b, Use of a computer-aided reconstruction system to examine the three-dimensional architecture of cyanobacteria, *J. Ultrastruct. Res.* **84**:73–82.

Nierzwicki-Bauer, S. A., Balkwill, D. L., and Stevens, S. E., Jr., 1984, The use of high-voltage electron microscopy and semi-thick sections for examination of cyanobacterial thylaoid membrane arrangements, *J. Microsc. (Oxford)* **133**:55–60.

Pease, D. C., 1960, *Histological Techniques for Electron Microscopy*, Academic Press, New York.

Pellegrini, M., 1980a, Three-dimensional reconstruction of organelles in *Euglena gracilis* Z. I. Qualitative and quantitative changes of chloroplasts and mitochondrial reticulum in synchronous photoautotrophic culture, *J. Cell Sci.* **43**:137–166.

Pellegrini, M., 1980b, Three-dimensional reconstruction of organelles in *Euglena gracilis* Z. II. Qualitative and quantitative changes of chloroplasts and mitochondrial reticulum in synchronous cultures during bleaching, *J. Cell Sci.* **46**:313–340.

Remsen, C. C., Watson, S. W., Waterbury, J. B., and Trüper, H. G., 1968, Fine structure of *Ectothiorhodospira mobilis* Pelsh, *J. Bacteriol.* **95**:2374–2391.

Research Resources Information Center, 1981, *CARTOS: Modeling Nerves in Three Dimensions*, Division of Research Resources, National Institutes of Health, NIH Publication 81-2289.

Research Resources Information Center, 1983, *Biotechnology Resources: A Research Resources Directory*, Division of Research Resources, National Institutes of Health, NIH Publication 83-1430.

Roth, J., 1983, The colloidal gold marker system for light and electron microscopic cytochemistry, in: *Techniques in Immunocytochemistry*, Vol. 2 (G. R. Bullock and P. Petrusz, eds.), pp. 217–284, Academic Press, New York.

Schötz, F., Bathelt, H., Arnold, C.-G., and Schimmer, O., 1972, Die Architektur und Organisation der *Chlamydomonas*-Zelle, Ergebnisse der Elektronenmikroskopie von Serienschnitten und der daraus resultierenden driedimensionalen Rekonstruktion, *Protoplasma* **75**:229–254.

Sjöstrand, F. S., 1958, Ultrastructure of retinal rod synapses of the guinea pig eye as revealed by three-dimensional reconstructions from serial sections, *J. Ultrastruct. Res.* **2**:122–170.

Sjöstrand, F. S., 1967, *Electron Microscopy of Cells and Tissues*, Academic Press, New York.

Sobel, I., Levinthal, C., and Macagno, E. R., 1980, Special techniques for the automatic computer reconstruction of neuronal structures, *Annu. Rev. Biophys. Bioeng.* **9**:347–362.

Spurr, A. R., 1969, A low-viscosity epoxy resin embedding medium for electron microscopy, *J. Ultrastruct. Res.* **26**:31–43.

Stevens, J. K., Davis, T. L., Friedman, N., and Sterling, P., 1980, A systematic approach to reconstructing microcircuitry by electron microscopy of serial sections, *Brain Res. Rev.* **2**:265–293.

Tyler, S., 1981, Another multiple-grid holder—this one especially for staining serial ultrathin sections, *Trans. Am. Microsc. Soc.* **100**:322–325.

Van Iterson, W., and Aten, J. A., 1976, Nuclear and cell division in *Bacillus subtilis:* Cell development from spore germination, *J. Bacteriol.* **126**:384–399.

Van Iterson, W., Michels, P. A. M., Vyth-Dreese, F., and Aten, J. A., 1975, Nuclear and cell division in *Bacillus subtilis:* Dormant nucleoids in stationary-phase cells and their activation, *J. Bacteriol.* **121**:1189–1199.

Veen, A., and Peachey, L. D., 1977, TROTS: A computer graphics system for three-dimensional reconstruction from serial sections, *Comput. Graphics* **2**:135–150.

Vivier, E., and Petitprez, A., 1972, Etude du système vacuolaire de l'hématozoaire *Anthemosoma garnhami* à l'aide des coupes sérieés et de reconstitutions tridimensionnelles, *J. Ultrastruct. Res.* **41:**219–237.

Ware, R. W., and LoPresti, V., 1975, Three-dimensional reconstruction from serial sections, *Int. Rev. Cytol.* **40:**325–440.

Waters, H., and Hunt, P., 1980, The *in vivo* three-dimensional form of a plant mycoplasma-like organism by the analysis of serial ultrathin sections, *J. Gen. Microbiol.* **116:**111–131.

Chapter 6

Electron Microscopy of Nucleic Acids

Claude F. Garon

Laboratory of Pathobiology
Rocky Mountain Laboratories
National Institute of Allergy and Infectious Diseases
Hamilton, Montana 59840

1. INTRODUCTION AND HISTORY

New techniques of molecular cloning and subcloning have stimulated renewed interest in the structural analysis of nucleic acid molecules by electron microscopy (EM). The specialized procedures for visualizing both naturally occurring nucleic acid molecules from cells or viruses and laboratory-constructed heteroduplexes and R-loop molecules could, and perhaps should, be a part of the technical capability of every fully equipped EM laboratory. While requirements for specialized equipment are minimal, this technique does require a separate set of interpretive skills including an appreciation of the fact that individual molecules are assayed without the benefit of the selective pressure present in other techniques, as well as a clear understanding of the relationship between preparative procedure and image quality. It is this latter relationship that can be most troublesome and one that I hope to deal with effectively in this chapter. My aim, then, rather than providing a comprehensive review of all of the modifications and permutations of the basic technique that have appeared in the literature over the years, is to describe a broad range of procedures that have, in my laboratory, most often provided useful and consistent structural data on nucleic acids. Since sample requirements for successful structural analysis of nucleic acid molecules often exceed those of other techniques, it will also be my aim to provide detailed purification, handling, and storage procedures that will provide samples of sufficient quantity and quality for ultrastructural analysis. Finally, I intend to "troubleshoot" the procedures as I go on to define those factors that are potential sources of problems and emphasize those procedures that most often provide crisp and unambiguous results. An often overlooked advantage of the EM analysis of nucleic acid molecules is the speed with which accurate qualitative analysis may be performed. Not even new mini-gel procedures offer the ability to quick scan an appropriately prepared sample to determine, literally within minutes, molecule size, concentration, homogeneity, and intactness. These same samples, often involving extremely small quantities of material, may be reprocessed later for greater precision and statistical analysis.

While EM in general has often played a significant role in genetic studies, that role was expanded appreciably with the description by Kleinschmidt and Zahn (1959) of a reliable and innovative procedure for visualizing nucleic acid molecules. Earlier procedures where DNA molecules were applied directly to grid surfaces inevitably resulted in tangled or aggregated forms of extremely low contrast, which made it difficult to evaluate a high percentage of molecules in a given field. The idea of coating nucleic acid strands with a basic globular protein to make them at once mechanically spreadable as part of a monomolecular film, larger, and more easily stainable is one that has served us well. Figure 1 demonstrates the images produced by double-stranded DNA photographed at identical magnifications but mounted for EM with (panel A) or without (panel B) cytochrome *c* coating. The contribution to diameter of the protein coat is clearly

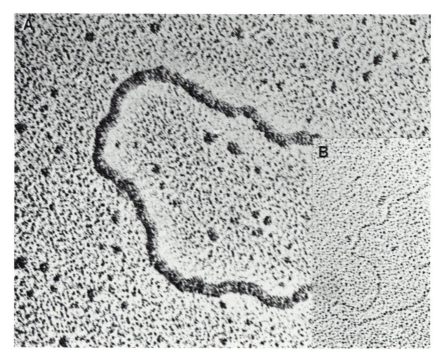

FIGURE 1. Double-stranded DNA molecules mounted for EM with (panel A) and without (panel B) cytochrome *c*. Molecules were photographed at identical magnification levels.

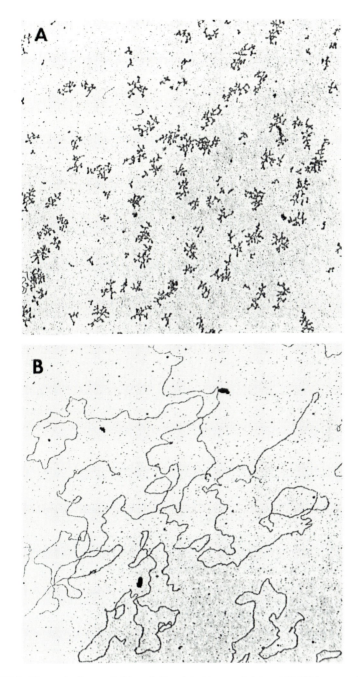

FIGURE 2. Linear, single-stranded (panel A) and double-stranded (panel B) DNA mounted for EM using the aqueous procedure. Molecules were rotary shadowed with platinum–palladium.

illustrated. It is this nucleic acid–protein interaction that is at the heart of all procedures to be described here. Newer modifications of this basic technique that were to follow have not, in my view, diminished the value and ingenuity of this original idea.

Although Kleinschmidt and Zahn's technique provided a simple and reliable method for examining double-stranded nucleic acids in the EM, little information could be gained about single strands. Under the mounting conditions described, single-stranded material aggregated into tight "bushes" due to intrastrand annealing. Figure 2 illustrates identical nucleotide sequences in both extended, double-stranded and collapsed, single-stranded forms. While the technique was, and still is, valuable for measuring double-strand lengths or for distinguishing (with certainty) between single- and double-stranded regions, information about the length of single-strand material is not obtainable from these images. However, new techniques, first described by Westmoreland et al. (1969), dealt with that problem simply by adding a chemical denaturant to the spreading solutions. Under their mounting conditions, double strands are stable, but the random base interactions in single strands are denatured, thereby extending the strands. Distinction between single and double strands is often based either on a subtle difference in strand thickness or on a small residual amount of intrastrand interaction that results in a slightly "kinky" appearance to the single-stranded regions. The ability to measure the contour lengths of single-stranded molecules or single-stranded regions of duplex molecules was now possible. Many modifications of these techniques have appeared in the literature; however, all may be placed into one of two categories: aqueous or formamide. These two major variations of the basic protein monolayer technique will be described in detail below.

2. SAMPLE MOUNTING TECHNIQUES

The mounting of nucleic acid molecules for EM analysis may be conveniently described in four steps: (1) the formation of a nucleic acid–protein complex; (2) the mechanical spreading of this complex onto a fluid surface in order to form a suitable monomolecular layer at the fluid–air interface; (3) the transfer of the monolayer to a solid support medium; and (4) contrast enhancement and viewing.

Parlodion (Cat. No. 19231, Ted Pella, Inc., Tustin, CA 92680)-coated 200- to 400-mesh copper grids (Cat. No. 14120, Ladd Research Ind., Burlington, VT 05402) are prepared each day for use in either aqueous or formamide procedures. Oven-dried (90°C for 24 hr) strips are dissolved in EM-grade butyl acetate (Cat. No. 3854, Polysciences Inc., Warrington, PA 18976) to give a 2.7–3.0% solution. The strips take several days at room temperature with gentle shaking to dissolve completely, but the solution is useful for 1 month or more thereafter when stored at room temperature in a tightly closed, dark bottle. New solutions are prepared when expanding pin holes appear in the film upon electron bean exposure. Rapid tearing of the film generally denotes an excessively thin coating rather than a faulty Parlodion solution. Grids are rinsed individually in reagent-grade acetone prior to placing them on a stainless steel screen beneath the deionized water surface of a drainable stainless steel trough. The initial drop of Parlodion solution is allowed to dry on the surface of the fluid in a 55°C oven for at least 1 min, is picked up on the tip of a Pasteur pipette, and discarded. This serves to clean the surface of the deionized water. The second drop is dried for a similar period of time before the water is drained slowly from the bottom of the trough allowing the film to settle gently upon the grids. The stainless steel screen is removed, covered with a glass petri dish, and allowed to dry at 55°C for 35 min. After cooling to room temperature, the grids are used without further treatment.

In general, spreading solutions must be made from refrigerated stocks immediately prior to use. The spreading solution, usually a total of 50 μl, contains the nucleic acid, cytochrome c, and appropriate buffers and salt. The hypophase solution (120 ml fills our spreading trough) serves as the support fluid upon which the protein monolayer is formed. Spreading is done in a lighted, Plexiglas hood (Cat. No. 22400, Ladd Research Ind.) to protect the monolayer from dust and disruption by air currents. Successful spreading of nucleic acid–protein monolayers has, however, been accomplished on bench tops.

2.1. Aqueous Procedure

The aqueous procedure remains in our hands one of the most useful and reliable methods for visualizing double-stranded DNA or RNA molecules in the EM. Stock reagents are more stable. Results appear less dependent on minor variation in technique, making them more consistent and reproducible. The hypophase solution for aqueous spreading consists of 120 ml of 0.25 M ammonium acetate. This volume completely fills a 100 × 15-mm square plastic petri dish (Cat. No. 4021, Lab Tek Products, Naperville, IL 60540). Teflon bars (120 mm in length) provide a nonwettable surface in the front of the dish and a convenient microscope slide support at the rear. Plain glass slides, which are stored in sulfuric acid–dichromate cleaning solution until use, are washed exhaustively in deionized water and allowed to air-dry immediately prior to use. A typical aqueous spreading solution (50 μl total volume) containing 35 μl of 1 mM EDTA, 5 μl of 5 M ammonium acetate, 5 μl of nucleic acid sample (concentration 1–5 μg/ml), and 5μl of cytochrome c (1 mg/ml; Cat. No. 250345; *Candida krusei;* Calbiochem, La Jolla, CA 92037) is drawn up in a capillary pipette and gently extruded onto a glass ramp approximately 1 cm above the 0.25 M ammonium acetate hypophase solution. The monolayer is aged for 50–60 sec before a Parlodion-coated grid is touched briefly to the surface of the hypophase solution at a point approximately one grid's diameter from the glass–fluid interface. The grid is immediately transferred to a 10^{-5} M uranyl formate (Leberman, 1965) staining solution freshly prepared each day in 90% ethanol. The grid is stained for 30 sec, transferred to a 90% ethanol wash for 10 sec, blotted with filter paper, and allowed to air-dry. All solutions and stains should be filtered (pore size 0.2 μm; Cat. No. 120-0020, Nalge Co., Rochester, NY 14602) after carefully removing any detergents or wetting agents that might have been incorporated into the filter membrane by washing with 90% ethanol. Grids may be viewed either before or after further contrast enhancement by rotary shadowing.

2.2. Formamide Procedure

Under the spreading conditions described above, single strands collapse into tight aggregates due to random base-to-base intrastrand interaction. These interactions may be discouraged and the single strands extended by the addition of any of a number of denaturants. Ones we have tested include formaldehyde, urea, dimethyl sulfoxide, caffeine, formamide, methyl mercury, sodium perchlorate, glyoxal, heat, and pH. The one most often chosen [first introduced in nucleic acid hybridization by Bonner *et al.* (1967); first EM application described by Westmoreland *et al.* (1969)] is formamide. Formamide (Cat. No. FX0420, MCB Manufacturing Chemists, Inc., Cincinnati, Ohio 45212) may be deionized by stirring for 1 hr at room temperature with 50 g/liter Dowex AG 501-X8 resin (Cat. No. 142-6424, Bio-Rad Laboratories, Richmond, CA 94804) followed by brief centrifugation to remove resin and storage in dark bottles at −20°C until use. Thawed bottles may be stored at 4°C for weeks. Formamide is so useful in these protocols: (1) because it is effective over a wide range of concentrations, and (2) because it enhances rather than reduces contrast over that range. At low concentrations of formamide, single strands are extended to some degree and duplex molecules are not visibly affected. At higher concentrations, single strands may be well extended, difficult to distinguish from duplexes in the same field, and partially homologous regions of an appropriate molecule (a heteroduplex molecule, for example) may show areas of strand separation. However, native or duplex molecules under these spreading conditions still appear unaffected. At very high formamide concentrations, all secondary structure may be removed from single-stranded DNA (often higher denaturing power is needed to remove secondary structures from RNA), and A–T-rich regions of native or duplex molecules may denature. Most duplex DNA molecules may be completely denatured in 98% formamide at room temperature at low concentrations of salt.

The effect of formamide on the thermal stability of native or reassociated nucleic acid molecules (expressed as reduction in T_m) has been calculated to be in the range of 0.60–0.72°C per each 1% increase in formamide concentration (McConaughy *et al.*, 1969; Bluthman *et al.*, 1973). Of equal importance is the finding that use of formamide produced no loss of specificity of reannealing. Reannealing can be carried out, therefore, in a solution containing 50% formamide at an optimal temperature 30°C lower with considerably less thermal breakage. However, while it is true that the addition of formamide to the basic protein monolayer spreading technique gives us additional information about single strands, allows us to detect regions of partial homology, and permits the mapping of denaturation profiles, the price may be a considerable loss of reliability and consistency. Image quality may be affected by all manner of procedural variation and, at certain formamide concentrations, the maintenance of reasonable differentiations between single and double strands may be a problem. Thus, formamide mounting protocols must be carefully controlled and contour lengths stringently calibrated by the inclusion of appropriate internal standards. The goal of the procedure to follow is not only adequate buffering of the formamide, which tends to become acidic, but provision for the proper balance of salt, temperature equivalents, and formamide concentration necessary to achieve the desired levels of denaturation during spreading while maintaining hybridized regions. Proper differentiations between single and double strands must be maintained as well.

A stock spreading solution (1 ml total volume) is prepared each day and stored on ice until use. It consists of 700 μl of deionized formamide, 100 μl of 1 M Tricine buffer (pH 8.0), 50 μl of 5 M sodium chloride, 20 μl or 0.5 M EDTA, and 130 μl of deionized water. Immediately prior to spreading, 8 μl of nucleic acid sample (1–5 μl/ml) is added to 40 μl of the above stock solution and mixed gently, but thoroughly, with 1 μl of cytochrome c (1 mg/ml). The sample is taken up in a capillary pipette and extruded down a cleaned glass slide ramp, as before, onto the surface of a hypophase completely filling a square plastic petri dish. We have successfully used either deionized water, or 0.01 M TRIS-HCl (pH 8.0), 0.001 M EDTA as a hypophase. Since components of the spreading solution diffuse away from the monolayer and into the water hypophase quickly, causing partial disassociation of the nucleic acid–protein complex, the monolayer must be transferred to a Parlodion-coated grid within 20 sec. Water hypophases seem to provide better distinction between single and double strands, "TRIS–EDTA" hypophases better contrast. Grids are stained with uranyl formate and destained with 90% ethanol as before. While nucleic acid strands mounted by the formamide procedure are visible following staining, the contrast is considerably less than that obtained with the aqueous procedure.

2.3. Contrast Enhancement

Rotary shadowing of stained grids with heavy metals remains the method of choice for achieving reasonable contrast enhancement, while maintaining good distinction between single and double strands and sufficient monolayer stability for EM and measurement of contour length. Approximately 3 cm of platinum–

palladium (80 : 20 mixture, 8-mil diameter, Cat. No. 12211, E. F. Fullam Inc., Schenectady, NY 12301) is wrapped tightly around the tip of a 20-mil tungsten (Cat. No. 27-100, Ted Pella, Inc.) V-filament, pumped to a vacuum of 2.5 × 10^{-5} Torr before vaporizing for 30 sec at a "distance from grids" to "height above grids" ratio of 8 : 1. The grids, placed as close to the center as possible, are rotated at approximately 40 rpm during the shadowing procedure.

Additional levels of contrast may be achieved by viewing grids at the lowest (40 kV or less) accelerating voltages available and by the use of small (50 μm or less) objective apertures. Darkfield illumination provided by an annular aperture provides considerable added contrast. Finally, contrast enhancement may be achieved photographically by alterations in the initial development of microscope negatives or during the production of final photographic prints. Some have resorted to xerography for a considerable step up in contrast.

Contrast, however, is largely determined in these procedures by the size of the column of protein surrounding the nucleic acid molecule relative to the thickness of the protein monolayer. This, in turn, is dependent on salt, formamide, and cytochrome c concentrations as well as time spent by the monolayer on the disrupting surface of the hypophase. Often a redefinition of these factors is more rewarding than attempts to rescue low-quality preparations photographically. While grids may be stored in open dishes or grid storage boxes for several weeks, some loss in image quality is expected over that time period. For that reason, immediate photographic recording of nucleic acid molecules is to be encouraged.

3. SAMPLE PREPARATION REQUIREMENTS

Difficulties encountered with mounting molecules for EM using the techniques described above can often be traced to one of several problems with the nucleic acid sample itself. Purification procedures that result in inadequate removal of either bound protein or certain chemical agents from the final sample or that result in structural damage to the nucleic acid molecules themselves are major sources of problems. Often molecules that appear suitably clean and intact for analysis by gel electrophoresis or other techniques will either prove unsuitable for EM analysis or disrupt mounting protocols completely. Effective spreading is difficult in the presence of even residual amounts of such agents as ethanol, phenol, cesium chloride, SDS, Triton X or other detergents, high salt, or high sucrose. Most, if not all, of the above are common reagents used in the purification of nucleic acid molecules from cells or viruses. Careful removal of these substances must be attended to before microscopy of nucleic acids can begin.

Damage to nucleic acid molecules may result either from mechanical shear forces or from the action of ubiquitous, contaminating nucleases. Large molecules appear more susceptible to both. The solution to the former category is extraordinary care in handling at all stages of the procedure. Sonication, freeze–thaw procedures, small-bore pipetting, stirring, and shaking are to be avoided. Often the introduction of a viscous agent such as sucrose (20%) serves to protect large molecules from shear damage during handling. Sucrose, however, must be removed prior to mounting protocols. The damage to nucleic acid molecules by nucleases may be minimized by the use of sterile reagents and glassware, by the maintenance of low temperature and alkaline pH (8.0–8.5) whenever possible, and by the inclusion of at least millimolar levels of chelating agents, such as EDTA, into all purification, spreading, and storage reagents. Restriction endonuclease digestion protocols have often been identified as a source of considerably random nicking activity, particularly when excessive enzyme is added or when long incubation times are used to ensure limit digests. An EM nicking assay for the detection of contaminating nucleases in reagents has been described previously (Garon, 1981). Nucleic acid samples are routinely stored in sterile containers at 4°C in 0.01 M TRIS-HCl (pH 8.0), 0.001 M EDTA. All enzyme treatments and spreading protocols presented here are compatible with samples added in the above buffer.

4. DNA ANALYSES

4.1. Contour Length Measurements

Often the question of molecular size is the first question to be asked about a newly isolated nucleic acid sample. Since nucleotides cannot be counted directly in the EM by these procedures, we must rely on carefully calibrated contour length measurements and accurate comparison of these with the linear densities of similar (and similarly prepared) nucleic acid molecules of known molecular size. Circular molecules are useful internal calibration standards since circles are, by definition, complete, intact, and are readily distinguishable from linear unknown samples. Both single- and double-stranded circular viral genomes or plasmids are available commercially and are suitable for this purpose. The presence of dimeric or polymeric forms frequently observed in these preparations, however, should not become a source of ambiguity or confusion. Measurement of the contour lengths of 25–50 molecules in each unknown and standard category should provide a suitable molecular size comparison. Molecular clones that have been treated with the appropriate restriction endonuclease to free the recombinant insert fragment from that of the cloning vehicle may be measured without added calibration standards since fragments from cloning vehicles of known molecular size are present in the same field as the insert. Nucleic acid microscopy appears at least as accurate as calibrated gel analysis for small fragments (0.5–5.0 kb) and demonstrably more accurate than gel analysis for larger molecules.

Figure 3 shows supercoiled and relaxed forms of a plasmid molecule. It is obvious from this micrograph that supercoiled molecules are not suitable in their unrelaxed form for accurate contour length measurement. However, we have employed three methods to allow EM observation of these molecules. One uses the spreading forces involved in monolayer formation (allow 3–4 min instead of the usual 50 sec) to force open the twisted form of the supercoil. The second uses an amount of the intercalating chemical ethidium bromide (usually 1 µg/ml) incorporated into both the spreading reagents and hypophase to unwind the superhelical turns in the molecule (see Sebring *et al.*, 1974). Both of the above procedures, while opening supercoiled molecules for observation, produce linear densities, upon careful contour length measurement, that are far above normal factors. Attention to the variation in these calibration factors must be carefully controlled. The third method involves enzymatic nicking of supercoiled molecules under conditions that will produce one or a few nicks per molecule, thereby producing relaxed and measurable circular forms. A convenient, controllable method involves a 25-min room-temperature incubation of a solution containing 35 µl of DNA, 5 µl of 1.0 M TRIS-HCl (pH 8.0), 5 µl of 0.1 $MgCl_2$, and 5 µl of pancreatic DNase I (Cat. No. DPFF, Worthington Diagnostic Systems, Freehold, NJ 07728) freshly diluted in water to contain 0.2 unit of enzymatic activity.

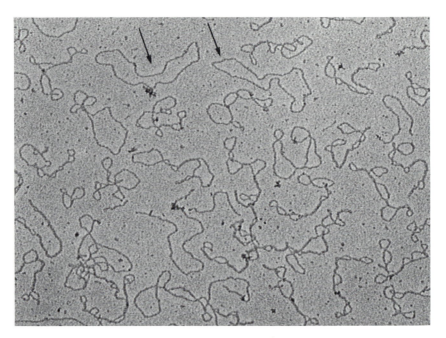

FIGURE 3. Covalently closed supercoiled and relaxed circular molecules (arrows) mounted for EM using the aqueous procedure.

The reaction is stopped by ice quenching and by the addition of 5 μl of 0.5 M EDTA. Molecules may be mounted for microscopy without further treatment using the methods described above. No change in linear density is noted with this procedure. This method may be used for heteroduplex analysis of supercoils since statistically supercoils are nicked but one time producing a single-stranded circle and a single-stranded linear form upon denaturation. It should be noted, however, that nicks are randomly placed.

4.2. Denaturation Profiles

A pattern of visible areas of strand separation or denaturation profile, may be produced along a molecule's length by a variety of chemical and/or physical treatments. Combinations of temperature, pH, formamide, urea, and formaldehyde have been used successfully. Denaturation begins in those areas of the nucleotide sequence that consist of the highest concentration of adenine (A) and thymine (T) residues and proceed at higher denaturing power to areas where guanine (G) and cytosine (C) predominate. Ultimately, the two strands will separate completely under high denaturation treatments. The pattern that is produced before complete denaturation takes place is an accurate ''footprint'' of the sequence arrangement for that particular molecule. Furthermore, these areas of strand separation may serve as physical reference points along the length of these molecules prepared for microscopy in this manner (Younghusband and Inman, 1974).

A convenient method for producing denaturation profiles is as follows: Since tris buffer cannot be used in reactions involving formaldehyde, nucleic acid samples should be dialyzed against a solution containing 0.1 M phosphate buffer (pH 7.8), 0.001 M EDTA. Formaldehyde (36%) is neutralized with 1.0 N sodium hydroxide to pH 7.8, placed in a capped tube in a boiling water bath for 10 min, and readjusted to pH 7.8 with 1.0 N sodium hydroxide. Dialyzed nucleic acid samples are reacted with 5% pH-adjusted formaldehyde for a 20-min period in a water bath at the appropriate temperature. At the end of that time period, the sample is ice quenched and mounted for EM using either the aqueous or formamide procedure. Samples may be dialyzed for a 1-hr period since the half-life of bound formaldehyde is on the order of 140 min.

4.3. Secondary Structure

The occurrence of inverted repeat sequences, hairpin and stem-loop (Figure 4) structures may be effectively demonstrated by EM following appropriate denaturation and reannealing conditions. A convenient protocol involves denaturation by mixing an equal volume of purified DNA and 0.2 N sodium hydroxide. Following an incubation period of 10 min at room temperature (all strands are demonstrably single-stranded by aqueous spreading), the pH is ad-

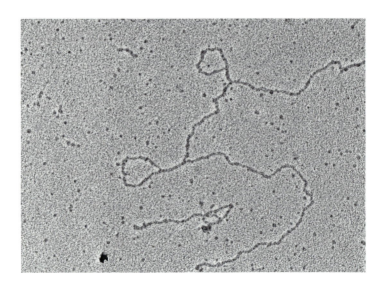

FIGURE 4. Human placental DNA following alkali denaturation and a brief period of reannealing. Note two characteristic stem-loop structures indicative of inverted repeat sequences. Molecule was mounted for EM by the formamide procedure as described.

justed to 8.5 by the addition of 2 M TRIS-HCl buffer. Formamide is added to give a 50% final concentration. Sodium chloride may be added to increase the stability of reannealed structures. Molecules may be mounted for EM using the formamide procedure as previously described after an incubation period of several minutes at 30–37°C.

4.4. Heteroduplex Formation

Heteroduplex molecules (i.e., duplexes composed of one strand from one source and one strand from a different source) are formed by mixing, denaturing, and reannealing related DNA strands under appropriate conditions of concentrations, time, and temperature equivalents (Davis *et al.*, 1971). Regions of the heteroduplex molecule that are identical in base sequence, or very nearly so, will form a duplex while regions of nonhomology will not anneal and remain single-stranded. These regions may be identified in the EM following, in most cases, formamide spreading or, in some instances, even aqueous mounting procedures. The detection limit for regions of noncomplementarity is generally agreed to fall in the range of 50–100 nucleotide pairs. Procedures for using single-strand-specific nucleases to increase sensitivity of detection to a few nucleotide pairs have been described (Bartok *et al.*, 1974).

Although the conditions affecting the reassociation of nucleic acids in solution are well defined (Wetmer and Davidson, 1968), optimum incubation conditions for the formation and subsequent mounting of specific heteroduplex molecules may require careful consideration. For example, molecules showing only limited homology or molecules that differ markedly in length may not form as efficiently or as stably as homoduplexes in the same solution. These molecules may be seen only transiently during the course of the incubation period. Therefore, these structures must be protected from subsequent reassortment by reannealing at low concentration or for shorter incubation times. Fortunately, the mounting procedures allow aliquots to be removed from the reannealing mixture for a rapid assessment of hybridization progress.

Complex heteroduplex structures (i.e., those showing many areas of strand separation) are invariably more difficult to spread unambiguously and are seen more often as tangled aggregates. Therefore, these structures may be more easily viewed following longer incubation periods at lower concentrations of DNA. Although statistically one would predict (with equal concentrations of the two DNA species in the reannealing mixture) a heteroduplex population approaching 50% of the annealed duplexes, due to factors such as the ones cited above the actual number of heteroduplex molecules present in a typical field may be considerably less.

A convenient protocol for the formation of heteroduplex molecules is as follows: a solution containing 440 μl of deionized formamide, 40 μl of 2.0 M TRIS-HCl (pH 8.0), and 10 μl of 5.0 M sodium chloride is prepared each day and stored in an ice bath until use. Ten microliters of a DNA mixture containing an equal amount of the two species to be compared is mixed with 10 μl of 0.2 N sodium hydroxide and incubated for 10 min at room temperature. After incubation, 24 μl of buffer formamide mixture is added, mixed gently but thoroughly, and the tube tightly capped. Heteroduplexes are formed by incubation at an appropriate temperature (usually 30–37°C) for one to several hours. Aliquots may be removed at various intervals, spread by the formamide procedure as previously described, to monitor reannealing progress. Figure 5 shows a typical heteroduplex molecule prepared using this method.

4.5. Molecular Clones

Recombinant DNA molecules prepared in either bacteriophage or plasmid cloning vehicles may be compared by heteroduplex analysis in one of two ways. Where possible, inserts may be cleaved out of the cloning vehicle by appropriate restriction enzymes and compared directly with no attempt made to remove bacteriophage or plasmid fragments (these fragments of known size may be used for internal calibration standards). Where only small regions of homology are expected, insert sequences may be stabilized by cleaving each recombinant clone at an identical restriction site present in the cloning vehicle but not the insert. Lambda clones may be heteroduplexed intact. An example of such experimental strategy is presented in Figure 6. Care must be exercised to ensure that homologous sequences are present in the same orientation within the cloning vector,

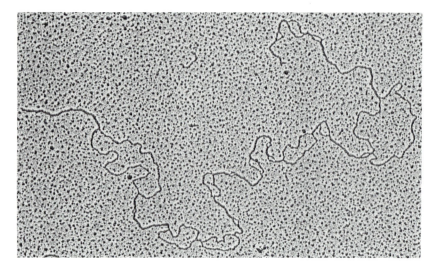

FIGURE 5. Heteroduplex comparison of the genomes of two highly oncogenic adenovirus serotypes mounted for EM using the formamide technique. Note areas of homology and heterology.

since opposite orientations of sequences that are completely homologous will not hybridize in these structures and appear as heterologous forms.

5. RNA ANALYSES

5.1. Contour Length Measurements

Double-stranded RNA is visually indistinguishable from double-stranded DNA and may be spread and measured in similar fashion (Figure 7). Although linear densities may vary slightly (Glass and Wertz, 1980), carefully calibrated experiments using duplex RNA standards should present no unexpected problems. Both aqueous and formamide mounting procedures may be used for these samples without modification. Extra care should be exercised in processing and handling duplex RNA samples especially noting sensitivity to alkali treatment and susceptibility to contaminating ribonucleases.

5.2. Secondary Structure

While double-stranded RNA molecules may be processed conventionally, mounting single-stranded RNA for accurate contour length measurement can present special problems. Efficient removal of secondary structure from RNA molecules in order to obtain accurate length data can sometimes tax our ability to denature. Applying conventional formamide spreading techniques to some sin-

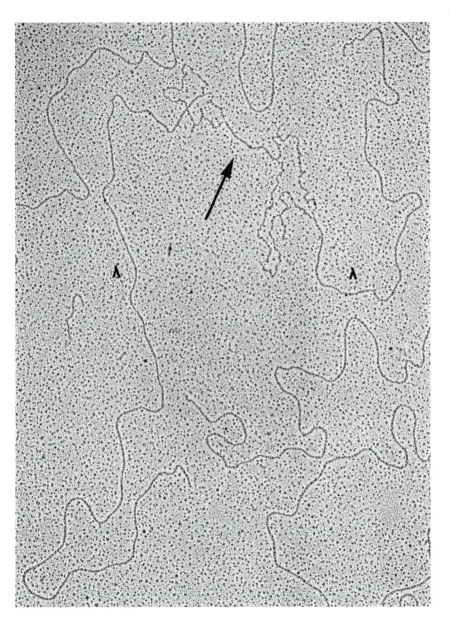

FIGURE 6. Heteroduplex molecule constructed with intact bacteriophage lambda recombinant clones. Duplex lambda arms are perfectly homologous; inserts show only a single, asymmetric area of homology (arrow).

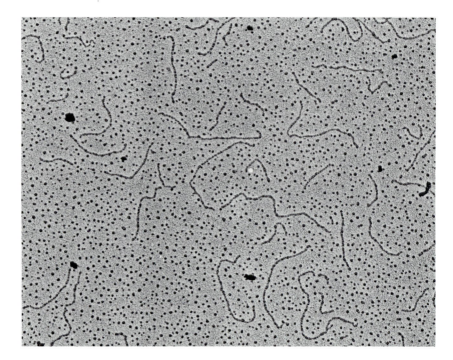

FIGURE 7. Reovirus double-stranded RNA fragments mounted for EM using the aqueous procedure.

gle-stranded RNA samples has left many molecules largely unextended. For that reason we have used a procedure incorporating urea and heat into a spreading protocol that has given good results in many systems. A urea/formamide reagent is prepared by dissolving 1.2 g of ultrapure urea (Cat. No. 2232, Polysciences Inc.) in 4.2 ml of deionized formamide. Five microliters of RNA is added to 40 μl of freshly prepared urea/formamide reagent. The mixture is heated to 53°C for 30 min and allowed to cool slowly to room temperature. Five microliters of cytochrome *c* stock [consisting of equal volumes of 1 mg/ml cytochrome *c* and 1.0 M TRIS-HCl (pH 8.0), 0.1 M EDTA] is added and spread on a hypophase of 0.1 M TRIS-HCl (pH 8.0), 0.001 M EDTA. Grids may be processed as previously described.

6. DNA–RNA INTERACTIONS

6.1. R-Looping Procedures

One of the most powerful forms of microscopic structural analysis in current use is the ability to accurately map the location of individual RNA transcripts along the length of a DNA molecule. The technique (Thomas *et al.*, 1976; White and Hogness, 1977) is made possible by a stringent set of incubation conditions that allow RNA to hybridize with complementary areas of DNA, causing an easily recognizable and characteristic displacement of the DNA strand. The resulting structure, called an "R-loop" (Figure 8), is mapped as an area of strand separation along a duplex DNA molecule defined by a duplex DNA–RNA hybrid arm and a single-stranded DNA arm. This DNA–RNA hybrid is formed during transient denaturation or "breathing" that is a consequence of the high formamide and temperature chosen for the incubation. Since DNA–RNA hybrids are slightly more stable than the DNA duplex under these conditions, R-loops may accumulate over the period of incubation and can be maintained during mounting procedures. For this reason, plus the use of long incubation periods, relatively impure RNA preparations may be used successfully in these procedures. A convenient reaction mixture for the formation of R-loops is made

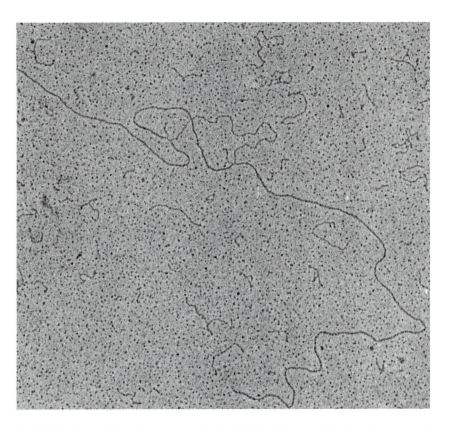

FIGURE 8. Typical R-loop showing duplex DNA–RNA hybrid arm and displaced single-stranded DNA arm. Branch points define length of hybridized transcript. Note unhybridized RNA molecules in background.

by mixing 70 μl of deionized formamide, 10 μl of 1.0 M Tricine buffer (pH 8.0), 5 μl of 5.0 M NaCl, 2 μl of 0.5 M EDTA, and 13 μl of the DNA–RNA mixture in a sealable vessel. Since the maximum rate of R-loop formation is within a degree or two of total denaturation for that DNA molecule, the choice of incubation temperature is critical. If the transcribed region of DNA never denatures, it never becomes available for hybridization with the RNA. If the DNA denatures completely, then the characteristic strand displacement structure is never seen. Therefore, the critical temperature is usually determined empirically for each R-loop experiment by raising the temperature of the reaction mixture incrementally until the conversion to single strands is observed in the EM by aqueous or formamide mounting procedures.

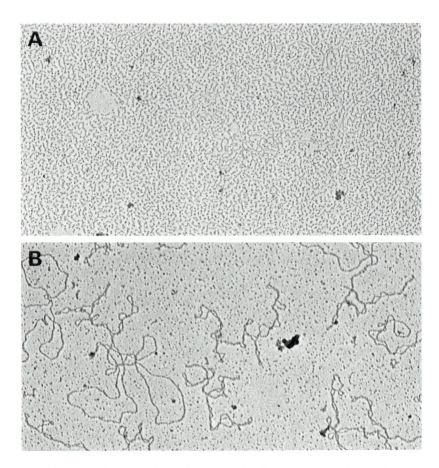

FIGURE 9. Removal of contaminating RNA molecules from DNA preparation by ribonuclease treatment. DNA molecules (panel A) are barely visible in overwhelming excess of RNA. Enzyme digestion does not result in damage to predominantly supercoiled plasmid molecules.

6.2. Enzymatic Procedures

Although poly(A) tails may occasionally be observed directly, Bender and Davidson (1976) have described a method for increased detection sensitivity by linking a visual marker to a hybridizable poly(T) strand. Since spreading and measuring of R-loop structures may be complicated by both partially hybridized and unhybridized RNA, we have proposed a method for removing this unhybridized material by a brief ribonuclease enyzme digestion performed immediately prior to the mounting procedures (Chan *et al.*, 1980). RNase digestion under these conditions (in the presence of formamide) neither destroys R-loops nor changes their size or configuration. Pancreatic Ribonuclease A (Cat. No. RAF, Cooper Biomedical, Malvern, PA 19355) is prepared in 0.3 mg/ml boiled stocks and stored at $-20°C$. Tubes are quick thawed and stored in an ice bath before use. Five microliters of the RNase stock added to a 50-μl reaction mixture and incubated for 10 min at 37°C is sufficient to bring unreacted single-stranded RNA to fragments that are below the level of detection in the EM (Figure 9). This method may also be used to remove residual, and obscuring, levels of RNA from DNA samples without fear of damage to DNA molecules.

7. DIGITIZING AND COMPUTER ANALYSIS

Having successfully purified, prepared, mounted, shadowed, and photographed nucleic acid molecules, some quantitation or measurement of the structures generated is invariably desired. That desire may be satisfactorily fulfilled by "equipment" as simple and inexpensive as a waxed string laid along the contour of a molecular strand and compared with the markings on a ruler or by equipment as complex as computerized image analyzers (Lipkin *et al.*, 1979). For most of us, the satisfactory solution lies between those extremes. We have obtained satisfactory, though tediously acquired, data with a mechanical map measurer (Cat. No. 1718, Dietzgen Corp., Switzerland). However, a considerable step up in both convenience and accuracy is obtainable with one of the commercially available computerized digitizer systems. Most work by registering the distance traversed by a cursor over the surface of a photographed molecule. The contour length data obtained may be displayed on a screen, printed on paper, or sent to memory for further calibration, statistical analysis, and/or storage. Our own laboratory has combined a similar digitizer with a projection system that allows us to measure images on the surface of a table-mounted screen, thereby eliminating the need for photographic printing. Programs have evolved in our laboratory over the years for handling individual values such as measurements of viral genomes or plasmids or for handling a number of subparts or regions of a molecule such as a heteroduplex or R-loop. Several of these computer programs (which are written in BASIC and were contributed to by many people) are available and may be obtained without charge from the author. Simple statistical analyses of the data groups, together with a reasonable capability to output data as histograms and schematics illustrating structure, are useful parts of these programs as well. The ability to sort, delete, insert, alter, group, and electronically file accumulated data is a common but important feature of our data handling system.

8. SUMMARY AND CONCLUSIONS

The technology for visualizing single strands and the introduction of the heteroduplex analysis method by Davis and Davidson (1968) and by Westmoreland et al. (1969) have permitted the growth of EM analysis of nucleic acids much beyond the simple definitions of size and conformation. Using appropriately denatured and reannealed samples, direct genetic comparisons have been made between members of related viral groups based on the visual identification of characteristic structures representing heterology, deletion, or substitution (Brack, 1981). Further improvements, mostly in the ability to titrate formamide concentrations in spreading solutions and hypophases, have also allowed a more precise look at regions of partial homology or genetic drift (Davis and Hyman, 1971).

We have suggested some areas of molecular biology where nucleic acid microscopy has made a contribution to the development of sound biological principles. Perhaps nowhere else in science does observed structure so often suggest function as in the ultrastructural analysis of nucleic acid molecules by EM. The potential for defining features of replication, transcription, or recombination often exists and should be fully exploited. It is our hope that these procedures will serve those investigators who want to use established procedures to explore their biological problems in different ways. Furthermore, it is our hope that this compilation of techniques and protocols will serve as a springboard to the development of new techniques and procedures that will enhance the usefulness and increase the sensitivity of this unique experimental approach.

9. REFERENCES

Bartok, K., Garon, C. F., Berry, K. W., Fraser, M. J., and Rose, J. A., 1974, Specific fragmentation of adenovirus heteroduplex DNA molecules with single strand specific nucleases of *Neurospora crassa*, *J. Mol. Biol.* **87**:437–449.

Bender, W., and Davidson, N., 1976, Mapping of poly (A) sequences in the electron microscope reveals unusual structure of type C oncornavirus RNA molecules, *Cell* **7**:595–607.

Bluthmann, H., Bruck, D., Hubner, L., and Schoffski, A., 1973, Reassociation of nucleic acids in solutions containing formamide, *Biochem. Biophys. Res. Commun.* **50**:91–97.

Bonner, J., Kung, G., and Bekhor, I., 1967, A method for the hybridization of nucleic acid molecules at low temperature, *Biochemistry* **6**:3650–3653.

Brack, C., 1981, DNA electron microscopy, *CRC Crit. Rev. Biochem.* **10**(2):113–169.

Chan, H. W., Garon, C. F., Chang, E. H., Lowy, D. R., Hager, G. L., Scolnick, E. M., Repaske, R., and Martin, M. A., 1980, Molecular cloning of the Harvey Sarcoma Virus circular DNA intermediates. II. Further Structural Analyses, *J. Virol.* **33**:845–855.

Davis, R. W., and Davidson, N., 1968, Electronmicroscopic visualization of deletion mutations, *Proc. Natl. Acad. Sci. USA* **60**:243–250.

Davis, R. W., and Hyman, R. W., 1971, A study in evolution: The DNA base sequence homology between coliphages T7 and T3, *J. Mol. Biol.* **62**:287–301.

Davis, R. W., Simon, M., and Davidson, N., 1971, Electron microscope heteroduplex methods for mapping regions of base sequence homology in nucleic acids, in: *Methods in Enzymology*, Vol. 21 (L. Grossman and K. Moldave, eds.), pp. 413–428, Academic Press, New York.

Garon, C. F., 1981, Electron microscopy of nucleic acids, in: *Gene Amplification and Analysis*, Vol. 2 (J. G. Chirikjian and T. S. Papas, eds.), pp. 573–589, Elsevier/North-Holland, Amsterdam.

Glass, J., and Wertz, G. W., 1980, Different base per unit length ratios exist in single-stranded RNA and single-stranded DNA, *Nucleic Acids Res.* **8**:5739–5751.

Kleinschmidt, A. K., and Zahn, R. K., 1959, Über desoxyribonucleinsauremolekeln in protein-mischfilmen, *Z. Naturforsch. Teil B Anorg. Chem. Org. Chem. Biochem. Biophys. Biol.* **14b**:770–779.

Leberman, R., 1965, Use of uranyl formate as a negative stain, *J. Mol. Biol.* **13**:606.

Lipkin, L., Lemkin, P., Shapiro, B., and Sklansky, J., 1979, Preprocessing of electron micrographs of nucleic acid molecules for automatic analysis by computer, *Comput. Biomed. Res.* **12**:279–289.

McConaughy, B. L., Laird, C. D., and McCarthy, B. J., 1969, Nucleic acid reassociation in formamide, *Biochemistry* **8**:3289–3295.

Sebring, E. D., Garon, C. F., and Salzman, N. P., 1974, Superhelix density of replicating SV40 DNA molecules, *J. Mol. Biol.* **90**:371–379.

Thomas, M., White, R. L., and Davis, R. W., 1976, Hybridization of RNA to double-stranded DNA: Formation of R-loops, *Proc. Natl. Acad. Sci. USA* **73**:2294–2298.

Westmoreland, B. C., Szybalski, W., and Ris, H., 1969, Mapping of deletions and substitutions in heteroduplex DNA molecules of bacteriophage lambda by electron microscopy, *Science* **163**:1343–1348.

Wetmer, J. G., and Davidson, N., 1968, Kinetics of renaturation of DNA, *J. Mol. Biol.* **31**:349–370.

White, R. L., and Hogness, D. S., 1977, R-loop mapping of the 18S and 28S sequences in the long and short repeating units of *Drosophila melanogaster* DNA, *Cell* **10**:177–192.

Younghusband, H. B., and Inman, R. B., 1974, The electronmicroscopy of DNA, *Annu. Rev. Biochem.* **43**:605–619.

Chapter 7

Freeze-Substitution of Fungi

Harvey C. Hoch

Plant Pathology Department
New York State Agricultural Experiment Station
Cornell University
Geneva, New York 14456

1. INTRODUCTION

To critical cytologists, well-preserved cells are of the utmost importance. It is difficult, if not impossible, to accurately interpret cell structure and relate function to cell components if the cells have not been adequately preserved. For mycologists or cell biologists interested in fungal ultrastructure, cell preservation achieved with chemical fixatives at room temperatures has served as the basis for nearly all structural interpretations to date. Some fungi, particularly members of the Phycomycetes, appear to be preserved well enough with conventional fixation protocols that serious fixation artifacts are not usually recognized. Most members of the Ascomycotina and Basidiomycotina (and Fungi Imperfecti), in contrast, are not well fixed. The cytoplasm of these fungi often appears so muddled that one cannot easily distinguish between mitochondria, Golgi, and endoplasmic reticulum (cf. Brushaber and Jenkins, 1971; Hammill, 1974; Hoch, 1977a; Collinge and Markham, 1982; O'Donnell and McLaughlin, 1984). Unfortunately, mycological cytologists have become accustomed to this quality of preservation and too often either do not recognize the gross cytoplasmic distortions or tend to ignore such artifactual problems and interpret cell structure as best they can.

2. FIXATION METHODS USED FOR FUNGI

Mycologists generally have used cell fixation protocols developed first by mammalian cell cytologists. Fixation by OsO_4, either in vapor form or in aqueous solutions, was one of the first methods recognized by electron microscopists as being effective in preserving cytoplasmic details not previously retained. Subsequently, Luft (1956) introduced potassium permanganate as a fixative that contrasted membranes very well, although it destroyed many other cell components. Mollenhauer (1959) later reported the usefulness of $KMnO_4$ in preserving plant cells. Many studies of fungal ultrastructure soon followed where $KMnO_4$ was employed (cf. Girbardt, 1961; Moore and McAlear, 1962; Bracker, 1968). In 1963, Sabatini and co-workers introduced the aldehydes as superior fixatives for animal cells. These chemicals were soon incorporated into the fixation schedules for plant and fungal cells. Nearly a decade passed in which most fungal microscopists used both $KMnO_4$ and aldehyde–OsO_4 fixations, cautiously comparing the results of each (cf. Grove et al., 1970; Hawker and Beckett, 1971). By the early 1970s, very few studies employed $KMnO_4$, giving way to glutaraldehyde–OsO_4 as the fixative of choice. Throughout this period of time, the quality of fixation of fungal cells gradually improved and many fixation artifacts were eliminated. Some artifacts, however, became so entrenched in the literature that they remain unrecognized as such. The lomasome, for example, was observed so ubiquitously in "well-fixed" fungal cells (Bracker, 1967; Grove et al., 1970; Heath and Greenwood, 1970; Brushaber and Jenkins, 1971; Hoch and Mitchell, 1972) (as well as other cell types) that even today it occasionally is treated as a real organelle (Bojović-Cretić and Vujičić, 1980). The toroidal swelling of the dolipore septum of many Homobasidiomycetes is another example. With the use of glutaraldehyde–OsO_4 fixation schedules, the structure has been recognized as being electron-lucent with a diffused fibril-like content (Setliff et al., 1972; Thielke, 1972; Beckett et al., 1974; Khan and Kimbrough, 1979). Careful correlative light and electron microscopic examination has shown that the swelling is probably swollen (enlarged) as a result of the fixation process (Hoch and Howard, 1981). Many fungal cytologists have recognized that certain cell structures so abundant in other organisms were not easily observed in fungal cells. F-actin, for example, has been documented in only a few fungal preparations, although it undoubtedly is ubiquitously present. The cytoplasmic organization of the hyphal tip, which has been the subject of intense study for decades, when examined by electron microscopy appears to be missing structural components responsible for its organization and maintenance. These examples of artifacts resulting from chemical fixations illustrate the need for improvement in the quality of cell fixation beyond that attainable with conventional glutaraldehyde–OsO_4.

Freeze-substitution provides an alternative to chemical fixation. Publication of the report by Howard and Aist (1979) using freeze-substitution protocols to improve fungal ultrastructure in *Fusarium acuminatum* was chiefly responsible for the reintroduction of this fixation method for fungal material. It was first used in fungal studies by Zalokar (1959, 1966) who wished to determine more precisely the localization of enymze activities and organelles in *Neurospora,* and later by Hereward (1976) who studied *Lipomyces.* As a cytological technique, freeze-substitution has been employed in studies involving mammalian cells much more frequently than for plant cell studies. The primary thrust of this chapter is directed toward providing the reader with a detailed protocol for using freeze-substitution as a means of preserving fungi, although the methods should be equally applicable to other microorganisms. The methods described are those that I have found to be quite satisfactory and easily implemented.

3. FREEZE-SUBSTITUTION

Freeze-substitution is very simple conceptually and equally simple in technique. Briefly, it involves immobilizing the cellular components (and cell function) in microseconds by freezing. In order to have material worthy of examining, the water of the cell must freeze in a vitreous state, or at least with ice crystals smaller than the resolution of the microscope employed. Once frozen, the water in the cell is exchanged (substituted), usually with an organic fluid such as acetone, ethanol, methanol, or tetrahydrofuran. Generally, the substituent also contains a chemical fixative, such as OsO_4, to minimize lipid loss and to further fix the cell's components. The cells are then brought to room temperature and embedded conventionally for sectioning.

The freezing operation is perhaps the most crucial part of the freeze-substitution procedure. To achieve satisfactory freezing of the cytoplasm, a variety of techniques have been developed. They include: (1) spray freezing, whereby the specimens (usually small cells or particles) suspended in liquid droplets are sprayed into a quenching agent (e.g., propane) (Bachmann and Schmitt, 1971; Pfaller and Rovan, 1978); (2) cryojet freezing, which employs a jet of propane sprayed onto the specimen from one or both sides (Moor *et al.,* 1976; Müller *et al.,* 1980; Pscheid *et al.,* 1980; Knoll *et al.,* 1982); (3) quenching of specimens by rapidly plunging them into a quenching agent (e.g., propane, Freon, or liquid N_2 slush) (Feder and Sidman, 1958; Howard and Aist, 1979; Elder *et al.,* 1982; Escaig, 1982; Hoch and Staples, 1983a); and (4) slammers. This last method was advanced by Fernández-Morán (1960) and Van Harreveld and Crowell (1964) and was further perfected by Dempsey and Bullivant (1976), Heuser *et al.* (1979), and Escaig (1982). The procedure essentially involves slamming the specimen against a copper block cooled with liquid helium or with liquid nitrogen (Boyne, 1979; Phillips and Boyne, 1984). Because of the ultralow temperature attainable with liquid helium and of the excellent heat exchange properties of the copper surface, cells theoretically can be frozen satisfactorily to depths greater than those possible with propane or Freon 23. Slammers in general have the disadvantage of not being amenable to processing a large number of samples, primarily because of the time needed to warm the plunger, mount the specimen, and clean the cooling block. The expense and complexity of the equipment are also limiting to many researchers, although some units have been designed for relative simplicity and lower cost (Heath, 1984). Liquid helium is expensive and usually must be obtained in minimum quantities of 30 liters or more. There also has been some concern that the first several micrometers of specimens is often smashed before becoming frozen. This, of course, is more of a disadvantage with single-celled microorganisms than with multicelled tissues. It should be noted that in using a model system of hydrated albumin, Bridgman and Reese (1984) observed that "the best quality of freezing obtained by immersion in propane mixtures was equivalent to the best quality obtained by slam-freezing." Others (Elder *et al.*, 1982) have also come to the same conclusion. Thus, the advantages of slammers may not always outweigh the benefits of quenching agents, particularly where expenses are involved. Cells generally can be frozen without gross ice crystal formation to a depth of 10–30 μm depending on the freezing protocol (Handley *et al.*, 1981). This, in itself, limits one to the outer regions of tissues or large cells. Fortunately for the mycologist, most fungal cells are less than 10 μm in diameter.

Once the cells are frozen, the vitreous ice must be substituted with a suitable solvent. Acetone is the most widely used substituent; however, methanol, ethanol, 2-methoxyethanol, diethyl ether, glycol, and others have been used (cf. Pearse, 1968; Pease, 1973; Mazzone *et al.*, 1979; Ornberg and Reese, 1981a,b). Various chemicals are usually added to the substituents to further stabilize the cellular constituents. Osmium tetroxide (2–5%) is most widely used. Uranyl acetate is added frequently for additional membrane preservation and enhanced contrast. Tannic acid has been used for added contrast, especially when used prior to osmication. It may also help protect microtubules from OsO_4-induced degradation. Other primary fixatives such as glutaraldehyde and acrolein are commonly employed; however, if they are not followed with OsO_4 fixation, poor preservation of lipid bodies and membranes will result.

The temperature at which substitution is carried out is important. The lower the temperature, the slower is the substitution process. Also, the temperature must be maintained above the melting point of the substituent. Acetone, for example, melts at about −95°C, whereas methanol and ethanol melt at −97 and −117°C, respectively. Most commercially available ultralow-temperature freezers operate with a lower temperature limit of −85°C, although models are available with lower attainable temperatures. It is important to recognize that recrystallization of water can occur if the temperature is raised before the cells have been completely substituted (Baud, 1952; Feder and Sidman, 1958; Elder *et al.*, 1982; Franks, 1982; Steinbrecht, 1982; Bridgman and Reese, 1984).

The time required for all water to be substituted depends on the substituent used, the size of the specimen, the substitution temperature, and properties of the cells. In our earlier studies, fungal cells were substituted for 2 to 4 weeks (Howard and Aist, 1979; Hoch and Howard, 1980). More recently, the time has been shortened to less than 2 days, a substitution time used by Zalokar (1966). This would be a good starting time, but workers should empirically determine the appropriate time for their specimens. Barlow and Sleigh (1979) determined that 3 µl of ice dissolved in methanol ($-70°C$) in 2–3 hr, whereas it took 20 hr to dissolve the same volume of frozen water in acetone. Similar results have been observed by others (Feder and Sidman, 1958; Zalokar, 1966). Incomplete substitution will result in poor specimen preservation and would be recognized by a distribution of ice crystals throughout all cells, indicating recrystallization had occurred (Elder *et al.*, 1982; Bridgman and Reese, 1984).

Most often, substitution temperatures are maintained at about $-85°C$ in commercially available ultralow-temperature freezers. Dry ice (solid CO_2) and immersion-type coolers using an ethanol or acetone bath (Heath, 1984) also have been used. Ornberg and Reese (1981a) have used a system whereby the frozen specimens are placed in scintillation vials containing substituent previously frozen with liquid N_2. The vials are kept in a well-insulated container and allowed to gradually warm to room temperature. The temperature is monitored using a dummy vial and thermocouple probe.

Freeze-substituted specimens generally have been embedded in conventional resin media. I routinely use an Epon–Araldite mixture (see Section 4.4) because it allows for very thin embedments (0.5 mm thick or less). Spurr's medium can also be used; however, it tends to be more brittle than Epon–Araldite and frequently cracks during removal from the glass substrates (molds). Spurr's medium also does not allow the sectioned specimen to be stained as readily as Epon–Araldite embedments. Other embedment media, including methacrylates and Vestopal (Pease, 1973), have been employed and, as noted in Chapter 10, low-temperature embedment with the Lowicryl media may provide useful alternatives for preservation of various cellular components. For more detailed information regarding development of freeze-substitution as a technique, additional freezing methodologies, substituents, and adjuvant fixatives, theory of temperature transfer from the specimen during cooling, and so on, it is suggested that the following be consulted as starting references: Pease (1973), Rebhun (1972), Pearse (1980), Dempsey and Bullivant (1976), Nei (1976), Boyne (1979), Howard and Aist (1979), Müller *et al.* (1980), Pscheid *et al.* (1980), Elder *et al.* (1982), Phillips and Boyne (1984).

4. RECOMMENDED PROTOCOL FOR PRESERVING FUNGI USING QUENCHING METHODOLOGIES

Slammers, propane sprayers, and quenching by plunging into eutectic fluids are the methods most commonly used for cryofixing cells. Quenching is probably the most convenient and adaptable method for use with filamentous fungi. It is also the least expensive to operate on a routine basis. Because quenching will likely be the method used by most fungal cell biologists, the following protocol is developed primarily for quenching. The protocol is also one that does not utilize glycerol or DMSO as cryoprotectants to reduce ice damage (Rebhun, 1972) since these agents most likely will have some morphological and/or physiological effect on the cells in question.

4.1. Growth and Preparation of Specimens

Since it is necessary to freeze specimens with a minimum amount of extracellular water, and in as thin a cell layer as possible, growth and support conditions generally need to be modified from the more conventional culture conditions (viz., growth on agar in petri dishes). For most filamentous fungi, nutrient- and water-permeable membranes laid over agar growth media have provided satisfactory supports. The support membranes (e.g., dissected dialysis tubing, cellophane) should be as thin as possible, yet they should be rigid enough to be physically supportive, especially during the manipulation procedures (e.g., plunging) for freezing. DuPont cellophane, type 215 PD-62 (E.I. DuPont de Nemours & Co., Wilmington, DE), has proven to be very satisfactory. It is preferable to cut the membranes into small pieces (ca. 3 × 7 mm) and sterilize them by autoclaving in the same nutrient medium (minus agar) on which the fungal cells will be grown. The membrane pieces are placed aseptically on the agar surface at predetermined distances from the growing colony margin. Once the fungal hyphae have grown onto the membranes (preferably about three-quarters across the membrane), a sharp razor blade or scalpel is used to sever the hyphae around the membranes. Next, for convenience, the membranes are either aligned and organized in a new noncolonized location on the agar surface of the original petri plate or they are transferred to new petri plates containing a similar agar medium. Because even the slightest disturbance can cause a shift and reorganization of cellular components, it is advisable to leave colonized membranes undisturbed in their new position for 15 to 60 min prior to freezing.

Some fungi, especially members of the Zygomycetes, do not adhere well to cellulose-based membranes. Consequently, utmost care must be taken in severing the hyphae and in manipulating the membranes. Alternately, polycarbonate membranes with pores (e.g., Nuclepore filters, Nuclepore Corp., Pleasanton, CA) seem to allow these fungi to become more firmly attached, possibly through partial anchorage into the pores. Fungi of other classes grow equally well on these surfaces; however, there are distinct disadvantages to these membranes. The pores of the membrane render the preparation too refractive for routine preliminary light microscopic observations of the embedments (Howard and Aist, 1979). In addition, the polycarbonate does not section well.

In special situations, aluminum foil (0.018 mm thick) or polyethylene provides an excellent support membrane (Hoch and Staples, 1983a). Fungi (e.g., rusts, powdery mildews) that grow from spores and do not require exogenous nutrients during the time course in which they are studied can be handled quite satisfactorily on these membranes. With the rust for example, uredospores are dusted onto 3 × 7-mm pieces of membrane to a density of 2000–3000 spores/cm². The spore-ladden polyethylene membranes are placed on the surface of 2.5% aqueous agar in petri dishes; the aluminum foil pieces are slightly inserted into the agar at one corner. The membranes may be very lightly, but evenly, atomized with water (or a nutrient solution) or they may be left alone to absorb atmospheric moisture from the closed container. The time for germination, however, will likely be longer without misting. Prior to cutting the aluminum foil and deposition of the spores, a uniform layer of tetrafluorethylene release agent is sprayed onto the reverse (dull) side (e.g., Miller–Stephenson Co., Danbury, CT). This facilitates the subsequent removal of the aluminum pieces following polymerization of the embedding medium. Also, alunimum foil is best cut by pushing a single-edge razor blade into the foil placed on a hard, smooth surface (e.g., glass) rather than using scissors or a slicing action.

As an alternative to placing the spore-ladden membranes on agar surfaces (spore side up), the membranes (polyethylene, aluminum foil, or other hydrophobic membranes) can be placed spore side down onto large drops of water, liquid growth media, or other test solutions. When it is time for the freezing operation, the membranes are grasped by forceps and carefully removed from the liquid, thus allowing nearly all of the water to roll off the hydrophobic membranes. They are then quickly frozen. The question one is attempting to answer about the cell's ultrastructure must be taken into consideration using this approach since the cells obviously will be disturbed briefly prior to the freezing operation.

For cells grown in liquid culture (e.g., yeasts, zoospores, bacteria, algae), as well as for studies involving isolated organelles, use of a wire loop may be best for handling these items during freezing. Lancelle et al. (1985) used a copper wire loop (3-mm diameter) coated with a thin Formvar film to support colonies of the actinomycete Frankia during freezing. Similar loops were used by R. J. Howard (personal communication) to suspend yeast-phase cells of Histoplasma capsulatum and by Barlow and Sleigh (1979) to suspend various protozoans. Depending on the composition of the liquid medium, it may be best to augment it with 0.05% agar or bovine serum albumin to provide an interlacing network of filaments that help retain the cells during substitution and embedding. Generally, the wire loop is dipped into the cell suspension and excess medium removed with a piece of filter paper until a very thin liquid film remains. The Formvar-filmed loops (Lancelle et al., 1985) appear advantageous in that removal of too much medium does not cause the medium film to break. The wire loops with cells are plunged into the cryofluid as decribed below for membranes. The dehydrated (substituted) films of the loops will yield a disk of material to be further processed.

Other specialized specimen holders that sandwich the cells between gold, copper, or titanium disks have been devised for use with plunging or propane jet freezing (Müller et al., 1980; Pscheid et al., 1980; Handley et al., 1981; Escaig,

1982; Knoll *et al.*, 1982). These holders, however, are probably only useful for yeastlike cells, bacteria, and isolated organelles (Wiemken *et al.*, 1979), since the cells invariably must be disturbed when they are placed in the holder.

4.2. Freezing Equipment and Procedures

The basic equipment for quench-freezing is relatively simple and inexpensive. It consists of a Dewar flask, freezing well, stirrer, thermocouple to monitor the temperature, and forceps. The freezing well consists of a copper reservoir, 3.5 cm across by 2.5 cm deep [made from a $1\frac{1}{8}$-inch (2.85 cm) Cu plumbing cap soldered (95/5 solder) to a $1\frac{1}{8}$- to 1-inch (2.54 cm) Cu plumbing reducer] (Figure 1). The freezing well is made so that it slips (via the reducer) snugly over a copper pipe 2.9 cm across by 22 cm long positioned in a 1-liter stainless steel Dewar flask (Aladdin Industries, Inc., Nashville, Tenn.) containing liquid N_2 when in use. Two brass rods, 0.45×2 cm, soldered (95/5 solder) upright on the inside base of the copper freezing well, were so positioned that they retained a stainless steel wire mesh basket to one side (one-half of the well diameter) (Figure 1). The baskets were made from either 0.051, 0.075, or 0.132 (mesh openings, inches) stainless steel wire mesh, properly shaped, heated, and set onto a 2-mm-thick polyethylene sheet. Excess polyethylene was later trimmed away. The other half of the well was used to submerge a copper–constantan thermocouple and a stirring impeller (1-cm diameter) connected to a variable-

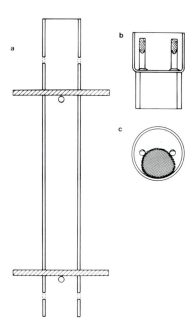

FIGURE 1. Freezing well and support pipe. (a) Copper support pipe is 2.9 cm across and 22 cm long. Four brass rods (hatched area) retain the support pipe in the center of the Dewar when in use. (b) Copper freezing well (3.5 cm across and 2.5 cm deep) is soldered to a copper plumbing reducer so that the unit slips snugly over the support pipe. Two brass rods, soldered to the inside base of the well, retain the sample basket (cross-hatched area of c).

speed motor by a flexible shaft. Temperature was monitored with an Omega trendicator (Model 400 A-TC, Omega Engineering, Inc., Stanford, CT).

Specimens may be frozen in any of several cryogenic fluids such as Freon 12, 22, 23, propane, isopentane, among others (Rebhun, 1972; Nei, 1976; Costello and Corless, 1978; Elder *et al.*, 1982). Their melting points and cooling (heat transfer) capacities vary, but it has generally been shown that propane provided the fastest cooling rates of commonly available hydrocarbons. In the past, propane has been my choice as the quenching fluid, although currently I prefer a mixture of propane and ethane, as discussed below. Strict precautions for the use of these inflammable hydrocarbons should be maintained as discussed in Section 7.

To prepare the apparatus (set up in a fume hood) for freezing, liquid N_2 is added to the Dewar flask containing the freezing well and support pipe. A plastic (polypropylene) lid previously should have been laid over the freezing well to prevent condensation of water and oxygen (see Section 7). Once the equipment has reached a stable temperature following nucleate boiling of the liquid N_2 (Phillips and Boyne, 1984), propane (99.5%; e.g., from Air Products, Tamagua, PA) is condensed in the freezing reservoir (Figure 2). Next, a wire mesh basket is placed in the reservoir, and the propane is stirred vigorously to prevent temperature gradients from forming, and also to help prevent solidification. This procedure generally allows one to supercool the propane to -189 to $-191°C$, several degrees below its melting point of $-186°C$. Alternatively, and preferably, propane is added to the freezing reservoir to a predetermined level (ca. 21 ml) to which ethane (99%) is added to form a propane/ethane mixture of approximately 89/11%. This mixture does not solidify in the liquid N_2 bath and a working temperature of $-194°C$ can be easily maintained. A temperature of $-195°C$ can be attained, depending on the design of the freezing well. Preservation by freezing of specimens in this mixture seems to be equal to or better than that obtained with propane alone. A similar mixture (75/25%) was used by Bridgman and Reese (1984), although the quenching temperature was not reported.

Freezing of the membrane-supported cells is accomplished by very quickly plunging them into the quenching fluid with a pair of curved No. 7 DuMont tweezers (Figures 3 and 4). The membranes are thrust in the quenching fluid so that they enter the fluid slightly off from a vertical insertion (insertion angle about 105°C) with the cell side entering first. It is most convenient to have a series of four or five tweezers lined up for use so that a large number of membranes can be frozen in a short period of time. The first used pair of tweezers will have warmed up before it is rotated into use again. The membranes are plunged into the quenching fluid in the wire mesh baskets, and released. Each basket will easily accommodate 10–30 membranes. For identification of various treatments or of different organisms in the same basket, the membranes can be "marked" by precisely clipping corners or by notching in a keyed pattern. Once a basket is filled with membranes, it is removed, shaken once to quickly drain

FIGURES 2–5. Condensation of cryogenic fluid and freezing operation. (2) Propane, followed by ethane is condensed in the bottom of the precooled freezing well. (3, 4) The membrane-supported cells are plunged very quickly, via rapid wrist action, into the sample basket and released. (5) Once all samples are frozen, the sample basket is transferred to liquid N_2. They are later transferred to the substitution fluid.

excess quenching fluid, and stored temporarily in liquid N_2 (Figure 5) before transferring to the substitution fluid. Once all specimens have been frozen, the removable freezing reservoir is placed in the rear of the hood using a pair of tongs so that inflammable vapors can safely escape. The fume hood door should be positioned for maximum hood efficiency.

4.3. Substitution Equipment and Procedures

Substitution equipment and supplies consist of an ultralow-temperature freezer (viz., capable of maintaining $-85°C$) (e.g., Forma Model 8307, Forma Scientific, Marietta, Ohio), 25 ml of substitution fluid (preferably 1–2% OsO_4 and 0.05% uranyl acetate in anhydrous acetone) contained in 100-ml polypropylene containers with snap-on polyethylene lids (Falcon, Oxnard, Calif.), forceps, and cloth gloves. Procedures for preparation of the substitution fluid vary, but I prefer to dissolve the uranyl acetate, with the help of ultrasonic sound, in Baker-brand HPLC-grade acetone. HPLC acetone is suggested since it usually contains less than 0.09% water. This container and a similar one containing pure OsO_4 only are first cooled to $-85°C$. Then the uranyl acetate–acetone mixture is poured into the OsO_4 container. The OsO_4 dissolves in about 1 hr at $-85°C$. A small pinhole should be placed in the container lid to relieve expansion pressure when the substitution fluid is eventually brought to room temperature.

The frozen specimens are removed from the liquid N_2 (in the ultralow-temperature freezer) and quickly placed into the cooled substitution fluid. Forty-eight hours of substitution is routinely used for a container holding three baskets of up to 20 membranes each. Substitution times at $-85°C$ as short as 4 hr have been used with excellent results, but the time will ultimately depend on the amount of water to be substituted, the nature of the cells, and the sample size. Additional changes of substitution fluid are not generally required. Following substitution at $-85°C$, the containers with specimens are placed at $-20°C$ for 1–2 hr, then at 4°C for 1 hr, and, finally, at room temperature (ca. 21°C) for 30–60 min. The baskets and specimens are next rinsed by transferring them through two 50-ml lots of HPLC acetone, 15 min each. The substitution fluid should have remained clear and yellow. If only a few specimens (e.g., total of 20) have been passed through the substitution fluid, it is returned to the ultralow-temperature freezer for reuse.

4.4. Embedment Schedule and Procedure

The specimens, processed in the same or similar baskets, are then infiltrated with an Epon–Araldite resin mixture (e.g., Araldite 6005, 20 ml; Epon 812, 25 ml; DDSA, 60 ml; DMP-30, ca. 3 drops per 5 ml resin mixture) via 10% increments 1 hr each, through 100% resin. It is important to add the resin catalyst (DMP-30) immediately before each change in the resin series to help prevent cell collapse as discussed in Section 7.

The membranes are next placed cell side down onto a glass plate (5 × 7.5 × 0.35 cm) previously coated with the tetrafluorethylene release agent and buffed to remove excess release agent. A similar glass plate to which aluminum foil-backed Teflon (TF-12, Lamart Corp., Clifton, NJ) was attached, is lowered onto the membrane samples creating a sandwich with a 0.5-mm-thick resin embedment. Following polymerization, the glass plates are snapped apart. The aluminum foil and polyethylene membranes are removed with a razor blade and a pair of fine-tipped tweezers leaving the fungal cells firmly attached to the embedment adhering to the tetrafluorethylene-treated glass. A very thin layer of Epon–Araldite is applied to these areas and polymerized to improve the optical properties of the newly exposed surfaces. The polymerized embedment is finally removed from the tetrafluorethylene-treated glass plate with the help of a razor blade. Polycarbonate and cellophane membranes are left embedded and sectioned along with the cells.

4.5. Selection of Specimens

Generally, one is interested in the ultrastructural aspects of specific cells, cell interactions, or other site-specific phenomena. For these reasons, but also to ensure that the specimens of interest are free of freeze damage (at the light microscopic level), the material is examined with phase-contrast light microscopy. Critical examination can be obtained by applying immersion oil between the resin embedment and supporting glass slide, and also above for 63× or 100× oil immersion objectives. Selected cells are marked and/or photographed, excised from the embedment, and mounted on blank specimen stubs for sectioning.

4.6. Staining of Sectioned Material

Most fungal material fixed by freeze-substitution and embedded in Epon–Araldite stains well with conventional uranyl acetate and lead citrate procedures. Generally, aqueous 1% uranyl acetate is applied at room temperature for 15 min followed with lead citrate, prepared according to Venable and Coggeshall (1965), for 4 min. If excessive lipid extraction has occurred due to the acetone substitution fluid and to poor osmication, then prolonged staining time or elevated staining temperature may be necessary. Alternatively, barium premanganate (Hoch, 1977b) or uranyl acetate mixed with isobutyl alcohol (Howard, 1981) can be used to lend some contrast to poorly staining material.

5. TYPICAL RESULTS AND CRITERIA FOR GOOD PRESERVATION

The perfectly preserved cell is elusive. Killing and preservation inevitably introduce artifacts. Rebhun (1972), recognizing that an artifactless preserved cell is unattainable, considers ". . . the production of interpretable artifacts . . ." to be the goal of the cytologist. In this way, cell structure and function can be more accurately interpreted. It has not been easy to recognize fixation artifacts until a better standard has been developed. Many fixation artifacts induced by $KMnO_4$ or OsO_4 were not readily recognized until aldehyde fixation protocols were introduced. Based on our knowledge of living cell structure obtained from high-resolution light microscopy and from ultrastructural studies with glutaraldehyde/OsO_4-fixed specimens, it is generally recognized that freeze-substitution can provide a state of cell preservation most representative of the living cell at the moment before fixation—at least in terms of organelle shape and position within the cell. Stabilization of cell components via chemicals and substitution fluids following freezing can still result in substantial artifact production. Acetone, for example, can extract lipids from membranes and lipid bodies. Osmium tetroxide is known to cause degradation of microfilaments (Maupin-Szamier and Pollard, 1978), at least in aqueous environments.

Under optimal conditions, most cells on a support membrane will be well fixed; however, since varying degrees of freeze damage occur, it is prudent to examine the cells with light microscopy prior to selection for electron microscopy. Such light microscopic examination of freeze-substituted and embedded cells can be used to discern specimens with gross freeze damage, e.g., ice crystals, cell rupture, breakage, and cell collapse. Examination with $63\times$ or $100\times$ oil immersion objectives can be used to select cells that would have the highest probability of being free of freeze damage due to ice crystal formation. Such damage is discerned as coarse to fine reticulate areas (Figures 6 and 7). Cell rupture, e.g., the outward flow of cytoplasm (Figure 8), has been observed to occur in 1–5% of cells, although the frequency varies with the fungus cell type. The cause of such cell rupture is not known with certainty, but obviously it occurs in the microseconds before the cell is completely frozen. Cell breakage appears as a sharp separation of the cell (i.e., the hypha) (Figure 9) and most likely occurs in the subsequent stages of manipulation for plastic infiltration. Cell collapse is a problem only reported for walled (fungal) cells (Figure 10). It occurs during the plastic infiltration steps and is probably the result of the wall layer not being freely permeable to the resin monomers. Retention of water-soluble polysaccharides or other wall components normally washed away during conventional fixation possibly accounts for the relative wall impermeability. This problem has been overcome, for the most part, by infiltrating the specimens with 10% increments of resin and by adding the catalyst (e.g., DMP-30) just prior to each change to ensure minimal polymerization before infiltration. Use of Spurr's resin does not seem to alleviate this problem.

Freeze-substituted and resin-embedded fungal hyphae provide excellent material for light microscopic study. Most cell organelles are clearly discerned with either phase-contrast or Nomarski interference-contrast optics. In some fungi, e.g., the rusts, organelles such as nuclei and mitochondria are not visible *in vivo* with phase-contrast optics due to the high lipid content and, hence, high light

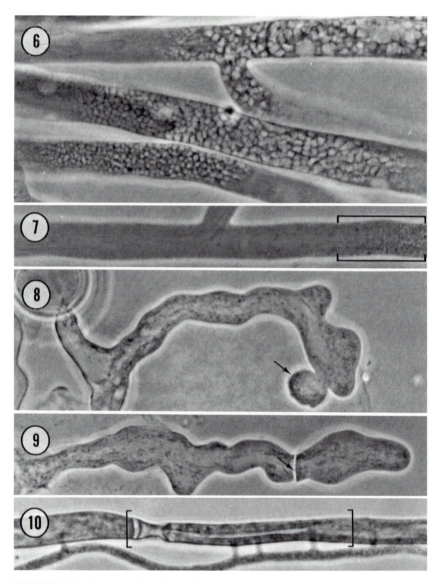

FIGURES 6–10. Examples of specimen damage, observed with light microscopy, that occur as a result of preparing fungi by freeze-substitution protocols. (6, 7) Gross and fine, barely perceivable (within area of brackets) ice crystal damage in hyphae of the basidiomycete *Laetisaria arvalis*. Both ×1600. (8) Ruptured cells have extruded cytoplasm (arrow). *Uromyces phaseoli*. ×1200. (9) Frozen and/or dehydrated cells are subject to breaks (arrow). *U. phaseoli*. ×1200. (10) Cells frequently exhibit a collapsed appearance (bracketed area) if they are not infiltrated with plastic resin mixtures in small increments of increasing concentrations. *Phycomyces blakesleeanus* parasitized by *Pythium acanthium*. ×680.

Table I
Some Features of Fungal Cells Fixed by Freeze-Substitution

	Well preserved	Poorly preserved
Membranes	Smooth, even profile with tripartite (dark–light–dark) layer; plasmalemma at hyphal apex is also smooth in profile with only a few, occasional vesicle-fusion profiles	Undulating or rough profile; granular appearance; broken segments; if not completely osmicated, only cristal membranes of mitochondria, plasmalemma, and apical vesicle membranes are preserved
Nuclear envelope	Smooth profile, membranes evenly spaced (parallel)	Undulating profile, membranes poorly resolved if not well osmicated (appear extracted)
Nucleoplasm	Fine, even granular appearance (appearance may vary depending on condition of chromatin and nucleus)	Frequently appears coarsely granular; most sensitive region to ice crystal damage; thus, it is the region in cell least likely to be well preserved
Lipid droplets	Smooth profile, no granularity, low to high electron density depending on osmication, lipid–cytoplasm interface appears as single-layered membrane	No electron density, indicating extraction or lack of osmication
Microtubules	Straight or gently curving smooth profile, width uniform; protofilaments and microtubule-associated side filaments may be observed	Collapsed; appearing as intermediate filaments; rough profile, granular
Vacuoles	Contents vary, but usually fine to moderately granular, with an evenly distributed content of moderate electron density; profile smooth	May appear with ice crystals of various size, contents coarsely granular
Vesicles (e.g., apical vesicles)	Spherical in shape, smooth profile, contents of very fine granularity with low to high electron density	Coarsely granular, rough surface profile
Mitochondria	Smooth organelle and membrane profile; evenly spaced outer and inner membranes; cristae easily discernible, even in higher fungi; matrix finely granular and moderately electron dense	Organelle profile rough, a "clear" zone may surround mitochondria, perhaps indicative of shrinkage
Endoplasmic reticulum	Easily discerned, even in higher fungi, smooth profile, membranes generally parallel to each other, rough ER with clearly resolved ribosomes	
Microvesicles	30- to 40-nm diameter, frequently polyhedral shaped, in filasomes, a filamentous coating composed of 5- to 8-nm-wide filaments	Coarsely granular appearance
Ground cytoplasm	Very finely granular and filamentous; moderate opacity, depending on staining	More coarsely granular; if not osmicated completely, appears extracted
Microfilaments (e.g., F-actin)	Clearly preserved with straight profiles; have been observed in nuclei, cytoplasm, and associated with microvesicles (as filasomes); filaments 6–8 nm in width	Poorly preserved and difficult to discern, frequently distorted in profile, only seen when in bundles

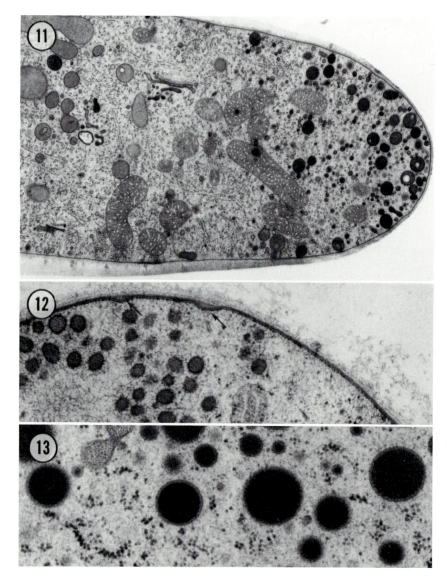

FIGURES 11–13. Ultrastructurally, freeze-substituted hyphal apices exhibit sperical apical vesicles with a smooth profile. The plasmalemma is also smooth in profile with only occasional fusion profiles (arrows, Figure 12) of apical vesicles. (11) *Pythium ultimum.* ×14,600. (12) *Laetisaria arvalis.* From Hoch and Howard (1980). ×52,500. (13) *Phycomyces blakesleeanus.* ×51,700.

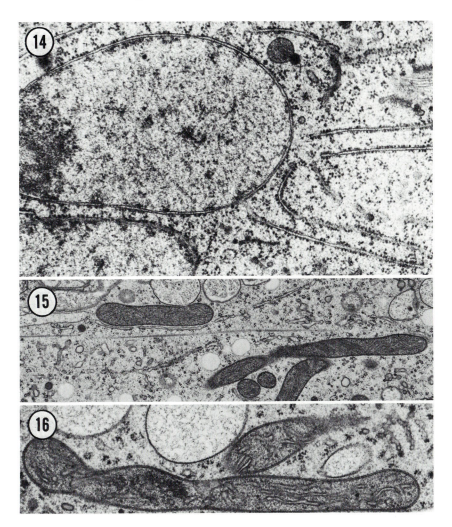

FIGURES 14–16. Nuclei, mitrochondria, and endoplasmic reticulum exhibit smooth membrane profiles. The organelles are rounded with parallel double membrane profiles. (14) *Endothia parasitica.* From Newhouse *et a.*, (1983). ×30,000. (15) *Uromyces phaseoli.* ×20,000. (16) *Rhizoctonia solani.* ×30,000.

refraction. Such cell components are, however, clearly seen in freeze-substituted rust cells where the refractive index of the cell and embedding medium are nearly the same (Figures 7–9).

At the ultrastructural level, well-preserved freeze-substituted hyphal cells have well-organized cell components with smooth profiles (Howard and Aist, 1979; Hoch and Howard, 1980; Hoch and Staples, 1983a). Some criteria for well-preserved cells are summarized in Table I. In general, membrane-bounded organelles, as well as the plasmalemma, are smooth in profile (Figures 11–13). Double membrane-bounded organelles, e.g., nuclei and mitochondria, generally exhibit a parallel and smooth profile of the two membranes (Figures 14–17). Lipid bodies should show some opacity, depending on the amount of extraction, fixation, and/or staining (Figure 15). Vacuoles should be uniformly filled with a slightly granular content (Figures 18–20), although vacuoles in some fungi (e.g., rust fungi) have more dispersed contents.

Ice crystal damage not discerned at the light microscopic level can be much more troublesome at the electron microscopic level of observation. Large crystal damage is obvious, but excluding this damage, how does one determine whether the cell is well preserved or if finer degrees of damage are present? Various sites within the cell may be used to assess the quality of freezing. Aside from the obvious voids created by ice crystals in the cytoplasm (Figure 21), microtubules, nuclei, and mitochondria exhibit varying degrees of sensitivity to freeze damage and can be used to determine the presence of more subtle damage. The cytoplasm, in general, may appear to be well preserved, although microtubules

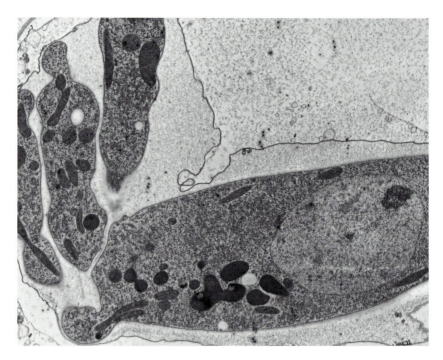

FIGURE 17. Haustorium and immature lobes of *Erysiphe graminis* frozen by "slamming" infected leaf tissue against a copper block. The cytoplasm is well preserved with rounded organelles having a smooth membrane profile. From H. Dahmen and J. Hobot (unpublished). ×900.

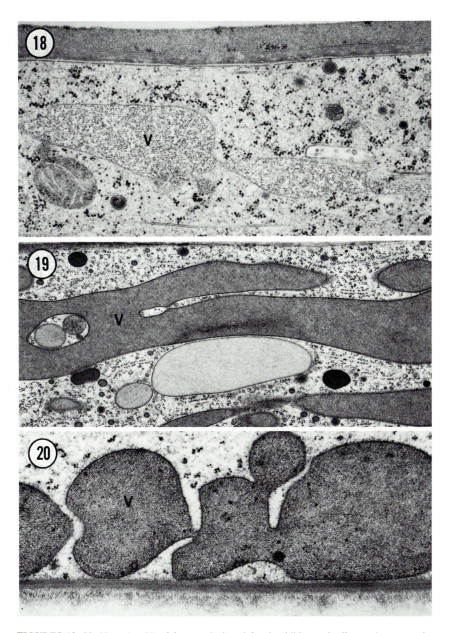

FIGURES 18–20. Vacuoles (V) of freeze-substituted fungi exhibit evenly dispersed contents that frequently appear finely granular, depending on the fungal species. (18) *Phycomyces blakesleeanus.* ×30,000. (19) *Pythium ultimum.* ×20,000. (20) *Laetisaria arvalis.* From Hoch and Howard (1980). ×30,000.

(nuclear and cytoplasmic) can exhibit damage as a collapsed appearance (Figure 22). Mitochondria appear to be slightly more susceptible to damage than microtubules. They show a roughened outline (Figures 21 and 23) as opposed to a very smooth, delimited profile (Figures 11, 15, and 16). The nucleoplasm in many cell types seems to be the most sensitive indicator of freeze damage caused by ice crystals (Zalokar, 1966; Phillips and Boyne, 1984). All other cell components can appear well preserved, yet the nuclear matrix may be poorly preserved—exhibiting a coarsely granular texture (Figure 24). Carmeron et al. (1984), using NMR relaxation times for water, noted that nuclei of various cell types examined had either more water and/or that it was less ordered than in the cytoplasm. They correlated this with the occurrence of larger ice crystals in the nuclei than in the cytoplasm. Such observations coincide with our observations that the nucleoplasm is the last cell region to be well frozen.

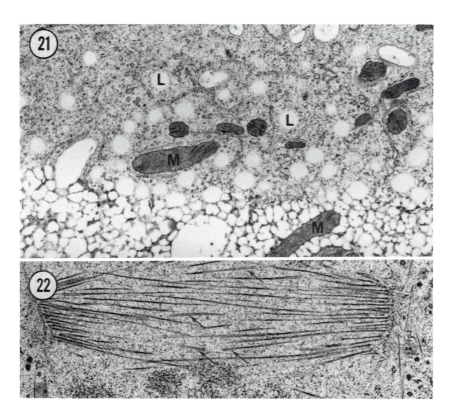

FIGURES 21, 22. Freeze damage at the ultrastructural level appears as clear voids formed by ice crystals [not to be confused with lipid bodies (L)] and as mitochondria (M) with roughened profiles. Microtubules (Figure 22), when damaged, appear collapsed (compared to noncollapsed portions, arrows). Both, *Uromyces phaseoli*. ×20,000 and ×29,100, respectively.

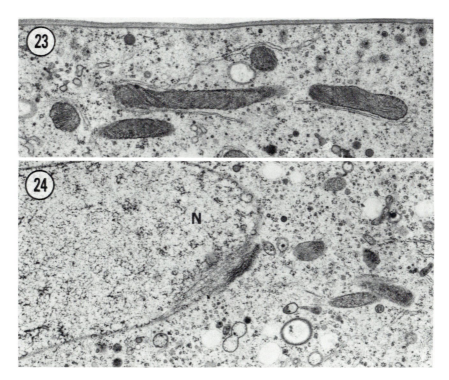

FIGURES 23, 24. The cytoplasm, in general , can be well frozen, but yet mitochondria can show freeze damage as illustrated by the roughened profiles. Nuclei (N) (Figure 24) are the cell component most sensitive to freeze damage, with the nucleoplasm exhibiting a fine-to-coarse appearance. Both, *Uromyces phaseoli.* ×26,900 and ×20,000, respectively.

6. ADVANTAGES OF USING FREEZE-SUBSTITUTION

The foremost goal when using freeze-substitution, as with all cryofixation techniques, is to achieve excellent cell preservation as a result of "freezing" (stopping) all cell activities in microseconds. Such a procedure is not expected to allow postfixation movement of cell components, organelle redistribution, or membrane blebbing as occur in conventionally fixed material processed at room temperature.

For most situations, freeze-substitution does not require close monitoring of many of the chemical and physical parameters associated with the fixation and rinse solvents used in conventional fixations. Composition of the buffer system (i.e., pH, osmolarity, ions) used in conventional fixation protocols generally is of no concern with freeze-substitution. Similarly, undesirable reaction products such as the precipitates that can occur in glutaraldehyde–OsO_4–phosphate-buffered specimens do not occur with freeze-substitution. Because freeze-substituted cells are not exposed to an aqueous environment during the fixation and rinse steps, water-soluble components such as extracellular mucilages and proteins are often better retained. This is especially evident in the extracellular wall region where large or additional "wall layers" are seen (Howard and Aist, 1979; Hoch and Howard, 1980; Hoch and Staples, 1983a). The outermost region of the extracellular matrix possibly constitutes many polysaccharides or glycoproteins that are normally rinsed away during conventional fixation procedures. Similarly, vacuolar contents, many of which are water soluble, are retained during freeze-substitution (Wiemken *et al.*, 1979; Hoch and Howard, 1980). Vacuoles of conventionally fixed cells usually appear empty or mostly empty with clumped, electron-opaque material (cf. Grove *et al.*, 1970; Hoch, 1977a; Hoch and Howard, 1980).

Several structures or profiles commonly observed in cells preserved by conventional fixation that are most likely artifacts, are not present in material preserved by freeze-substitution. Lomasomes and mesosomes in actively growing fungi and bacteria are not present following cryogenic preservation (Ebersold *et al.*, 1981; Hoch and Staples, 1983a). Likewise, membrane "blisters" and blebbing are not seen in frozen cells (Bretscher and Whytock, 1977; Stolinski *et al.*, 1978; Hay and Hasty, 1979; Pfenninger, 1979; Hoch and Howard, 1980). The extreme apical tips of hyphae do not show extensive "vesicle fusion profiles" when preserved by freeze-substitution (Figures 11 and 12). Some organelles of certain fungi are preserved with improved clarity following freeze-substitution. Mitochondria, Golgi bodies (Figures 25–27), and endoplasmic reticulum are especially well preserved in ascomycetous and basidiomycetous fungi. The organization and shape of cell structures such as F-actin, as well as organelle profiles, are much better preserved with freeze-substitution. Certain actinlike filaments are straighter following freeze-substitution (Figures 28–30) as compared to those following conventional procedures (Hoch and Howard, 1980; Hoch and Staples, 1983a,b).

Microfilaments, presumably F-actin, have been observed for the first time surrounding microvesicles (filasomes) in fungi fixed using freeze-substitution procedures (Figures 23, 27, and 28) (Hoch and Howard, 1980; Hoch and Staples, 1983a; Howard, 1981). The size of cells and cell components following freeze-substitution is probably more comparable to the *in vivo* state than following conventional fixation. Howard and Aist (1979) and Hoch and Howard (1980) noted clear differences in the size of apical vesicles, microvesicles, and microtubules fixed by the two procedures. The toroidal swelling of dolipore septa of the basidiomycetous fungi, *Laetisaria arvalis* and *Rhizoctonia solani*, were about 60% larger in specimens fixed by conventional protocols as compared to those preserved by freeze-substitution (Hoch and Howard, 1981). Using light microscopy, similar increases in the size of the toroidal swelling were seen in cells during the course of glutaraldehyde fixation. The implication was that conventional fixation procedures caused a size increase. It is noted, however,

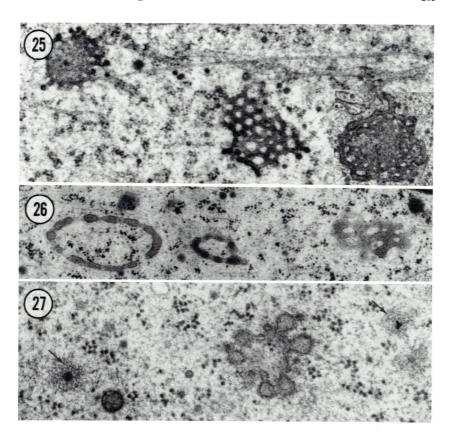

FIGURES 25–27. Golgi bodies of fungi are well preserved by freeze-substitution protocols. Golgi bodies of members of Ascomycotina and Basidiomycotina are notorious for being poorly preserved with conventional procedures. Filasomes (arrows, Figure 27), microvesicles surrounded by actin, are likewise well preserved using freezing methodologies. (25) *Fusarium solani.* From J. R. Aist (unpublished). ×40,600. (25 inset) *Endothia parasitica.* ×48,000. (26) *Phycomyces blakesleeanus.* ×30,000. (27) *Laetisaria arvalis.* From Hoch and Howard (1980). ×60,000.

that substitution in some fluids may cause a shrinkage of certain cell types (Barlow and Sleigh, 1979).

7. DISADVANTAGES OF USING FREEZE-SUBSTITUTION

The most obvious problem of employing freeze-substitution for cell preservation is the occurrence of freeze damage to the cells. The degree of ice crystal formation depends to a large extent on the size of the specimen and the osmolarity of the cytoplasm. Different fungal genera vary in their susceptibility to ice crystal formation. Oomycetes (e.g., *Pythium, Saprolengia*) can be frozen quite successfully (Figures 11 and 31) (Heath, 1984; Heath *et al.*, 1984); however, considerably more ice crystal damage has been noted when frozen under the same conditions as those used to preserve satisfactorily various Ascomycotina (e.g., *Endothia, Fusarium, Neurospora*), Basidiomycotina (e.g., *Uromyces, Puccinia, Laetisaria, Rhizoctonia*), and Actinomycetes (e.g., *Frankia*) (Aist, unpublished; Hoch, unpublished; Howard and Aist, 1979, 1980; Hoch and Howard, 1980; Howard, 1981; Hoch and Staples, 1983a; Newhouse *et al.*, 1983; Lancelle *et al.*, 1985). The Zygomycetes, *Phycomyces, Mucor,* and *Gilbertella,* also froze without substantial ice crystal damage.

Freeze-substitution is most successful when applied to single cell layers (e.g., fungi grown on membranes or cells held in thin aqueous films) and to cells

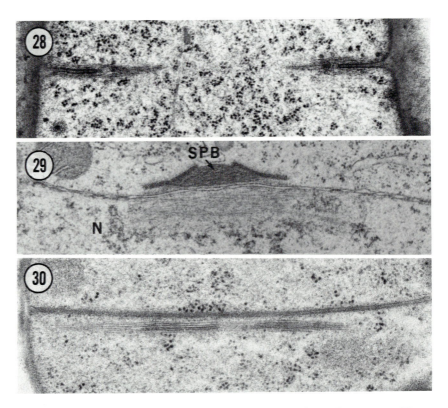

FIGURES 28–30. Microfilaments (mostly F-actin) are preserved in fungi adjacent to septa (Figures 28 and 30), associated with filasomes (Figure 28), and in the nucleoplasm subjacent the spindle pole body (SPB). N, nucleus. (28) *Laetisaria arvalis.* From Hoch and Howard (1980). ×36,800. (29) *Uromyces phaseoli.* From Hoch and Staples (1983a). ×40,000. (30) *Frankia* sp. From Lancelle *et al.* (1985). ×60,000.

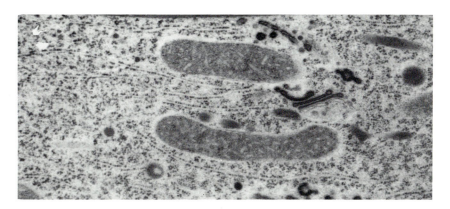

FIGURE 31. Well-frozen cells frequently exhibit poorly preserved organelle membranes such as those of the endoplasmic reticulum, outer mitochondrial membrane, and nuclear envelope (not shown) if adequate osmication has not been achieved. *Pythium ultimum.* ×30,000.

10 μm or less in thickness. Adequate freezing rates for ultrastructural studies of multicelled layers, such as hymenial layers, ascocarps, or mycelial strands (rhizomorphs), cannot be easily achieved with current methods employing cryogenic fluids and quenching techniques because these specimens are too thick.

Loss of lipids, particularly from membranes, can occur if insufficient osmication has occurred before the material is warmed to room temperature (Figure 31). Van Harreveld *et al.* (1965) indicated that osmication does not occur significantly at −80°C. Bridgman and Reese (1984) believe that the lowest temperature at which membranes show signs of being fixed is between −40°C and −50°C. These authors also noted that more polar substituents such as methanol and ethanol cause more loss of membranes (and extraction of cytoplasm) than do acetone or tetrahydrofuran. Small cytoplasmic voids, appearing as an obvious absence of organelles, especially a lack of ribosomes, has been noted with alcohol substituents (J. R. Aist, personal communication).

Collapse of freeze-substituted walled hyphal cells can occur when they are infiltrated with the embedding medium (Figure 10). This artifact presumably occurs because much more extracellular material is conserved and does not allow ready entry of the unpolymerized resin into the cell. Monitoring of hyphae, using light microscopy, indicated that cell collapse occurred above the 60% resin level.

Freezing of the cells with hydrocarbons presents certain dangers not normally encountered with conventional fixation protocols. Propane, ethane, and certain other hydrocarbons used as the quenching agent are inflammable and require special handling precautions. First, they should be used only in a well-ventilated fume hood. Second, it is theoretically possible for propane, ethane, and the others to interact with condensed oxygen and explode, even at the low temperatures used during freezing. For this reason, many reseachers do not work with propane below about $-180°C$ since oxygen condenses at about $-184°C$. While we routinely freeze our specimens at $-193°C$ or lower, the freezing well is kept covered with a lid to minimize condensation of oxygen until propane/ethane is added. Furthermore, the hydrocarbon quenching agent is not used for more than about 10 min at which time the freezing well is removed from the Dewar and placed in the rear of the hood for evaporation. Several papers alluding to the precautionary use of these hydrocarbons are available and *must* be consulted (cf. Stephenson, 1954; Feder and Sidman, 1958; Rebhun, 1972; Elder *et al.*, 1982). Fluorohydrocarbons (e.g., Freon 22, 23) are not inflammable and can be substituted for propane. However, they do not yield as good freezing rates and therefore cell preservation as propane, but they may be adequate in many instances.

The equipment required for propane-quenched, freeze-substituted material is not particularly expensive, although it may be beyond the means of some researchers. Ultralow-temperature freezers (2 m^3) sell for about $3000. The other necessary equipment costs between $5 and $400, depending on whether or not they are custom-made. Alternatively, a very inexpensive "freezing Dewar" can be constructed by placing a pipette canister into a block of Styrofoam. A solid CO_2 (dry ice) chamber (ca. $-79°C$) can be used in lieu of the ultralow-temperature freezer for only a few dollars.

8. CELL SYSTEMS TO BE STUDIED USING FREEZE-SUBSTITUTION

The types of problems best approached with freeze-substitution are those involving small cells and organisms that can be cultured as single cell layers. Because most cells can be frozen without apparent freeze damage only 10–15 μm deep, cell size presents some limitation. Tissues can be frozen, particularly with "slammers," but still the useful zone, free of freeze damage, is limited. However, I am impressed with the quality of fixation and lack of ice damage. Studies that involve elucidation of spatial organization of cell components are well adapted to freezing methodologies. Certain cell components, such as F-actin in fungi, are not well preserved, if at all, by conventional fixation methods. Freeze-substitution has been shown to enhance preservation of those filaments. Retention of water-soluble components such as carbohydrates, ions, and proteins by freeze-substitution makes these studies amenable to quantitative examination. Consequently, better localization of cell components via cytochemistry and ion probe (e.g., Chandra *et al.*, 1984) analysis can be achieved. Obviously, very rapid cell movements (i.e., flagellar movements, fusion profiles of vesicles) in cells of *Venturia inaequalis* and *Erysiphe graminis* in apple and wheat leaves, respectively (Figure 17; Dahmen and Hobbot, unpublished), frozen with the slammer described by Escaig (1982), also can be more accurately studied.

I have presented considerable detail regarding the methods that I use to freeze-substitute fungi. Hopefully this will not discourage its use. As with any new laboratory technique, the first time through may be overwhelming. To the contrary, the procedure is simple. Freeze-substitution will not likely replace conventional protocols as a means of fixing all fungal specimens; however, because it usually provides greatly improved cell preservation, it has become the method of choice for many specimens.

ACKNOWLEDGMENTS. I am pleased to acknowledge the helpful discussions and advice of Drs. J. R. Aist, I. B. Heath, and R. J. Howard during the developmental phases of freeze-substitution protocols for fungi. Dr. Howard, especially, is acknowledged for his role in reintroducing freeze-substitution to the field of mycology. I am also very grateful to Mr. T. M. Bourett and Dr. B. E. Tucker for their help in preparing and reviewing the manuscript. Some of the results reported herein were made possible, in part, by grants from the National Science Foundation (DCB-8315713) and the Whitehall Foundation.

9. REFERENCES

Bachmann, L., and Schmitt, W. W., 1971, Improved cryofixation applicable to freeze-etching, *Proc. Natl. Acad. Sci. USA* **68**:2149–2152.

Barlow, D. I., and Sleigh, M. A., 1979, Freeze-substitution for preservation of ciliated surfaces for scanning electron microscopy, *J. Microsc. (Oxford)* **115**:81–95.

Baud, C. -A., 1952, La fixation "par substitution," *Bull. Microsc. Appl.* Ser. 2 **2**:158–160.

Beckett, A., Heath, I. B., and McLaughlin, D. J., 1974, *An Atlas of Fungal Ultrastructure,* Longman, London.

Bojović-Cvetić, D., and Vujičić, R., 1980, Membranous aggregates in hyphal tips of *Aspergillus flavus,* Arch. Microbiol. **126**:245–249.

Boyne, A. F., 1979, A gentle, bounce-free assembly for quick-freezing tissues for electron microscopy: Application to isolated torpedine ray electrocyte stacks, *J. Neurosci. Methods* **1**:353–364.

Bracker, C. E., 1967, Ultrastructure of fungi, *Annu. Rev. Phytopathol.* **5**:343–374.

Bracker, C. E., 1968, The ultrastructure and development of sporangia in *Gilbertella persicaria, Mycologia* **60**:1016–1067.

Bretscher, M. D., and Whytock, S., 1977, Membrane-associated vesicles in fibroblasts, *J. Ultrastruct. Res.* **61**:215–217.

Bridgman, P. C., and Reese, T. C., 1984, The structure of cytoplasm in directly frozen cultured cells. I. Filamentous meshworks and the cytoplasmic ground substance, *J. Cell Biol.* **99**:1655–1668.

Brushaber, J. A., and Jenkins, S. F., Jr., 1971, Lomasomes and vesicles in *Poria monticola, Can. J. Bot.* **49**:2075–2084.

Cameron, I. L., Hunter, K. E., Ord, V. A., and Fullerton, G. D., 1984, Ordered water in cells: Ice crystal size as an indicator of water content and water proton NMR relaxation times. *J. Cell Biol.* **99**:33a.

Chandra, S., Harris, W. C., Jr., and Morrison, G. H., 1984, Distribution of calcium during interphase and mitosis as observed by ion microscopy, *J. Histochem. Cytochem.* **32:**1224–1230.

Collinge, A. J., and Markham, P., 1982, Hyphal tip ultrastructure of *Aspergillus nidulans* and *Aspergillus giganteus* and possible implications of Woronin bodies close to the hyphal apex of the latter species, *Protoplasma* **113:**209–213.

Costello, M. J., and Corless, J. M., 1978, The direct measurement of temperature changes within freeze-fracture specimens during rapid quenching in liquid coolants, *J. Microsc. (Oxford)* **112:**17–37.

Dempsey, G. P., and Bullivant, S., 1976, A copper block method for freezing non-cryoprotected tissues to produce ice-crystal-free regions for electron microscopy, *J. Microsc. (Oxford)* **106:**251–270.

Ebersold, H. R., Cordier, J., and Luthy, P., 1981, Bacterial mesosomes: Method dependent artifacts, *Arch. Microbiol.* **130:**19–22.

Elder, H. Y., Gray, C. C., Jardine, A. G., Chapman, J. N., and Biddlecombe, W. H., 1982, Optimum conditions for cryoquenching of small tissue blocks in liquid coolants, *J. Microsc. (Oxford)* **126:**45–61.

Escaig, J., 1982, New instruments which facilitate rapid freezing at 83K and 6K, *J. Microsc. (Oxford)* **126:**221–229.

Feder, N., and Sidman, R. L., 1958, Methods and principles of fixation by freeze-substitution, *J. Biophys. Biochem. Cytol.* **4:**593–600.

Fernandez-Morin, H., 1960, Low-temperature preparation techniques for electron microscopy of biological specimens based on rapid freezing with liquid helium. II. *Ann. N.Y. Acad. Sci.* **85:**689–713.

Franks, F., 1982, The properties of aqueous solutions at subzero temperatures, in: *Water: A Comprehensive Treatise*, Vol. 7 (F. Franks, ed.), pp. 215–334, Plenum Press, New York.

Girbardt, M., 1961, Licht-und electronenmikroskopische untersuchungen an *Polystictus versicolor.* II. Die Feinstruktur von Grundplasma und Mitochondrien, *Arch. Mikrobiol.* **39:**351–359.

Grove, S. N., Bracker, C. E., and Moore, D. J., 1970, An ultrastructural basis for hyphal tip growth in *Pythium ultimum, Am. J. Bot.* **57:**245–266.

Hammill, T. M., 1974, Electron microscopy of phialides and conidiogenesis in *Trichoderma saturnisporum, Am. J. Bot.* **61:**15–24.

Handley, D. A., Alexander, J. T., and Chien, S., 1981, The design and use of a simple device for rapid quench-freezing of biological samples, *J. Microsc. (Oxford)* **121:**273–282.

Hawker, L. E., and Beckett, A., 1971, Fine structure and development of the zygospore of *Rhizopus sexualis* (Smith) Callen, *Philos. Trans. R. Soc. London Ser. B* **263:**71–100.

Hay, E. D., and Hasty, D. L., 1979, Extrusion of particle-free membrane blisters during glutaraldehyde fixation, in: *Freeze Fracture: Methods, Artifacts, and Interpretations* (J. E. Rash and C. S. Hudson, eds.), pp. 59–66, Raven Press, New York.

Heath, I. B., 1984, A simple and inexpensive liquid helium cooled 'slam freezing' device, *J. Microsc. (Oxford)* **135:**75–82.

Heath, I. B., and Greenwood, A. D., 1970, The structure and formation of lomasomes, *J. Gen. Microbiol.* **62:**129–137.

Heath, I. B., Rethoret, K., and Moens, P. B., 1984, The ultrastructure of mitotic spindles from conventionally fixed and freeze-substituted nuclei of the fungus *Saprolegnia, Eur. J. Cell Biol.* **35:**284–295.

Hereward, F. V., 1976, Nuclear membrane breakdown at mitosis in *Lipomyces lipofer* shown by freeze-substitution, *Protoplasma* **87:**113–119.

Heuser, J. E., Reese, T. S., Dennis, M. J., Jan, Y., Jan, L., and Evans, L., 1979, Synaptic vesicle exocytosis captured by quick freezing and correlated with quantal transmitter release, *J. Cell Biol.* **81:**275–300.

Hoch, H. C., 1977a, Mycoparasitic relationships. III. Parasitism of *Physalospora obtusa* by *Calcarisporium parasiticum, Can. J. Bot.* **55**:198–207.

Hoch, H. C., 1977b, Use of permanganate to increase the electron opacity of fungal walls, *Mycologia* **69**:1209–1213.

Hoch, H. C., and Howard, R. J., 1980, Ultrastructure of freeze-substituted hyphae of the basidiomycete *Laetisaria arvalis, Protoplasma* **103**:281–297.

Hoch, H. C., and Howard, R. J., 1981, Conventional chemical fixations induce artifactual swelling of dolipore septa, *J. Exp. Mycol.* **5**:167–172.

Hoch, H. C., and Mitchell, J. E., 1972, The ultrastructure of *Aphanomyces euteiches* during asexual spore formation, *Phytopathology* **62**:149–160.

Hoch, H. C., and Staples, R. C., 1983a, Ultrastructural organization of the nondifferentiated uredospore germling of *Uromyces phaseoli* variety *typica, Mycologia* **75**:795–824.

Hoch, H. C., and Staples, R. C., 1983b, Visualization of actin *in situ* by rhodamine-conjugated phalloin in the fungus *Uromyces phaseoli, Eur. J. Cell Biol.* **32**:52–58.

Howard, R. J., 1981, Ultrastructural analysis of hyphal tip cell growth in fungi: Spitzenkorper, cytoskeleton and endomembranes after freeze-substitution, *J. Cell Sci.* **48**:89–103.

Howard, R. J., and Aist, J. R., 1979, Hyphal tip cell ultrastructure of the fungus *Fusarium:* Improved preservation by freeze-substitution, *J. Ultrastruct. Res.* **66**:224–234.

Howard, R. J., and Aist, J. R., 1980, Cytoplasmic microtubules and fungal morphogenesis: Ultrastructural effects of methyl benzimidazole-2-ylcarbamate determined by freeze-substitution of hyphal tip cells, *J. Cell Biol.* **87**:55–64.

Khan, S. R., and Kimbrough, J. W., 1979, Ultrastructure of septal pore apparatus in the lamellae of *Nematoloma puiggarii, Can. J. Bot.* **57**:2064–2070.

Knoll, G., Oebel, G., and Plattner, H., 1982, A simple sandwich-cryogen-jet procedure with high cooling rates for cryofixation of biological materials in the native state, *Protoplasma* **111**:161–176.

Lancelle, S. A., Torrey, J. G., Hepler, P. K., and Callaham, D. A., 1985, Ultrastructure of freeze-substituted *Frankia* strain HFPCc13, the actinomycete isolated from root nodules of *Casuarina cunninghamiana, Protoplasma* **127**:64–72.

Luft, J. H., 1956, Permanganate—A new fixative for electron microscopy, *J. Biophys. Biochem. Cytol.* **2**:799–802.

Maupin-Szamier, P., and Pollard, T. D., 1978, Actin filament destruction by osmium tetroxide, *J. Cell Biol.* **77**:837–852.

Mazzone, R. W., Durand, C. M., and West, J. B., 1979, Electron microscopic appearances of rapidly frozen lung, *J. Microsc. (Oxford)* **117**:269–284.

Mollenhauer, H. H., 1959, Permanganate fixation of plant cells, *J. Biophys. Biochem. Cytol.* **6**:431–436.

Moor, H., Kastler, J., and Müller, M., 1976, Freezing in a propane jet, *Experientia* **32**:805.

Moore, R. T., and McAlear, J. H., 1962, Fine structure of mycota. 7. Observations on septa of Ascomycetes and Basidiomycetes, *Am. J. Bot.* **49**:86–94.

Müller, M., Meister, N., and Moor, H., 1980, Freezing in a propane jet and its application in freeze-fracturing, *Mikroskopie* **36**:129–140.

Nei, T., 1976, Review of the freezing techniques and their theories, in: *Recent Progress in Electron Microscopy of Cells and Tissues* (E. Yamada, V. Mizuhira, K. Kurosumi, and T. Nagano, eds.), pp. 213–243, University Park Press, Baltimore.

Newhouse, J. R., Hoch, H. C., and MacDonald, W. L., 1983, The ultrastructure of *Endothia parasitica:* Comparison of virulent with a hypovirulent isolate, *Can. J. Bot.* **61**:389–399.

O'Donnell, K. L., and McLaughlin, D. J., 1984, Postmeiotic mitosis, basidiospore development, and septation in *Ustilago maydis, Mycologia* **76**:486–502.

Ornberg, R. L., and Reese, T. S., 1981a, Beginning of exocytosis captured by rapid-freezing of *Limulus amebocytes, J. Cell Biol.* **90**:40–54.

Ornberg, R. L., and Reese, T. S., 1981b, Localization of calcium in quick frozen cells by freeze-substitution in tetrahydrofuran and low temperature embedding, *J. Cell Biol.* **91**:387(a).

Pearse, A. G. E., 1968, *Histochemistry: Theoretical and Applied,* Vol. 1, Churchill Livingstone, Edinburgh.

Pease, D. C., 1973, Substitution techniques, in: *Biological Electron Microscopy* (J. K. Koehler, ed.), pp. 35–66, Springer-Verlag, Berlin.

Pfaller, W., and Rovan, E., 1978, Preparation of resin embedded unicellular organisms without the use of fixatives and dehydration media, *J. Microsc. (Oxford)* **114**:339–351.

Pfenninger, K. H., 1979, Subplasmalemmal vesicle clusters: Real or artifact?, in: *Freeze Fracture: Methods, Artifacts, and Interpretations* (J. E. Rash and C. S. Hudson, eds.), pp. 71–80, Raven Press, New York.

Phillips, T. E., and Boyne, A. F., 1984, Liquid nitrogen-based quick freezing: Experiences with bounce-free delivery of cholinergic nerve terminals to a metal surface, *J. Electron Microsc. Tech.* **1**:9–29.

Pscheid, P., Schudt, C., and Plattner, H., 1980, Cryofixation of monolayer cell cultures for freeze-fracturing without chemical pre-treatments, *J. Microsc. (Oxford)* **121**:149–167.

Rebhun, L. I., 1972, Freeze-substitution and freeze-drying, in: *Principles and Techniques of Electron Microscopy: Biological Applications,* Vol. 2 (M. A. Hayat, ed.), pp. 2–49, Van–Nostrand Reinhold, Princeton, New Jersey.

Sabatini, D. D., Bensch, K., and Barnett, R. J., 1963, Cytochemistry and electron microscopy: The preservation of cellular structure and enzymatic activity by aldehyde fixation, *J. Cell Biol.* **17**:19–58.

Setliff, E. C., MacDonald, W. L., and Patton, R. F., 1972, Fine structure of the septal pore apparatus in *Polyporus tomentosus, Poria latemarginata,* and *Rhizoctonia solani, Can. J. Bot.* **50**:2559–2563.

Steinbrecht, R. A., 1982, Experiments on freezing damage with freeze-substitution using moth antennae as test objects, *J. Microsc. (Oxford)* **125**:187–192.

Stephenson, J. L., 1954, Caution in the use of liquid propane for freezing biological specimens, *Nature (London)* **174**:235.

Stolinski, C., Breathnach, A. S., and Bellairs, R., 1978, Effect of fixation on cell membrane of early embryonic material as observed on freeze-fracture replicas, *J. Microsc. (Oxford)* **112**:293–299.

Terracio, L., and Schwabe, K. G., 1981, Freezing and drying of biological tissues for electron microscopy, *J. Histochem. Cytochem.* **29**:1021–1028.

Thielke, C., 1972, Die dolipore der basidiomyceten, *Arch. Mikrobiol.* **82**:31–37.

Van Harreveld, A., and Crowell, J., 1964, Electron microscopy after rapid freezing on a metal surface and substitution fixation, *Anat. Rec.* **149**:381–386.

Van Harreveld, A., Crowell, J., and Malhotra, S. K., 1965, A study of extracellular space in central nervous tissue by freeze-substitution. *J. Cell Biol.* **25**:117–137.

Venable, J. H., and Coggeshall, R., 1965, A simplified lead citrate stain for use in electron microscopy, *J. Cell Biol.* **25**:407–408.

Wiemken, A., Schellenberg, M., and Urech, K., 1979, Vacuoles: The sole compartments of digestive enzymes in yeast (*Saccharomyces cerevisiae*), *Arch. Microbiol.* **123**:23–35.

Zalokar, M., 1959, Growth and differentiation of *Neurospora* hyphae, *Am. J. Bot.* **46**:602–610.

Zalokar, M., 1966, A simple freeze-substitution method for electron microscopy, *J. Ultrastruct. Res.* **15**:469–479.

Chapter 8

Freeze-Fracture (-Etch) Electron Microscopy

Russell L. Chapman
Department of Botany
Louisiana State University
Baton Rouge, Louisiana 70803

and

L. Andrew Staehelin
Department of Molecular, Cellular, and Developmental Biology
University of Colorado
Boulder, Colorado 80309

1. INTRODUCTION

Freeze-fracture electron microscopy (FEM) is a unique and powerful research technique that provides a triad of major advantages for the microbiologist interested in ultrastructure. First, because FEM is a replicating technique, it can provide both face-on as well as cross-sectional views of cellular components exposed by the fracturing of frozen cells. Thus, the investigator can visualize the morphology and distribution of specific components both within the plane and on both surfaces of membranes. In addition, it is often possible to perceive three-dimensional relationships that would require tedious reconstruction if studied in thin-section transmission electron microscopy. Second, even when chemical pretreatment is used during processing of samples, FEM provides an alternate method of subsequent processing and thus allows comparative analysis of thin-section electron microscopy data and FEM data. Third, when coupled with a purely physical specimen fixation method (i.e., cryofixation), FEM can completely eliminate the need for chemical fixation and is thus extremely valuable for the identification of artifactual structures in electron microscopy (EM) images. Furthermore, because of the rapid rate at which all cellular activities can be halted by cryofixation, FEM can be used to investigate rapid cellular processes that cannot be preserved in chemically fixed samples. From the start, FEM has offered the promise of an ''artifact-free'' glimpse at the ultrastructure of cells. Even if the ultimate goal is not a routine achievement, FEM has been and remains an important research tool in biology in general, and in microbiology in particular. This chapter will provide both a general overview of FEM and more specific coverage of the techniques applicable to microorganisms. It will not reiterate theoretical foundations of FEM covered by the numerous books and reviews available (many of which are cited). It will, however, provide both an introduction to the technique and a summary of some recent developments in the field. Many of the more specific comments relate to operation of the Balzers freeze-etch apparatus (Balzers High Vacuum Corp., Principality of Liechtenstein and Hudson, NH), but most methodological aspects of this chapter are also applicable to other instruments.

2. A BRIEF HISTORY

The research upon which FEM of biological specimens is founded may date back as far as the 1940s (Sjöstrand, 1943) and includes studies such as those of Hall (1950), Merryman (1950), and Merryman and Kafig (1955). However, Steere's study of fractured frozen samples of crystallized tobacco mosaic virus (Steere, 1957; Steere and Schaffer, 1958) is generally cited as the starting point in the successful application of the technique in biology. Replicas of cracked red blood cells produced by Haggis (1961) were another of the early successes. However, it was above all the developmental work of Moor *et al.* (1961) in Switzerland that transformed the technique and its associated instrumentation from a ''prototype'' state to that of a routine EM method. Indeed, the basic design parameters of the original instrument are still retained to a large extent even in the most modern Balzers commercial freeze-etch apparatus. Several simple designs, such as the freeze-fracture device of Bullivant and Ames (1966), were developed in the 1960s, but none of these simple instruments, although sold through commercial outlets, have withstood the test of time. More recent developments of the freeze-etch technology have focused more on improvements in specimen preparation than on major changes in the preparation of the replicas from the frozen samples. However, the introduction of features to reduce the contamination of fracture faces by water and oil vapors should not go unmentioned.

Although the development of these refinements continues (e.g., Müller and Pscheid, 1981), by the 1970s freeze fracture was already a well-established tool among biologists. Controversy about how membranes fractured (i.e., along the surface or through the middle) had been resolved (e.g., Branton, 1966; Branton and Deamer, 1972) and a standardization of nomenclature for the faces and surfaces of fractured membranes had been devised (Branton *et al.*, 1975; see also Section 3.5 for a discussion of how this nomenclature is applied to both prokaryotic and eukaryotic cell membranes). Over the years a variety of artifacts, the bugaboo of any newly developed technique, have been recognized and cataloged (see, e.g., Bullivant, 1977; Sleytr and Robards, 1977, 1982; Böhler, 1979; Rash and Hudson, 1979). Sleytr and Robards (1982) concluded their review of the artifact problem in freeze-fracture replication with the assertion ''that the early hopes that freeze-fracture and etching techniques might provide artefact-free images because chemical fixation can be replaced by (undamaging) cryofixation have not been fulfilled.'' Fortunately, increased understanding of the artifacts has led to the elimination of many of them. Further, even those artifacts that cannot be avoided need not interfere with our understanding of the cell biology of microorganisms as long as the artifacts are recognized as such and appropriate precautions are taken in the interpretation of micrographs. Numerous comprehensive reviews of the general techniques of freeze fracture (e.g., Moor, 1969, 1971; Koehler, 1972; Bullivant, 1973; Southworth *et al.*, 1975; Stolinski and Breathnach, 1975; Sleytr, 1978; Willison and Rowe, 1980; Gilkey and Staehelin, 1986) are available and should be consulted by the reader interested in more detail than can be provided in this chapter.

Microorganisms and viruses are well suited for examination by FEM, and indeed much of the early research was devoted to the ultrastructure of bacteria (e.g., Herrmann and Staehelin, 1965; Remsen, 1966; Remsen and Lundgren, 1966; Giesbrecht, 1968; Lickfeld, 1968; Nanninga, 1968; Holt and Leadbetter, 1969), cyanobacteria (Jost, 1965; Jost and Matile, 1966), fungi (e.g., Matile *et al.*, 1965), algae (Staehelin, 1966; Branton and Southworth, 1967), yeast (Moor and Mühlethaler, 1963), and, of course, tobacco mosaic virus (Steere, 1957; Steere and Schaffer, 1958). Freeze-fracture studies of protozoans apparently began somewhat later (e.g., see Janisch, 1972; Plattner *et al.*, 1973; Satir *et al.*, 1973; Speth and Wunderlich, 1973). Because of their small size, microorganisms and suspension culture cells from higher organisms are also well suited for recently developed techniques of ultrarapid freezing (see Sections 3.2.2b–e) and can be processed without excision or trimming (cf. the processing of tissues), thus allowing the investigator to avoid artifacts that could be induced during such preparation. Exciting new observations on microorganisms will undoubtedly accompany the further development of freeze fracture of unfixed, uncryoprotected, ultrarapidly frozen samples. The goal of an artifact-free glimpse at the ultrastructure of microorganisms may be within reach.

3. METHODS AND RECOMMENDATIONS

3.1. Pretreatment

Optimal preservation of specimen structure requires that samples be frozen under conditions that keep ice crystal size at less than a few hundred angstroms. This requires the use of either cryoprotectants or ultrarapid freezing techniques to reduce the number of ice nuclei that are formed and the rate of growth of these nuclei during freezing. Parameters that are of critical importance for the quality of freezing are freezing point, recrystallization point, specimen size, and rate of heat transfer. The interval between the freezing point and the recrystallization temperature is known as the critical temperature interval because this interval has to be passed as quickly as possible if ice crystal formation, and thus specimen damage, is to be avoided (Figure 1).

The critical temperature interval can be reduced by experimental lowering of the freezing temperature and by increasing the recrystallization temperature. This can be achieved by reducing the free water available for ice crystal formation, either by dehydrating the specimen, by infiltrating it with a water-binding cryoprotectant, or by placing the sample under high pressure. Many microorganisms contain natural cryoprotectants (e.g., hydrophilic carbohydrates), but, with the exception of spores and the like, none escape ice crystal formation entirely when conventional dip-freezing methods are employed. Glycerol was the first cryoprotectant to be used in conjunction with FEM (Moor and Mühlethaler, 1963), and has remained the most commonly used cryoprotectant to date (others include ethylene glycol and DMSO). Nevertheless, glycerol has been shown to cause a variety of major and minor artifacts, such as the swelling of mitochondria, vesiculation of the endoplasmic reticulum, and clumping of intramembrane particles (e.g., see Rash and Hudson, 1979; Gilkey and Staehelin, 1986). To circumvent some of these problems, specimens are often prefixed with

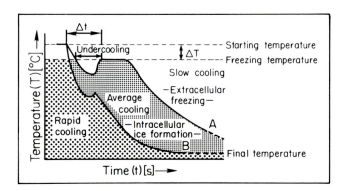

FIGURE 1. Cooling-rate curve. As a liquid cools, its temperature falls below the freezing point (undercooling), until the first crystals form on contaminants ("nuclei," nucleation centers) in it. The released crystallization heat warms the liquid to the freezing temperature. Only then does the temperature go down to the final temperature (curve A). Freezing processes with rates above this curve lead to extracellular freezing due to the slow progress. Freezing processes with rates below this curve progress at an average speed and lead to intracellular ice formation and thus to destruction of the cell structure. Only when more heat is extracted from the sample during cooling than is released by the crystallization heat, is the freezing rate almost linear (curve B) and vitrification takes place. Adapted from Freeze-Etching, a Balzers Report by W. Niedermeyer.

glutaraldehyde (e.g., 1–2% for 15–60 min) before glycerination, but this negates some of the potential advantages of cryofixation and FEM over more conventional thin-section EM specimen preparation methods.

Additional problems arise when deep-etching of specimens is desired to expose macromolecules and true membrane surfaces below the fracture plane; the concentration of solute molecules in the specimen medium must be less than 10 mM. Thus, no glycerol or other nonsubliming cryoprotectant may be used when samples are to be etched. Carefully formulated buffered solutions containing less than 10 mM total solutes should be used for the rinses prior to freezing to reduce possible extraction artifacts caused by distilled water (Staehelin, 1976). Alternatively, specimens may be stabilized by chemical fixatives before being subjected to the water washes.

The second method of eliminating ice crystal formation is to accelerate passage through the critical temperature interval (Figure 1) by various methods of ultrarapid freezing (see Sections 3.2.2b–e) without pretreatment of the samples with a cryoprotectant. Ultrarapid freezing also nearly instantaneously stops even the most rapid of cellular processes (e.g., vesicle–plasma membrane fusion events) and is clearly the method of choice when it can be suitably employed (see below).

3.2. Freezing

3.2.1. Specimen Supports and Mounting of Samples

A variety of specimen supports are available from the manufacturers of freeze-etch machines; for example, gold–nickel alloy "hats" of about 2.6-mm diameter are available from Balzers Corporation. Some investigators may wish to slightly roughen the smooth surfaces with emery cloth to provide better adhesion of the specimen. In our laboratories, similar hat-shaped supports are punched from 0.1-mm-thick copper foil (diameter ca. 3 mm, rim ca. 0.5 mm wide, step height ca. 0.3 mm). The flat specimen surface is roughened with emery cloth (a slightly coarser grit is used for the copper hats). The copper specimen supports are easily bent, but far less expensive than gold supports and can therefore be produced in large quantities. Either type of support should be cleaned well (e.g., sonicated in acetone or methanol/HCl, rinsed thoroughly with distilled water, and dried) immediately before use. Specimen supports can be marked on the lower surface with felt-tip pens to color code samples. Supports made from cardboard or cellulose acetate have poor thermal conductivity and should be avoided.

Mounting of the sample will, of course, vary with the nature of the sample. For many microorganisms and cell suspension cultures, a small droplet (ca. 1 μl) of the concentrated cells is placed on the roughened area of the specimen support. It is important that the fluid sample not flow over the entire surface and onto the rim of the hat. Too large a sample will interfere with or prevent loading into the specimen holder. Similarly, if a filamentous specimen or piece of fungal tissue is to be mounted, a small (ca. 0.5 to 1 mm^3) sample must be excised. In all cases, it is essential that sample excision, mounting, and subsequent freezing are carried out as quickly as possible to avoid potential dehydration artifacts. Such artifacts occur more frequently than commonly appreciated because of the large surface-to-volume ratio of the samples. If the minimum time required for mounting specimens is too great to avoid dehydration artifacts, the environment around the specimen should be experimentally modified to minimize changes. For example, if there is a chance that a sensitive specimen is dehydrating during the excision and mounting of a suitably small sample, the processing should be carried out in a humid chamber. Such a chamber can be easily constructed from a glass dish "coated" with moistened tissue paper on its inside and covered with plastic with a small hole (4 × 4 cm) to provide tweezer access to the humid interior of the chamber. If several samples are being prepared and frozen in rapid succession, one should be careful not to use cold forceps for handling the samples prior to freezing, as cold forceps can cause some slow partial precooling or prefreezing of the sample. It is also undesirable to mount several samples first and then freeze them after the last sample has been mounted. Freezing of a given sample should be done as soon as possible after mounting to reduce dehydration and other artifacts.

Cell suspensions can be concentrated by centrifugation into a pellet of controlled density to ensure that the small volume mounted will provide an adequate number of cells. Possible effects of centrifugation on the cells must be taken into consideration, and in some cases other concentration methods may have to be adopted to avoid breaking or distorting cells. High densities of living cells may also lead to anoxic conditions and related artifacts. Some types of concentrated cell suspensions may not be suitably dispersed onto specimen supports by means of a pipette. In such cases, a dissecting needle can be used to remove a small sample from the pellet and to transfer it to the holder. Under all conditions, care must be taken to prevent dessication and other possible damaging changes of the sample. Small samples concentrated by centrifugation are often rediluted by small volumes of medium left on the centrifuge tube wall. The problem can be eliminated by wiping the interior wall of the tube with a rolled piece of filter paper immediately after the supernatant has been decanted.

A useful alternate method for concentrating cultured plant cells that avoids many of the above-mentioned pitfalls is the following. A piece of nylon cloth of appropriate mesh size is placed over a petri dish (or other vessel) and some of the cell suspension is poured in. When cells are to be mounted on the support, a portion of the cloth is raised from the dish and medium, and a small amount of the cell slurry is removed from the cloth with a dissecting needle or small spatula. This method avoids the time and physical force associated with centrifugation, and keeps the cells relatively undisturbed and aerated until the moment they are to be mounted on the support. This method also allows the investigator some degree of control over the relative density of the sample.

If for any reason the frozen samples tend to break off the specimen support during subsequent stages of specimen preparation, one should check the surface of the specimen support to be sure that it has been adequately roughened and cleaned. Traces of fatty substances can greatly reduce the strength of adhesion between frozen samples and specimen supports. To stabilize the attachment of very large algal cells and/or tissue pieces, such samples can be mounted with a paste of yeast cells suspended in the same medium as the sample.

If a double replica is to be prepared, a sandwich type of specimen between two supports has to be used. Because the complete "sandwich" of support–sample–support must fit into a specialized holder (Figure 2), sample thickness may be limited by the dimensions of the holder. This is often the case with samples to be frozen in the propane jet freezer (see Section 3.2.2d), which

FIGURE 2. A freeze-fracture double replica device for the production of complementary replicas. Three different types of specimen carriers are shown in the three specimen slots. Courtesy of Balzers Corporation.

requires samples less than 70 μm thick. To prevent crushing of sandwiched samples, one or more spacers generally must be used. A typical preparation would involve placing a spacer on the specimen surface of a support, placing the small sample in the center of the spacer, and placing a second support on top of the specimen (Figure 3). Copper slot grids (ca. 70 μm thick) or gold slot grids (ca. 30 μm thick) work well but reduce the specimen area somewhat. A modified thin-section EM grid from which about two-thirds of the inner mesh has been removed provides more area for sample and even greater control over the spacing. Due to the thinness of the grids (12–15 μm), several can be mounted on top of each other. Although EM grids are too soft to simply be placed in a punch for the preparation of a central hole, the inner mesh can be removed by special techniques such as electrode discharge boring. If microorganisms or plant suspension culture cells are relatively large (ca. 70 μm in diameter), there is a chance that the fracture will follow the cell surfaces and that few cells will be cross-fractured through central interior regions. Under such conditions the plane of fracture can be directed, to some extent, through the center of the sample by mounting 30-μm gold slot grids on the surfaces of each specimen support with Formvar. The samples are loaded as described above and usually the fracture will occur between the two spacer grids and will split the cells in half rather than follow the cell surface. To locate complementary portions of the two replicas, gold marker grids can be used as spacers provided that the sample is smaller than the spaces between the grid bars (Neushul, 1970).

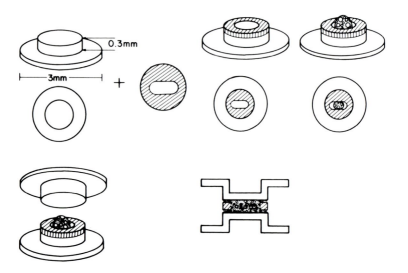

FIGURE 3. A schematic representation of the preparation of a specimen ''sandwich'' for use with the double replica device (and the propane jet freezer).

3.2.2. Freezing Methods

Five major means of freezing samples can be employed: (1) dipping or plunging the sample into a coolant, (2) spraying the sample into a coolant, (3) spraying cooled propane onto the sample, (4) spraying liquid nitrogen under high pressure onto the sample, and (5) bringing the sample rapidly into contact with a cold metal block. These methods can facetiously, but not inaccurately, be called "dipping," "spraying," "blasting," "bombing," and "slamming," respectively. Dipping is the most common and readily available means of preparing microorganisms for FEM. The other methods are more involved, but can yield superior freezing results and enable workers to circumvent the use of chemical fixatives and cryoprotectants.

3.2.2a. Plunge Freezing. This most commonly employed method involves plunging a specimen (up to 1 mm thick) into a coolant such as Freon 12, 22, or propane held just above its freezing point ($-160°C$ to $-190°C$, depending on the coolant). The typical freezing rate of such samples is between -200 and $-1100°C$/sec (Plattner and Bachmann, 1982). When used in conjunction with 20–35% glycerol as a cryoprotectant, this method can yield very nicely frozen samples. A small (ca. 2 cm deep) metal receptacle is mounted in a liquid nitrogen-filled Dewar flask (Figure 4) and filled with gaseous Freon or propane, which condenses onto the cold metal receptacle during filling. Left unattended, the Freon or propane usually solidifies within about 5 min. The frozen Freon or propane can, however, be reliquefied by melting with a metal rod, and the sample, held with metal forceps (at room temperature), is plunged into the liquid Freon and kept there for a few seconds. It is very important that the plunging of the sample be as quick as possible because deleterious partial freezing can occur as the sample is moved through the cold gaseous zone above the liquid Freon. The sample is then rapidly transferred to an open liquid nitrogen Dewar flask for short-term storage. Samples can also be held for long-term storage in a liquid nitrogen storage Dewar. For this purpose the frozen samples are next transferred to a precooled small can with a perforated lid and an attached string or handle for retrieval purposes. When large numbers of samples have to be stored, it is advantageous to use specially designed storage containers that remain submerged in liquid nitrogen even when the nitrogen level is low (Krah *et al.*, 1973).

3.2.2b. Spray-Freezing. Despite the general usefulness of the dipping technique, the rate of freezing is usually not fast enough to prevent detectable ice crystal formation in uncryoprotected specimens. Plattner *et al.* (1973) employed the spray-freezing technique developed by Bachmann and Schmitt (1971) to freeze bacteria, algae, and protozoans. Although the actual rate of freezing cannot be measured, it is estimated that the small sample droplets are frozen at rates between $-20,000$ and $-100,000°C$/sec. At this rapid rate of freezing,

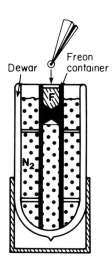

FIGURE 4. A typical Dewar setup for dip or plunge freezing in Freon. Adapted from Freeze-etching, a Balzers High Vacuum Report—EM-Preparation Technique by H. Moor.

excellent structural preservation of cells can be obtained without the use of cryoprotectants. Solutions of macromolecules or cells up to 40 μm in diameter can be prepared by this method, but larger cells often exhibit signs of mechanical damage due to the shearing forces to which they are exposed during the formation of the fluid droplets in the airbrush used to spray the sample into the liquid propane. Following the spray-freezing, the propane is evaporated (at −85°C), and the droplets of frozen sample (ca. 5–40 μm in diameter) are combined with cold (−85°C) *n*-butylbenzene to form a paste, which can be transferred with a cold platinum wire onto precooled specimen supports. Samples are transferred to liquid nitrogen and then processed routinely. Special caution must, of course, be taken in handling pathogenic microorganisms for which spray-freezing may be an unacceptably risky procedure (see also Lickfeld, 1976). Although the rapid cooling effected by this technique makes it an important means of sample preparation, the somewhat tedious handling procedures (including spraying and mounting), the limited size of cells that can be frozen without damage (10 μm or less), and the generally low yield of useful sample images (due to droplet density and smearing of the solidified *n*-butylbenzene over the specimen area caused by the microtome cutting) are major drawbacks.

3.2.2c. Cold Metal Block Freezing. Slamming or slam-freezing utilizes a cold (i.e., liquid nitrogen or liquid helium temperature) metal block with which a sample is rapidly brought into contact, either by hand or mechanically. Although initial studies on the technique may date back to the early 1940s, successful routine application began in the 1960s and 1970s (Van Harreveld and Crowell, 1964; Van Harreveld *et al.*, 1965; Heuser *et al.*, 1976). Samples are mounted on a metal carrier designed to fit on the stage of a freeze-etch apparatus. The carrier is mounted on the sample holder of the freezing device and is backed by a cushion of rubber, plastic foam, or fixed tissue (e.g., liver or lung). The cushion prevents crushing of the sample against the metal block. The very rapid cooling rate achievable (up to $-50,000°C/sec$), the availability of commercial apparatus (e.g., the "Slammer," Polaron Instruments, 2293 Amber Drive, Hatfield, PA 19440; the "Gentleman Jim," Quick Freezing Devices, 112 East Gitting Street, Baltimore, MD 22230; the HITEK M9500 "Fast-Freeze," EMTEK Scientific, 3880 Industrial Way, Benicia, CA 94519; a device available from J. Heuser, Department of Physiology, Washington University School of Medicine, St. Louis, MO 63110; and a device available from J. Escaig, 105 Boulevard Raspail, 75006 Paris, France), and the fact that relatively large tissue samples can be processed make the cold metal block freezing technique very useful. It must be noted that, because the freezing is unilateral, excellent freezing is limited to a depth of 10–15 μm from the sample surface. Also, for many microorganisms and cell suspension cultures, jet freezing may provide simpler handling and, in some cases, may also be more reproducible.

3.2.2d. Propane Jet Freezing. Propane jet freezing has been routinely used in our laboratories to provide excellent freezing of unfixed, uncryoprotected samples of ca. 40- to 70-μm thickness. The technique is relatively new (Moor *et al.*, 1976; Müller *et al.*, 1980) and requires a jet freezer in which gaseous propane is condensed and allowed to cool to near liquid nitrogen temperature [a commercial jet freezer is available from Balzers Corp.; the design of our jet freezer is described in Gilkey and Staehelin (1986)]. A sample and spacer are loaded onto a hat-shaped specimen carrier (as described above) and a second specimen carrier is placed upside down on top of the sample. The sandwich is quickly loaded into a specimen holder and the latter is rapidly inserted into the freezer. Liquid propane under about 10 atm pressure is sprayed onto both sides of the specimen sandwich. The sample is sprayed or blasted from both sides, which increases the depth of adequate ultrarapid freezing to about 40 μm vis-à-vis about 10–15 μm in unilaterally cooled specimens as in the cold metal block freezer. The frozen sample is quickly transferred to a Dewar of liquid nitrogen. If gold alloy specimen carriers (Balzers) are used, the central area of the "hats" are thinned to about 75–125 μm to increase the rapidity of freezing. If copper specimen carriers are punched from foil, a thin (< 50 μm thick) sheet is used. As described for the dipping method, care must be taken to avoid allowing the specimen to dry out during the loading of the specimen carriers. Propane jet freezing is limited to small samples (maximum diameter 2 mm, maximum thickness 40–70 μm) but it is simple to use, and it provides freezing rates of up to $-50,000°C/sec$. The technique is especially appropriate for the study of many microorganisms and suspension culture cells, which can be processed from undisturbed cells in medium to ultrarapidly frozen samples within seconds with a minimum of handling and no chemical fixation or cryoprotection.

3.2.2e. High-Pressure Freezing. Sample size limitations of propane jet freezing (and other ultrarapid freezing techniques) will limit or preclude the study of even some microbiological specimens. Under atmospheric pressure, no existing system can provide ultrarapid freezing to a greater sample depth than the maximal 40 μm provided by the propane jet freezer. This problem can be partly overcome by freezing under high pressure, due to the fact that high pressure (2000 atm) not only can reduce the freezing point of water to $-20°C$, but can also slow down ice crystal nucleation and growth. Thus, by suppressing ice crystal formation by means of high pressure, more time is gained for the freezing process (Riehle and Hoechli, 1973; Moor *et al.*, 1980; Wolf *et al.*, 1981). In theory at least, spherical samples up to 1.0 mm in diameter could be ultrarapidly frozen without pretreatment. The potential benefits of this technique are exciting, but two major concerns must be addressed. First, the apparent lack of deleterious effects of the high pressure used must be confirmed (pressure buildup to 2100 atm occurs in 20 msec before freezing is initiated); and second, the routine, successful operation of the prototype high-pressure freezers must be established. The first commercial high-pressure freezer (Balzers HPM 010) became available in January 1985; however, it will probably take several years to fully evaluate the potential of this method for biological studies.

3.3. Freeze-Fracturing, Deep-Etching, and Replica Processing

3.3.1. Freeze-Fracturing

The two major modes of fracturing are by cutting (actually, chipping) the sample with a liquid nitrogen-cooled blade to produce a single replica (Figure 5) or by tearing the sample apart in special holders to form a double replica. In the former, a razor blade or piece thereof typically is used. These inexpensive knives must be well cleaned (e.g., with acetone) and tightly mounted on the microtome arm. To ensure good thermal contact, the knife-holding clamp should be coated with grease (e.g., Apiezon N, Apiezon Products). For hard samples, Balzers Corporation also produces special steel knives. Various standard procedures are followed to evacuate, cool, and load the work chamber. In the Balzers device the microtome knife arm can be cooled to $-196°C$ to serve as a cryopump and to reduce water vapor contamination [see also Müller and Pscheid (1981) for a more in-depth treatment of this topic]. To prevent water vapor condensation on the specimen holder during transfer of the frozen specimen to the cold stage, modern instruments allow for premounting of the frozen samples under liquid nitrogen into clamp-on holders that can be transferred to the precooled specimen stage under dry nitrogen counterflow conditions. This mode of transfer reduces water vapor artifacts and speeds up subsequent pumping down of the chamber. The fracturing can begin when the vacuum reaches $< 2 \times 10^{-6}$ bar. The cutting speed and amount of knife advance must be determined empirically for the specific device and sample. The general rule is to begin with a slow speed and

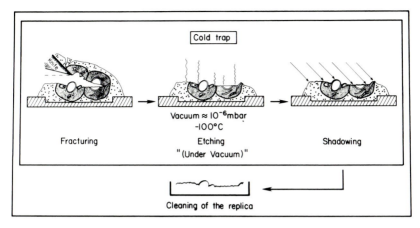

FIGURE 5. Steps in preparing a single replica by fracturing the sample.

small advance in order to avoid shattering or dislodging the specimen. At −100°C the samples are chipped rather than cut; scratching, which is to be avoided due to associated melting artifacts, occurs when a knife becomes dull. The appearance of the specimen surface indicates the quality of fracturing. Well-fractured specimens appear pearly, scratched surfaces appear mirrorlike. The precise stage temperature during the fracturing may also have to be determined empirically, and potential drift in the temperature gauges of older instruments must be kept in mind. If no deep-etching of the sample is required, the stage is kept at about −110°C and the carbon–platinum shadowing is begun as soon as possible after the final pass of the knife. An excellent discussion of the theory and practice of shadowing can be found in Willison and Rowe (1980).

Preparation of double replicas requires a special holder in which one or more samples are loaded (Figure 2). The samples (as described earlier) consist of two specimen supports separated by the specimen and spacer. The frozen specimen is torn apart as the special holder is pulled apart. Shadowing is completed as for single replicas.

There are several sources of artifacts during the fracturing and shadowing steps. Many (e.g., water vapor contamination) can be eliminated or minimized by optimizing critical specimen processing steps, but others cannot and must be considered in the interpretation of replicas (see Section 3.4). Several mini-reviews dealing with artifacts and the interpretation of freeze-fracture (-etch) images in the book edited by Rash and Hudson (1979) and comparable works are essential readings for the novice.

3.3.2. Deep-Etching

In both single replicas and double replicas, the fracturing process exposes internal fracture *faces* of membranes that are exposed by splitting along the

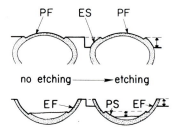

FIGURE 6. Exposure of fracture faces and true surfaces of the plasma membrane of a red blood cell ghost frozen in water. (Left) After fracturing, protoplasmic face (PF) and exoplasmic face (EF) are exposed. (Right) After subsequent etching, i.e., after ice sublimation, the exoplasmic surface (ES) and protoplasmic surface (PS), previously concealed underneath the ice, become visible.

central hydrophobic plane of their lipid bilayer. If the true membrane or cell wall surfaces are to be exposed, the fracturing has to be followed by an "etching" step during which ice surrounding the structures of interest is allowed to sublimate (Figure 6). Some sublimation may occur following routine fracturing. If larger etched surfaces are to be exposed, the specimen temperature should be raised to between -95 and $-100°C$ and the liquid nitrogen-cooled microtome arm (or cold trap) placed directly above the specimen for the full duration of the etching. The optimal duration of deep-etching must be determined empirically, and can range from a few seconds to 30 min. As previously noted, chemical cryoprotectants such as glycerol have a very low vapor pressure at low temperature; thus, specimens that are to be "deep-etched" should not be treated with such agents before freezing. Best results are usually obtained with samples suspended in distilled water, provided that the samples can withstand such treatment without damage. If a buffer has to be used, the total concentration of solutes should be kept below 10 mM for plunge-frozen samples. Heuser and co-workers (Hirokawa and Heuser, 1981) often add 15% methanol as a volatile cryoprotectant to their samples to improve the quality of freezing in their slam-freezer. It should be noted that even small traces of solutes can give rise to spurious filamentous (artifactual) structures in etched, ultrarapidly frozen samples (Miller *et al.*, 1983).

3.3.3. Replicas

3.3.3a. Shadowing. Immediately after fracturing or deep-etching, the sample is shadowed, first at an angle (ca. 30 to 55°, usually 45°) with platinum–carbon (or tungsten–tantalum in some systems) and then from directly above with pure carbon. The platinum–carbon shadow provides contrast and the three-dimensional relief seen in the EM; the pure carbon shadow holds the replica together and provides the rigidity needed for additional processing. Generally, the platinum–carbon layer is about 2 nm thick, and the carbon film, about 5–20 nm. The thickness of the shadows and support films can be accurately controlled and reproduced with a quartz-crystal film thickness monitor. Variations of the shadowing angles and shadow thicknesses may be required for special applications (e.g., detection of macromolecules requires a small shadowing angle of 5 to 10°; replicas of very rigid-walled cells may tend to break up during processing and therefore often require a thicker carbon shadow). A piece of white filter paper positioned at the base of the specimen stage may be used to monitor the platinum–carbon shadowing and the carbon deposition.

Shadowing requires use of either resistance-heating evaporation of the shadow materials or electron-gun evaporation. The former method is the conventional means of shadowing and involves preparation of carbon rods and platinum wire coils [see Staehelin (1980) for a detailed description]. The preparation of the carbon rods and wire coils, together with determination of the correct procedures (e.g., timing, current levels) for producing a good shadow render this aspect of freeze fracture more of an art than a science. The development of electron-gun shadowing has improved the efficiency and consistency of the shadowing process. Combined with additional automation of the procedure on new models of freeze-etch devices, the electron-gun shadowing has significantly simplified the process of getting excellent replicas. The electron gun, which bombards the shadowing metal with electrons, can evaporate high-melting-point metals and provides more reproducible results than the resistance-heating systems. Although the specific parameters for setting up an electron gun will vary with the specific equipment used, the cleanliness of the gun is a general concern. In our experience it has been appropriate to clean the unit and readjust the height of the anode after every second or third run. Because radiation from the gun during the initial heating can damage the fracture face, we have added a shutter with a 4-sec time-delay switch in front of the unit (Staehelin, 1980). Rotary shadowing (Margaritis et al., 1977) can provide additional information on the morphology of linear molecules and repetitive structures for which unilateral shadowing causes a loss of information. A small shadowing angle (7 to 30°) is usually employed for freeze-fracture specimens. However, when highly extracted, deep-etched samples rich in filaments are to be examined, a high shadowing angle (50 to 60°) tends to give the best results (e.g., Hirokawa and Heuser, 1981).

3.3.3b. Cleaning. After shadowing and the venting of the chamber, the replica can, if necessary, be strengthened by the application of 0.5% collodion or 2% Lexan (Steere and Erbe, 1983). The strengthening material is dissolved away after the replica has been cleaned of biological material. The specimen is gently slipped off the support at an oblique angle onto the surface of distilled water or other cleaning solution. To reduce general fragmentation of replicas of some samples, it may be necessary to float the replica off onto the same buffer used for the specimen freezing. If the replica does not float off freely, the entire sample can be left in the solution and, if necessary, the replica can be teased off the support (e.g., by gently agitating it with a stream of solution from a pipette tip). For most biological samples, undiluted commercial bleach works well as the initial cleaning solution (for 30 min to 12 hr). Stronger cleaning solutions can be used if necessary (e.g., following distilled water washes, transfer to 70% H_2SO_4, fuming HNO_3, 40% CrO_3, or other cleaning solutions is possible). Mild warming (60°C) during the cleaning period in bleach or acid often enhances removal of the biological material. Small platinum wire loops are used to transfer the replicas from one solution to the next (n.b., distilled water washes between the different cleaning fluids are necessary to minimize mixing-induced damage to the replicas). Because the replica will be carried in the meniscus within the loop, it is helpful to have loops of various sizes on hand. White porcelain Coors spot plates or small petri dishes (on a white surface) are generally used to hold the various cleaning solutions and distilled water. If small pieces of replicas do not float, they can be pipetted gently. The biological-material-free replicas should be rinsed several times by transfers to distilled water before mounting on an EM grid. If the replica sinks and is too large to be pipetted, the solutions can be changed in the vessel. After the final rinse, the replica can be brought to the surface by teasing with a platinum loop or by adding some 1 : 1 chloroform–methanol. If a large replica is curled up, it sometimes can be unfolded by replacing the water with alcohol or acetone and then transferring the replica to a clean distilled water surface. The process will either uncurl or break up the replica (unfortunately, often the latter). The replicas are picked up from below or above on coated or uncoated thin-section EM grids. Uncoated grids can be dipped in a "grid glue" of 1.5% polybutene in xylene to ensure good adhesion (n.b., excess solution can be removed from the grid and forceps with a piece of filter paper), and allowed to dry. If uncoated grids are used for pickup from below, a brief pretreatment of the grid with 0.5% NaOH or dilute wetting agent is useful to reduce hydrophobicity of the grid. If the replica is picked up from above, it will sit in a small droplet of water on the grid surface. A small piece of filter paper should be used to draw down the droplet of water slowly and lodge the replica on the center of the grid. Methods used in cleaning replicas of plant tissue (Platt-Aloia and Thomson, 1982) may also be useful with replicas of microorganisms.

3.4. Artifacts and Interpretation

Numerous articles on the cause and recognition of artifacts have been cited (see Section 2), and a volume such as that edited by Rash and Hudson (1979) is an invaluable aid in the analysis of replicas. Such a complete and detailed coverage of the subject is beyond the realm of this chapter but a brief overview of the major types of artifacts is appropriate. Böhler (1979) distinguishes six major classes of artifacts based on the preparation step during which the artifact can be introduced. The artifacts are classified as follows: (1) those caused by cryoprotectants, (2) freezing artifacts, (3) cutting or fracturing artifacts, (4) sublimation and condensation artifacts, (5) replication artifacts, and (6) insufficient cleaning of the replica artifacts. We would like to add to this list artifacts associated with the handling and processing of the specimen prior to freezing (aside from those related to cryoprotectants). These are often the most difficult artifacts to avoid and to recognize. Also, as all the other processing steps are improved, the handling and processing prior to freezing becomes the most limiting step for further progress in our understanding of biological structures. Although possible artifacts caused by chemical fixation have already been mentioned, there are also changes in the specimens that can occur during the harvesting and handling of the specimen. For microorganisms and suspension culture cells, harvesting of the cells may involve abnormal conditions, which may induce changes at the subcellular level. Centrifugal force, temporary anoxia, partial dehydration, and abnormal cell densities may cause artifacts that, although not major, could nevertheless preclude an artifact-free glimpse at the microorganism's ultrastructure. Because ultrarapid freezing arrests very rapid cellular processes and fixes sensitive components in place, sources of minor perturbation of cell processes during harvesting and handling may become increasingly significant. We should not forget that the quality of the final image is ultimately dependent on the quality of the sample at the time of the initial rapid-freeze fixation. At best, the subsequent steps can preserve what was stabilized during the initial fixation event. They cannot, however, eliminate artifactual changes that occur as the sample is prepared for rapid freezing, and they cannot improve on the general quality of specimen preservation.

Of the six classes of artifacts cited by Böhler, the first two can largely be eliminated by ultrarapid freezing; skilled application of freeze-fracture techniques can significantly reduce most of the last three (except for those associated with deep-etched samples). Some of the artifacts such as plastic deformation are inherent in the physics of the frozen sample and the system. Fortunately, plastic deformation is recognizable and often a minor impediment to interpretation of good replicas. Under ideal conditions the routine resolution of FEM is 3 to 4 nm, but resolutions of 1.0 to 1.5 nm may be obtained under special conditions including the use of heavy metals (e.g., tungsten and tantalum), ultrahigh vacuums, and liquid helium temperatures (Gross, 1979; Niedermeyer, 1982).

3.5. Freeze-Fracture (-Etch) Nomenclature

In 1975 Branton and 13 coauthors proposed a mnemonic labeling scheme for the fracture faces and surfaces of cellular membranes (illustrated in Figure 7), which now has been adopted by nearly all researchers who employ FEM. In this scheme, any cytoplasmic membrane that can be split by freeze-fracturing is subdivided into two halves (leaflets), a "P"-half and an "E"-half. The half closest to the cytoplasm, nucleoplasm, chloroplast stroma, or mitochondrial matrix is designated the "*p*rotoplasmic" half, thus P-half; in turn, the half closest to the *e*xtracellular space, *e*xoplasmic space, or *e*ndoplasmic space is designated the E-half. The concept of exoplasmic space includes the interior of

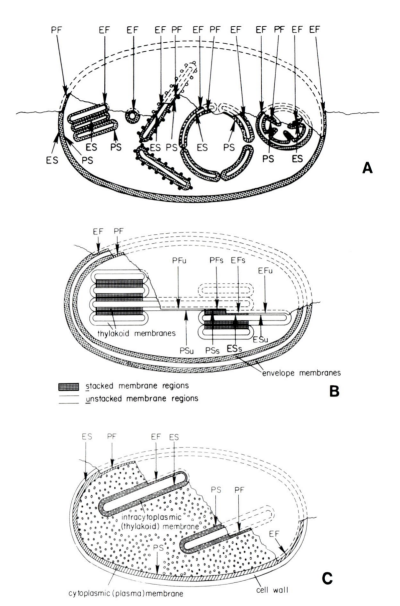

FIGURE 7. (a) Diagrammatic illustration of the Branton *et al.* (1975) nomenclature for freeze-fractured (-etched) membranes of eukaryotic cells. Labels for some of the frequently studied membrane fracture faces and surfaces are shown. The dark line through the cell and some of its organelles traces the course of a hypothetical fracture prior to etching. See text for details. (b) Diagram illustrating the freeze-fracture and freeze-etch nomenclature as applied to a chloroplast with stacked (s) and unstacked (u) thylakoid membranes. See text and Staehelin (1976) for details. (c) Diagram illustrating how the Branton *et al.* (1975) nomenclature applies to the cellular membranes of prokaryotic cells (example: cyanobacterial cell with internal membranes). Note, however, that the Branton *et al.* (1975) nomenclature does *not* apply to cell wall layer fracture faces that are occasionally revealed in freeze fracture replicas of gram-negative bacteria and some algae.

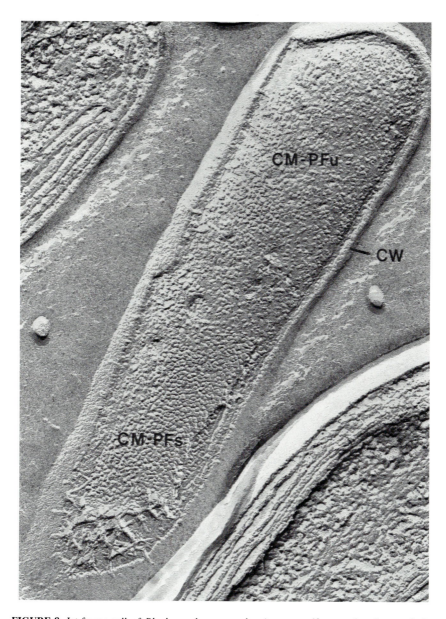

FIGURE 8. Jet-frozen cell of *Rhodopseudomonas palustris,* a nonsulfur, purple, photosynthetic bacterium. The cytoplasmic (plasma) membrane (CM) is differentiated into two distinct regions, a photosynthetic membrane domain labeled CM-PFs due to the fact that the differentiation is caused by membrane stacking with an underlying photosynthetic membrane, and a nonphotosynthetic membrane domain (CM-PFu), which carries a normal complement of transport complexes. The arrowheads point to tubular connections between cytoplasmic membrane and the underlying continuous photosynthetic membranes. The smooth, turgid appearance of the cytoplasmic membrane indicates optimal freezing. CW, cell wall. ×125,000. Photo: A. Varga; see Varga and Staehelin (1983) for details.

endocytotic vacuoles, phagosomes, primary and secondary lysozomes, ordinary plant vacuoles, and Golgi vesicles. The endoplasmic space includes the lumen of the endoplasmic reticulum, the space between the two membranes of the nuclear envelope, and the cisternae of Golgi stacks.

Following this definition of the half-membrane leaflets, the scheme proposed to call the fracture faces of the two halves revealed by membrane splitting, PF (for P-face) and EF (for E-face). Finally, the true membrane surfaces were

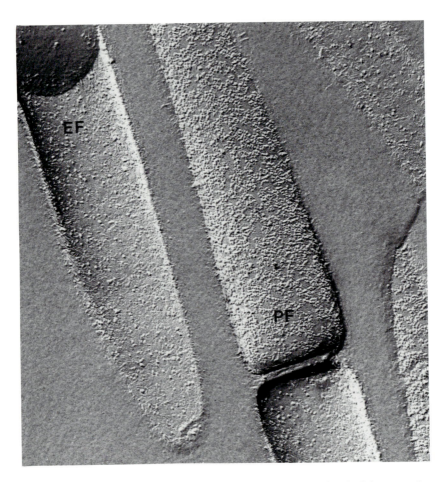

FIGURE 9. P-face and E-face views of the cytoplasmic (plasma) membrane of the green, thermophilic bacterium *Chloroflexus aurantiacus* grown under aerobic (nonphotosynthetic) conditions. The cells were fixed by addition of 1% glutaraldehyde to the growth medium at 53°C. Glycerol (final concentration 30%) was added slowly over a 30-min period, starting 15 min after onset of fixation. Freezing was by hand-dipping into Freon 12 held at its melting point [see Sprague *et al.* (1981) for details]. ×65,000.

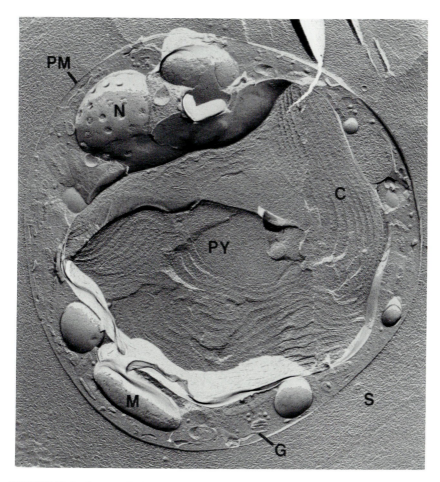

FIGURE 10. Jet-frozen cell of the red alga *Porphyridium cruentum* frozen in its growth medium. Note the smooth contours of the plasma membrane (PM) and of the internal membrane systems, as well as the very finely dispersed state of the slime (S) in the surrounding medium (slightly etched to reveal the slime filaments). All of these features indicate optimal structural preservation of the sample (i.e., virtual vitrification of the specimen during the freezing). N, nucleus; C, chloroplast; PY, pyrenoid; M, mitochondrion; G, Golgi. ×20,000.

designated ES (the hydrophilic surface of the E-half) and PS (the hydrophilic surface of the P-half). In chloroplasts the scheme has been further refined to take into account the differentiation of the membranes into stacked (appressed) and unstacked (nonappressed) membrane regions (Figure 7b; Staehelin, 1976).

   Although the Branton *et al.* (1975) nomenclature paper did not specifically diagram how the nomenclature should be applied to prokaryotic cells, the same P- and E-membrane halves can be recognized in bacteria as in eukaryotic cells.

Thus, the labeling scheme can be equally well used for eukaryotic and prokaryotic cells as shown in Figure 7c (see also Giddings and Staehelin, 1979;
Varga and Staehelin, 1983). This general applicability of the Branton *et al.*
(1975) nomenclature to all cytoplasmic membrane systems (one exception: the
central membrane of the triple chloroplast envelope membrane system of the
Euglenophyceae and Dinophyceae) appears to have been lost on some microbiologists, who have recently felt compelled to invent two partly new nomenclatures for labeling bacterial cytoplasmic membranes and cell wall layers
(Schmid *et al.*, 1980; Rodgers and Davey, 1982). The problems with both of
these nomenclatures is that in trying to accommodate the cell envelope layers of
gram-negative bacteria into their schemes, the authors have had to partly redefine

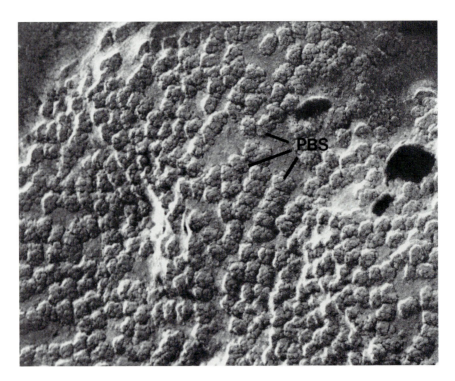

FIGURE 11. True surface view of an isolated chloroplast membrane of the red alga *Porphyridium
cruentum* revealed by freeze-etching (5-min etch at −100°C). The isolated membranes were fixed
with 1% glutaraldehyde before being washed with distilled water to preserve the phycobilisomes
(PBS) on the membrane surface. The "subunits" of the phycobilisomes correspond to end-on views
of the phycobiliprotein-containing rod elements. To illustrate the three-dimensional structure of the
phycobilisomes more clearly, the first half of the Pt/C shadow was employed in a unidirectional
mode, the second in a rotational mode. In addition, the micrograph has been printed in reverse
contrast, which is often beneficial for roughly contoured structures revealed by deep-etching.
×120,000.

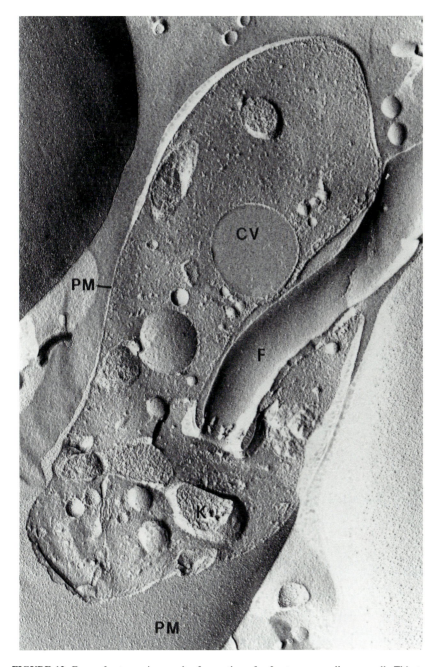

FIGURE 12. Freeze-fracture micrograph of a portion of a *Leptomonas collosoma* cell. This trypanosomatid flagellate was preserved by jet-freezing in growth medium. Note the smooth, turgid appearance of all membrane systems, including the contractile vacuole (CV), flagellum (F), plasma membrane (PM), and the kinetoplast (K). ×36,000. Photo: J. Linder; see Linder and Staehelin (1979) for details.

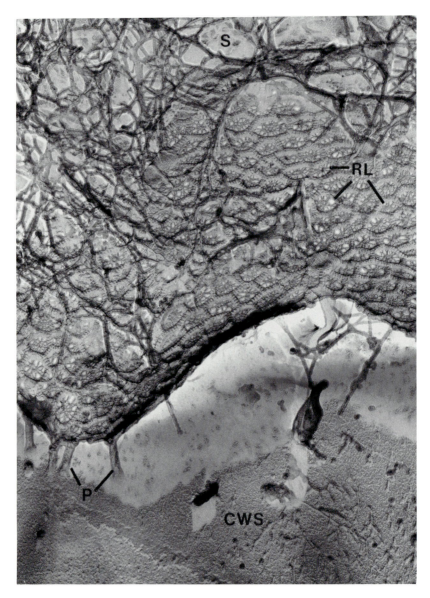

FIGURE 13. Cell wall ornamentations of the green alga *Scenedesmus quadricauda* revealed by deep-etching. The cell wall surface (CWS) is seen in the bottom half of the picture. The elaborate reticulate layer (RL) consists of hexagonal and pentagonal elements and its ribs exhibit faint striations. Each hexagonal (pentagonal) element contains a core consisting of finely interwoven filaments that radiate toward the ribs. The reticulate layer is supported by props (P) that are anchored to the cell wall surface. Slime (S) filaments are seen external to the reticulate layer. ×45,000. See Staehelin and Pickett-Heaps (1975) for details.

the Branton *et al.* (1975) nomenclature for cytoplasmic membranes [e.g., by proposing that a P-leaflet can give rise to a P-face (PF) and an E-surface (ES), which obviously does not make sense]. In our opinion, the Branton *et al.* (1975) nomenclature is perfectly adequate for labeling *all cytoplasmic membranes* of prokaryotic cells and should therefore not be changed. It does not, however, address the problems of labeling of *cell wall layers* as found in gram-negative bacteria and in many eukaryotic algae. Since cell wall layers are not directly related to the leaflets of cytoplasmic membranes, it would seem desirable to refrain from using PF, PS, EF, and ES labels for fracture faces and surfaces of *extracellular* layers. We urge those researchers interested in the morphology of extracellular layers to *jointly* approach the problem of developing a *new* nomenclature for such layers. Because neither of the authors of this review has worked on bacterial cell wall layer morphology, we feel that it would be inappropriate for us to propose such a nomenclature.

4. TYPICAL RESULTS

FEM studies of microorganisms are published in a large number of major scientific journals and typical results of the application of the technique abound. For those who are just beginning FEM studies of microorganisms, a comparison of their results with published FEM micrographs remains an excellent means of assessing the quality of the replicas. Rather than only direct the reader to the published literature and conclude this chapter without inclusion of any micrographs, we have selected a few views of microorganisms from work in our laboratories. The micrographs (Figures 8–13) also illustrate the results of ultra-rapid freezing with an advanced prototype propane jet freezer developed in the laboratory of L. A. Staehelin. The reader should note the absence of any visible ice crystals in the specimens and the smooth, turgid appearance of the various cellular membranes.

ACKNOWLEDGMENTS. The authors wish to acknowledge Balzers Corporation, Hudson, New Hampshire for permission to modify and use several figures. The technical assistance of M. Dewit, M. Henk, and T. Roller is gratefully acknowledged. The authors wish to thank M. Henk and S. Matthews for editorial corrections and suggestions. Supported by NIH Grant GM-18639 to L.A.S.

5. REFERENCES

Bachmann, L., and Schmitt, W. W., 1971, Improved cryofixation applicable to freeze-etching, *Proc. Natl. Acad. Sci. USA* **68:**2149–2152.
Böhler, S., 1979, Artifacts and defects of preparation in freeze-etch technique, in: *Freeze Fracture:*

Methods, Artifacts, and Interpretations (J. E. Rash and C. S. Hudson, eds.), pp. 19–29, Raven Press, New York.

Branton, D., 1966, Fracture faces of frozen membranes, *Proc. Natl. Acad. Sci. USA* **55**:1048–1056.

Branton, D., and Deamer, D. W., 1972, Membrane structure, *Protoplasmatologia* II/E/1.

Branton, D., and Southworth, D., 1967, Fracture faces of frozen *Chlorella* and *Saccharomyces* cells, *Exp. Cell Res.* **47**:648–653.

Branton, D., Bullivant, S., Gilula, N. B., Karnovsky, M. J., Moor, H., Muehlethaler, K., Northcote, D. H., Packer, L., Satir, B., Satir, P., Speth, V., Staehelin, L. A., Steere, R. L., and Weinstein, R. S., 1975, Freeze-etching nomenclature, *Science* **190**:54–56.

Bullivant, S., 1973, Freeze-etching and freeze-fracturing, in: *Advanced Techniques in Biological Electron Microscopy* (J. Koehler, ed.), pp. 67–112, Springer-Verlag, Berlin.

Bullivant, S., 1977, Evaluation of membrane structure facts and artefacts produced during freeze-fracturing, *J. Microsc. (Oxford)* **111**:101–116.

Bullivant, S., and Ames, A., 1966, A simple freeze-fracture replication method for electron microscopy, *J. Cell Biol.* **29**:435–447.

Giddings, T. H., and Staehelin, L. A., 1979, Changes in thylakoid structure associated with the differentiation of heterocysts in the cyanobacterium, *Anabaena cylindrica, Biochim. Biophys. Acta* **546**:373–382.

Giesbrecht, P., 1968, Zur Darstellung der DNS von Bakterien and plastischer biologischer Strukturen mit Hilfe Gefrieratzung, *Zentralbl. Bakteriol. Parasitenkd. Infektionskr. Hyg. Abt. I Orig.* **207**:198–205.

Gilkey, J. R., and Staehelin, L. A., 1986, Advances in ultrarapid freezing for the preservation of cellular ultrastructure, *J. Electron Microsc. Tech.* **3**:177–210.

Gross, H., 1979, Advances in ultrahigh vacuum freeze fracturing at very low specimen temperature, in: *Freeze Fracture: Methods, Artifacts, and Interpretations* (J. Rash and C. S. Hudson, eds.), pp. 127–139, Raven Press, New York.

Haggis, J. H., 1961, Electron microscope replicas from the surface of a fracture through frozen cells, *J. Biophys. Biochem. Cytol.* **9**:841–852.

Hall, C. E., 1950, A low temperature replica method for electron microscopy, *J. Appl. Phys.* **21**:61–62.

Herrmann, C., and Staehelin, L. A., 1965, Licht- und elektronenmikroskopische Untersuchungen an *Pediococcus cerevisiae, Schweiz. Brau. Rundsch.* **76**:76–79.

Heuser, J. E., Reese, T. S., and Landis, D. M. D., 1976, Preservation of synaptic structure by rapid freezing, *Cold Spring Harbor Symp. Quant. Biol.* **40**:17–24.

Hirokawa, N., and Heuser, J. E., 1981, Quick-freeze, deep-etch visualization of the cytoskeleton beneath surface differentiations of intestinal epithelial cells, *J. Cell Biol.* **91**:399–409.

Holt, S. C., and Leadbetter, E. R., 1969, Comparative ultrastructure of selected aerobic spore-forming bacteria: A freeze-etching study, *Bacteriol Rev.* **33**:346–378.

Janisch, R., 1972, Pellicle of *Paramecium caudatum* as revealed by freeze-etching, *J. Protozoul.* **19**:470–472.

Jost, M., 1965, Die ultrastruktur von *Oscillatoria rubescens* D.C., *Arch. Mikrobiol.* **50**:211–245.

Jost, M., and Matile, P., 1966, Zur Charakterisierung der Gasvacuolen der Blaualge *Oscillatoria rubescens, Arch. Mikrobiol.* **53**:50–58.

Koehler, J., 1972, The freeze-etching technique, in: *Principles and Techniques of Electron Microscopy*, Vol. 2 (M. Hyatt, ed.), pp. 53–98, Van Nostrand–Reinhold, Princeton, New Jersey.

Krah, S., Staehelin, L. A., and Nettesheim, G., 1973, A new type of storage container for freeze-etch specimens, *J. Microsc. (Oxford)* **99**:349–352.

Lickfeld, K. G., 1968, Der frostgeätzte Bakterienkern: Ein Beitrag zur Klärung seiner Tertiärstruktur, *Z. Zellforsch.* **88**:560–564.

Lickfeld, K. G., 1976, Transmission electron microscopy of bacteria, in: *Methods in Microbiology*, Vol. 9 (J. R. Norris, ed.), pp. 127–176, Academic Press, New York.

Linder, J. C., and Staehelin, L. A., 1979, A novel model for fluid secretion by the trypanosomatid contractile vacuole apparatus, *J. Cell Biol.* **83:**371–382.

Margaritis, L., Elgsaeter, A., and Branton, D., 1977, Rotary replication for freeze-etching, *J. Cell Biol.* **72:**47–56.

Matile, P., Jost, M., and Moor, H., 1965, Intrazelluläre Lokalisation proteolytischer Enzyme von *Neurospora crassa, Z. Zellforsch.* **68:**205–216.

Merryman, H. T., 1950, Replication of frozen liquids by vacuum evaporation, *J. Appl. Phys.* **21:**68.

Merryman, H. T., and Kafig, E., 1955, The study of frozen specimens, ice crystals, and ice crystal growth by electron microscopy, *Res. Rep. Nav. Med. Res. Inst.*, Natl. Nav. Med. Ctr. **13:**529–544.

Miller, K. R., Prescott, C. S., Jacobs, T. L., and Lassignol, N. L., 1983, Artifacts associated with quick-freezing and freeze-drying, *J. Ultrastruct. Res.* **82:**123–133.

Moor, H., 1969, Freeze-etching, *Int. Rev. Cytol.* **25:**391–412.

Moor, H., 1971, Recent progress in the freeze-etching technique, *Philos. Trans. R. Soc. London Ser. B* **261:**121–131.

Moor, H., and Mühlethaler, K., 1963, Fine structure in frozen-etched yeast cells, *J. Cell Biol.* **17:**609–628.

Moor, H., Mühlethaler, K., Waldner, H., and Frey-Wyssling, H., 1961, A new freezing ultramicrotome, *J. Biophys. Biochem. Cytol.* **10:**1–13.

Moor, H., Kistler, J., and Müller, M., 1976, Freezing in a propane jet, *Experientia* **32:**805.

Moor, H., Bellin, G., Sandri, C., and Akert, K., 1980, The influence of high pressure freezing on mammalian nerve tissue, *Cell Tissue Res.* **209:**201–216.

Müller, M., Meister, N., and Moor, H., 1980, Freezing in a propane jet and its application in freeze-fracturing, *Mikroskopie* **36:**129–140.

Müller, W., and Pscheid, P., 1981, An improved freeze-fracturing procedure preventing contamination artifacts at fracturing temperatures below 163 K (−110°C) in an unmodified Balzers unit, *Mikroskopie* **39:**143–148.

Nanninga, N., 1968, Structural features of mesosomes (chondriods [sic]) of *Bacillus subtilis* after freeze-etching, *J. Cell Biol.* **39:**251–263.

Neushul, M., 1970, A freeze-etching study of the red alga *Porphyridium, Am. J. Bot.* **57:**1231–1239.

Niedermeyer, W., 1982, Freeze-fracturing at low temperatures. I. A device for fracturing biological specimens at 77–10 K under high vacuum, *J. Microsc. (Oxford)* **125:**307–318.

Platt-Aloia, K. A., and Thomson, W. W., 1982, Freeze-fracture of intact plant tissue, *Stain Technol.* **57:**327–334.

Plattner, H., and Bachmann, L., 1982, Cryofixation: A tool in biological ultrastructural research, *Int. Rev. Cytol.* **79:**237–304.

Plattner, H., Schmitt-Fumian, W. W., and Bachmann, L., 1973, Cryofixation of single cells by spray-freezing, in: *Freeze-Etching Techniques and Applications* (E. L. Benedetti and P. Favard, eds.), pp. 81–100, Societé Francaise de Microscopie Electronique, Paris.

Rash, J. E., and Hudson, C. S. (eds.), 1979, *Freeze Fracture: Methods, Artifacts, and Interpretations,* Raven Press, New York.

Remsen, C. C., 1966, The fine structure of frozen-etched *Bacillus cereus* spores, *Arch. Mikrobiol.* **54:**266–275.

Remsen, C. C., and Lundgren, D. G., 1966, Electron microscopy of the cell envelope of *Ferrobacillus ferrooxidans* prepared by freeze-etching and chemical fixation techniques, *J. Bacteriol.* **92:**1765–1771.

Riehle, U., and Hoechli, M., 1973, The theory and technique of high pressure freezing, in: *Freeze-Etching Techniques and Applications* (E. L. Benedetti and P. Favard, eds.), pp. 31–61, Societé Francaise de Microscopie Electronique, Paris.

Rodgers, F. G., and Davey, M. R., 1982, Freeze-etch nomenclature for procaryotic bacteria, *Micron* **13:**419–424.

Satir, B., Schooley, C., and Satir, P., 1973, Membrane-fusion in a model system: Mucocyst secretion in Tetrahymena, J. Cell Biol. 56:153–176.

Schmid, E. N., Sleytr, U. B., and Lickfeld, K. G., 1980, Nomenclature of frozen-etched bacterial envelopes, J. Ultrastruct. Res. 71:22–24.

Sjöstrand, F. S., 1943, Electron-microscopic examination of tissues, Nature (London) 151:725–726.

Sleyr, U. B., 1978, Gefrierbruch-Abdruckmethoden: technische Entwicklungen and Interpretation, Mikroskopie 34:2–5.

Sleytr, U. B., and Robards, A. W., 1977, Plastic deformation during freeze-cleavage: A review, J. Microsc. (Oxford) 110:1–25.

Sleytr, U. B., and Robards, A. W., 1982, Understanding the artefact problem in freeze-fracture replication: A review, J. Microsc. (Oxford) 126:101–122.

Southworth, D., Fisher, K., and Branton, D., 1975, Principles of freeze fracturing and etching, in: Techniques of Biochemical and Biophysical Morphology, Vol. 2, (D. Glick and R. Rosenbaum, eds.), pp. 247–282, Wiley, New York.

Speth, V., and Wunderlich, F., 1973, Membranes of Tetrahymena. II. Direct visualization of reversible transitions in biomembrane structure induced by temperature, Biochim. Biophys. Acta 291:621–628.

Sprague, S. G., Staehelin, L. A., and Fuller, R. C., 1981, Semiaerobic induction of bacteriochlorophyll synthesis in the green bacterium Chloroflexus aurantiacus, J. Bacteriol. 147:1032–1039.

Staehelin, L. A., 1966, Die Ultrastruktur der Zellwand und des Chloroplasten von Chlorella, Z. Zellforsch. Mikrosk. Anat. 74:325–350.

Staehelin, L. A., 1976, Reversible particle movements associated with unstacking and restacking of chloroplast membranes in vitro, J. Cell Biol. 71:136–158.

Staehelin, L. A., 1980, Freeze-fracture and freeze-etch techniques, in: Handbook of Phycological Methods. Developmental and Cytological Methods (E. Gantt, ed.), pp. 355–365, Cambridge University Press, London.

Staehelin, L. A., and Pickett-Heaps, J. D., 1975, The ultrastructure of Scenedesmus (Chlorophyceae). I. Species with the "reticulate" or "warty" type of ornamental layer, J. Phycol. 11:163–185.

Steere, R. L., 1957, Electron microscopy of structural detail in frozen biological specimens, J. Biophys. Biochem. Cytol. 3:45–60.

Steere, R. L., and Erbe, E. F., 1983, Supporting freeze-etch specimens with "lexan" while dissolving biological remains in acids, Proc. 41st Annu. Meet. Electron Microsc. Soc. Am. p. 618.

Steere, R. L., and Schaffer, F. L., 1958, The structure of crystals of purified Mahoney poliovirus, Biochim. Biophys. Acta 28:241–246.

Stolinski, C., and Breathnach, A. S., 1975, Freeze-Fracture Replication of Biological Tissues: Techniques, Interpretation and Applications, Academic Press, New York.

Van Harreveld, A., and Crowell, J., 1964, Electron microscopy after rapid freezing on a metal surface and substitution fixation, Anat. Rec. 149:381–385.

Van Harreveld, A., Crowell, J., and Malhotra, S. K., 1965, A study of extracellular space in central nervous tissue by freeze-substitution, J. Cell Biol. 25:117–137.

Varga, A. R., and Staehelin, L. A., 1983, Spatial differentiation in photosynthetic and non-photosynthetic membrances of Rhodopseudomonas palustris, J. Bacteriol. 154:1414–1430.

Willison, J. H. M., and Rowe, A. J., 1980, Replica, shadowing and freeze-etching techniques, in: Practical Methods in Electron Microscopy (A. M. Glauert, ed.) Vol. 8, North-Holland, Amsterdam.

Wolf, K. V., Stockem, W., and Wohlfarth-Bottermann, K. E., 1981, Cytoplasmic actomyosin fibrils after preservation with high pressure freezing, Cell Tissue Res. 217:479–495.

Chapter 9

Preparation of Freeze-Dried Specimens for Electron Microscopy

Thomas H. Giddings, Jr., and George P. Wray

Department of Molecular, Cellular, and Developmental Biology
University of Colorado
Boulder, Colorado 80309

1. INTRODUCTION

The application of freeze-drying to electron microscopy (EM) involves rapid freezing of the specimen followed by the removal of the frozen water by controlled sublimation. In the simplest case, freeze-drying is followed directly by observation in a transmission electron microscope. The freeze-drying step can also be incorporated into a wide range of more complex processing protocols, which we will refer to briefly, directing the reader to recent reviews and some examples from the literature. Our emphasis will be on the methodology of freeze-drying itself and on the results that can be expected from this technique. Today there are many preparative techniques to choose from, including critical point-drying, freeze-substitution, freeze-etching, and standard fixation and embedding. All are capable of yielding excellent morphological preservation when appropriately applied. We will discuss freeze-drying in the context of these methods in order to identify its unique capabilities.

As an introduction to this subject, we can consider the following: When specimens are prepared by freeze-drying, chemical fixation is optional. When prior chemical fixation is omitted, fixation is achieved by the initial rapid freezing of the sample. Fixation by rapid freezing is both faster and more "comprehensive" than chemical fixation; it is complete within milliseconds rather than seconds or minutes (Fitzharris *et al.*, 1972; Gilkey and Staehelin, 1986; Mersey and McCully, 1978) and all molecules are immobilized rather than just those that can be cross-linked by the first chemical reagent (e.g., glutaraldehyde). The distribution of small molecules, such as ions and organic monomers, as well as macromolecules can remain essentially unchanged from the living state. Since the sample is dried without exposure to fixative or solvents, unlike critical point-drying and freeze-substitution, there is little or no opportunity for subsequent diffusion of soluble compounds. Avoidance of chemical fixatives and solvents can also be advantageous in preserving enzymatic activity for cytochemical procedures and antigenicity for immunolabeling. Freeze-drying has been employed in more strictly morphological studies as a control for potential artifacts arising from exposure to fixatives and solvents.

Freeze-drying relates closely to the subject matter of some of the other chapters of this book. When freeze-drying or "etching" is limited to the upper exposed surface of a sample and is followed by metal shadowing, removal of the biological material, and visualization of the metal replica rather than the sample itself, the process is considered freeze-etching, which is dealt with in Chapter 8. We will, however, consider the use of metal shadowing to enhance contrast and preservation in samples that have been completely freeze-dried. Electron microscopy of frozen-hydrated samples requires specialized equipment, i.e., an electron microscope equipped with a low-temperature stage; it will be dealt with here very briefly as a point of comparison with the preservation obtained by freeze-drying.

2. HISTORICAL ORIGINS

Freeze-drying as a preparative method for electron microscopy was first developed in the 1940s as an alternative to air-drying, before the advent of current fixation procedures. Wyckoff (1946) was the first to describe the freeze-drying and metal shadowing of viruses, bacteria, and some plant and animal tissues for observation in the electron microscope. Williams (1953), working with tobacco mosaic virus and red blood cell ghosts, further developed the technique of freeze-drying and shadowing. The quality of morphological preservation in both cases was superior to that obtained by air-drying. These early reports dealt with many of the concerns that still deserve attention when freeze-drying is undertaken today. They include observations that: residual salts can obscure detail (Wyckoff, 1946); faster freezing improves preservation; freeze-drying above a certain temperature will have deleterious effects; sublimation must be thoroughly completed before the sample is warmed up; condensation can occur and damage ultrastructure if samples are removed from the vacuum while still cold; and some collapse can occur but can be minimized by prior fixation (Williams, 1953).

3. METHODS AND INSTRUMENTATION

Since there are many different applications for freeze-drying, there are also many variations in the technique. Nevertheless, the processes of rapid freezing and controlled sublimation are common to all of these procedures and are the source of the most significant problems. Damage from the formation of ice crystals is the most serious threat to good structural preservation in freeze-drying. Other considerations include keeping the sample surface free of contamination and preventing rehydration after freeze-drying. We will consider these requirements in detail in order to provide a rationale for the design of equipment and procedures discussed below.

3.1. The Formation of Ice Crystals, "Freezing Damage"

All cryofixation techniques, including freeze-drying, freeze-substitution, and freeze fracture, encounter the problem of ice crystal formation. When aqueous samples are frozen, ice crystals form and expand, excluding all nonaqueous components of the sample into an area of increasing solute concentration, called a eutectic. After freeze-drying, during which the crystals of pure ice are sublimed away leaving the eutectic behind, this damage is easily recognized by the appearance of clear spaces within the sample and distortion of surrounding structures. In whole cells for example (Porter and Anderson, 1982), voids in the cytoplasm will be present, membrane contours can be distorted, and in extreme cases, organelles or whole cells can be crushed. Freezing damage is more problematic in freeze-drying than freeze fracture because the most common (and most effective) cryoprotectants, such as glycerol or sucrose, must be avoided since they are not volatile enough to be sublimed away.

Samples are susceptible to ice crystal formation as long as they are fully or partially hydrated, but the most critical time is during the initial freezing process. Several excellent reviews on the theory and practice of rapid freezing for preservation of biological ultrastructure have been published (e.g., Costello and Corless, 1978; Gilkey and Staehelin, 1986; Plattner and Bachmann, 1982; Terracio and Schwabe, 1981). The key arguments include the following: The size of the ice crystals that form during a freezing process will vary as a function of the cooling rate. Faster cooling rates increase the rate of ice nucleation within the sample and thus increase the number but decrease the size of the ice crystals that are formed. A faster cooling rate also means that recrystallization, the growth of larger crystals at the expense of smaller ones, will be minimized because the sample will traverse the thermal range between the onset of ice nucleation and the minimum temperature at which recrystallization occurs more rapidly. Factors that influence the rate of heat extraction from a given area within the sample include the rate of heat extraction from the surface of the sample and the rate of heat flow from that area to the surface. The former is determined by the freezing method employed, summarized below. The latter depends mainly on the size and geometry of the sample, as well as the depth of the area of interest from the rapidly cooled surface (Gilkey and Staehelin, 1986).

Whether it is possible to prevent ice crystal formation completely by vitrifying the sample is controversial (Terracio and Schwabe, 1981). True vitrification implies that the water molecules are solidified without any reorganization into crystals. It is known that vitreous ice can be produced by condensing water vapor onto a cold surface at low pressure (Dowell and Rinfret, 1960). Vitrification of liquid water, however, has been well documented in only a few of the many reports in the literature. Bruggeller and Mayer (1980) showed by differential thermal analysis that they had vitrified micrometer-sized droplets of water suspended in emulsion form in ethane, a liquid that is a much better thermal conductor than water. Dubochet and co-workers have reported vitrification of very thin aqueous samples including suspensions of virus particles and bacteria (Adrian *et al.*, 1984; Dubochet *et al.*, 1983; McDowall *et al.*, 1983). Electron diffraction of the aqueous phase of these samples yielded a diffuse ring pattern characteristic of an amorphous solid rather than crystalline ice. Most other reports claim vitrification on the grounds that no evidence of ice crystal damage is visible in electron micrographs of the sample. However, the absence of visible evidence of ice crystal formation is not proof that a sample has been vitrified, only that the crystals are too small to distort ultrastructure sufficiently to be recognized.

Regardless of whether vitrification occurs during rapid freezing, it follows from the above discussion that the faster the rate of heat extraction from a given region of the sample, the better the ultrastructural preservation of that region. This was demonstrated by Schwabe and Terracio (1980), who measured actual cooling rates achieved by a variety of freezing techniques and found that increased cooling rates correlated with the reduction of ice crystal damage visible after freeze-drying. Based on the experience of many laboratories, the goal of "ultrarapid freezing," to cool the sample rapidly enough to control the growth of ice crystals to a level that is not detectable in the electron microscope, is reasonable and attainable (Gilkey and Staehelin, 1986; Schwabe and Terracio, 1980; Terracio and Schwabe, 1981).

3.2. Freezing Techniques

There are five general methods, each applied with innumerable individual variations, currently in use for rapid freezing for ultrastructural preservation. For comprehensive and detailed discussions of these methods, the reader is referred to the recent review by Gilkey and Staehelin (1986) and Chapter 8 in this book. To summarize very briefly, the most common freezing technique consists of plunging the sample into a cryogenic liquid, usually propane or Freon maintained just above its freezing point (Costello and Corless, 1978). Methods that have been developed for faster cooling rates include "slam-freezing," in which the sample is brought into contact with a cold metal block; jet-freezing, in which the cryogenic liquid is squirted onto the sample at high velocity; spray-freezing, in which fine droplets of sample are injected into the cryogenic liquid; and high-pressure freezing, in which ice crystal formation is suppressed by the high pressure. The choice of freezing method is influenced by the size and type of sample. This is discussed in Section 3.5. Once frozen, samples can be maintained in stable form under liquid nitrogen until being transferred to the freeze-drying apparatus.

3.3. Sublimation

Freeze-drying, or sublimation, takes place because water molecules equilibrate between the solid (ice) phase and the vapor phase. As the temperature of the ice increases, so does the vapor pressure of water in contact with it. When the water vapor is continuously removed by either a vacuum system, a desiccant, or a flow of dry gas, the vapor pressure of water is maintained below its equilibrium level, and a net loss of water molecules from the surface occurs. For pure water ice, 10°C can make about an order of magnitude difference in the sublimation rate (Davey and Branton, 1970). At −100°C, the rate is about 100 nm/min while at −90°C it increases to about 1000 nm/min.

Clearly, some adjustment must be made when estimating how long to freeze-dry a biological sample. Extracellular ice may freeze-dry at close to the same rate as pure water ice. However, sublimation from within intact cells and membrane-limited organelles takes much longer. Using low-temperature high-voltage electron microscopy to monitor the process of freeze-drying, we have observed that the interior of an intact cell will dry much more slowly than the extracellular space. Sublimation from organelles within the cell, such as mitochondria, takes even longer than the cytoplasm in general (Fotino and Giddings, 1985). A cultured mammalian cell, with a maximum thickness of 5 μm at the nucleus, can take 48 hr to dry completely at −90°C, for an average rate of approximately 100 nm/hr.

For fast and efficient freeze-drying, one generally wishes to select the highest temperature at which optimal structural preservation is retained. Unfortunately, we know of no systematic study of the effect of drying temperature on the quality of the resulting ultrastructure. It is known, nevertheless, that the sample must not be allowed to warm up to its recrystallization temperature, T_r, while totally or partially hydrated. Even if ice damage was prevented or controlled to an acceptable level during the initial freezing, damage can still occur during freeze-drying as a result of recrystallization. T_r for pure water ice is about −40°C (Mackenzie, 1977); for solutions, it can be somewhat lower. One freeze-fracture study suggested that yeast can be held at −50°C without evidence of recrystallization (Bank, 1973). Some types of whole mounts, such as virus particles and protein crystals, have been successfully freeze-dried at −35°C (Smith, 1980). It is possible in this case that the very thin aqueous layer that the particles were embedded in was completely etched away as the sample warmed up to the intended drying temperature, i.e., that the actual freeze-drying took place at a lower temperature. Larger samples, such as cultured mammalian cells, have generally been dried at −80 to −95°C (Porter and Anderson, 1982; Wray and Giddings, unpublished). Since the T_r of most biological samples is not known, a practical suggestion is to freeze-dry initially at low temperatures, perhaps −95°C, and then experiment with warmer temperatures.

3.4. Design Principles for a Freeze-Dryer

There are two major functions required of a freeze-dryer. One is to maintain the sample at the desired temperature. The other is to ensure that the vapor pressure of all volatile compounds, including water leaving the sample surface, is low enough in the vicinity of the sample that net accumulation on the cold sample is not favored. This can be accomplished with a flow of cold, dry gas or with a vacuum chamber. Freeze-drying in the presence of atmospheric pressure is slowed by the short mean free path of water molecules leaving the specimen surface, which results in the formation of a humid atmosphere around the sample (Zingsheim, 1984). Vacuum-based systems appear to yield consistently superior results and are in more general use.

Whether one is purchasing a commercial freeze-dryer, building one, or adapting a freeze-etch device, it is important to ensure that conditions within the device are optimized for freeze-drying. One of the most important requirements is that contamination of the sample surface must be avoided. Possible sources of contamination in a vacuum-based freeze-drying system are the sample, the system itself, and leaks. Leaks in the system will admit water and other vapors into the chamber that will condense on all cold surfaces including the sample. It is therefore wise to periodically check for leaks especially if a buildup of "frost" is noticed on cold surfaces. Helium leak detectors (mass spectrometers sensitive only to helium) are probably the best instruments for this purpose. The pumping system itself can be a source of contaminants, backstreaming diffusion pump oil being the major one. Adequate water-cooled baffles between the pump and the chamber can usually solve this problem. Non-oil-based pumps such as turbomolecular pumps avoid the problem altogether. Cold shrouds placed close to the sample can reduce contamination from all sources. Condensable molecules will be in equilibrium between the condensed and vapor phase. If the shroud is the coldest surface, condensables will tend to accumulate there. By positioning the shroud in such a way that molecules do not have a clear path to the sample, condensation on the samples can be minimized.

The second major component of a freeze-drying system is the specimen temperature control. If one desires maximum versatility in the selection of a freeze-drying temperature, then liquid nitrogen appears to be the most appropriate coolant. Since some samples require long freeze-drying at low temperatures, the freeze-dryer should be able to maintain sample temperature with a minimal rate of cryogen consumption. The stage temperature can be maintained at the desired level by a resistance heater driven by a thermocouple-based temperature controller. These components are all commercially available. Efficient designs, i.e., those that consume a minimum of liquid nitrogen, can be based on an understanding of heat flow within the device.

A schematic diagram of an efficient freeze-drying apparatus is shown in Figure 1. The liquid nitrogen is the ultimate heat sink in this system. Heat sources include radiation from warm surfaces of the apparatus, conduction through materials attached to the heat sink, and the specimen stage heating element. The objective is to control the flow of heat in the system to maintain the sample at the desired freeze-drying temperature with a minimal amount of heat transferred to the liquid nitrogen heat sink. A liquid nitrogen Dewar is attached to the specimen stage, eliminating the need for transfer lines. The Dewar can be a high-efficiency commercial Dewar or one constructed of concentric layers of a poor thermal conductor, such as stainless steel. In the latter case, the vacuum system provides the Dewar's insulation and the Dewar in turn improves the vacuum by cryopumping. If heat leaks from the external environment are controlled in this way, the remaining major source of heat is the sample stage heater. Heat flow from the sample stage to the heat sink must be controlled to avoid wasting liquid nitrogen. In the design depicted, this is accomplished by arranging the heating element in direct contact with the stage and placing a barrier to heat flow between the stage and the liquid nitrogen. In this way, heat supplied by the controlled element works to maintain the specimen temperature rather than boiling liquid nitrogen. The barrier can consist of a thermal insulator of appropriate thickness so that the stage will equilibrate at about $-120°C$ without external heat applied. The stage temperature can then be warmed to the freeze-drying range with a minimum of applied heat.

The specimen carrier should be designed to provide the best possible thermal contact between the specimen and the stage (Figures 2 and 3). The specimen carrier shown in Figure 3 was designed primarily to freeze-dry samples frozen directly on EM grids. The grids are clamped firmly into the carrier, which in turn fits snugly in the stage. Loading is done under liquid nitrogen. The carrier has sufficient mass to prevent significant warming during sample insertion into the freeze-dryer. It is important to test the actual specimen temperature relative to the stage temperature with a low-mass thermocouple placed on a grid, simulating an actual specimen. Even in a well-designed specimen holder, thermal contact may not be as good as one would expect. Grid temperatures can easily be $5-10°C$ warmer than the stage temperature.

The apparatus should incorporate an easy-to-operate specimen insertion/removal mechanism that allows specimen exchanges while the system is under vacuum and the stage is at low temperature. A double-sealed air lock chamber such as the one incorporated in the Fullam freeze-drying unit will accomplish this (Figure 1). If the gas volume in the air lock chamber is very small, it is not necessary to prepump it.

3.5. Specimen Preparation

Freeze-drying can be an intermediate step in such a wide variety of procedures, that handling before, during, and after the freeze-drying step is dictated

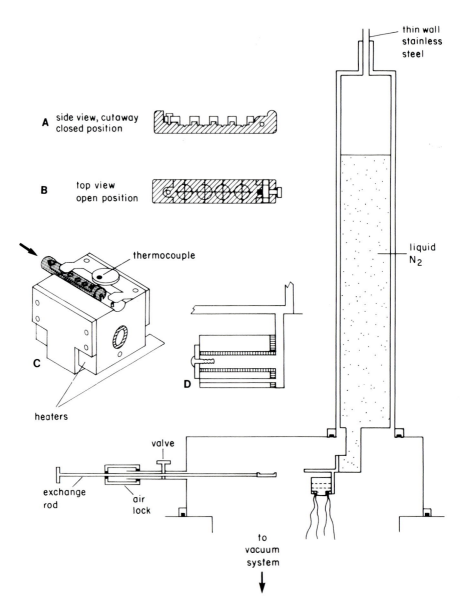

FIGURE 1. Schematic diagram of a freeze-drying apparatus built in our laboratory. Insets: (A, B) Specimen holder for EM grids, made of copper. (C) Specimen stage with grid holder (shaded) partially inserted. The stage is solid copper. Two foil heaters are clamped onto the sides by aluminum plates. A cylindrical plastic or stainless steel insulator slows heat flow to the heat sink. (D) Enlarged cross-section view of the specimen stage attached to a copper rod that in turn is attached to copper bars in contact with the liquid nitrogen heat sink. Shaded areas represent insulators.

FIGURE 2. Sample stage (designed by E.F. Fullam, Inc., Schenectady, NY) of the freeze-dryer depicted in Figure 1. A thermocouple attached to the top of the stage is monitored by an automatic temperature controller that drives two heating elements. The specimen carrier is shown in place.

to a large extent by requirements for prior or subsequent processing. We will begin with a discussion of techniques for freeze-drying whole mounts for direct visualization in a transmission electron microscope.

Attention must first be given to handling the specimen before it is fixed, either by the freezing step or by conventional chemical fixatives. For whole mounts, it is best to prepare the sample directly on EM grids. In the case of cultured animal cells (Porter and Anderson, 1982), the procedure involves picking up Formvar-coated grids on glass coverslips, carbon coating, glow discharging, and placing the coverslip with grids in a culture dish covered with a suspension of cells. This is the same procedure employed for critical point-drying this type of sample (Wolosewick and Porter, 1979a,b). The medium and growth conditions are manipulated to induce the cells to settle on the coverslip (some will be over grids) in a monolayer and to spread out forming thin, flat margins.

The normal growth medium is too rich in solutes to permit direct freeze-drying; it would leave behind a heavy residue. The procedure of Porter and Anderson (1982) of washing the cells briefly in isotonic (0.16 M) ammonium acetate buffer, replaces the growth medium with a volatile salt that will sublime to a large extent during the freeze-drying process. After rinsing in the volatile buffer, the sample is blotted quickly on buffer-moistened filter paper to leave a minimal aqueous layer, then frozen by plunge freezing in liquid propane. Reduction of the aqueous layer to a minimal depth is critical for two reasons: first, it permits an adequate freezing rate throughout the sample so that damaging ice crystals are not formed; and second, it will reduce the amount of nonvolatile

FIGURE 3. Specimen carrier. A hinged locking clamp holds EM grids in shallow wells. The carrier is treaded on one end for attachment to the insertion rod.

solutes that may remain in the extracellular phase and that could leave deposits on the cells.

Other types of whole mounts, viruses, bacterial cells, or isolated organelles for example, can be processed in a similar fashion. Grid preparation is usually the same, with Formvar and carbon coating. Glow discharging is recommended to promote even wetting of the grid by the sample. This is critical for obtaining a thin aqueous layer of uniform thickness prior to freezing. If necessary, materials such as polylysine can be applied to promote adhesion. As in negative staining, the sample can be applied by simply bringing the grid into contact with a suspension of the sample. Once again, the sample must be washed in a dilute medium, ideally not more than 10 mM in total solutes, or in a volatile buffer. If the sample is not stable in low-ionic-strength solutions or volatile buffers, then it is necessary to employ a chemical fixation, followed by washing in distilled water. There are other instances when prior application of a fixative may be appropriate. If dissection of tissue is required or any prolonged handling in unsuitable environments, fixation will probably be required to prevent damage prior to the freezing step. In addition, fixation, followed by thorough rinsing in distilled water or a dilute, volatile cryoprotectant such as 10% methanol, will extract unfixed, soluble components and improve the imaging of remaining fixed structures.

In our experience, samples thin enough to be viewed as whole mounts, including those prepared for high-voltage EM, are thin enough to be ultrarapidly frozen, i.e., without the formation of visible ice crystals, by simple plunging into liquid propane maintained below $-185°C$. After a few seconds in the propane, grids are transferred quickly into liquid nitrogen for storage. It is advisable to carefully blot off residual propane using filter paper immersed in the liquid nitrogen. Because samples mounted on grids are very fragile, the faster freezing techniques such as cold metal block freezing and propane jet freezing are not generally employed.

Of course, many structural questions cannot be answered by whole mount preparations. Intact tissues and often single cells must be sectioned in order to visualize internal structure. It is then necessary to freeze the larger sample and obtain frozen sections for freeze-drying, or to freeze-dry the tissue for subsequent embedding and conventional sectioning (see below). Since nonvolatile cryoprotectants cannot be used, a major difficulty in processing larger pieces of tissue is achieving an adequate freezing rate. The formation of large ice crystals is not only destructive of ultrastructure, but also makes collection of intact frozen sections extremely difficult (Dubochet et al., 1983). The use of a volatile cryoprotectant after fixation in order to suppress ice crystal formation has been employed for freeze-etching (Heuser and Kirschner, 1980; Schiller and Taugner, 1980) and may facilitate the cutting of frozen sections (Giddings, unpublished observations) but to our knowledge has not been extensively applied to freeze-drying. Those using cryoprotectants should be aware that such solutes can significantly lower the temperature of recrystallization (Nei, 1973), necessitating a lower freeze-drying temperature. Once obtained, frozen sections can be pressed onto coated grids (see Seveus, 1978) and freeze-dried. Due to the many difficulties of frozen sectioning and the marginal quality of the morphological preservation, freeze-drying of frozen sections has, to date, found utility only in cytochemical applications where other methods of tissue preparation are unacceptable, e.g., X-ray microanalysis of diffusible compounds.

There are several alternatives to direct visualization of the freeze-dried sample in the electron microscope. For example, a number of methods have been developed using freeze-drying to visualize suspensions of macromolecules, viruses, membranes, and various other small particles. The methods involve settling the particles on a smooth, flat surface; followed by rapid freezing, sublimation of the surrounding ice, and metal shadowing in a freeze-etch device. This technique was refined and extensively applied by Heuser (1983) who mixed a suspension of sample with a suspension of mica flakes, followed by etching and shadowing. After replication the sample is handled as a freeze-etch replica, i.e., the biological material and the mica flakes are digested away and the replica viewed in the electron microscope. A variation on this method involves settling the sample on a larger flat substrate such as a 5- to 10-mm^2 piece of mica, freeze-drying completely, an shadowing. In this case, the sample and replica are viewed together (Kistler and Kellenberger, 1977; Nermut and Frank, 1971; Studer et al., 1981). It is also possible to float the replica off the mica substrate and clean off the biological sample as is done with freeze-etch samples (Dewit and Staehelin, personal communication).

In addition to metal shadowing, it is possible to stabilize freeze-dried material by exposure to osmium vapor. Freeze-dryers have been constructed that permit exposure of the sample to osmium vapor immediately following freeze-drying, avoiding any exposure to air prior to fixation (Coulter and Terracio, 1977). The sample can then be viewed without embedding resin, in the case of whole mounts or freeze-dried sections, or embedded in resin for subsequent sectioning, in the case of tissue (e.g., Terracio and Schwabe, 1981). This technique, in conjunction with fast freezing by a cold metal block method, has been used to preserve antigenicity and localization in tissues for immunolabeling (Dudek et al., 1982).

3.6. The Freeze-Drying Step

The appropriate sublimation protocol depends entirely on the sample in question. Whole cells, with their intact plasma membranes and organelle membranes, take much longer at a given temperature than a suspension of nonmembranous particles. We have routinely freeze-dried at about −90 to −95°C to avoid recrystallization. At these temperatures, it is necessary to freeze-dry about 48 hr to ensure complete dehydration of all of the organelles in cells that are about 5 μm thick at the nucleus. This was confirmed by imaging the cells during the freeze-drying process by low-temperature high-voltage EM. Thin suspensions of particles in pure water ice should freeze-dry at a rate of 1000 nm/min at −90°C according to the results of Davey and Branton (1970). As a rule, however, we allow overnight freeze-drying in these cases to ensure complete dehydration. With efficient freeze-drying systems, this does not have to consume large volumes of liquid nitrogen. If the sample is to be imaged without further stabilization, it should be warmed to room temperature or slightly above to prevent condensation of water from the air onto the sample. Rehydration from humidity is a common problem and care must be taken to keep the sample dry until it is imaged. Carbon coating immediately after removal from the freeze-dryer not only stabilizes the sample for subsequent observation in the electron microscope, but also seems to provide some protection from the damaging effects of rehydration.

4. EXAMPLES OF RESULTS

The results that can be obtained by observation of freeze-dried and frozen-hydrated specimens in the transmission electron microscope can be divided into those that are purely morphological and those that are cytochemical.

4.1. Morphological Studies

4.1.1. Whole Mounts

Virus particles were among the first specimens successfully studied by the freeze-drying method (Williams, 1953; Wyckoff, 1946). An example of a viral plant pathogen prepared by freeze-drying and metal shadowing in a vacuum device (Roberts and Duncan, 1981) is shown in Figure 4. The resulting preservation and resolution were sufficient to give an indication of the packing geometry of the coat proteins. In general, preservation of structural detail at high resolution in freeze-dried and shadowed samples was superior to what was obtained for air-dried and shadowed preparations (Kistler and Kellenberger, 1977; Smith, 1980; Williams, 1953). Some viruses are better preserved by this method than by negative staining, which can cause disassembly or collapse (Roberts and Duncan, 1981). In addition, shadowed preparations can give a better idea of the three-dimensional surface shape of the capsid than negative staining can. Numerous examples of freeze-dried and shadowed viruses can be found in the literature.

Larger specimens, such as whole bacterial cells, isolated organelles, and eukaryotic cells, can be prepared as freeze-dried whole mounts for EM if the accelerating voltage of the electron microscope to be used is adequate to penetrate the specimen without significant loss of resolution. High-voltage (1000 kV) EM centers dedicated to biological studies have been available for several years and

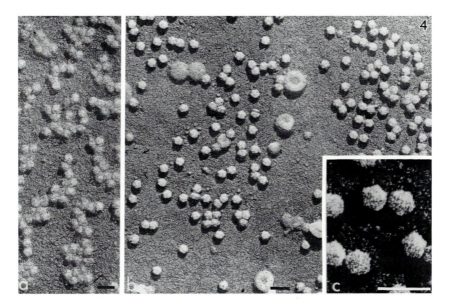

FIGURE 4. Lucerne transient streak virus metal shadowed after air-drying (a) or freeze-drying (b, c). Bar = 50 nm. From Roberts and Duncan (1981).

are open to visiting scientists. Intermediate-voltage (200–500 kV) microscopes are becoming increasingly common.

An instructive example of the application of freeze-drying to the study of whole bacterial cells can be found in the work of Todd *et al.* (1984). The structure and organization of cell surface pili in colonies of *Neisseria gonorrhoeae* were studied by freeze-drying, critical point-drying, and freeze-etching. Rapid freezing followed by freeze-drying *in vacuo* at a controlled temperature of −100°C for 24 hr and observation in a high-voltage electron microscope revealed the presence of a network of branched, rod-shaped structures, about 30 nm in diameter, forming a three-dimensional lattice between cells (Figure 5). Very similar images were obtained following fixation and critical point-drying (see Chapter 3). This suggests that the structures could not be dismissed as artifacts of either method. After negative staining, however, unbranched pili with a characteristic diameter of 7 nm were observed. Neither the branched thick rods nor the unbranched pili were observed in nonpiliated control cultures. The apparent disparity between the freeze-dried and negative-stained images was resolved by freeze-etch EM. This technique revealed individual pili adhering to the outer surface of the cell that coalesced into large bundles. The bundles but not single pili left the cell surface, branched, and formed the lattice that interconnects the cells in the colony. Thus, in this case, the image produced by freeze-drying was

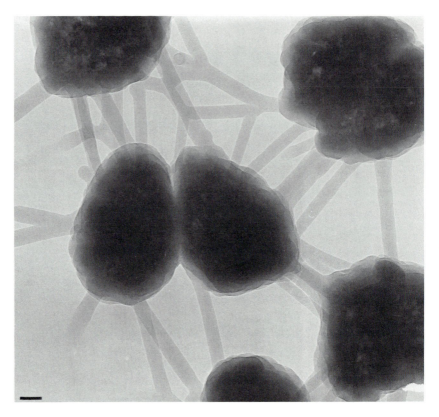

FIGURE 5. Piliated *Neisseria gonorrhoeae* prepared by freeze-drying without staining. Imaged at 1000 kV. Bar = 0.1 μm. From Todd *et al.* (1984).

accurate, but failed to resolve the substructure of individual pili within the larger bundles. This loss of high-resolution information in freeze-dried samples that have not been stabilized in some way by heavy metals, will also be apparent in subsequent examples and constitutes a limitation on the effectiveness of present freeze-drying technology.

Cultured animal cells have also been imaged following freeze-drying. Porter and Anderson (1982) freeze-dried PtK2 cells (a line of kidney cells derived from the marsupial *Potorus tridactylis*) that had been cultured directly on coated EM grids. The cells can be induced to settle in monolayers on the grids and flatten so that much of the cell's cytoplasm is in relatively thin margins, less than 2 μm thick (see Section 3.5 for discussion of the methods). Comparisons were made between these unfixed cells and others that were fixed prior to freeze-drying or critical point-drying. No significant differences were found. We have repeated these experiments in the course of developing a more efficient freeze-dryer. Some examples of the results are shown in Figure 6. Like Porter and Anderson (1982), we found that freeze-dried cells exhibited very good overall preservation, although most grids will have a mixture of well-preserved cells and others that have been damaged by ice crystals or by air-drying. In the case of unfixed, freeze-dried cells, the contrast obtained is remarkable in that it is derived solely from electron scattering by the biological material; no added heavy metals are present. An example showing mitochondrial cristae is shown in Figure 6b.

Some shrinkage is frequently observed around mitochondria and outside the nuclear envelope. Observations of the process of freeze-drying (Fotino and Giddings, 1985) indicate that this shrinkage occurs in the latter stages of dehydration although it could represent a latent effect of inadequate freezing (Gilkey and

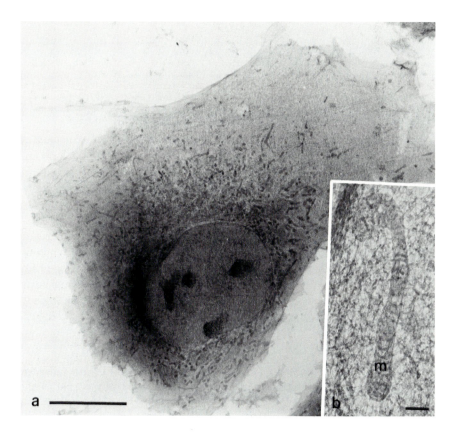

FIGURE 6. (a) Normal rat kidney (NRK) cell cultured on an EM grid, fixed with glutaraldehyde, rinsed with distilled water, and freeze-dried. Bar = 10 μm. (b) Mitochondrion (m) in the cytoplasm of an NRK cell that was freeze-dried without prior chemical fixation. Bar = 0.2 μm. Both micrographs taken at 1000 kV.

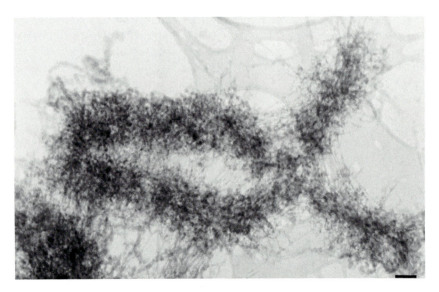

FIGURE 7. Freeze-dried metaphase chromosome isolated from a human lymphocyte. Imaged at 1000 kV. Bar = 0.2 μm. Chromosomes isolated by Dr. Gunter Bahr and Walter Engler.

Staehelin, 1986). The membranes that surround these organelles slow the rate of sublimation so that the interior of the organelles remains hydrated after the cytoplasm has completed dehydration. When the organelles finally do dehydrate, their diameter may decrease by 25%. The cytoplasmic lattice surrounding them often contracts at the same time, leaving a clear halo.

Isolated organelles or other subcellular fractions can also be processed as whole mounts by rapid freezing followed by freeze-drying. We have imaged isolated human lymphocyte metaphase chromosomes, sea urchin embryo mitotic spindles, and reassembled microtubules by freeze-drying and high-voltage EM. An example of the results obtained by freeze-drying isolated human lymphocyte chromosomes is shown in Figure 7. The isolated chromosomes were attached to coated EM grids, washed in distilled water, and frozen without prior chemical

---

→

FIGURE 8. Mitotic spindle from a sea urchin embryo, freeze-dried (without chemical fixation) and imaged in a high-voltage (1000 kV) electron microscope equipped with a cold stage. Dividing cells were lysed on coated EM grids and washed in 10 mM PIPES buffer containing 10 μM taxol to stabilize microtubules (preparation by Dr. E. D. Salmon). Bar = 5 μm.
FIGURE 9. Microtubules in a spindle prepared as in Figure 8. Although reasonably well preserved, the microtubules lack visible substructure and have variable outside diameters. Bar = 0.1 μm.
FIGURE 10. Isolated sea urchin microtubules freeze-dried and imaged by low-temperature high-voltage EM after fixation and staining with uranyl acetate. Excess stain was removed by rinsing in distilled water prior to freezing to give a ''positive'' stain. Microtubules isolated by Dr. John Scholey. Bar = 0.1 μm.

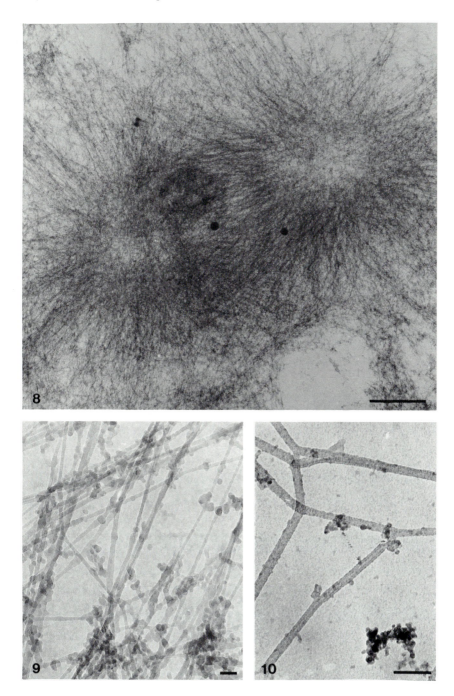

fixation. The structure is generally similar to what had been obtained by fixation and critical point-drying (Bahr and Engler, personal communication). There is relatively little indication of collapse onto the support film as judged by measurement of the depth visible in stereo pairs. Single chromatin fibers exhibit adequate contrast at 1000 kV, even though no heavy metal staining had been applied.

Microtubules are clearly visible in the freeze-dried mitotic spindle shown in Figures 8 and 9. Material between microtubules is visible in these preparations and similar to filamentous material seen in sections (Salmon and Segall, 1980). The lysed cell was washed in weakly buffered taxol to stabilize microtubules. The chromosomes were not well preserved by this treatment.

Examples of freeze-dried isolated microtubules are shown in Figure 10. Microtubules that were fixed and stained with uranyl acetate exhibit fairly well-preserved shapes, and protofilaments can be resolved. Unfixed unstained microtubules, freeze-dried from 10 μM taxol in dilute buffer, show irregular diameter and no indication of subunit structure (similar to those observed in whole spindles, Figure 9). This appears to be a characteristic feature of biological samples freeze-dried without application of heavy metal stain or coating.

4.1.2. Freeze-Dried Cryosections

Specimens too large to be viewed in whole mount preparations may be frozen as blocks, then sectioned by cryoultramicrotomy. Since the techniques are more difficult, require specialized equipment, and often yield results of inferior quality in terms of morphological preservation, freeze-drying of cryosections has been applied mainly for the purpose of cytochemical localization (described below).

4.2. Cytochemical Localization

Freeze-drying takes advantage of the rapid spatial fixation within cells of all molecules, both high and low molecular weight, that can be obtained by rapid freezing. Components that are soluble or unstable in fixatives or organic solvents have no opportunity to diffuse from their *in vivo* location and losses of enzymatic activity or antigenicity are minimized. The methods of detection include X-ray microanalysis and electron energy loss spectroscopy for elemental analysis review by Moreton, 1981), autoradiography (e.g., Johnson and Bronk, 1979), and immunolabeling. There are many examples in the literature in which ionic distributions have been mapped by elemental analysis following freeze-drying. Karp *et al.* (1982) were able to localize K, Na, and Cl ions in freeze-dried cryosections of skeletal muscle (see also Somlyo *et al.*, 1977). As shown in Figure 11, the quality of morphological preservation achieved by these authors was exceptional. Autoradiography of freeze-dried cryosections has been used to localize [22]Na uptake by mouse kidney cells (Baker and Appleton, 1976) and

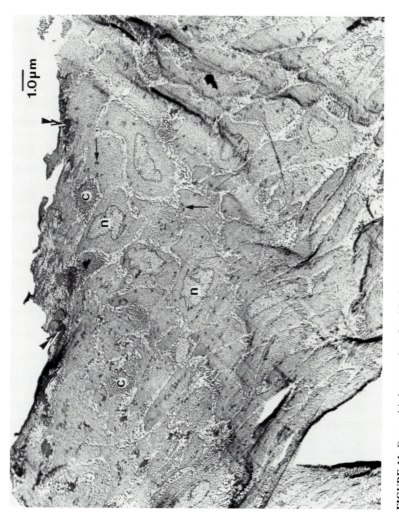

FIGURE 11. Freeze-dried cryosection of rapidly frozen rabbit vascular smooth muscle tissue. Visible ultrastructure includes internal elastic lamellae (double arrowheads), mitochondria (small arrows), nuclei (n), and collagen bundles (c). From Karp *et al.* (1982).

uptake of a tritiated sugar by intestinal epithelia (Johnson and Bronk, 1979). Freeze-drying followed by embedding usually results in superior ultrastructural preservation if sectioning is required and still eliminates exposure of the sample to liquid fixatives and all of the solvents except the resin itself. Burry and Lasher (1978) localized [^{3}H]-γ-aminobutyric acid uptake by cultured neuronal cells by autoradiography of sections prepared in this way. This type of processing has also been used to preserve antigenicity and localization for immunolabeling (Dudek et al., 1982).

4.3. Frozen-Hydrated Specimens

Perhaps the closest approximation of visualizing living specimens is achieved by imaging in the frozen-hydrated state. Some remarkable examples of results obtained with this technology are shown in Figures 12 and 13 (see Adrian et al., 1984; Dubochet et al., 1983). Morphological preservation appears to be superior to images of comparable samples that have been freeze-dried. Since an EM equipped with a high-resolution cold stage is required for this work, virtually all of these images have been produced by laboratories devoted to the techniques and instrumentation of cryomicroscopy. Cold stages are now becoming available from most EM manufacturers and several independent firms. Hopefully, these successful attempts will be reproducible with commercially available equipment in a wider range of laboratories. Aside from the EM modifications that are required, the techniques associated with sample preparation are essentially the same as those encountered in freeze-drying.

4.4. Potential Artifacts of the Freeze-Drying Method

Part of the motivation for undertaking freeze-drying as a preparative procedure is that steps that are known to introduce certain types of artifacts, such as fixation with glutaraldehyde and osmium, can be avoided. However, freeze-drying can also introduce artifacts. Since nonvolatile cryoprotectants cannot be used, damage from ice crystal formation may occur. Samples thin enough to be imaged as whole mounts including viruses (e.g., Roberts and Duncan, 1981; Smith, 1980; Williams, 1953), bacteria (Todd et al., 1984), or cultured mammalian cells (Porter and Anderson, 1982) have been frozen and freeze-dried without significant ice damage. Larger samples usually exhibit freezing damage more than 10–20 μm from the surface. Cryosectioning, even of optimally frozen tissue, introduces considerable damage in the form of chatter and deformation of structures (Dubochet et al., 1983).

An additional source of artifacts arises from the fact that soluble components of the cell and surrounding medium will remain, either attached to other structures or forming structures of their own. Miller et al. (1983) described latticelike structures produced by freeze-drying buffers or salt solutions. Clearly,

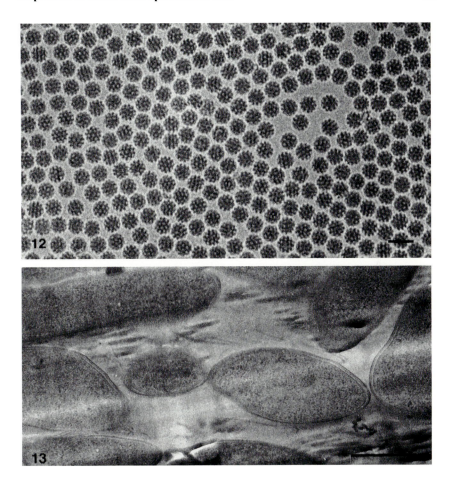

FIGURE 12. Frozen-hydrated Semliki Forest viruses. Bar = 0.1 μm. Courtesy of Dr. J. Dubochet. See Adrian *et al.* (1984) for details.

FIGURE 13. Frozen-hydrated section of *E. coli* cells displaying excellent preservation of the cytoplasm and cell envelope. Bar = 0.5 μm. From Dubochet *et al.* (1983).

all nonvolatile compounds in a sample will remain after freeze-drying and contribute to the image. For extracellular spaces, this can be minimized by using a volatile buffer such as ammonium acetate and reducing the extracellular fluid to a minimum before freezing. The extent to which intracellular soluble compounds contribute to apparent cytoskeletal structures is difficult to assess. However, the observation that the lattice structure of cytoplasm in freeze-dried cells is essentially identical to what is observed in fixed and critical-point-dried cells (Porter and Anderson, 1982) suggests that the image is not likely to be an artifact of either method.

A disturbing concern is the inability to resolve fine detail in unstained biological samples that have been imaged directly following freeze-drying. For example, compare the images of unstained microtubules in Figure 9 to those of stained microtubules in Figure 10. The results of Todd *et al.* (1984) suggest a similar loss of high-order structure in bacterial pili when samples are freeze-dried or critical point-dried. The fine structure of the bundled pili was preserved, however, by metal shadowing after freeze-etching. Thus, sublimation of surrounding ice probably does not directly cause the loss of structural information. Instead, it appears that freeze-dried samples are very sensitive to damage in the electron beam especially when viewed free of embedding matrix and without stabilization by heavy metals applied either as a ''stain'' or as a ''shadow.'' This loss of higher-order structure is not due to absorption of water or any other consequence of exposure to atmosphere since the same phenomenon occurs when samples are freeze-dried directly in the cold stage of an EM. Possibly low-electron-dose techniques could reduce this effect.

5. CONCLUDING REMARKS

5.1. Appropriate Applications of Freeze-Drying

Hopefully, we have shown that freeze-drying equipment and procedures need not be prohibitively expensive. Nevertheless, given the wide range of alternative methods for preparing samples for EM, when is it appropriate to employ freeze-drying? In many cases, cytochemical localization of diffusible compounds and ions cannot be achieved by any other method (frozen-hydrated specimens may be an alternative if one has access to instruments equipped with a cold stage). The ability of freeze-drying to preserve antigenicity for immunolabeling has been established (e.g., Dudek *et al.*, 1982).

In terms of morphological preservation, the following can be considered as potential alternatives: critical point-drying, freeze-substitution, and deep-etching (shadowing). Conventional critical point-drying, like freeze-drying, yields a resinless whole mount or section. Resinless samples have numerous advantages related to the absence of the electron-scattering matrix. These include the ability to image structures of similar (low) electron density to common embedding materials; and the ability to penetrate much thicker samples at a given accelerating voltage without loss of resolution. These methods, particularly when combined with stereo imaging, can yield an excellent overall view of cell structure. However, critical point-drying involves exposure to solvent and chemical fixatives. Freeze-drying can be used in conjunction with critical point-drying as a control for possible artifacts of exposure to the fixatives and solvents. In addition, like all methods involving rapid freezing as a first step, the time resolution, or speed of fixation is much greater so that fast processes on a molecular scale can be captured. Conventional freeze-substitution, although achieving rapid freeze-fixation, involves exposing the sample to fixatives and solvents and does not result in a resinless sample. However, it is possible to combine freeze-substitution with subsequent critical point-drying to achieve a resinless specimen as Porter and Anderson (1982) have done.

Deep-etching as described by Heuser and Kirschner (1980) can yield spectacular-looking micrographs and incorporates many of the same advantages as simple freeze-drying. However, there are significant differences. Deep-etching has been employed successfully to image cytoskeletal elements. In such preparations, the cells are first heavily extracted to remove most soluble components, including ions, organic monomers, and many soluble macromolecules including some proteins and membranes. Thus, much of the native cell structure has been removed from the image. Freeze-drying also has the advantage of being able to visualize greater depth, so that the architecture of whole cells can be presented in single micrographs or stereo pairs.

5.2. Need for Further Analysis

Perhaps due to the sporadic use of freeze-drying and the diversity of purposes to which it has been applied, there have been few if any systematic studies of the quality of ultrastructural preservation that it delivers. As it now stands, we can say that freeze-drying is a relatively simple technique that is free of many of the potential artifacts of other methods. It can produce a resinless sample suitable for three-dimensional imaging. Its applications to cytochemical problems are well established. Its prospects as an imaging technique are intriguing and require more exploration.

ACKNOWLEDGMENT. This work was supported by NIH Grant RR-00592 from the Division of Research Resources.

6. REFERENCES

Adrian, M., Dubochet, J., Lepault, J., and McDowall, A. W., 1984, Cryo-electron microscopy of viruses, *Nature (London)* **308**:32–36.

Baker, J. R., and Appleton, T. C., 1976, A technique for electron microscope autoradiography (and X-ray microanalysis) of diffusible substances using freeze-dried fresh frozen sections, *J. Microsc. (Oxford)* **108**:307–315.

Bank, H., 1973, Visualization of freezing damage. II. Structural alterations during warming, *Cryobiology* **10**:157–170.

Bruggeller, P., and Mayer, E., 1980, Complete vitrification in pure liquid water and dilute aqueous solutions, *Nature (London)* **288**:569–571.

Burry, R. W., and Lasher, R. S., 1978, Freeze-drying of unfixed monolayer cultures for electron microscope autoradiography, *Histochemistry* **58**:259–272.

Costello, M. J., and Corless, J. M., 1978, The direct measurement of temperature changes within freeze-fracture specimens during rapid quenching in liquid coolants, *J. Microsc. (Oxford)* **112**:17–37.

Coulter, H. D., and Terracio, L., 1977, Preparation of biological tissue for electron microscopy by freeze-drying, *Anat. Rec.* **187**:477–493.

Davey, J. G., and Branton, D., 1970, Subliming ice surfaces: Freeze-etch electron microscopy, *Science* **168**:1216–1218.

Dowell, L. G., and Rinfret, A. P., 1960, Low temperature forms of ice as studied by X-ray diffraction, *Nature (London)* **188**:1144–1148.

Dubochet, J., McDowall, A. W., Menge, B., Schmid, E. N., and Lickfeld, K. G., 1983, Electron microscopy of frozen-hydrated bacteria, *J. Bacteriol.* **155**:381–390.

Dudek, R. W., Childs, G. V., and Boyne, A. F., 1982, Quick-freezing and freeze-drying in preparation for high quality morphology and immunocytochemistry at the ultrastructural level: Application to pancreatic beta cell, *J. Histochem. Cytochem.* **30**:129–138.

Fitzharris, T. P., Bloodgood, R. A., and McIntosh, J. R., 1972, The effect of fixation on the wave propagation of the protozoan axostyle, *Tissue Cell* **4**:219–225.

Fotino, M., and Giddings, T. H., 1985, Ultrastructural visualization of unfixed and unstained whole mounts by high-voltage electron microscopy at low temperatures. *J. Ultrastruct. Res.* **91**:112–126.

Gilkey, J. C, and Staehelin, L. A., 1986, A critical evaluation of the use of ultrarapid freezing techniques for the preservation of cellular ultrastructure, *J. Electron Microsc. Tech.* 1986.

Heuser, J. E., 1983, Procedure for freeze-drying molecules adsorbed to mica flakes, *J. Mol. Biol.* **169**:155–195.

Heuser, J. E., and Kirschner, M. W., 1980, Filament organization revealed in platinum replicas of freeze-dried cytoskeletons, *J. Cell Biol.* **86**:212–234.

Johnson, I. T., and Bronk, J. R., 1979, Electron microscope autoradiography of a diffusible intracellular constituent, using freeze-dried frozen sections of mammalian intestinal epithelium, *J. Microsc. (Oxford)* **115**:187–194.

Karp, R. D., Silcox, J. C., and Somlyo, A. V., 1982, Cryoultramicrotomy: Evidence against melting and use of a low temperature cement for specimen orientation, *J. Microsc. (Oxford)* **125**:157–165.

Kistler, J., and Kellenberger, E., 1977, Collapse phenomena in freeze-drying, *J. Ultrastruct. Res.* **59**:70–75.

McDowall, A. W., Chang, J.-J., Freeman, R., Lepault, J., Walter, C. A., and Dubochet, J., 1983, Electron microscopy of frozen hydrated sections of vitreous ice and vitrified biological samples, *J. Microsc. (Oxford)* **131**:1–9.

Mackenzie, A. P., 1977, Non-equilibrium freezing behavior of aqueous systems, *Philos. Trans. R. Soc. London Ser B* **278**:167–189.

Mersey, B., and McCully, M. E., 1978, Monitoring of the course of fixation of plant cells, *J. Microsc. (Oxford)* **114**:49–76.

Miller, K. R., Prescott, C. S., Jacobs, T. L., and Lasignal, N. L., 1983, Artifacts associated with quick-freezing and freeze-drying, *J. Ultrastruct. Res.* **82**:123–133.

Moreton, R. B., 1981, Electron-probe X-ray microanalysis: Techniques and recent applications in biology, *Biol. Rev.* **56**:409–461.

Nei, T., 1973, Growth of ice crystals in frozen specimens, *J. Microsc. (Oxford)* **99**:227–233.

Nermut, M. V., and Frank, H., 1971, Fine structure of influenza A2(Singapore) as revealed by negative staining, freeze-drying and freezer-etching, *J. Gen. Virol.* **10**:37–51.

Plattner, H., and Bachmann, L., 1982, Cryofixation: A tool in biological ultrastructural research, *Int. Rev. Cytol.* **79**:237–304.

Porter, K. R., and Anderson, K. L., 1982, The structure of the cytoplasmic matrix preserved by freeze-drying and freeze-substitution, *Eur. J. Cell Biol.* **29**:83–96.

Roberts, I. M., and Duncan, G. H., 1981, A simple device for freeze-drying electron microscope specimens, *J. Microsc. (Oxford)* **124**:295–303.

Salmon, E. D., and Segall, R. R., 1980, Calcium-labile mitotic spindles isolated from sea urchin eggs (*Lytechinus variegatus*), *J. Cell Biol.* **86**:355–365.

Schiller, A., and Taugner, R., 1980, Freeze-fracturing and deep-etching with the volatile cryoprotectant ethanol reveals true membrane surfaces of kidney structures, *Cell Tissue Res.* **210**:57–69.

Schwabe, K. G., and Terracio, L., 1980, Ultrastructural and thermocouple evaluation of rapid freezing techniques, *Cryobiology* **17**:571–584.

Seveus, L., 1978, Preparation of biological material for X-ray microanalysis of diffusible elements, *J. Microsc. (Oxford)* **112**:269–279.

Smith, P. R., 1980, Freeze-drying specimens for electron microscopy, *J. Ultrastruct. Res.* **72**:380–384.

Somlyo, A. V., Shuman, H., and Somlyo, A. P., 1977, Elemental distribution in striated muscle and the effects of hypertonicity: Electron probe analysis of cryo sections, *J. Cell Biol.* **74**:828–857.

Studer, D., Moor, H., and Gross, H., 1981, Single bacteriorhodopsin molecules revealed on both surfaces of freeze-dried and heavy metal-decorated purple membranes, *J. Cell Biol.* **90**:153–159.

Terracio, L., and Schwabe, K. G., 1981, Freeze-drying of biological tissues for electron microscopy, *J. Histochem. Cytochem.* **29**:1021–1028.

Todd, W. J., Wray, G. P., and Hitchcock, P. J., 1984, Arrangement of pili in colonies of *Neisseria gonorrheae*, *J. Bacteriol.* **159**:312–320.

Williams, R. C., 1953, A method of freeze-drying for electron microscopy, *Exp. Cell Res.* **4**:188–199.

Wolosewick, J. J., and Porter, K. R., 1979a, Microtrabecular lattice of the cytoplasmic ground substance: Artifact or reality?, *J. Cell Biol.* **82**:114–139.

Wolosewick, J. J., and Porter, K. R., 1979b, Preparation of cultured cells for electron microscopy, in: *Practical Tissue Culture Applications* (K. Maramorosch and H. Hirumi, eds.), pp. 59–85, Academic Press, New York.

Wyckoff, R. W. G., 1946, Frozen-dried preparations for the electron microscope, *Science* **104**:36–37.

Zingsheim, H. P., 1984, Sublimation rates of ice in a cryoultramicrotome, *J. Microsc. (Oxford)* **133**:307–312.

Chapter 10

Low-Temperature Embedding

Barbara L. Armbruster

Division of Infectious Diseases
Washington University Medical School
St. Louis, Missouri 63110

and

Edward Kellenberger

Abteilung Mikrobiologie
Biozentrum, Universität Basel
CH-4056 Basel, Switzerland

1. INTRODUCTION

1.1. General Considerations

The examination of cell structure by means of electron microscopy is severely limited by several factors: (1) preparation artifacts as living specimens are subjected to the rigors of chemical fixation, dehydration in organic solvents, and embedment in various plastics and resins (most involving high-temperature curing); (2) distortion generated during sectioning; and (3) beam damage during observation. Suitable preparation methods must be developed before the pursuit of high-resolution data from thin-sectioned material can generate meaningful results. Sjöstrand (1976) has considered the parameters necessary to minimize conformational changes: (1) peptide chain freedom of movement is lessened by inter- and intramolecular cross-linking, (2) low temperature should reduce the extent of conformational changes (i.e., rearrangement of the peptide chains) that are due to the actions of organic liquids, (3) complete dehydration should be avoided to maintain hydration shells, and (4) the specimen environment should be polar.

The development of the Lowicryl resins was a direct outgrowth of the work of Sjöstrand and colleagues on the appearance of mitochondrial membranes after exposure to various fixation and embedment regimes (Sjöstrand, 1976; Sjöstrand

Present address for BLA: Monsanto Company, St. Louis, Missouri 63167.

and Barajas, 1968; Sjöstrand and Kretzer, 1975). Under the low-denaturation regime (fixation in 1% or less glutaraldehyde, dehydration with ethylene glycol, and low-temperature embedment in Vestopal, cured with ultraviolet light at low temperature), the contrast of cristal membranes was reversed. Mitochondria showed well-defined inner and outer membranes, the inner membrane being compact with no intracristal space and having stained regions in the center of the cristal membrane, or composed of two distinct membrane units (Sjöstrand and Kretzer, 1975). The exact conditions leading to the "broad band," "double line," or intermediate phases could not be determined until resins whose physicochemical properties could be independently adjusted were developed (the theoretical basis of these problems is discussed in Kellenberger et al., 1980).

The three main groups of resins, the epoxys, the polyesters, and the methacrylates, were investigated in this respect by Carlemalm and Villiger, joined later by Acetarin and others in the Basel laboratory. By virtue of their low viscosity and of the possibility to cure the resins at low temperatures, the acrylate–methacrylate mixtures became the focus of experimentation (Carlemalm et al., 1982a). The potential of polyesters might justify further exploration (Acetarin, 1981), whereas no formulations of epoxys were found to be both adequate for low-temperature work and preservation of tissue.

Four Lowicryl formulations presently available were the result of this work. As discussed below, they share several properties with previous methacrylate embedding media but many deleterious characteristics have been avoided by introducing strong cross-linking.

1.2. Development of Methacrylate-Based Resins

Studies using methacrylate-embedded specimens were primarily conducted in the 1950s. Newman et al. (1949) introduced the technique, discussing major problems with polymerization damage. The use of benzoyl peroxide as an initiator and polymerization at 40–50°C for 6–8 hr produced swollen blocks, the phenomenon being termed "explosion damage." Birbeck and Mercer (1956) suggested that these explosions were due to osmotic swelling of tissue as unpolymerized monomer dissolved into polymer within the tissue. Studies by phase-contrast microscopy (Borysko, 1956) confirmed that most of the serious distortions occurred during polymerization. In addition, shrinkage might also be observed: when resin polymerized faster outside than inside the cell, the monomer might be extracted out and thus cause cells to shrink considerably. Explosion or shrinkage of cells in the embedding material also depends strongly on the type of fixation and other pretreatments of the cell, because their products might interfere in different ways with the initiation and catalysis of the curing reactions. Borysko and Sapranauskas (1954) overcame these problems by using prepolymerized resin, thus matching the viscosity (and in turn the rate of polymerization) inside and outside the tissue. Alternatively, cross-linkers could be added to the resin mixture (Watson, 1963). Polymerization of methacrylates at high temperature (Borysko, 1956) was preferred because the tissue was then usually stronger and resisted damage during resin infiltration. Low-temperature infiltration with polymerization by means of ultraviolet light allowed a longer time for osmotic damage to occur.

Another disadvantage in using the early methacrylate formulations was the tendency of the embedment to melt and to distort the biological material during examination in the electron beam. Using metal-shadowed sections of embedded material coated both before and after first observation, Ryter and Kellenberger (1958) showed that non-cross-linked methacrylate mixtures melted during exposure to the electron beam, causing distortions and loss of image fidelity.

1.3. Development of Hydrophilic Resins and Lowicryl

The difficulties presented above instigated the interdependent development of embedding resins based on epoxys (Glauert et al., 1956; Glauert and Glauert, 1958; Spurr, 1969) and polyesters (Kellenberger et al., 1956). The most used representatives bf these resins (Epon, Spurr's, Vestopal) are neither very hydrophilic nor of a polar nature. It is generally believed—but was never really demonstrated—that hydrophilic, polar media should preserve macromolecular structure better than hydrophobic ones. It was reasoned that hydrophilic media should be more similar to water and thus be less deleterious to structural features maintained in an aqueous environment. It was also thought that hydrophilic embedding media should be more suited for histochemical reactions and particularly for specific labeling with antibodies and lectins.

Hydrophilic resins are, by definition, either completely miscible with water or, at least, are able to dissolve a large amount of water. Concentrations of water-miscible resins in a graded series could be used as dehydrating agents rather than alcohols or other organic solvents, thus eliminating the need for intermediate solvents. In addition, hydrophilic resins are in general able to polymerize in the presence of 20–30% water. In all known cases, blocks with such a high water content could not be successfully sectioned.

In the 1960s, several research groups focused their efforts on the development of "water-miscible" resins. Only Durcupan (Stäubli, 1963) is an epoxy-based resin and has to be polymerized at high temperature. All the others, such as glycol methacrylate (Bernhard, 1969; Rosenberg et al., 1960), hydroxypropyl methacrylate (Leduc and Holt, 1965), and mixtures of the latter (Leduc and Bernhard, 1967), were methacrylate-based. Since these resins were not cross-linked, all specimens embedded suffered from explosion or shrinkage and beam-induced melting. For an extensive review including additional resins of hydrophilic or hydrophobic nature, we refer the reader to Luft (1973).

None of these early methacrylate resins were designed particularly for low-temperature work. In 1982, Carlemalm et al. (1982a) published two new formulations of cross-linked acrylic–methacrylic resins designed for low-temperature infiltration and polymerization. Since one resin is highly polar and the other highly nonpolar, an experimental embedding system to examine the effect of polarity on preservation is feasible. In agreement with Rosenberg *et al.* (1960), Carlemalm's group also found that commercially available methacrylates had very variable properties and had to be purified. This necessarily led to the commercialization of two methacrylate mixtures designated Lowicryl K4M and HM20. The polar K4M was shown to be extremely suitable for immunolabeling by means of the protein A–gold technique (Roth *et al.*, 1980, 1981; Roth, 1982, 1983a,b) while the other, HM20, with its high hydrogen content is particularly suited for observation of unstained (without osmium, lead or uranyl salts) biological material with ratio-contrast (reviewed by Carlemalm *et al.*, 1985b).

2. TECHNICAL PROCEDURES

2.1. Composition and Properties of Resins

Briefly, the Lowicryl resins were designed to use the positive properties of methacrylates–acrylates and minimize polymerization damage and instability in the electron beam. The resins are stabilized by the addition of a cross-linker, triethyleneglycol dimethacrylate, at concentrations between 5 and 15 mole%. Within this range the resins have a stability almost equivalent to Epon in the electron beam. Higher concentrations of cross-linker can lead to brittle blocks, while lower concentrations produce unstable sections. The problems of explosion and shrinkage are linked to polymerization rate (see Section 1.2). The combination of photoinitiators, such as benzoin methylether and benzoin ethylether, with low temperature and indirect ultraviolet irradiation allows controlled and uniform polymerization. No effects of shrinkage are seen in specimens polymerized correctly, for loss of volume is restricted to the resin at the top of the container. One also has to keep in mind that a volume shrinkage of 10% corresponds to a linear shrinkage of only 3.5%.

Four Lowicryl formulations are currently available, the polar K4M and K11M and the nonpolar HM20 and HM23. K11M and HM23 were recently introduced for work at still lower temperatures; only limited experience is available concerning them (Carlemalm *et al.*, 1985b). We will therefore restrict our discussion to the two formulations that have benefitted from several years of use. K4M is a polar, low-viscosity formula whose main component is hydroxypropyl methacrylate (Carlemalm *et al.*, 1982a). It is miscible with most polar dehydrating agents, including ethylene glycol, and blocks containing up to 6% water can be sectioned without too much difficulty. In contrast to Epon, the polymerization reaction of K4M is not disturbed by the presence of water. The chemical requirements for achieving the polarity of K4M restrict its use to above −40°C; below this temperature its viscosity increases rapidly with decreasing temperature. Polymerization with ultraviolet light is, however, possible down to about $-50°C$. HM20 is a nonpolar (hydrophobic) resin of low density (1.1 g/ml) and very low viscosity that can be used down to $-60°C$. Polymerization must take place at $-50°C$ or above, for reasons of chemical reactivity. It is miscible with most polar dehydration agents, though not sufficiently with ethylene glycol or dimethylformamide.

2.2. Means to Achieve Low Temperatures

The most convenient way to achieve low temperatures is to use commercially available equipment (Balzers Union). Obviously, different types of freezers can be used to achieve low temperatures as well as freezing mixtures. Further details about improvised equipment are described in the instructions accompanying the Lowicryl kits.

2.3. Fixation

Specimens—small tissue pieces, cell suspensions, or monolayers of cells—can be chemically fixed with various aldehydes or rapidly frozen by plunging into liquid propane, by the propane jet or against a cold copper block. Osmication of specimens should be avoided when one intends to polymerize the resin with ultraviolet irradiation, since the brownish-black osmium reaction product screens off the UV light from the interior of the specimen, resulting in incomplete polymerization.

2.4. Dehydration Procedures

Chemically fixed specimens can be dehydrated by solvents in conjunction with the progressive lowering of temperature (PLT) technique or conventionally dehydrated either at room temperature or in a cold room. Rapidly frozen specimens may undergo lyophilization or freeze-substitution.

A typical dehydration schedule for PLT combined with K4M can be as follows:

30% ethanol	0°C	30 min
50% ethanol	-20°C	60 min
70% ethanol	-35°C	60 min
95% ethanol	-35°C	60 min
100% ethanol	-35°C	60 min
100% ethanol	-35°C	60 min

It must be pointed out that the concentration of solvent and corresponding temperature have to be such that the specimen never freezes, i.e., the temperature in a subsequent dehydration step is not allowed to be lower than the freezing point of the preceding step. With this in mind, one can design PLT dehydration schemes for any solvent compatible with K4M or HM20. If desired, one can obviously use higher temperatures or conventional room-temperature dehydration.

2.5. Infiltration

Preparation of resin: avoid breathing fumes or contact of the resin with the skin. Mix cross-linker and monomer in a tared vial. Stir resin mixture gently with a glass rod, turn the capped vial upside down a few times or bubble dry nitrogen gas through the mixture. In comparison to highly viscous embedding media like Epon, the Lowicryls need much less stirring. Note also that oxygen, introduced through too vigorous stirring, inhibits the curing reaction. Soft blocks are the consequence of such overzealous mixing. The initiator is added to the cross-linker and monomer and thoroughly distributed.

Resin mixtures

K4M	
Cross-linker A	2.7 g
Monomer B	17.3 g
Initiator C	0.1 g
HM2O	
Cross-linker D	3.0 g
Monomer E	17.0 g
Initiator C	0.1 g

These mixtures produce medium-hard blocks. If harder blocks are needed, more cross-linker should be added. For HM20, the cross-linker concentration may be varied from 5 to 17 wt% (1.0 to 3.4 g/20 g resin); in the case of K4M, the cross-linker concentration may be varied from 4 to 18 wt% (0.8 to 3.6 g/20 g resin).

A typical low-temperature infiltration schedule for K4M is:

Resin : solvent	Temperature	Time
1 : 1	−35°C	60 min
2 : 1	−35°C	60 min
pure resin	−35°C	60 min
pure resin	−35°C	12 hr

The temperature at which infiltration occurs may vary as can the length of any infiltration step. HM20 can be used down to −60°C. The viscosity of K4M increases rapidly below $-35°C$; it is possible to infiltrate at lower temperature if one prolongs the infiltration times. Generally it is advisable to agitate the specimens during infiltration. Infiltration can also be performed conventionally at room temperature.

2.6. Polymerization

Samples can be placed in BEEM or gelatin capsules that have been oven-dried. Capsules should be suspended in wire holders to ensure uniform exposure to UV irradiation. The UV source must have a wavelength of about 360 nm, and fluorescent tubes such as Philips TLD 15W05 or a Black-Ray lamp (Model ULV-56, UVP, Inc.) are suitable.

For polymerization at low temperature, a chamber lined with aluminum foil can be made to fit into a freezer or cold room. A reflecting screen must be placed between the UV source and the capsules to ensure diffuse illumination only. Specimens should be placed approximately 30 cm away from the UV lamp. Slits for ventilation should be cut into the top and bottom of the side of the polymerization chamber. Small UV sources may be used, but the lamp-to-specimen distance must be shortened for adequate polymerization. A trial run using capsules entirely filled with resin is recommended to test the apparatus. If the blocks shrink or cavities appear in the resin, the rate of polymerization is too high and the lamps must be moved farther away from the blocks or the intensity of the UV source has to be reduced. To embed specimens, gelatin capsules are filled almost to the top with precooled resin. Samples are transferred with wide-mouth Pasteur pipettes into the capsules, the capsules are closed and allowed to cool to the desired temperature. All of these steps should take place in the cold to minimize condensation of water in the samples. Samples are polymerized at low temperature $(-30$ to $-50°C)$ for at least 24 hr with UV light. Capsules are then placed under UV light at room temperature for 2–3 days to improve the sectioning properties of the blocks. As an alternative, Fryer et al. (1983) have proposed replacing the room-temperature polymerization by an exposure to sunlight for 2–3 weeks.

2.7. Chemical Polymerization of Lowicryl Resins (Carlemalm et al., 1982a, Appendix)

For embedments at 70°C or above, no activators are necessary. At 60°C, add 0.2% (by weight) dibenzoylperoxide to K4M (0.5% to HM20). It is recommended that capsules to be polymerized should be placed in holes drilled in a metal plate, which acts as a heat sink during the chemical reaction.

Room- and low-temperature $(-35°C)$ chemical polymerization are as follow:

		HM2O	K4M
Room temp.	Dibenzoylperoxide	0.08%	0.05%
	Dimethylparatoluidine	0.05%	0.03%
−35°C	Dibenzoylperoxide	1%	0.4%
	Dimethylparatoluidine	0.6%	0.25%

Note: The resin should be divided into two parts, one with the peroxide, the other with the activator. Place the resins at the desired temperature and mix. Polymerization starts immediately, unlike photoinitiated polymerization. The reaction speed can be increased by embedment in an inert atmosphere (CO_2 or N_2) or under vacuum.

2.8. Rapid (4 hr) Method for K4M Embedments (Altman *et al.*, 1984)

Fixation is by 4% paraformaldehyde or 2% glutaraldehyde in buffer. Wash with buffer, followed by immersion in 50% dimethylformamide (DMF), 75% and 90% solvent (all for 10 min), K4M:DMF (1:2) for 15 min, K4M:DMF (1:1) for 20 min, 100% K4M for 25 min followed by 100% K4M for 30 min. All steps take place at room temperature. Polymerization of resin and included specimens takes place in BEEM capsules at 4°C with UV light (GE 15-watt black light, 12 cm, 45 min or less). Specimens so embedded were employed for immunolocalization studies. The authors state that labeling levels for rapidly embedded specimens were higher than those of the standard protocol or embedment in BSA.

2.9. Sectioning

Blocks should be trimmed to a pyramid with a trimming machine or with glass knives on the microtome. All faces should be clean and free of debris. The sides of the pyramid should be trimmed at a 30° angle from the face of the block. HM20 is a hydrophobic, highly cross-linked resin and sections easily with glass or diamond knives. K4M is hydrophilic and care must be taken not to wet the block face during sectioning. Ultramicrotome sectioning speeds of 2–5 mm/sec give the best results.

2.10. Section Staining

For purposes of preliminary observation of specimen sectioning or the state of preservation, sections may be placed unstained in the microscope operating at low accelerating voltage with a small objective aperture. The contrast is too low for conventional imaging, and sections must be stained with heavy metal salts for photographic imaging. Normally, sections are stained with saturated aqueous or alcoholic solutions of uranyl acetate; these may be followed by Reynolds's lead citrate (1963) or Millonig's lead stain (1961). A recommended sequence (W. Villiger, unpublished data) is as follows:

	HM2O	K4M
Saturated aqueous uranyl acetate	35 min	5–10 min
Millonig's lead stain (Method II)	1–3 min	1–3 min

3. RESULTS

3.1. Preservation of Protein Crystal Structure

In the past 5 years the Lowicryl resins have been used for a variety of applications, the first of which was a test on aldehyde-fixed protein crystals (Carlemalm et al., 1980, 1982a; Garavito et al., 1980). Crystals as a model system were suggested by the work of Douzou (1977) and Petsko (1975) on the preservation of protein crystal structure at subzero temperatures by replacing the crystal mother liquor with an organic aqueous mixture of adequately low freezing point. Crystals were ideal for such work because structural preservation can be monitored quantitatively by X-ray diffraction. While the "unsharpness" of the diffraction spots gives indications about the general disorder of the crystal, the intensity pattern of the spots reflects directly the average mass distribution within the unit cell of the lattice, i.e., in this case the conformation of the protein molecule. If the polypeptide chains of the protein were to rearrange as a consequence of exposure to organic solvents, this should become detectable. While the purpose of the work of Douzou and Petsko was to decrease thermal "vibrations" within the crystal, and by that to preserve its order, that of Garavito and Carlemalm was to study conformational changes and general disorder resulting from fixation, dehydration, infiltration, and polymerization. The authors used what is now known as the stepwise PLT procedure, in which the temperature at each step is adjusted to that of the solvent water mixtures such that no ice can form within the crystal (or within the tissue block in biological work). The results of Garavito and Carlemalm (Carlemalm et al., 1982a) with aspartate aminotransferase (AAT) and catalase (CAT) crystals are summarized as follows: no loss of resolution (up to 0.28 nm) was found in glutaraldehyde-fixed AAT or CAT crystals, and preservation during dehydration was dependent on temperature and solvent. At room temperature, ethylene glycol and hydroxyethyl methacrylate better preserved AAT crystal structure than did acetone, methanol, and ethanol. At low temperature ($-35°C$), polar solvents were more successful in preserving molecular order than less polar solvents. For CAT crystals the reverse was seen, and polar solvents caused greater disorder, ethylene glycol being less effective than acetone. Low temperature coupled with medium to highly polar solvents allowed greater retention of molecular order. Resin infiltration and polymerization caused little damage: K4M-embedded AAT (at $-35°C$) showed better diffraction maxima (up to 0.8 nm) while HM20-embedded CAT crystals maintained up to 0.8-nm diffraction maxima. The authors concluded that very polar solvents preserved AAT best, nonpolar solvents worked best for CAT crystals, molecular disorder arose during dehydration, and low temperature minimized this problem. Sectioned CAT crystals embedded in HM20 showed a regular lattice that contained a 14% compression due to the sectioning process.

The preliminary resolution achieved on sections—as tested by optical diffraction on micrographs—was not better than 2.7 nm. Better results have been published by others (Lange *et al.*, 1979; Hoppe, 1974) on other crystals and embedding procedures. The sudden decrease of resolution upon sectioning is most disturbing with respect to the potential use of sections for high-resolution work. It must be decided experimentally which of the limitations (stain distribution or mechanical distortion due to sectioning) is predominant. The hypothesis that staining might cause misinformation was demonstrated with septate junctions of the testis of *Drosophila melanogaster* fixed in glutaraldehyde and embedded in HM20 (Garavito *et al.*, 1982; Carlemalm and Kellenberger, 1982). Unstained sections observed by ratio-contrast imaging in STEM clearly showed transmembrane proteins while the same material stained with uranyl acetate viewed in TEM showed a lack of staining on initially hydrophobic parts. The lack of adequately equipped STEMs has not allowed many investigators to pursue the study of proteins in biological membranes by these methods.

3.2. Preservation of Cellular Components

Of particular concern was and still is the question of the "extraction" of matter. An apparent lack of matter in regions of the cell where a high concentration was expected is generally interpreted as an extraction that occurs during the procedures leading to embeddings. Until now, no attempt was made to decide experimentally whether this is really so or if the apparent absence is only the consequence of tissue lacking stain affinity. Unambiguous experimental results are available only for the extraction of lipids. Even with the use of cryoprotectants such as dimethyl sulfoxide and glycerol and infiltration with prepolymerized glycol methacrylate, Cope (1968) still found cell damage and extraction in embedded mouse tissue. A further study (Cope and Williams, 1968) using macrophages fed with tritiated glyceryl oleates demonstrated that glycol methacrylate removed 100% of the neutral glyceride in cells unless cells were osmicated or processed at low temperature. Even at low temperature ($-20°C$), glycol methacrylate embedments retained no more lipid than conventionally processed cells. These results were confirmed and extended by recent work by Weibull *et al.* (1980, 1983, 1984). These authors used *Acholeplasma* cells that have a particular need for fatty acids for growth. The fatty acids end up exclusively in the plasma membrane. By means of radioactively labeling the fatty acids, one achieves highly specific lipid labeling of the membrane. To summarize, these authors found that methanol, acetone, and the Lowicryls extracted lipids at temperatures above $-70°C$. Extraction was prevented by osmication, which acts even with saturated lipids. As a consequence of these findings, Carlemalm and Acetarin have developed resins of the Lowicryl type for use at still lower temperatures (Carlemalm et al., 1985b). A resin in which lipids seem not to be extracted is based on melamine (Bachhüber and Frösch, 1983; Shinagawa and Shinagawa, 1978). Unfortunately, this resin is cured at 80°C and has proven unsuitable for immunolabeling.

We feel it is important to emphasize here that most biological membranes because of their protein content are recognizable with or without lipids being present. Until now, it is not possible to visualize reproducibly the functionally so important transmembrane proteins by any of the usual heavy metal stains. The fine structural details of the biological membrane are not correlated with the presence or absence of lipids.

An important discovery that supports the use of low-temperature procedures concerns the preservation of DNA. It is well known that aqueous solutions of nucleic acids and of most proteins start to precipitate out of solution when water-miscible organic liquids become admixed. This precipitation occurs at concentrations of organic liquids between 40 and 80%. The purpose of fixation by aldehydes and osmium tetroxide is essentially to minimize the size of aggregates that occur during dehydration and/or any further steps. DNA-containing plasms are particularly suited for such studies (Kellenberger, 1962; Schreil, 1964; Kellenberger et al., 1981). Carlemalm, Villiger, Hobot, and one of us (E.K.) discovered that the lower the temperature of the dehydrants, the finer were the DNA aggregates. With cryosubstitution they were so fine that attempts to find a fixative for DNA that does not contain heavy metals could be discontinued. This question will be discussed further below.

3.3. Z or Ratio-Contrast Imaging

The theory for this imaging mode is derived from the work of Crewe et al. (1975) designed for the imaging of single atoms. The first use of ratio-contrast to image unstained sections of biological material embedded in Lowicryl (Carlemalm and Kellenberger, 1982) demonstrated the potential of this technique. A ratio-contrast signal is generated in the STEM by dividing the elastic by inelastic electron signals. The contrast so generated reflects mainly the chemical composition of the object, whereas in conventional imaging the contrast depends essentially upon differences either of thickness or densities of the specimen.

The images created by ratio-contrast from unstained sections of cells not pretreated with heavy metals, show in STEM a much better definition with ratio-contrast than with darkfield (Carlemalm and Kellenberger, 1982; Carlemalm *et al.*, 1985a). According to these authors, this "unsharpness" has not only an optical cause due to inelastic electrons, but is mainly due to the distorted surface of sections. Indeed, with microtomes the matter is cleaved into slices; the variable properties of matter with respect to plastic flow and tensile strength result in a surface relief that easily reaches 5 nm. In conventional imaging of unstained sections the relief is the main source of contrast, while with ratio-contrast the matter-specific differences *within* the section are responsible. Staining with uranyl acetate penetrates many micrometers deep into the block; slices are therefore stained throughout. As a consequence, section staining, which was shown to result in heavy metal deposits of the same order of weight as the biological macromolecular components themselves (Bashong *et al.*, 1984 see also Carlemalm *et al.*, 1985a), is therefore also emphasizing the interior of the slice over the relief. The morphology and location of this deposit being unknown, ratio-contrast imaging of unstained sections is required before we can gain further understanding. In the first comparisons, it was shown that the hydrophobic parts of integral membrane proteins are not stained (Carlemalm and Kellenberger, 1982; Garavito *et al.*, 1982).

Imaging by ratio-contrast at 80–100 kV can be applied optimally to sections between 30 and 40 nm thick. If they are thicker, multiple scattering becomes predominant and destroys the effects (Reichelt and Engel, 1984). In addition to largely eliminating the deleterious effects of relief in the observation of unstained, sectioned material, ratio-contrast also allows for precise determinations of concentrations (Reichelt *et al.*, 1985).

In order to demonstrate convincingly the influence of the relief and as an extension of ratio-contrast imaging, bacteriophage T4 absorbed to *E. coli* cell envelopes were embedded in a "negative" contrast resin consisting of tributyl-stannyl-methacrylate, methyl methacrylate, and ethylene glycol dimethacrylate (Carlemalm *et al.*, 1982b). Calculations on contrast formation predicted that such an incorporation of tin into the surrounding resin should provide great contrast to specimens when viewed with ratio-contrast in the STEM. Contrast was not adequate in CTEM operating in brightfield mode or in STEM by elastic and inelastic darkfield modes. High contrast was only obtained with ratio-contrast imaging. Resolution was limited to 1–3 nm by granularity, probably induced by beam damage. The results suggested that the addition of low amounts of heavy metals to frozen-hydrated specimens may also generate "negative" contrast in ratio-contrast imaging. A new tin-containing resin that is more beam resistant has been developed (Acetarin *et al.*, 1986).

3.4. Section Surface Labeling

Great interest in K4M was generated when it was shown that on-section immunolabeling with the protein A–gold procedure was easy and reproducible

(Roth, 1982, 1983b, 1984). The reason for efficient section surface labeling is debatable and still under investigation. It was found (Carlemalm *et al.*, 1985a), that the surfaces of resin sections are rough. Metal shadowing of sections of *Histoplasma capsulatum* embedded in K4M and Spurr's resin illustrates this point (Figures 1 and 2). Lowicryl sections had a relief as great as 5 nm, while Spurr's sections were 3–5 times smoother. The following explanation seems plausible: Epon, epoxys, and Spurr's are the best glues or cements presently known. The resin indeed covalently links with the specimen, which results in the well-known observation that epoxy-mended objects rarely break a second time along the first fracture. This would imply that the surfaces of antigens embedded in Epon would not become exposed unless the section is etched (Erlandsen *et al.*, 1979). Methacrylates like the Lowicryls are not very good glues; it seems obvious then that the antigenic surfaces would become exposed by the sectioning process and allow reaction with antibodies. Besides this advantageous way of cleavage of K4M, it seems quite certain that low temperature leads to an increased reactivity of the antigen with antibodies; this was experimentally studied on microtubules of *Leishmania* (Armbruster *et al.*, 1983a; see Section 3.9).

In many other cases, labeling was also achieved on K4M sections with colloidal gold coated with lectins (Roth, 1983a) and with toxins (Schwab and Thönen, 1978). The reader is referred to Chapter 17 for further information on these topics.

3.5. Bacteria

When *E. coli* cells are fixed in osmium, several structurally defined layers can be identified in the cell envelope (Beveridge, 1981). The most external layer is the outer membrane, which is composed of lipids and (glyco)proteins. Underneath lies the peptidoglycan layer, which may be separated from the outer membrane by an intermediate layer. The periplasmic space lies between the peptidoglycan and cytoplasmic membrane. Isolated cell envelopes fixed in glutaraldehyde and embedded in HM20 did not show this organization (Armbruster *et al.*, 1982). Instead only three bands were observed. The study was continued utilizing freeze-substitution in conjunction with low-temperature embedding methods (Hobot *et al.*, 1984). In bacteria that were freeze-substituted as well as in bacteria that were glutaraldehyde-fixed, low-temperature-dehydrated and K4M-embedded (Figure 3), the cell envelope had a constant width and was divided into cytoplasmic membrane, an intermediate layer called the periplasmic gel, and outer membrane. A distinct "empty" periplasmic space was absent in these preparations. These findings agreed with the published results of other freeze-substitution studies (Amako *et al.*, 1983), examinations of freeze-etched bacteria (Bayer and Remsen, 1970) and frozen-hydrated cryosections of bacteria (Dubochet *et al.*, 1983). In all these preparations, the distance between inner and outer membranes was constant. The observation of unstained sections by ratio-contrast in STEM demonstrated

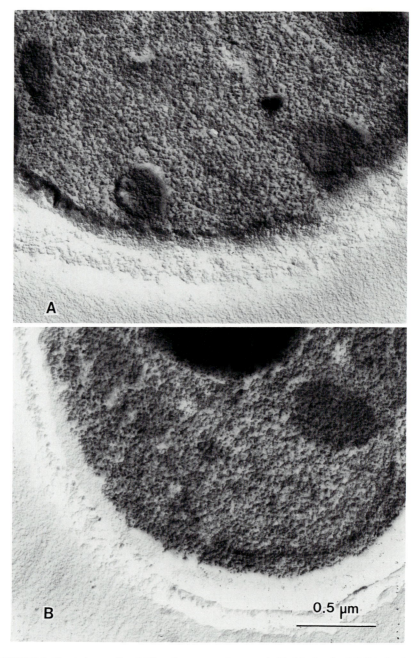

FIGURE 1. Yeast-phase cells of *Histoplasma capsulatum* embedded in Lowicryl K4M showing section relief. In the untreated section (A), the surface relief is more pronounced over the embedded cell in comparison to the surrounding resin. Panel (B) illustrates excess relief, i.e., separation of the cell wall from the cytoplasm resulting from section being etched with saturated aqueous sodium metaperiodate.

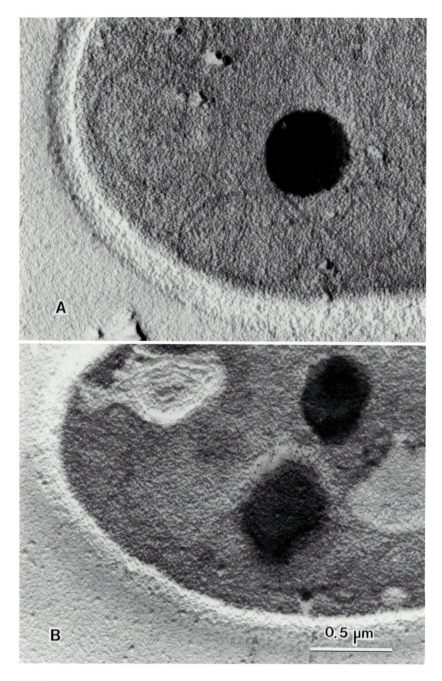

FIGURE 2. Yeast-phase cells of *H. capsulatum* embedded in Spurr's resin. Membrane profiles and the cell wall periphery are clearly demarcated in the untreated section (A) whereas such delineations are lost following etching procedures (B). These specimens proved unreactive to immunoreagents.

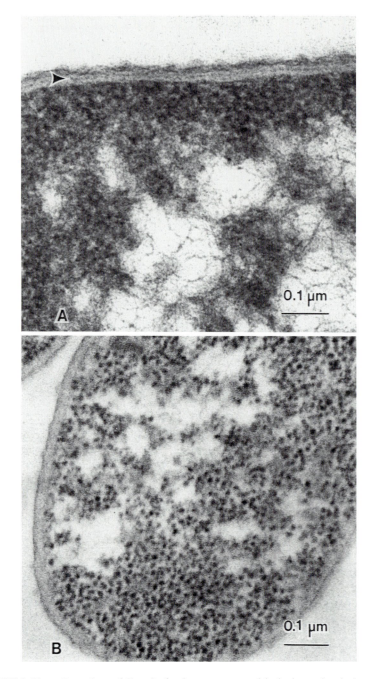

FIGURE 3. The cell envelope of *E. coli* after low-temperature dehydration and embedment. No periplasmic space is visible between inner and outer membranes (arrowhead) viewed by TEM (A). In unstained section viewed by STEM (B), the periplasmic space is also absent. Photographs courtesy of J. A. Hobot.

matter throughout the whole space between inner and outer membranes and showed no change of concentrations (Hobot et al., 1984). These observations suggested that the peptidoglycan is not a compacted structure as was assumed previously but rather very hydrated and loose as are the cellulose walls of plants. Equilibrium centrifugations in differential density gradient media [penetrating (CsCl) and nonpenetrating (Percoll)] showed a water content of more than 80%. These experiments were made with isolated peptidoglycan (obtained by shaking bacteria with glass beads in the Mickle apparatus). The "Mickle" envelopes were composed of the outer membrane and an attached layer that corresponds by measurement to the outer membrane plus half of the periplasmic gel of intact cells. It was concluded that the periplasmic gel layer was probably more cross-linked in the area just under the outer membrane (corresponding to the peptidoglycan) and less so toward the cytoplasmic membrane. In freeze-substituted preparations of plasmolyzed cells, no adhesion sites—"Bayer bridges"—between the outer membrane and the cell body were seen. The results suggested that the difference between gram-positive and gram-negative bacteria might be due to the amount of cross-linking in the periplasmic gel, linkages existing throughout the gel in the case of gram-positive bacteria.

The microscopic examination of the structure of prokaryotic DNA has not shed much light on its in vivo state. Part of the difficulty lies in the effect of fixatives on the appearance of the nucleoid. Aldehyde fixation usually produces dispersed nucleoids, whereas osmium fixation creates a more localized, confined nucleoid (Hobot et al., 1985). Both fixatives immediately render the plasma membranes permeable such that the intracellular potassium and frequently the magnesium leak out (Moncany, 1982) as was already shown for eukaryotic cells (Morel et al., 1971; Arborgh et al., 1976). The change of internal ionic strength and composition would influence the repartition between ribosomes and the DNA-containing plasma of the nucleoid. Rapid freezing before any chemical fixation should preserve the native state of exponentially growing cells. Cryosubstituted E. coli in Figure 4 showed "ribosome-free spaces" containing most of the DNA. The "spaces" corresponded to the phase-contrast image of nucleoids in living bacteria (Hobot et al., 1985).

A more important question concerns the fine structure or organization of bacterial DNA. After cryosubstitution, no compact chromatin masses were present in E. coli that could be compared to the compact metaphase chromosomes of eukaryotes. Apparently, in prokaryotes, chromatin is always in a dispersed state, comparable to that of interphase chromatin in eukaryotes. The authors concluded that the compact masses found "swimming" in a nucleoid after conventional "fixation" with aldehyde or osmium were the consequence of solvent-induced aggregation of the unfixed DNA plasma (Kellenberger et al., 1981). Only the ribosome-containing cytoplasm would eventually become fixed so as to withstand aggregation in the organic liquids. It was also observed (E. Carlemalm and E. Kellenberger, unpublished) that aggregation of the DNA plasma in organic

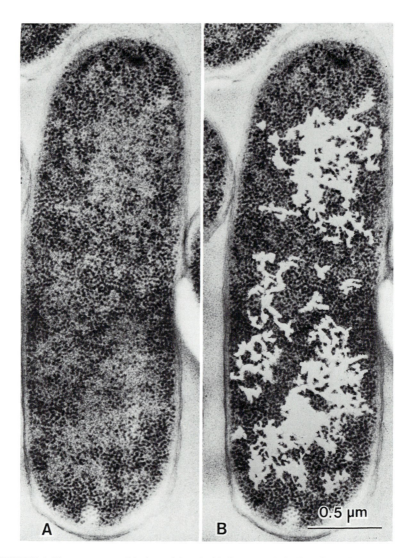

FIGURE 4. The appearance of the bacterial nucleoid after cryosubstitution of *E. coli*. Bacteria were substituted in 3% glutaraldehyde in acetone at −90°C and embedded at −35°C in K4M. The ribosome-free regions in cell (A) are shaded in cell (B) for better spatial definition. Photographs courtesy of J. A. Hobot.

liquids became finer with lower temperature. These observations confirmed previous data showing that bacterial nucleoids need conditions of processing different from those for eukaryotic chromosomes in order to be preserved from aggregation in organic liquids (Kellenberger *et al.*, 1981). One has to conclude (Hobot *et al.*, 1985)that much more experimental data are needed before extrapolating the nucleosomic structure of eukaryotes to the noneukaryotic nuclei.

3.6. Isolated Organelles

In order to study the variation of the appearance of mitochondrial cristae under different fixation and embedment regimes, mitochondria from isolated hepatocytes and intact liver were embedded at low temperature in Lowicryl (Armbruster *et al.*, 1982). Since short exposures to glutaraldehyde were recommended in previous studies (Sjöstrand, 1976; Sjöstrand and Barajas, 1968), the effect of fixation time on cristal morphology was first examined. No significant change in cristal width or morphology was seen when fixation times were varied between 5 and 60 min (Armbruster *et al.*, 1982). The cristae did show a globular substructure similar to that previously reported (Sjöstrand, 1977; Sjöstrand and Barajas, 1968). In the absence of osmium, membranes showed reversed contrast, a phenomenon also reported in the above-cited references.

Embedments of isolated yeast mitochondria in K4M may be useful in elucidating transport mechanisms and localizing particular proteins in these organelles. Preliminary work by one of us (B.L.A.) in collaboration with R. Hay (Department of Biochemistry, Biozentrum, University of Basel) has shown that it is possible to locate proteins in the outer membrane of mitochondria by means of protein A–gold and suitable antibodies (Figure 5A). Similar work by Bendayan and Shore (1982) localized an outer membrane protein and matrix enzyme in intact hepatocytes.

The structure of chloroplast thylakoids subjected to low temperature was distinctly different from that seen in conventionally prepared material (Weibull *et al.*, 1980). Sjöstrand and Kretzer (1975) reported that the membranes of algal chloroplasts stained weakly after a short glutaraldehyde fixation, exhibiting a particulate substructure. Methacrylate embedment mixtures could only be used effectively below −25°C and caused lipid extraction in room-temperature embedments. The thylakoid membranes of pea and spinach chloroplasts embedded at −35°C in the Lowicryl resins had a particulate substructure, the membranes appearing light between the stained partition gaps (Weibull *et al.*, 1980). After exposure to hydroxypropyl methacrylate, photosystem II activity disappeared whereas photosystem I activity remained after infiltration.

Under carefully controlled conditions, it is possible to isolate membrane-depleted nuclei and metaphase chromosomes from associated cytoplasm in Chinese hamster ovary tissue culture cells (Wunderli *et al.*, 1983). Such prepara-

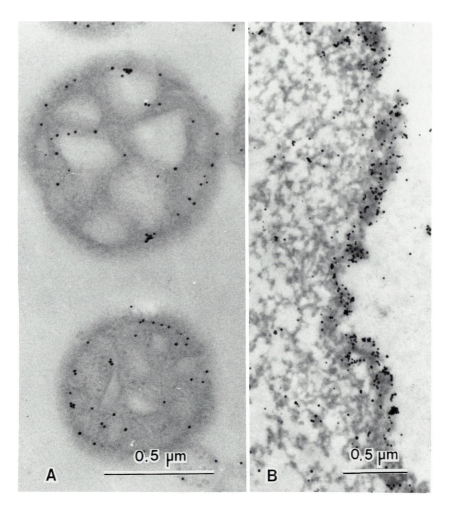

FIGURE 5. Immunolocalization in isolated organelles. Antibodies to an outer membrane protein of yeast mitochondria (A) and histone 2B in Chinese hamster ovary nuclei (B) recognize specific antigens in these specimens prepared at low temperature.

tions of chromatin are sensitive to the concentration of magnesium in the medium, becoming condensed upon the addition of 5 mM magnesium chloride (compare Figure 6A,C to Figure 6B,D). Immunolocalization studies by Armbruster et al. (1983b) on interphase and metaphase chromatin structures have shown that actin and tubulin are associated with chromatin throughout the cell cycle. Histone 2B is particularly concentrated in the periphery of these interphase nucleoids (Figure 5b).

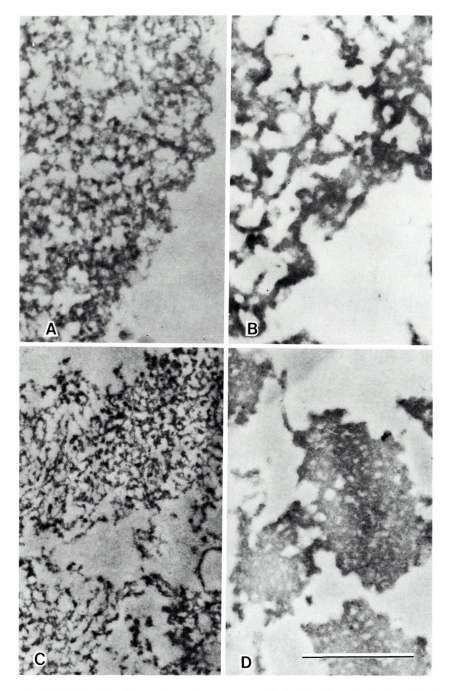

FIGURE 6. The effect of magnesium on the appearance of isolated membrane-depleted nuclei and metaphase chromosomes. In both interphase (A, B) and metaphase (C, D) preparations from Chinese hamster ovary cells, chromatin condenses upon the addition of 5 mM magnesium chloride.

3.7. Dinoflagellates

These protists are of special interest to cell biologists because their chromosomes are observable throughout the divisional cycle (Herzog and Soyer, 1983; Thomas and Cox, 1973; Giesbrecht, 1965; de Haller *et al.*, 1964). Fixation and dehydration produce nearly the same effects as for the nucleoids of bacteria (de Haller *et al.*, 1964). The fibrillar fine structure is also very comparable to that of the nucleoids of prokaryotes. Low-temperature techniques were applied only very recently to these microorganisms (Michel-Salamin *et al.*, 1984). Thin sections of the dinoflagellate *Prorocentrum micans* E. were made from specimens that were (1) conventionally fixed with or without osmium, embedded in either Epon or Lowicryl at $-35°C$; (2) rapidly frozen followed by cryofracture or freeze-substitution with osmium–acetone, then Epon embedment; or (3) sectioned for frozen-hydrated viewing. Specimens subjected to freeze fixation had chromosomal thin fibrils of 2–4 nm while chemically fixed specimens showed fibril diameters of 5–8 nm. Regular chromosomal fibril distribution was seen only in physically fixed specimens; chemical fixation produced subunit collapse. Using ratio-contrast imaging and unstained sections, the nucleofibrils could be clearly imaged. This image contrast may result from the aforementioned collapse or from different rates of mass loss between resin and specimen.

3.8. *Paramecium*

The process by which *Paramecium* extrudes its trichocysts has been the focus of study of several laboratories. Although it has been shown that the decondensation of the crystalline matrix of trichocysts is controlled by a Ca^{2+} influx (Bilinski *et al.*, 1981), actin (Livolant, 1980) and calmodulin (Rauh and Nelson, 1981) have also been reported in trichocysts. Immunolocalization studies with K4M-embedded *Paramecium* (Kersken *et al.*, 1984) showed that actin and calmodulin were not present in significant levels in intact cells *in situ*. Microfilament bundles did surround the trichocyst top and appeared to originate beneath the attachment site of the trichocyst on the plasma membrane (Plattner *et al.*, 1982). Figure 7A demonstrates the presence of microfilament fasciae in *Paramecium*. Discharged trichocysts did show a positive reaction for calmodulin as did structurally impaired trichocysts within cells and trichocyst membranes. The authors concluded that decondensation of the internal crystalline matrix during exocytosis was due to a rearrangement of this protein's conformation and was not influenced by the presence of contractile or regulatory proteins.

3.9. *Leishmania*

Microtubulelike structures under the cytoplasmic membrane of trypanosomatids have been frequently observed (Angelopoulos, 1970), but their classification as microtubules was only determined from their shape and estimated

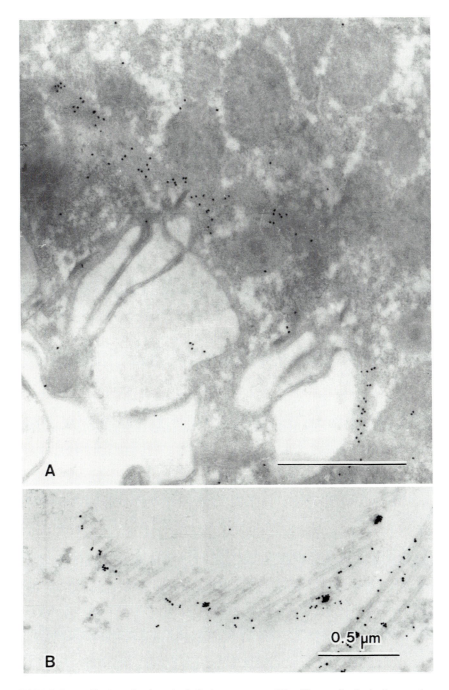

FIGURE 7. Localization of actin and tubulin in protozoans. Microfilament fasciae in *Paramecium* (A) and microtubules of isolated cortical arrays from *Leishmania* (B) are labeled by means of protein A–gold. A, courtesy of H. Plattner.

diameter. Bordier *et al.* (1982) isolated and analyzed the major protein of the pellicular membrane of *Leishmania tropica* and identified the components as α- and β-tubulin by immunological and biochemical comparison to subunits of mammalian brain tubulin. Immunolocalization studies by means of antitubulin antibodies and protein A–gold on K4M-embedded sections of pellicular arrays also indicated the presence of tubulin (Figure 7B).

Cortical microtubule arrays were used in a study to determine the effect of the temperature at which dehydration and embedment take place and resin polarity on the preservation of tubulin antigenicity (Armbruster *et al.*, 1983a). The experimental scheme included three temperature regimes for dehydration, embedment in hdyrophilic K4M or hydrophobic Epon, and pre- and postembedding immunolabeling. Qualitative and quantitative assessment of labeling were made using purified antitubulin antibodies and protein A–gold. Specimens embedded in K4M at 20°C showed levels of labeling comparable to those of Epon-embedded specimens. A temperature-dependent variation in background labeling made accurate quantitative evaluation of the 20°C and 0°C preparations difficult. Labeling levels decreased by a factor of four between specimens dehydrated by the PLT technique and those dehydrated only at 0°C, suggesting that denaturation took place during solvent exchange.

4. PERSPECTIVES

None of the embedding materials available today is ideal in all of its properties. Several do afford the investigator good conditions for particular

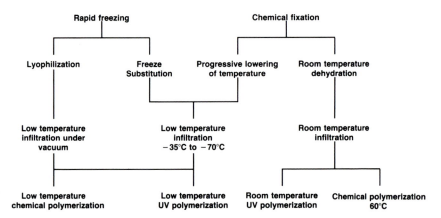

FIGURE 8. Versatility of low-temperature embedding. The flexibility of the Lowicryl system is illustrated by the broad range of temperatures at which the resins remain fluid for specimen penetration. The effect of fixation, solvent, and polarity of the embedding environment can be addressed with such an experimental system.

specimens, and the reader is referred to Hayat (1970) for an expanded comparison of the merits and liabilities associated with various media.

The Lowicryl resins can be employed in various ways, as illustrated in Figure 8. In combination with frozen or chemically fixed tissues, these resins allow the investigator a broad choice of temperature regimes, infiltration schedules, and methods for resin polymerization. The flexibility of the system allows experimentation and adjustment of conditions to the requirements of particular tissues.

Several areas of research will be pursued in the future: since unstained material can be viewed by STEM, perhaps we will clarify the action of heavy metal stains. Smaller metal tags will increase the precision of immunolocalization. Two new acrylic embedding resins for ultralow-temperature work are currently being developed in the Basel laboratory. Such efforts will provide the means to explore questions of cell structure and function.

ACKNOWLEDGMENTS. This work was supported by USPHS Grants 5T 32 AI07015, 5T 32 AI16228, 1F 32 AI07047 (individual award to B.L.A.) and the Swiss National Science Foundation.

5. REFERENCES

Acetarin, J. D., 1981, Nouvelles recherches sur les resines d'inclusion pour la microscopie electronique, Thesis, Univ. Louis Pasteur, Strasbourg, France.

Acetarin, J. D., Villiger, W., and Carlemalm, E., 1986, A new heavy metal containing resin for low temperature embedding and imaging of unstained sections of biological material, *J. Electron Micro. Tech.* (submitted).

Altman, L. G., Schneider, B. G., and Papermaster, D. S., 1984, Rapid embedding of tissues in Lowicryl K4M for immunoelectron microscopy, *J. Histochem. Cytochem.* **32:**1217–1223.

Amako, K., Murata, K., and Umeda, A., 1983, Structure of the envelope of *Escherichia coli* observed by the rapid-freezing and substitution fixation method, *Microbiol. Immunol.* **27:**95–99.

Angelopoulos, E., 1970, Pellicular microtubules in the family Trypanosomatidae, *J. Protozool.* **17:**39–51.

Arborgh, B., Bell, P., Brunk, U., and Collins, V. P., 1976, The osmotic effect of glutaraldehyde during fixation: A transmission electron microscopy, scanning electron microscopy and cytochemical study, *J. Ultrastruct. Res.* **56:**339–350.

Armbruster, B. L., Carlemalm, E., Chiovetti, R., Garavito, R. M., Hobot, J. A., Kellenberger, E., and Villiger, W., 1982, Specimen preparation for electron microscopy using low temperature embedding resins, *J. Microsc. (Oxford)* **126:**77–85.

Armbruster, B. L., Garavito, R. M., and Kellenberger, E., 1983a, Dehydration and embedding temperatures affect the antigenic specificity of tubulin and immunolabeling by the protein A–colloidal gold technique, *J. Histochem. Cytochem.* **31:**1380–1384.

Armbruster, B. L., Wunderli, H., Turner, B. M., Raška, I., and Kellenberger, E., 1983b, Immunocytochemical localization of cytoskeletal proteins and histone 2B in isolated membrane-depleted nuclei, metaphase chromatin, and whole Chinese hamster ovary cells, *J. Histochem. Cytochem.* **31:**1385–1393.

Bachhüber, K., and Frösch, D., 1983, Melamine resins, a new class of water-soluble embedding media for electron microscopy, *J. Microsc. (Oxford)* **130:**1–9.

Bashong, W., Bashong, C., Wurtz, M., Carlemalm, E., Kellenberger, C., and Kellenberger, E., 1984, Preservation of protein structures for electron microscopy by fixation with aldehydes and/or osmium tetroxide. *Eur. J. Cell Biol.* **35:**21–26.

Bayer, M. E., and Remsen, C. C., 1970, Structure of *Escherichia coli* after freeze-etching, *J. Bacteriol.* **101:**304–313.

Bendayan, M., and Shore, G., 1982, Immunocytochemical localization of mitochondrial proteins in the rat hepatocyte, *J. Histochem. Cytochem.* **30:**139–147.

Bernhard, W., 1969, A new staining procedure for electron microscopical cytology, *J. Ultrastruct. Res.* **27:**250–265.

Beveridge, T. J., 1981, Ultrastructure, chemistry and function of the bacterial wall, *Int. Rev. Cytol.* **72:**229–317.

Bilinski, M., Plattner, H., and Matt, H., 1981, Secretory protein decondensation as a distinct Ca^{+2}-mediated event during the final steps of exocytosis in *Paramecium* cells, J. Cell Biol. **88:**179–188.

Birbeck, M. S. C., and Mercer, E. H., 1956, Applications of an epoxide embedding medium to electron microscopy, *J. R. Microsc. Soc.* **76:**159–161.

Bordier, C., Garavito, R. M., and Armbruster, B., 1982, Biochemical and structural analyses of microtubules in the pellicular membrane of *Leishmania tropica*, *J. Protozool.* **29:**560–565.

Borysko, E., 1956, Recent developments in methacrylate embedding. I. A study of the polymerization damage phenomenon by phase contrast microscopy, *J. Biophys. Biochem. Cytol.* **2:**3–14 (Suppl.).

Borysko, E., and Sapranauskas, P., 1954, A new technique for comparative phase-contrast and electron microscope studies of cells grown in tissue culture, with an evaluation of the technique by means of time-lapse cinematographs, *Bull. Johns Hopkins Hosp.* **95:**68–79.

Carlemalm, E., and Kellenberger, E., 1982, The reproducible observation of unstained embedded cellular material in thin sections: Visualization of an integral membrane protein by a new mode of imaging for STEM, *EMBO J.* **1:**63–67.

Carlemalm, E., Garavito, R. M., and Villiger, W., 1980, Advances in low temperature embedding for electron microscopy, in: *Proceedings of the 7th European Congress on Electron Microscopy,* Vol. 2 (P. Brederoo and W. de Priester, eds.), pp. 656–657.

Carlemalm, E., Garavito, R. M., and Villiger, W., 1982a, Resin development for electron microscopy and an analysis of embedding at low temperature, *J. Microsc. (Oxford)* **126:**123–143.

Carlemalm, E., Acetarin, J. D., Villiger, W., Colliex, C., and Kellenberger, E., 1982b, Heavy metal-containing surroundings provide more "negative" contrast by Z-imaging in STEM than with conventional modes, *J. Ultrastruct. Res.* **80:**339–343.

Carlemalm, E., Colliex, C., and Kellenberger, E., 1985a, Contrast formation in electron microscopy of biological material, in: *Advances in Electronics and Electronphysics* (P. W. Hawkes, ed.), pp. 269–334, Academic Press, New York.

Carlemalm, E., Villiger, W., Hobot, J. A., Acetarin, J. D., and Kellenberger, E., 1985b, Low temperature embedding with Lowicryl resins: Two new formulations and some applications, *J. Microsc. (Oxford)* **140:**55–63.

Cope, G. H., 1968, Low-temperature embedding in water-miscible methacrylates after treatment with antifreezes, *J. R. Microsc. Soc.* **88:**235–257.

Cope, G. H., and Williams, M. A., 1968, Quantitative studies on neutral lipid preservation in electron microscopy, *J. R. Microsc. Soc.* **88:**259–277.

Crewe, A. V., Langmore, J. P., and Isaacson, M. S., 1975, Resolution and contrast in the scanning transmission electron microscope, in: *Physical Aspects of Electron Microscopy and Microbeam Analysis* (M. Siegel and D. R. Beaman, eds.), pp. 47–62, Wiley, New York.

de Haller, G., Kellenberger, E., and Rouiller, C., 1964, Etude au microscope electronique des plasmas contenant de l'acide desoxyribonucleique. III. Variations ultrastructurales des chromosomes d'*Amphidinium*, *J. Microsc. (Paris)* **3:**627–642.

Douzou, P., 1977, *Cryobiochemistry: An Introduction*, Academic Press, New York.

Dubochet, J., McDowall, A. W., Menge, B., Schmid, E. N., and Lickfeld, K. G., 1983, Electron microscopy of frozen-hydrated bacteria, *J. Bacteriol.* **155**:381–390.

Erlandsen, S. L., Parson, J. A., and Rodning, C. B., 1979, Technical parameters of immunostaining of osmicated tissue in epoxy sections, *J. Histochem. Cytochem.* **27**:1286–1289.

Fryer, P. R., Wells, C., and Ratcliffe, A., 1983, Technical difficulties overcome in the use of Lowicryl K4M electron microscopy embedding resin, *Histochemistry* **77**:141–143.

Garavito, R. M., Carlemalm, E., and Villiger, W., 1980, Low temperature embedding of spermatids, in: *Proceedings of the 7th European Congress on Electron Microscopy*, Vol. 2 (P. Brederoo and W. de Priester, eds.), pp. 658–659.

Garavito, R. M., Carlemalm, E., Colliex, C., and Villiger, W., 1982, Septate junction ultrastructure as visualized in unstained and stained preparations, *J. Ultrastruct. Res.* **80**:344–353.

Giesbrecht, P., 1965, Über das Ordnungsprinzip in den Chromosomen von Dinoflagellaten und Bakterien, *Zentralbl. Bakteriol. Parasitenkd. Infektionskr. Hyg.* **196**:516–519.

Glauert, A. M., and Glauert, R. H., 1958, Araldite as an embedding medium in electron microscopy, *J. Biophys. Biochem. Cytol.* **4**:191–194.

Glauert, A. M., Rogers, G. E., and Glauert, R. H., 1956, A new embedding medium for electron microscopy, *Nature (London)* **178**:803.

Hayat, M. A., 1970, *Principles and Techniques of Electron Microscopy*, Vol. 1, Van Nostrand–Reinhold, Princeton, New Jersey.

Herzog, M., and Soyer, M. O., 1983, The native structure of dinoflagellate chromosomes and their stabilization by Ca^{2+} and Mg^{2+} cations, *Eur. J. Cell Biol.* **30**:33–41.

Hobot, J. A., Carlemalm, E., Villiger, W., and Kellenberger, E., 1984, Periplasmic gel: New concept resulting from the reinvestigation of bacterial cell envelope ultrastructure by new methods, *J. Bacteriol.* **160**:143–152.

Hobot, J. A., Villiger, W., Escaig, J., Maeder, M., Ryter, A., and Kellenberger, E., 1985, The shape and fine structure of the nucleoid observed on sections of ultra-rapid frozen and cryosubstituted bacteria, *J. Bacteriol.* **162**:960–971.

Hoppe, W., 1974, Towards three-dimensional "electron microscopy" at atomic resolution, *Naturwissenschaften* **61**:239–249.

Kellenberger, E., 1962, The study of natural and artificial DNA-plasms by thin sections, in: *The Interpretation of Ultrastructure*, Vol. 1, pp. 233–249, Symp. Int. Soc. Cell Biol., Berne.

Kellenberger, E., Schwab, W., and Ryter, A., 1956, L'inclusion d'un copolymere du groupe des polyesters comme materiel d'inclusion en ultramicrotomie, *Experientia* **12**:421–422.

Kellenberger, E., Carlemalm, E., Villiger, W., Roth, J., and Garavito, R. M., 1980, *Low Denaturation Embedding for Electron Microscopy of Thin Sections*, Chemische Werke Lowi, G.m.b.H., Waldkraiburg.

Kellenberger, E., Carlemalm, E., Stauffer, E., Kellenberger, C., and Wunderli, H., 1981, In vitro studies of the fixation of DNA, nucleoprotamine, nucleohistone and proteins, *Eur. J. Cell Biol.* **25**:1–4.

Kersken, H., Tiggemann, R., Westphal, C., and Plattner, H., 1984, The secretory contents of *Paramecium tetraurelia* trichocysts: Ultrastructural cytochemical characterization, *J. Histochem. Cytochem.* **32**:179–192.

Lange, R. H., Blödern, J., Magdowski, G., and Trampish, H. J., 1979, Crystalline preparations of rhombohedral porcine insulin as studied by electron diffraction, *J. Ultrastruct. Res.* **68**:81–91.

Leduc, E. H., and Bernhard, W., 1967, Recent modifications of the glycol methacrylate embedding procedure, *J. Ultrastruct. Res.* **19**:196–199.

Leduc, E. H., and Holt, S. J., 1965, Hydroxypropyl methacrylate, a new water-miscible embedding medium for electron microscopy, *J. Cell Biol.* **26**:137–155.

Livolant, F., 1980, Mise en evidence d'actine dans les trichocystes de *Prorocentrum micans* (Dinoflagelle), *Biol. Cell* **39**:10a.

Luft, J. H., 1973, Embedding media—Old and new, in: *Advanced Techniques in Biological Electron Microscopy* (J. K. Koehler, ed.), pp. 1–34, Springer, Berlin.

Michel-Salamin, L., Gautier, A., Soyer-Gobillard, M. O., de Billy, F., Dubochet, J., McDowall, A. W., Kellenberger, E., and Carlemalm, E., 1984, Appearance of "arch-shaped" chromosomes in dinoflagellates as observed in thin sections and cryofractures, following various preparation procedures, in: *Proceedings 8th European Congress Electron Microscopy*, Vol. 3 (A. Csanady, P. Rohlich, and D. Szabo, eds.), pp. 1803–1804.

Millonig, G., 1961, A modified procedure for lead staining of thin sections, *J. Biophys. Biochem. Cytol.* **11**:736–739.

Moncany, M. L. J., 1982, Determination des conditions intracellulaires chez *E. coli:* Consequences biologiques de leur modification, Thesis, University of Basel.

Morel, F. M. M., Baker, R. F., and H. Wayland, 1971, Quantitation of human red blood cell fixation by glutaraldehyde, *J. Cell Biol.* **48**:91–100.

Newman, S. B., Borysko, E., and Swerdlow, M., 1949, New sectioning techniques for light and electron microscopy, *Science* **110**:66–68.

Petsko, G. A., 1975, Protein crystallography at sub-zero temperatures: Cryo-protective mother liquors for protein crystals, *J. Mol. Biol.* **96**:381–392.

Plattner, H., Westphal, C., and Tiggemann, R., 1982, Cytoskeleton–secretory vesicle interactions during the docking of secretory vesicles at the cell membrane in *Paramecium tetraurelia* cells, *J. Cell Biol.* **92**:368–377.

Rauh, J. J., and Nelson, D. L., 1981, Calmodulin is a major component of extruded trichocysts from *Paramecium tetraurelia, J. Cell Biol.* **91**:860–865.

Reichelt, R., and Engel, A., 1984, Monte Carlo calculations of elastic and inelastic electron scattering in biological and plastic materials, *Ultramicroscopy* **13**:279–294.

Reichelt, R., Carlemalm, E., Villiger, W., and Engel, A., 1985, Concentration determination of embedded biological matter by scanning transmission electron microscopy, *Ultramicroscopy* **16**:69–80.

Reynolds, E. S., 1963, The use of lead citrate at high pH as an electron-opaque stain in electron microscopy, *J. Cell Biol.* **17**:208–213.

Rosenberg, M., Bartl, P., and Lesko, J., 1960, Water-soluble methacrylate as an embedding medium for the preparation of ultrathin sections, *J. Ultrastruct. Res.* **4**:298–303.

Roth, J., 1982, The protein A–gold (pAg) technique—A qualitative and quantitative approach for antigen localization on thin sections, in: *Techniques in Immunocytochemistry*, Vol. 1 (G. R. Bullock and P. Petrusz, eds.), pp. 107–133, Academic Press, New York.

Roth, J., 1983a, Application of lectin–gold complexes for electron microscopic localization of glycoconjugates on thin sections, *J. Histochem. Cytochem.* **31**:987–999.

Roth, J., 1983b, The colloidal gold marker system for light and electron microscopy: Theory and application, in: *Techniques in Immunocytochemistry*, Vol. 2 (G. R. Bullock and P. Petrusz, eds.), pp. 217–284, Academic Press, New York.

Roth, J., 1984, Light and electron microscopic localization of antigenic sites in tissue sections by the protein A–gold technique, *Acta Histochem.* **29**:9–22.

Roth, J., Bendayan, M., Carlemalm, E., and Villiger, W., 1980, Immunocytochemistry in thin sections with the protein A–gold (pAg) technique, *Experientia* **36**:757.

Roth, J., Bendayan, M., Carlemalm, E., Villiger, W., and Garavito, R. M., 1981, Enhancement of structural preservation and immunocytochemical staining in low temperature embedded pancreatic tissue, *J. Histochem. Cytochem.* **29**:663–671.

Ryter, A., and Kellenberger, E., 1958, L'inclusion au polyester pour l'ultramicrotomie, *J. Ultrastruct. Res.* **2**:200–214.

Schreil, W. H., 1964, Studies on the fixation of artificial and bacterial DNA-plasms for the electron microscopy of thin sections, *J. Cell Biol.* **22**:1–20.

Schwab, M. E., and Thönen, H., 1978, Selective binding, uptake, and retrograde transport of tetanus toxin by nerve terminals in the rat iris, *J. Cell Biol.* **77**:1–13.

Shinagawa, Y., and Shinagawa, Y., 1978, Melamine resin as water-containing embedding medium for electron microscopy, *J. Electron Microsc.* **27**:13–17.

Sjöstrand, F. S., 1976, The problems of preserving molecular structure of cellular components in connection with electron microscopic analysis, *J. Ultrastruct. Res.* **55**:271–280.

Sjöstrand, F. S., and Barajas, L., 1968, Effect of modifications in conformation of protein molecules on structure of mitochondrial membranes, *J. Ultrastruct. Res.* **25**:121–155.

Sjöstrand, F. S., and Kretzer, F., 1975, A new freeze-drying technique applied to the analysis of the molecular structure of mitochondrial and chloroplast membranes, *J. Ultrastruct. Res.* **53**:1–28.

Spurr, A. R., 1969, A low-viscosity epoxy resin embedding medium for electron microscopy, *J. Ultrastruct. Res.* **26**:31–43.

Stäubli, W., 1963, A new embedding technique for electron microscopy, combining a water-soluble epoxy resin (Durcupan) with water-insoluble Araldite, *J. Cell Biol.* **16**:197–201.

Thomas, R. N., and Cox, E. R., 1973, Observations on the symbiosis of *Peridinium balticum* and its intracellular alga. I. Ultrastructure, *J. Phycol.* **9**:304–323.

Watson, M. L., 1963, Explosionfree methacrylate embedding, *J. Appl. Phys.* **34**:2507.

Weibull, C., Carlemalm, E., Villiger, W., Kellenberger, E., Fakan, J., Gautier, A., and Larsson, C., 1980, Low-temperature embedding procedures applied to chloroplasts, *J. Ultrastruct. Res.* **73**:233–244.

Weibull, C., Christiansson, A., and Carlemalm, E., 1983, Extraction of membrane lipids during fixation, dehydration and embedding of *Acholeplasma laidlawii*-cells for electron microscopy, *J. Microsc. (Oxford)* **129**:201–207.

Weibull, C., Villiger, W., and Carlemalm, E., 1984, Extraction of lipids during freeze-substitution of *Acholeplasma laidlawii*-cells for electron microscopy, *J. Microsc. (Oxford)* **134**:213–216.

Wunderli, H., Westphal, M., Armbruster, B., and Labhart, P., 1983, Comparative studies on the structural organization of membrane-depleted nuclei and metaphase chromosomes, *Chromosoma* **88**:241–248.

Chapter 11

High-Voltage Electron Microscopy

Sandra A. Nierzwicki-Bauer

Department of Biology
Rensselaer Polytechnic Institute
Troy, New York 12181

1. INTRODUCTION

High-voltage electron microscopy (HVEMY) is transmission or scanning–transmission electron microscopy that is performed at accelerating potentials much higher than those used by conventional transmission (CTEM) or scanning–transmission (CSTEM) instruments. HVEMY requires a specialized instrument, commonly referred to as a high-voltage electron microscope (HVEM), that can operate at accelerating potentials of 1 MV or higher. These HVEMs differ considerably from CTEMs in their physical characteristics, primarily as a result of the difficulties associated with production of such high voltages. HVEMs should not be confused with intermediate-voltage electron microscopes (IVEMs), which operate at 200–400 kV and are physically similar to CTEMs and CSTEMs. The discussion that follows is restricted to the HVEM because, as yet, there are insufficient data to evaluate the usefulness of IVEMs for the examination of microorganisms.

The first HVEMs were designed to satisfy the needs of metallurgists and materials scientists; it was assumed that such instruments would be of little use to biologists. Except for the 1.5-MV instrument at Toulouse, which was built in the hope that it would facilitate the examination of living cells (Dupouy *et al.*, 1960), this assumption remained true for many years. Biologists became interested in the HVEM only after they realized that it could produce high-resolution images of unusually thick specimens (i.e., specimens that are much thicker than those viewable with a CTEM). Interest in HVEMY among biologists then increased rapidly, and their needs are now considered routinely when new special features and functions of HVEMs are to be developed. This trend should eventually minimize the technical problems associated with HVEMY of biological samples, while leading to broadened applicability and increased productivity of the HVEM in biological research.

The HVEM has been used extensively in biological and biomedical research during the past 10–15 years (for reviews, see Glauert, 1974; Hawes, 1981; King *et al.*, 1980), the result being a significant contribution to our understanding of eukaryotic cell biology and tissue structure. Investigations involving HVEMY of microorganisms have been quite rare (specific examples are cited below), however, and most of these have been directed at eukaryotic rather than prokaryotic species. The purpose of this chapter is to stimulate electron microscopists who work with microorganisms (eukaryotic and prokaryotic) to make more extensive use of HVEMY in their research. Toward this goal, I describe the HVEM itself, explain in some detail how HVEMY can be applied to microorganisms, and illustrate the unique advantages of this approach with examples from the literature.

2. THE HIGH-VOLTAGE ELECTRON MICROSCOPE

2.1. General Physical Characteristics

The most noticeable difference between the HVEM and the CTEM is that the former is very much larger (Figure 1). A typical HVEM may be two or three stories high, weigh 25–30 tons, and occupy several thousand cubic feet of space. These massive proportions result primarily from (1) the size of the equipment required to generate potentials of 1 MV or higher; (2) the fact that an accelerator tank must be positioned directly above the column; (3) the fact that more powerful and, therefore, larger lenses are required to control the faster-moving electrons; and (4) the need to shield operators from the high levels of X-rays produced within the instrument.

2.2. Standard Features and Controls

Except for its size, the HVEM is quite similar to a typical CTEM; an HVEM operator encounters all of the controls and features normally found in CTEMs or CSTEMs. Only a few minor differences are noticeable, the most important of which involve the vacuum and image viewing/recording systems.

2.2.1. Vacuum System

HVEMs contain two vacuum systems, one for the accelerator and one for the column. These two systems can be isolated from each other via an "intersystems valve" located above the first condenser lens. The accelerator must be isolated from the column during any operation that might leak gas into the

FIGURE 1. Photograph of the AEI EM7 MkII high-voltage (1.2 MV) electron microscope. Note the large accelerator and generator tanks situated above the column of the instrument. Photograph courtesy of Dr. D. F. Parsons.

instrument (e.g., a specimen change), because even a slight deterioration in the accelerator vacuum might cause dangerous electrical arcing.

2.2.2. Image Viewing and Recording

A variety of means are available for viewing HVEM images, including phosphor-based devices (viewing screens), television systems, and charge-imaging devices (King *et al.*, 1980). Phosphor-based devices are most common, but some operators find focusing more difficult than with phosphor-based screens on CTEMs. Consequently, many HVEM facilities provide television images or other alternatives in the hope of alleviating this problem.

HVEM images are usually recorded on photographic emulsions, although the storage of digitized television images for image processing represents an interesting alternative. Photographic emulsions must be isolated from the column except during actual exposure to avoid X-ray fogging. X-ray fogging can be further reduced by the use of aluminum film cassettes and cut-film sheets in preference to glass plates. Exposures are usually made on conventional Kodak 4489 electron microscopy film, although Dupont "Lo-Dose" film (which is 22 times more sensitive to 1-MV electrons) can be used to help minimize beam damage to ultrasensitive specimens.

2.3. Special Features and Functions of Use in Biological Research

Some special features and functions of the HVEM (Table I) were designed specifically to meet the needs of biologists; others were originally developed for nonbiological applications and later modified for use with biological specimens. Those having the broadest range of applications in biological research are described below.

2.3.1. Single-Tilt, Rotation-Tilt, and Double-Tilt Stages

Most HVEMs provide a selection of sophisticated goniometer stages that enable the operator to orient a specimen in the electron beam very precisely throughout a wide range of positions. The most frequently used goniometer stages in HVEMs are the single-tilt stage, the rotation-tilt stage, and the double-tilt stage. Single-tilt stages allow the specimen to be tilted up to ±60° about a single horizontal axis, while rotation-tilt stages permit the specimen to be rotated about the vertical axis (often a full 360°) before it is tilted about a single horizontal axis. Double-tilt stages allow a specimen to be tilted about two perpendicular horizontal axes (up to ±60° in both cases), so that virtually any desired specimen orientation can be achieved. The primary advantage of all HVEM goniometer stages for biological investigations is that they permit one to produce stereo-pair micrographs of thick specimens, thereby greatly facilitating the interpretation of

Table I
Special Features and Functions of the HVEM That Are of Use in Biological Research

Feature/function	Application(s)	Reference(s)
Goniometer stages	Production of stereo pairs, tilting to improve image clarity (see text)	Glauert (1974), Hawes (1981), King *et al.* (1980)
Darkfield imaging	High-resolution studies on thin specimens (see text)	Glauert (1974), King *et al.* (1980)
High-voltage STEM	Electronic image processing, increased contrast, and so on (see text)	King *et al.* (1980)
Energy-dispersive X-ray analysis	Elemental analysis of thick sections or whole cells	Tivol *et al.* (1983)
Environmental chambers	Observation of hydrated cells	Glauert (1974), King *et al.* (1980), Turner *et al.* (1981)
Cold stages, including tilting cold stages	Observation of biological specimens at controlled temperatures	King *et al.* (1980)
Faraday cups, etc.	Accurate measurement of beam current for elemental analysis and other studies	King *et al.* (1980)
Electron energy-loss spectrophotometers (EELS)	Elemental analysis for low-atomic-number elements	Tivol *et al.* (1983)
Magnetic imaging filters	Improved penetration power and resolution	King *et al.* (1980)
Diffraction apertures	Electron diffraction studies	King *et al.* (1980)

3-D structural information within those specimens (Section 4.3). Goniometer stages also allow the investigator to orient a specimen in such a way as to maximize the clarity of certain structural features within it (Section 4.4).

2.3.2. Darkfield Imaging

Darkfield imaging can be achieved with HVEMs by use of either a tilted beam, a central beam stop (strioscopy), or electronic rotating hollow-cone illumination (Dubochet, 1973; King *et al.*, 1980). Dupouy (1968, 1973) has shown that darkfield methods can overcome signal-to-noise ratio problems in thin specimens that are examined at high voltages, thereby producing high-contrast images. Darkfield HVEMY of thin biological objects, such as thin sections (Dupouy, 1973; Massover, 1974) and isolated molecules (Massover, 1972; Massover *et al.*, 1973), also produces images with higher resolution than those obtained at conventional accelerating voltages (perhaps by reducing chromatic aberration).

Dupouy (1968, 1973) initially developed darkfield imaging techniques to examine bacterial flagella (Dupouy *et al.*, 1969) and bacterial cell walls (Dupouy, 1973). Little has been done since to develop these promising procedures further, although Mannella and Ratkowski (1983) have described efforts to adapt the HVEM for routine biological darkfield imaging. Perhaps their work will stimulate renewed interest in the technique.

2.3.3. High-Voltage Scanning–Transmission Electron Microscopy (HVSTEMY)

HVEMs can be modified to include a STEM mode by addition of scanning coils to raster the beam and an extra lens to reduce beam diameter (King *et al.*, 1980). HVSTEMY offers several distinct theoretical advantages over HVEMY. The most significant for biological research are (1) decreased radiation damage as a result of increased beam efficiency in the STEM mode, (2) the ability to enhance and improve contrast via electronic manipulation of detector signals, and (3) improved resolution and sensitivity with X-ray analysis due to the use of a smaller, more intense beam probe.

2.4. Availability of HVEMs for Biological Research

The increased use of HVEMs for biological research in recent years is, in part, a result of the increased availability of these instruments to qualified investigators. There are now three HVEM facilities in the United States (located in Albany, N.Y.; Boulder, Colo.; and Madison, Wisc.) that are dedicated to biological work. All are funded by the National Institutes of Health as part of the National Biotechnology Research Resources Program. As such, they may be used by any qualified investigator without cost after submission and approval of a simple project application. For further information on the three NIH-funded facilities, see Research Resources Information Center (1983).

Parsons and Ratkowski (1980) have compiled a full list of HVEM facilities and their services. This list includes more than 30 such facilities worldwide, many of which provide beam time to biologists.

3. THEORETICAL ADVANTAGES AND DISADVANTAGES OF HIGH ACCELERATING VOLTAGES

3.1. Advantages

3.1.1. Increased Resolution

High accelerating potentials provide improved resolving power because wavelength and optimum aperture angle decrease with increasing voltage. The improvement is partially offset by the increased spherical aberration of the larger lenses required to control high-energy electrons but, even so, an increase in accelerating voltage from 100 to 1000 kV produces a 2.5-fold increase in calculated resolving power (Meek, 1976). Calculated resolving powers are almost never realized with biological samples at any accelerating voltage, however, because resolution is ultimately limited by the amount of chromatic aberration produced as the electron beam interacts with the specimen. Still, the HVEM provides better resolution with biological samples than the CTEM does because beam–specimen interaction (and, therefore, chromatic aberration) decreases as accelerating voltage increases. Resolution is only marginally better when thin objects are viewed with HVEMY, and darkfield imaging techniques are required to compensate for the loss in contrast. The improvement is far more significant with thicker specimens, however, and one need not use special imaging techniques to take advantage of it.

3.1.2. Increased Penetrating Power

The penetrating power of the HVEM is considerably greater than that of the CTEM. Specimen thickness can be increased by a factor of 2.3 and still remain within a single-scattering regime if the accelerating voltage is increased from 100 to 1000 kV. It is this attribute of the HVEM that most significantly extends the instrument's applications beyond those of the CTEM by making it possible to examine denser materials, thicker specimens (Section 3.1.3), hydrated cells in environmental chambers, and so on.

3.1.3. Ability to View Thick Specimens

The ability to obtain high-resolution images of thick specimens, a result of increased penetrating power and reduced chromatic aberration, is probably the most significant and practical advantage of the HVEM (Glauert, 1974). It is possible to obtain a 20-fold increase in practical resolving power (i.e., resolving power as limited by chromatic aberration) with thick specimens by increasing the accelerating voltage from 100 to 1000 kV (Meek, 1976). The HVEM can thus be used to view 1-μm-thick sections at a resolution equal to that obtained when thin sections (50–60 nm) are viewed with a CTEM. Depending on the nature of the specimen, sections up to 5 μm in thickness may be viewed at 1000 kV with adequate resolution for most biological studies.

The ability to view thick specimens is desirable for many reasons, the most important of which may be summarized as follows:

1. Thick specimens provide improved sampling statistics because they contain a much greater volume of cell or tissue material than do thin sections. The increased sampling of cell material results in a more complete and, sometimes, more realistic image of the specimen and its substructures. Conventional thin

sections sometimes yield relatively misleading or inaccurate information by comparison (for examples, see Section 4 and 5, as well as Glauert, 1974; Hawes, 1981; King *et al.*, 1980).

2. Because thick specimens contain so much material, the investigator can obtain large amounts of quantitative structural information on entire cells in relatively short periods of time by looking at serial thick sections (Section 4.5) or critical-point-dried whole cells (Section 5).

3. HVEM images of thick specimens contain much 3-D information, especially if stereo pairs are made to facilitate their interpretation. This technique allows one to visualize 3-D spatial relationships directly, without extrapolation from the 2-D data contained in conventional thin sections or tedious reconstruction of serial thin sections.

3.2. Disadvantages

3.2.1. Specimen Damage

Considering the reduced interaction between the electron beam and the specimen, one might expect specimen damage in the HVEM to be less severe than in the CTEM. Yet, it now appears that biological specimens can incur considerable damage when they are examined by HVEMY and that it may be advisable to undertake precautions to prevent such damage. This aspect of HVEMY has been discussed in detail by King *et al.* (1980).

3.2.2. Cost

Admittedly, the cost of operating an HVEM does not directly affect many of its users because most of these instruments are located in facilities that are supported by government programs like the NIH Research Resources Program in the United States. Nevertheless, the initial cost of an HVEM is enormous ($1–5 million plus the cost of a building suitable for housing it), as are the expenses of maintenance and routine operation. These expenses must be viewed as a disadvantage when one is comparing the HVEM to the CTEM. The travel expenses that most investigators incur in order to use one of the small number of HVEMs available to biologists represent an additional disadvantage.

3.2.3. Reduction of Contrast

The major practical disadvantage of HVEMY of biological specimens is the reduction in image contrast that accompanies increased accelerating voltages. Ironically, this problem is caused by the same phenomenon that is responsible for better resolution of thick specimens at high voltages: decreased beam–specimen interaction. The lack of contrast can usually be overcome, but only through the use of modified staining methods, darkfield imaging, and so on.

4. EXAMINATION OF THICK SECTIONS

4.1. Introduction

The HVEM is most often used in biological research to examine relatively thick sections of plastic-embedded specimens. Studies using this approach have yielded a considerable amount of information on the 3-D architecture of cells, the ultrastructure of cellular features that are difficult to visualize with thin sectioning or other approaches, and the general relationship between structure and function in cells and tissues. Most of these investigations have dealt with eukaryotic tissue material (for reviews, see Glauert, 1974; Hawes, 1981; King *et al.*, 1980); only a few of them have been directed at some of the larger unicellular microorganisms (see below). Nevertheless, the latter studies have been highly productive and serve to illustrate the potential value of HVEMY as applied to microorganisms.

Image contrast and resolution must be maximized for each type of specimen when thick sections of plastic-embedded biological material are investigated with HVEMY. The approaches described below are applicable to a wide range of biological specimens but may require modification to yield optimal results with some types of samples. The following is not a detailed discussion of thick sectioning and HVEMY techniques but, rather, a general overview of the pertinent methods with a few specific recommendations (based on personal experience) concerning the use of these methods with microorganisms.

4.2. Preparation of Thick Sections

The preparation of thick sections for HVEMY is a reasonably straightforward task, being essentially a slight modification of the methodology used to prepare thin sections. The following factors should be considered when thick sections are produced: (1) the choice of fixation, (2) the choice of embedding resin, (3) cutting techniques, (4) the choice and preparation of specimen grids, and (5) methods for staining to enhance image contrast.

4.2.1. Fixation Procedures

All of the fixations commonly used to produce conventional thin sections can also be used to produce thick sections. The suitability of a specific protocol depends on the characteristics of the specimen and the goals of the investigation, just as it does in thin sectioning. A general fixation that preserves a wide variety of cell features (e.g., glutaraldehyde prefixation followed by OsO_4 postfixation) will suffice in many cases. The only drawback is that, because thick sections contain so much cell material, preservation and staining of too many cell features may produce images that are difficult to interpret because of excessive superimposition of different structures. General cell preservation is most likely to result in confusing thick-section images with prokaryotic microorganisms that have very dense cytoplasms (Glauert, 1974; Weibull, 1974). Eukaryotic microorganisms, in which the cytoplasm is somewhat less crowded, are less susceptible to this problem.

Few, if any, fixation procedures provide truly complete preservation of biological specimens; most protocols tend to preserve certain types of cell features better than others. As a result, some investigators may want to use a variety of different fixations to maximize the amount of information that can be obtained from the resulting sections. In addition, the varying ability of different fixations to preserve certain features can be used to avoid the confusing images sometimes produced by general fixations as described above. Highly selective fixations make it possible to preserve and densely stain the cell feature that is of primary interest, thereby eliminating (through lack of preservation) unwanted structures that would simply obscure one's view of this feature. Such an approach facilitates rapid elucidation of a specific feature's detailed ultrastructural characteristics (including 3-D characteristics) and often enables one to view sections thicker than 1 μm without incurring an unacceptable loss of resolution.

Selective staining can be used as an alternative to selective fixation for increasing the contrast of certain cell features (Glauert, 1974; Hawes, 1981). Most selective staining procedures for HVEMY involve the use of heavy metals that associate preferentially with specific features. The reagents are generally applied to the specimen during or after the fixation, but prior to dehydration and embedding. Examples of successful selective stains include OsO_4 impregnations (Poux et al., 1974; Rambourg et al., 1973, 1974), OsO_4–zinc iodide (ZIO) impregnations (Harris, 1979; Harris and Chrispeels, 1980), lead stains for acid phosphatase (Carasso et al., 1971; Favard et al., 1971), and silver impregnations (Favard and Carasso, 1973; Palay and Chan-Palay, 1973).

4.2.2. Embedding Resins

Epon and Araldite (or Epon–Araldite) epoxy resins have been used most frequently as embedding media for thick sectioning, although Spurr's resin (Spurr, 1969) and Maraglas have also been used with some success. The following formulation of Poly/Bed 812 (an Epon 812 substitute sold by Polysciences, Inc., Warrington, Pa.) is recommended for most routine applications:

Poly/Bed 812	31.45 g
DDSA (dodecenylsuccinic anhydride)	16.37 g
NMA (nadic-methyl anhydride)	12.43 g
DMP-30 [2,4,6-tri(dimethylaminomethyl)-phenol]	0.88 ml
(Curing time: 15 hr, minimum, at 65°C)	

This formula yields fairly soft blocks with excellent cutting properties for sections from 10 nm to 3 μm in thickness. The blocks should be fairly soft because (1) thick sections cut from softer blocks tend to have less compression and fewer striations and (2) softer blocks allow a greater number of consecutive sections to be cut from a single portion of the knife edge.

4.2.3. Cutting the Sections (Ultramicrotomy)

Any ultramicrotome can be used to cut sections up to several micrometers in thickness, and high-quality sections can be obtained with either glass or diamond knives. Diamond knives may be superior because (1) they can cut many consecutive, defect-free sections and (2) blocks of average hardness (such as those normally prepared for thin sectioning) can be cut without difficulty. The major drawback of the diamond knife for thick sectioning, however, is that its edge dulls far more rapidly than during thin sectioning. The routine use of diamond knives for thick sectioning, then, is feasible only in laboratories that can afford to replace or resharpen their diamond knives frequently.

When a glass knife is used to cut thick sections, it should be possible to cut ten or more high-quality sections without moving to a new portion of the knife edge if the specimen blocks are made relatively soft as described above. Only one to four sections can be obtained if blocks of average hardness are sectioned instead. The actual number of good sections obtained with any block will depend on the quality of the knife, the thickness of the sections, and the inherent properties of the embedded specimen. The sections will eventually begin to display tears, striations, and/or other defects in all cases and, when this happens, a new portion of the knife edge (or a new knife) must be used to prevent gouging of the block face. A problem frequently encountered when very thick sections (2–10 μm) are cut is that the block face becomes rough after only one or two sections have been cut. This problem cannot be avoided entirely, but it can be minimized if several thinner sections (0.25–0.5 μm) are cut after each thick section. The thin sectioning polishes the block face so that higher-quality thick sections are obtained.

Section thickness ranging from 0.25 to several micrometers can be obtained with reasonable accuracy by proper adjustment of the advance controls on most microtomes. (Specific instructions are usually given in the operation manual supplied with each instrument.) Section thickness can then be checked to some extent by observation of interference colors (Table II). It is important to ascertain the most effective section thickness for each new type of specimen examined, a quantity that depends on the fixation and staining techniques, as well as on the internal structural characteristics of the specimen itself. One can determine optimal thickness by preparing sections of different thickness and then comparing them in the HVEM. In a recent study of the photosynthetic membranes in a unicellular cyanobacterium (Nierzwicki-Bauer *et al.*, 1984), for example, 0.25-, 0.5-, and 1.0-μm sections (Figure 2) were compared with respect to how effectively they depicted complex membranous intersections. The 0.25-μm sections were found to be the most useful in this case because of the organism's very dense cytoplasm. The resolution obtained with the 0.5- and 1.0-μm sections was not inferior to that observed with the 0.25-μm sections. Rather, the thicker sections just contained too many overlapping dense cell features to be readily interpreted.

4.2.4. Choice and Preparation of Grids

Virtually any type of specimen grid can be used for HVEMY of thick sections. Standard square-mesh grids are probably used most often, but single-slot or folding grids may be preferable in some instances. Single-slot grids are useful for mounting ribbons of serial thick sections. They also permit the use of larger tilt angles because of the absence of obscuring grid bars. Folding grids are used when sections tend to detach from standard grids during staining or during examination in the microscope, a problem that occurs frequently with sections that are thicker than 1 μm. The sandwiching effect of the folding grid prevents detachment, but the sections may be less stable in the electron beam because folding grids are usually not coated with Formvar (see below). An alternative method for preventing detachment of thick sections has been described by Drummond (1950). Dipping the grids (any type) into a solution of 1% isobutylene in xylol prior to picking up the sections makes them slightly adhesive.

Table II
Interference Colors
of Araldite Thick Sections[a]

Section thickness (nm)	Interference color
300	Blue
340	Yellow
650	Red
710	Green
1200 or greater	Colorless

[a]From Locke and Krishnan (1971).

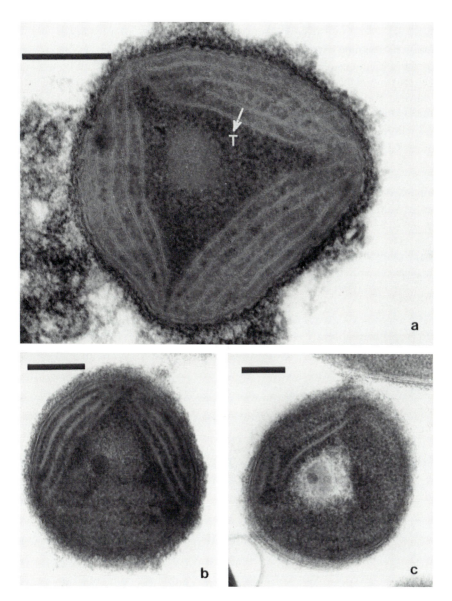

FIGURE 2. High-voltage electron micrographs of thick sections through the unicellular cyanobacterium *Agmenellum quadruplicatum*, comparing section thicknesses of 0.25 μm (a), 0.5 μm (b), and 1.0 μm (c). Bars = 0.5 μm. Note increased contrast and greater tendency of photosynthetic thylakoid membranes (T) to become obscured by darkly stained materials in the thicker sections. Reproduced from Nierzwicki-Bauer *et al.*, (1984), with copyright permission from Blackwell Scientific Publications, Ltd.

All grids to be used for thick sections and HVEMY should be cleaned thoroughly. Cleaning is best accomplished by ultrasonication in detergent, followed by multiple rinsing in distilled water and then acetone. A final glow-discharge cleaning is recommended as well. The cleaned grids can be used as they are for picking up sections or, as is more often done, they can be coated with Formvar. Formvar coatings provide additional physical support for the sections (thereby increasing stability in the electron beam) and facilitate adhesion of the sections to the grid. A 0.8% solution of Formvar in ethylene dichloride will produce relatively thick Formvar films (silver-to-light-gold interference color) that are suitable for most HVEMY applications. If the films do not adhere properly to the grids, adhesion can be improved by dipping the grids in 0.1% Formvar (in ethylene dichloride) and allowing them to air-dry before they are coated with 0.8% Formvar. Formvar-coated grids should be recoated with carbon (about 10nm) and glow-discharged. Carbon-coating strengthens the support film and increases its conductivity, thus reducing heat buildup in the electron beam. Glow-discharging reduces the hypdrophobicity of the support film, so that sections can be retrieved more easily.

The investigator usually mounts thick sections on grids by picking them up from below the water surface. Touching the edge of the grid to a piece of filter paper removes the excess water. When sections must be placed on grids in some precise orientation (e.g., when ribbons of serial sections are mounted), the "sol-gel" method of Anderson and Brenner (1971) is helpful.

4.2.5. Poststaining

4.2.5a. Section Staining. Mounted thick sections must be stained intensely if they are to produce sufficient image contrast when viewed at high accelerating voltages. A modified version of the procedure often used to poststain thin sections will serve this purpose. The general protocol is outlined in Table III.

Table III
General Procedure for Poststaining Thick Sections

Operation	Time	Temperature
Immerse grids in 2% uranyl acetate (aqueous)	1–4 hr[a]	50°C
Rinse grids with gentle stream of distilled water	1 min	RT[b]
Immerse grids in 0.4% lead citrate (aqueous)[c]	30–60 min[a]	37°C
Rinse grids with gentle stream of distilled water	1 min	RT
Remove excess liquid from grid and air-dry	—	RT

[a]Exact time depends on section thickness; see Table IV.
[b]RT, room temperature.
[c]Formulated according to Reynolds (1963).

Table IV
Poststaining Time for Thick Sections

Section thickness (μm)	Staining time (min)	
	Uranyl acetate[a]	Lead citrate[a]
0.25	60	30
0.5	75	40
1.0	90	50
1.5	120	60

[a]Stain formulas and concentrations as specified in Table III.

The grids are immersed in the staining solutions so that the two sides of the sections are stained simultaneously. The Hiraoka staining kit sold by Polysciences can be used to facilitate staining sections on both sides. Alternatively, the grids can be inserted (on edge) into dental wax and then covered with large droplets of stain solution. Because of the long staining times, the lead citrate staining should be done in a small, closed petri dish in which NaOH pellets have been placed around the droplets of stain solution. If stain precipitates appear on the sections in spite of these precautions, they can be removed by treatment of the grids with 0.05% nitric acid (Favard and Carasso, 1973).

Most of the differences between the procedure in Table III and the protocol normally used for thin sections (i.e., staining on both sides, longer times, higher temperatures) reflect the fact that stains penetrate thick sections very slowly. The penetration rates of aqueous solutions vary according to the characteristics of the stain itself. Uranyl acetate solutions penetrate epoxy resins more slowly (Locke and Krishnan, 1971) than do lead citrate solutions (Venable and Coggeshall, 1965). In practice, the optimal staining times are dictated by the thickness of the sections being stained. Recommended staining times for sections up to 1.5 μm in thickness are given in Table IV. Staining times for thicker sections can be determined by trial and error, using the method of Favard and Carasso (1973) to ascertain when penetration is complete.

The only disadvantage to poststaining with both uranyl acetate and lead citrate is that the resulting images may be confusing because many different cell features are stained to an equal density. A somewhat more selective staining effect can be achieved if either stain is used alone. Uranyl acetate tends to stain membranous features, whereas lead citrate selectively stains ribosomes and the cytoplasmic ground substance (Glauert and Mayo, 1973).

4.2.5.b. *En Bloc* and Block Staining. Good contrast can be obtained with thick sections if the specimens are stained *en bloc* prior to dehydration and embedding, either in place of or in addition to poststaining of the finished sections. Hot (60°C) alcoholic solutions (Locke and Krishnan, 1971; Locke *et al.*, 1971) penetrate effectively within approximately 24 hr. The most frequently used reagents are uranyl acetate in ethanol, uranyl acetate in methanol, and phosphotungstic acid in ethanol. Sequential staining with different reagents can also be employed (Favard *et al.*, 1971).

Pesacreta and Parthasarathy (1984) recently described a technique for postpolymerization block staining that greatly increased the contrast of thick-sectioned plant material. Polymerized blocks of epoxy resin (rather than the cut sections) were treated with 5% uranyl acetate (in 75% ethanol) for 36 hr at 60°C. This approach was advantageous because (1) there was no possibility of Formvar film deterioration caused by exposure to hot ethanolic solutions (see Hayat, 1970), and (2) there were fewer problems with chatter and tearing when the sections were cut, presumably because the hot ethanolic solution softened the block prior to cutting. Moreover, the contrasting of ribosomes, microtubules, microfilaments, and other detailed structures was far superior to that obtained by poststaining of finished sections.

4.3. Preparation and Usefulness of Stereo Micrographs

Micrographs of thick sections are often difficult to interpret because these sections contain so much cell material that structures situated at different depths within the section are frequently superimposed in the final photographic image. To fully understand the structural information contained in thick sections as revealed by the HVEM, then, one must produce stereo-pair micrographs of important regions of the specimen. These stereo pairs can be fused with a stereoscopic viewer (or, in the absence of a viewer, by adjustment of one's eyes) to produce a 3-D image. An experienced investigator can determine from such an image where different structures within the thick section are positioned with respect to one another in 3-D space. This knowledge, in turn, facilitates accurate interpretation of the micrographs and leads more rapidly to a successful elucidation of the subject's structural characteristics (compare monographic and stereographic views of Figures 3–8).

Stereo pairs are produced when a thick section is photographed from two different angles; one of the goniometer stages described above (Section 2.3.1) is used to achieve the required specimen orientations. The optimal tilt angles for each stereo pair depend on section thickness and the final magnification of the micrographs. These angles can be calculated from the data of Hudson and Makin (1970), or one can use the settings recommended by Beeston (1973) for micrographs to be examined with a 2× hand-held stereo viewer.

Aside from facilitating the interpretation of confusing thick-section images, stereo pairs are especially useful for studying the 3-D configuration of cell features that fit within a single section and the overall 3-D architecture of entire cells. Figure 3 depicts a stereo pair (from Hawes, 1981) of a 1-μm thick section through a fungal conidium, in which the 3-D structures of the mitochondria and

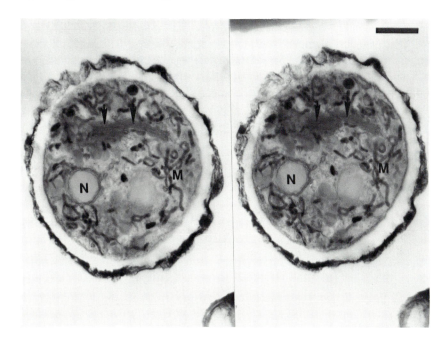

FIGURE 3. Stereo-pair high-voltage (1 MV) electron micrograph of a 1 -μm thick section through a fungal conidium (*Ceratocystis adiposa*) that has been stained by impregnation with zinc and OsO_4 (ZIO technique) in order to show the 3-D distribution of the mitochondria (M) and endoplasmic reticulum (arrows). N, nucleus. Bar = 2 μm. Reproduced from Hawes (1981), with copyright permission from Pergamon Press, Ltd.

endoplasmic reticulum are readily apparent. In contrast to conventional thin sections, which give the impression that the cells have many short mitochondria, thick sections reveal that the mitochondria are actually long and tubular and are sometimes shaped like rings.

Stereo pairs are also helpful for tracing the paths of fibrous or membranous features as they pass through a thick section. This approach has proven to be quite effective for the examination of photosynthetic membranes in unicellular (Nierzwicki-Bauer *et al.*, 1983, 1984) and filamentous cyanobacteria (Figure 4). Hawes (1981) also used thick-section stereo pairs to study membrane arrangements, in this case the thylakoids of bean leaf chloroplasts. Stereo pairs of 1-μm thick sections (Figure 5) showed clearly that the membranes twist as they pass from one stack of grana to another.

4.4. Use of Bidirectional Tilting to Clarify Structures

Thick sections have been found to be superior to thin sections for detecting and elucidating the detailed structure of membranous intersections and other three-dimensionally complex features. Thin sections may fail to detect very small, isolated fusion points between membranes because the chances that any given section will happen to pass through such a structure are quite small. Thick sections, because they contain 10–30 times as much cell material, are far more likely to include one or more fusion points within the plane of sectioning. Thin sections may fail to portray membrane intersections effectively in some instances because membranes in thin sections stand out clearly only when they are oriented almost parallel to the electron beam. Unless both membranes are so oriented, one of them will be indistinct, and the intersection may go unnoticed. The clarity of membranes in thick sections is also dependent on their orientation, but the

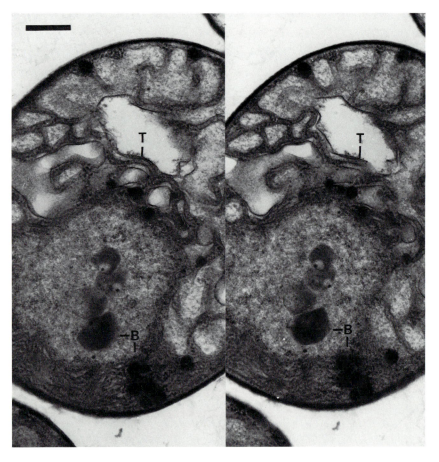

FIGURE 4. Stereo-pair high-voltage (1 MV) electron micrograph of a 0.25-μm thick section through a vegetative cell of the filamentous cyanobacterium *Anabaena* 7120. Bar = 0.5 μm. The 3-D distributions of the photosynthetic thylakoid membranes (T) and various inclusion bodies (B) are seen clearly.

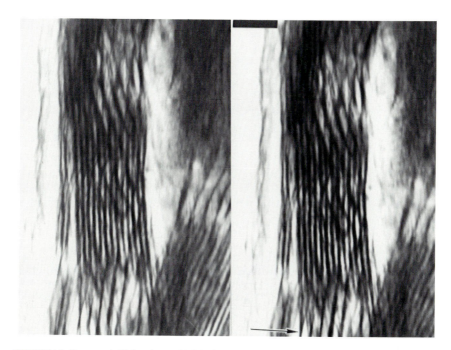

FIGURE 5. Stereo-pair high-voltage (1 MV) electron micrograph of a 1-μm thick section through a stack of grana in a zinc–OsO$_4$ (ZIO)-impregnated bean leaf chloroplast. Bar = 0.1 μm. The interrelationships between the membranes in the grana are seen clearly; note the twist in the membranes as they pass between adjacent grana (arrow). Reproduced from Hawes (1981), with copyright permission from Pergamon Press, Ltd.

improved sampling and greater depth of the thicker sections make it easier to recognize membranes that are not situated parallel to the beam, making it easier, in turn, to detect complex intersections. Glauert and Mayo (1973) were able to detect regions of fusion between the cisternae of the rough endoplasmic reticulum and the plasma membrane in chondrocytes with HVEMY of thick sections, even though such fusions were not seen in thin sections. Similarly, thick sections have revealed interconnections between microfibrils and neighboring connective tissue elements in eukaryotic cells (Ichimura and Hashimoto, 1982) and, more recently (D. L. Balkwill and S. A. Nierzwicki-Bauer, unpublished results), between the thylakoid and cytoplasmic membranes of filamentous cyanobacteria (Figure 6).

The ability of thick sections to reveal and clearly depict complex small structures like membranous interconnections can be enhanced by tilting of the sections about two perpendicular axes until the clearest possible views of the feature being examined are obtained. This bidirectional tilting is best accomplished with double-tilt goniometer stages, although rotation-tilt stages are also effective. Nierzwicki-Bauer *et al.* (1983, 1984) used the former to study com-

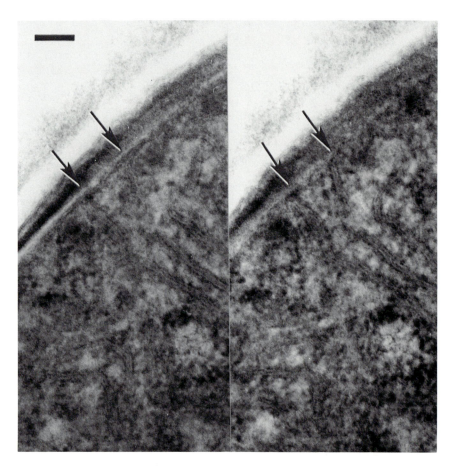

FIGURE 6. Stereo-pair high-voltage (1 MV) electron micrograph of a 1-μm thick section through a vegetative cell of the filamentous cyanobacterium *Mastigocladus laminosus,* illustrating contact (arrows) between the photosynthetic thylakoid membranes and the cytoplasmic membrane. Bar = 0.1 μm.

plex membranous features in a unicellular cyanobacterium and to demonstrate that this organism's thylakoids contacted the cytoplasmic membrane (Figure 7). Bidirectional tilting of thick sections should be equally effective if it is used to examine other complex microbial substructures.

4.5. Use of Serial Thick Sections to Obtain Quantitative Information and to Create 3-D Reconstructions

It is not very difficult to produce serial thick sections through microbial cells because relatively few sections are required to cover an entire cell, at least in

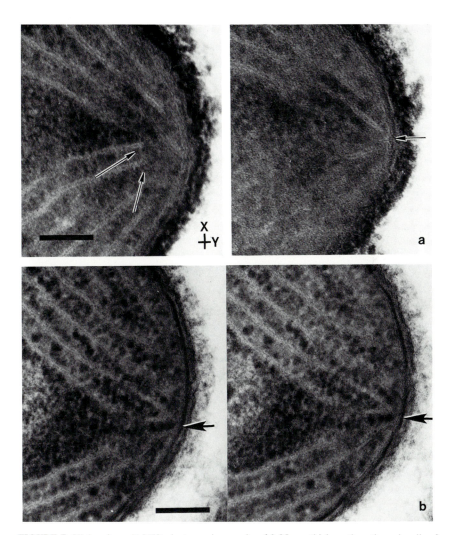

FIGURE 7. High-voltage (1 MV) electron micrographs of 0.25-μm thick sections through cells of the unicellular cyanobacterium *Agmenellum quadruplicatum,* illustrating the use of bidirectional tilting to clarify the structure of membranous features. Bar = 0.2 μm. (A) Use of tilting to study the nature of a complex membranous intersection. Left panel depicts an untilted section in which several thylakoid membranes appear to terminate independently of each other (arrows). Right panel depicts same section after being tilted +5° about X and −12° about Y. Thylakoid membranes now appear to extend to the cytoplasmic membrane (arrow), possibly intersecting at a common point. (b) Stereo-pair micrograph of a tilted section, depicting the use of tilting to maximize the clarity of a contact point (arrow) between the thylakoid and cytoplasmic membranes. Tilt angles: +8° about X for both halves of pair; −25° about Y for left half of pair; −19° about Y for right half of pair. (b) reproduced from Nierzwicki-Bauer *et al.* (1984), with copyright permission from Blackwell Scientific Publications, Ltd.

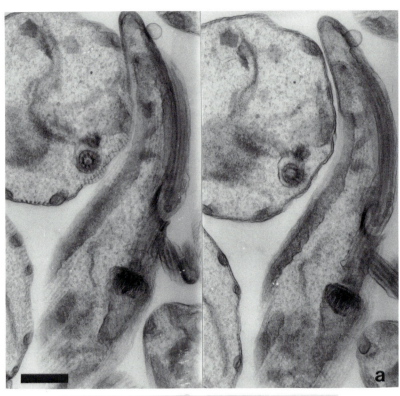

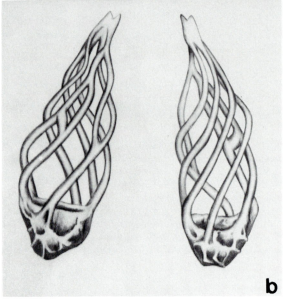

comparison to the number of serial thin sections needed to accomplish the same task. Such series of thick sections can be used to obtain quantitative structural information about whole cells quite rapidly. Serial 0.25-μm sections, for example, were used to determine the numbers and intracellular locations of several types of inclusion bodies in a unicellular cyanobacterium (Nierzwicki-Bauer *et al.*, 1983).

Serial thick sections may be used in place of thin sections for the production of manual or computer-aided 3-D reconstructions (see Chapter 5) of whole cells. Thick sections provide two important advantages over thin sections when these reconstructions are done: (1) much time is saved because fewer sections need be cut, and (2) a considerable amount of 3-D information that can be of assistance toward completing the reconstruction is contained within each section. Paulin (1975) used serial thick sections to study the configuration of mitochondria in trypanosome cells. This study revealed that each cell contains a continuity of mitochondrial extensions rather than a group of small, independent mitochondria. The nature of the extensions was seen clearly in single 0.5-μm sections (Figure 8a), and an artist's drawing of the complete mitochondrial system could be prepared from serial thick sections through entire cells (Figure 8b). Serial thick sections were used by Crang and Pechak (1978) to determine the overall 3-D intracellular organization of the black yeast *Aureobasidium pullulans*. Computer-aided reconstructions of cell tracings (Figure 9) showed that (1) large mitochondrial networks are situated about the cell periphery and around the nuclei, (2) vacuoles are located in the central region of the cell and are connected by channels, and (3) the nuclei are also in the center of the cell but are not connected.

4.6. HVEM in Combination with Light Microscopy

HVEMY studies involving thick sections can be facilitated by initial phase-contrast light-microscopic screening of serial sections for structures or areas of interest (Rieder and Bowser, 1983). This technique helps to minimize the amount of valuable HVEM beam time wasted by examination of useless sections. Similarly, fluorescence light microscopy of embedded cells in thick sections has been followed by ultrastructural characterization (with the HVEM) of labeled regions (Rieder and Bowser, 1985; Rieder et al., 1985).

←———

FIGURE 8. Use of HVEMY and serial thick sections to determine the 3-D configuration of the mitochondrion in the chondriome of the trypanosomatid *Blastocrithidia culicis*. (a) Stereo-pair high-voltage (1 MV) electron micrograph of a 0.5-μm thick section through *B. culicis*, showing portions of the mitochondrial extensions that were found to be present in this organism. The kinetoplast is at the viewer's lower right. Bar = 1.0 μm. (b) Artist's drawing of the complete mitochondrion of *B. culicis*, prepared with the aid of serial thick sections. Reproduced from Paulin (1975), with copyright permission from The Rockefeller University Press.

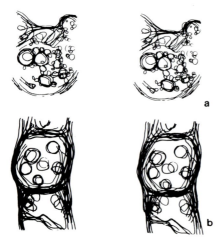

FIGURE 9. Stereo pairs made from line-drawing overlays of serial thick sections (0.25 μm; photographed by HVEMY at 1 MV) through cells of the black yeast *Aureobasidium pullulans*. (a) Vacuole profiles indicate that vacuoles are clustered in the center of the cell and are interconnected at various levels. (b) Nuclear profiles indicate that nuclei are not interconnected. Reproduced from Crang and Pechak (1978), with copyright permission from Springer-Verlag.

5. EXAMINATION OF WHOLE CELLS

5.1. Introduction

With the increased penetrating power of the HVEM making it possible to view thick specimens, it was inevitable that this instrument would eventually be used to study whole cells. Approaches frequently employed to do so include examination of (1) very thick sections (i.e., sections that actually include entire cells), (2) critical-point-dried whole cells, and (3) hydrated cells in special environmental chambers. To date, only the first two approaches have been used with a high degree of success. Nevertheless, examination of hydrated cells may yet become a feasible approach if some of the technical difficulties associated with this technique can be eliminated (for further information, see Glauert, 1974; King *et al.,* 1980).

Interpretation of the recorded images is perhaps the most important and most difficult aspect of using HVEMY to study whole cells. As with thick sections, judicious use of selective staining and/or stereo pairs can be helpful. It is also important to concentrate on one area or one level of the cell at a time when viewing low-magnification images, in order to avoid the confusion that can be caused by the great depth of field in whole-cell images.

HVEMY studies involving whole cells have been confined almost exclusively to the examination of air- or critical-point-dried eukaryotic cells that were either isolated and transferred to the grids or cultured directly on the grid support films (see Glauert, 1974; King *et al.,* 1980). Microbial cells would seem to be ideal subjects for such investigations but, to date, the results have been disappointing (Glauert, 1974). The main problem is that microbial cells often have very dense cytoplasms, thereby making it difficult to resolve features inside the cell. It should be possible to overcome this drawback by using selective staining or by growing the organisms under conditions that reduce the number of ribosomes in the cytoplasm (e.g., nitrogen limitation). Considering the impressive results that have been obtained by observation of cultured eukaryotic cells, efforts to make this approach more applicable to microorganisms should be given a high priority.

5.2. Embedded Specimens (Very Thick Sections)

Unlike many eukaryotic tissue cells, most microorganisms are small enough to fit entirely within the plane of a single thick section. Such cells can be examined as embedded specimens rather than as critical-point-dried samples, an approach that provides good specimen stability and compatibility with a wide variety of selective fixation and staining procedures. This method has served (S. A. Nierzwicki-Bauer and D. L. Balkwill, unpublished results) to reveal the 3-D distribution of photosynthetic membranes and inclusion bodies in cells of *Rhodospirillum rubrum* (Figure 10), as well as the number and arrangement of magnetic particles in the magnetotactic bacterium *Aquaspirillum magnetotacticum* (Figure 11). The images were relatively easy to interpret in both cases because these organisms do not have a dense cytoplasm. With microorganisms that possess denser cytoplasms, it will be necessary to deal with the problems described above to obtain equally understandable and effective images.

5.3. Critical-Point-Dried (CPD) Specimens

Critical point-drying is the technique used most often to prepare whole-cell specimens for HVEMY. The primary advantages of CPD specimens over embedded whole cells are that (1) the cells have never been exposed to dehydrating or embedding chemicals; (2) the absence of embedding resin decreases electron scattering, thereby permitting thicker specimens to be viewed with higher resolution; and (3) the preparation protocols are simpler.

5.3.1. Preparation of CPD Specimens

Preparation techniques for critical point-drying are not overly difficult. Cells can be grown on specimen grids initially or attached to the grids after growth elsewhere. If the cells are to be grown on the grids, the grids must be made of gold or some other nontoxic material (copper grids are not suitable). After coating with Formvar and carbon, the grids should also be sterilized by UV irradiation prior to inoculation (for examples and further information, see Byers and Porter, 1977; Porter and Tucker, 1981; Wolosewick and Porter, 1976, 1979). Cells that are not grown on grids can be attached to them through the use of polylysine (see De Souza and Benchimol, 1984).

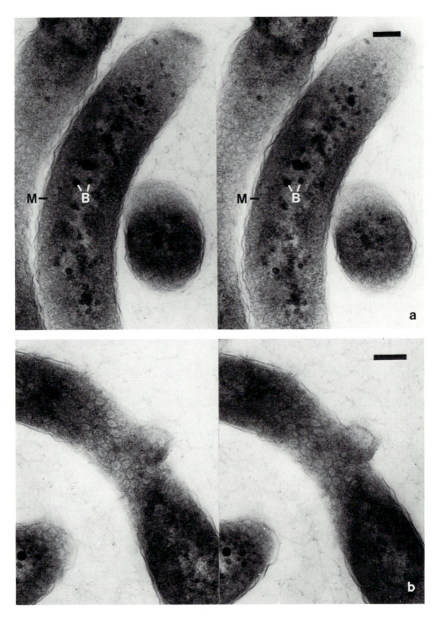

FIGURE 10. Stereo-pair high-voltage (1 MV) electron micrographs of 1-μm thick sections through cells of the photosynthetic bacterium *Rhodospirillum rubrum*. Bars = 0.2 μm. (a) Section area in which the entire cell body is contained within the plane of sectioning. Note the 3-D distribution of inclusion bodies (B) and appearance of the photosynthetic membranes (M). (b) Oblique section through the edge of a cell, in which only the portion of the cytoplasm containing the photosynthetic membranes is contained within the plane of sectioning. Note vesicular nature of the membranes.

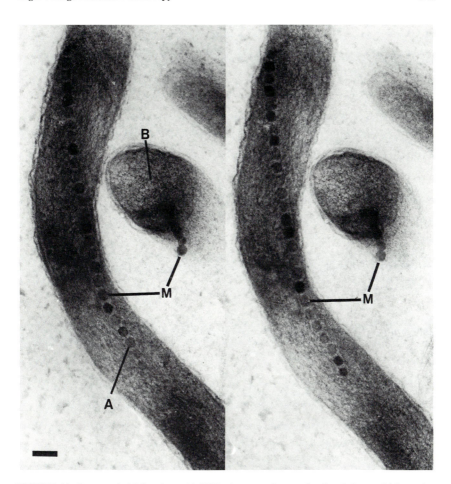

FIGURE 11. Stereo-pair high-voltage (1 MV) electron micrograph of a 0.5 μm thick section through cells of the magnetotactic bacterium *Aquaspirillum magnetotacticum*. Bar = 0.1 μm. Cell A contains a long chain of magnets (M) that appears to be quite straight (i.e., almost planar), even though the cell body is curving in and out of the plane of sectioning. The linear nature of the magnet chain can also be seen in cell B, which has been cross-sectioned.

A general procedure for the preparation of CPD whole cells for HVEMY is given in Table V. The CPD specimens should be coated with approximately 10 nm of carbon and then stored in a desiccator until they are viewed.

5.3.2. Observations and Results

Studies involving HVEMY of CPD whole microbial cells have not been numerous, but they have produced results that would have been difficult or even

Table V
General Procedure for Critical Point-Drying Whole Cells for HVEMY

Operation	Time	Temperature
Fix with 2% glutaraldehyde in 0.1 M cacodylate buffer (pH 7.3)	20 min	CGT[a]
Rinse with 0.1 M cacodylate buffer	1 min	CGT
Fix with 1% OsO_4 in 0.1 M cacodylate buffer	10 min	RT[b]
Soak in distilled water	30 min	RT
Dehydrate as follows:		
50% acetone	2 min	RT
75% acetone	2 min	RT
95% acetone	2 min	RT
100% acetone (repeat 2–3 times)	5 min	RT
Transfer to critical point dryer and flush with liquid CO_2 (repeat 4–5 times)	2 min	—
Critical point-dry[c]		

[a]CGT, cell growth temperature, i.e., the temperature at which the specimen is normally grown.
[b]RT, room temperature.
[c]Protocol depends on the specific instrument used.

impossible to obtain with other, more conventional approaches. These studies are described below to illustrate the range of information that can be obtained with this approach.

De Souza and Benchimol (1984) used HVEMY to investigate the structural organization of CPD whole trypanosomatids (protozoa). They found that intact whole cells were too thick for satisfactory observation of internal details, and it was therefore necessary to extract some of the cellular components with Triton X-100. This extraction rendered the cells more suitable for viewing. The degree to which cellular components were extracted could be controlled by adjustment of the concentration of the detergent. As a consequence, different aspects of the overall cellular architecture could be observed by examination of cells extracted with different concentrations (Figure 12). A similar approach was used by Loftus *et al.* (1984) to elucidate the structural relationships between different filamentous elements within platelets. In this study, it was found that extraction of living cells with Triton and other nonionic detergents removes the less stable cytoplasmic components while leaving behind a cytoskeleton of three major filament types.

Todd *et al.* (1984) examined piliated gonococcal cells prepared by critical point-drying and found that thick pilus structures branch and fuse to form an irregular 3-D lattice around the cells (Figure 13). This lattice was not apparent in epoxy-embedded specimens, and was destroyed during air-drying for subsequent negative staining. Cox and Juniper (1983) have examined whole CPD cells of the protonemata of the moss *Bryum tenuisetum*, as well as CPD cells of bacteria and cyanophytes. The prokaryotic cells they examined did not contain the "micro-

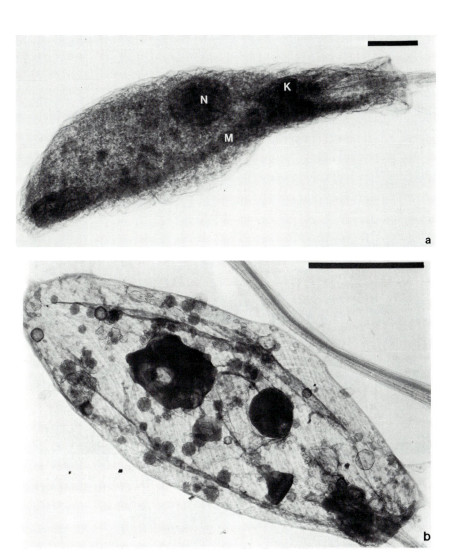

FIGURE 12. High-voltage (1 MV) electron micrographs of critical-point-dried whole cells of the trypanosomatid *Herpetomonas samuelpessoai,* following extraction with Triton X-100. Bars = 1 μm. (a) Cell treated with 0.05% Triton X-100 for 5 min. N, nucleus; K, kinetoplast; M, mitochondrion. (b) Cell treated with 0.1% Triton X-100 for 5 min. The organization of the subpellicular microtubules is visible. Reproduced from De Souza and Benchimol (1984), with copyright permission from Editrice Compositori.

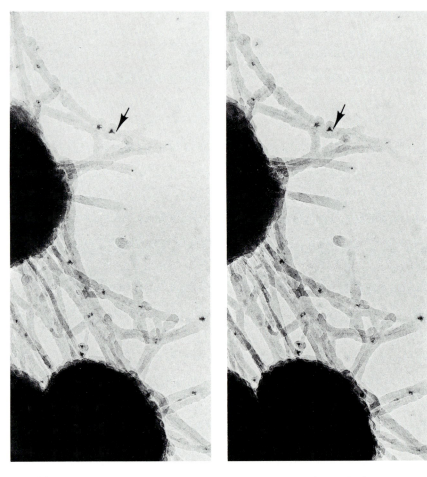

FIGURE 13. Stereo-pair high-voltage (1 MV) electron micrograph of critical-point-dried whole cells of the bacterium *Neisseria gonorrhoeae*, depicting an extracellular network of thick pilus structures that branch and rejoin. Electron-dense material (arrows) is often associated with pili in specimens processed by the critical point method. Micrograph from Todd *et al.* (1984). ×68,000.

trabecular lattices'' frequently observed in plant and animal cells. HVEMY of CPD whole cells has also been used to study the reproduction of viruses in eukaryotic tissue cells (Fonte and Porter, 1974; Gershenbaum *et al.*, 1974; Kilarski *et al.*, 1976).

A technique for sectioning polyethylene glycol-embedded material (Mazurkiewicz and Nakane, 1972) has been adapted to facilitate observation of resin-free sections in the electron microscope (Guatelli *et al.*, 1982; Wolosewick, 1980). This approach was used by Hawes *et al.* (1983) to study resin-free sections of plant tissues, and the resulting images were similar to those obtained from the edges of CPD whole cells. This technique might be useful for studying whole cells of small microorganisms or thick, resin-free sections through larger microorganisms. It also has the advantage of being highly compatible with cytochemical and immunocytochemical procedures.

6. CONCLUSIONS

HVEMY is a valuable alternative to the more traditional approaches that are commonly used to study the ultrastructure of prokaryotic and eukaryotic microorganisms. The principal advantage of HVEMY is that it allows one to obtain high-resolution images of unusually thick specimens. Thick sections or even entire microbial cells can be viewed at a resolution equal to that achieved with thin sections by the CTEM. The improved sampling statistics provided by thicker samples result in a more complete and, sometimes, more realistic image of the specimen. Such images also contain a considerable amount of 3-D information, the visualization of which can be facilitated by the production of stereo-pair micrographs. In practice, then, HVEMY of thick specimens can (1) facilitate accurate elucidation of cellular ultrastructure by avoiding the comparatively less complete and, occasionally, misleading information provided by randomly cut thin sections; (2) facilitate rapid elucidation of the 3-D relationships between different cell features or of the 3-D architecture of the entire cell; (3) allow one to obtain large amounts of quantitative structural information in a relatively short period of time; and (4) provide certain types of structural information that cannot be obtained as readily with any other technique.

HVEMY has not been applied extensively to the study of microbial ultrastructure. Nevertheless, the small number of investigations described to date have been highly productive in terms of yielding novel information. In some cases, the findings have even demonstrated that ultrastructural models based on thin sectioning were incorrect or incomplete. Such findings justify more widespread use of HVEMY to examine microbial subjects, and they imply that interesting and novel observations will continue to be made as an increasing variety of microorganisms are investigated in this way.

ACKNOWLEDGMENTS. I am indebted to Drs. David L. Balkwill and S. Edward Stevens, Jr., for their collaboration on some of the HVEMY work described in this chapter. I am also grateful to Dr. D. L. Balkwill for valuable advice and critical reading of the manuscript, and to Drs. M. Benchimol, R. E. Crang, W. De Souza, C. Hawes, D. F. Parsons, J. Paulin, and W. J. Todd for contributing micrographs. I wish to thank Drs. D. F. Parsons, A. J. Ratkowski, W. Tivol, D. Barnard, and S. Davilla for their advice and assistance in using the AEI EM7 MkII HVEM.

Some of the work described in this chapter was supported by United States Public Health Service Grant RR0219 awarded by the Division of Research Resources, Department of Health and Human Services, to support the New York State high-voltage electron microscope as a National Biotechnology Resource.

7. REFERENCES

Anderson, R. G. W., and Brenner, R. M., 1971, Accurate placement of ultrathin sections on grids: Control by sol-gel phases of a gelatin flotation fluid, *Stain Technol.* **46**:1–6.

Beeston, B. E. P., 1973, High voltage microscopy of biological specimens: Some practical considerations, *J. Microsc. (Oxford)* **98**:402–416.

Byers, H. R., and Porter, K. R., 1977, Transformation in the structure of the cytoplasmic ground substance in erythrophores during pigment aggregation and dispersion. I. A study using whole cell preparations in stereo high voltage electron microscopy, *J. Cell Biol.* **75**:541–558.

Carasso, N., Ovtracht, L., and Favard, P., 1971, Observation, en microscopie électronique haute tension, de l'appareil de Golgi sur coupes de 0.5 à 5 µ d'épaisseur, *C. R. Acad. Sci. Ser. D* **273**:876–879.

Cox, G., and Juniper, B. E., 1983, High-voltage electron microscopy of whole, critical-point dried plant cells—Fine cytoskeletal elements in the moss *Bryum tenuisetum*, *Protoplasma* **115**:70–80.

Crang, R. E., and Pechak, D. G., 1978, Serial section reconstruction of the black yeast, *Aureobasidium pullulans* by means of high voltage electron microscopy, *Protoplasma* **96**:225–234.

De Souza, W., and Benchimol, M., 1984, High voltage electron microscopy of critical point dried trypanosomatids, *J. Submicrosc. Cytol.* **16**:237–242.

Drummond, D. G. (ed.), 1950, The practice of electron microscopy, *J. R. Microsc. Soc.* **70**:1–141.

Dubochet, J., 1973, High resolution dark-field electron microscopy, in: *Principles and Techniques of Electron Microscopy,* Vol. 3 (M. A. Hayat, ed.), pp. 115–151, Van Nostrand–Reinhold, Princeton, New Jersey.

Dupouy, G., 1968, Electron microscopy at very high voltages, *Adv. Opt. Electron Microsc.* **2**:167–250.

Dupouy, G., 1973, Performance and applications of the Toulouse 3 million volt electron microscope, *J. Microsc. (Oxford)* **97**:3–28.

Dupouy, G., Perrier, F., and Durrieu, L., 1960, Microscopie électronique: L'observation de la matiére vivante au moyen d'un microscope électronique fonctionnant sous très haute tension, *C. R. Acad. Sci.* **251**:2836–2841.

Dupouy, G., Perrier, F., Enjalbert, L., Lapchine, L., and Verdier, P., 1969, Accroisement du contraste des images d'objects amorphes en microscopie électronique, *C. R. Acad. Sci. Ser. B* **268**:1341–1345.

Favard, P., and Carasso, N., 1973, The preparation and observation of thick biological sections in the high voltage electron microscope, *J. Microsc. (Oxford)* **97**:59–81.

Favard, P., Ovtracht, L., and Carasso, N., 1971, Observations de spécimens biologiques en microscopie électronique à haute tension. 1. Coupes épaisses, *J. Microsc. (Paris)* **12**:301–316.

Fonte, V. G., and Porter, K. R., 1974, Visualization in whole cells of herpes simplex virus using SEM and TEM, in: *Scanning Electron Microscopy 1974* (O. Johari and I. Corvin, eds.), pp. 827–834, Illinois Institute of Technology Research Institute, Chicago.

Gershenbaum, M. R., Shay, J. W., and Porter, K. R., 1974, The effects of cytochalasin B in Balb/373 mammalian cells cultured *in vitro* as observed by scanning and high voltage electron microscopy, in: *Scanning Electron Microscopy 1974* (O. Johari and I. Corvin, eds.), pp. 589–596, Illinois Institute of Technology Research Institute, Chicago.

Glauert, A. M., 1974, The high voltage electron microscope in biology, *J. Cell Biol.* **63**:717–748.

Glauert, A. M., and Mayo, C. R., 1973, The study of the three-dimensional structural relationships in connective tissues by high voltage electron microscopy, *J. Microsc. (Oxford)* **97**:83–94.

Guatelli, J. C., Porter, K. R., Anderson, K. L., and Boggs, D. P., 1982, Ultrastructure of the cytoplasmic and nuclear matrix of human lymphocytes observed using high voltage electron microscopy and embedment-free sections, *Biol. Cell* **43**:69–80.

Harris, N., 1979, Endoplasmic reticulum in developing seeds of *Vicia faba:* A high voltage electron microscope study, *Planta* **146**:63–69.

Harris, N., and Chrispeels, J. J., 1980, The endoplasmic reticulum of mung-bean cotyledons: Quantitative morphology of cisternal and tubular ER during seedling growth, *Planta* **148**:293–303.

Hawes, C. R., 1981, Applications of high voltage electron microscopy to botanical ultrastructure, *Micron* **12**:227–257.

Hawes, C. R., Juniper, B. E., and Horne, J. C., 1983, Electron microscopy of resin-free sections of plant cells, *Protoplasma* **115**:88–93.

Hayat, M. A., 1970, *Principles and Techniques of Electron Microscopy*, Vol. 1, p. 266, Van Nostrand–Reinhold, Princeton, New Jersey.

Hudson, B., and Makin, M. J., 1970, The optimum tilt angle for electron stereomicroscopy, *J. Phys. E.* **3**:311.

Ichimura, T., and Hashimoto, P. H., 1982, Three-dimensional fine structure of elastic fibers in the perivascular space of some circumventricular organs as revealed by high-voltage electron microscopy, *J. Ultrastruct. Res.* **81**:172–183.

Kilarski, W., Iwasaki, Y., Porter, K. R., and Koprowski, H., 1976, High voltage electron microscopy of human brain cells infected with vaccinia and parainfluenza 1 viruses, *J. Microsc. Biol. Cell.* **25**:81–86.

King, M. V., Parsons, D. F., Turner, J. N., Chang, B. B., and Ratkowski, A. J., 1980, Progress in applying the high-voltage electron microscope to biomedical research, *Cell Biophys.* **2**:1–95.

Locke, M., and Krishnan, N., 1971, Hot alcoholic phosphotungstic acid and uranyl acetate as routine stains for thick and thin sections, *J. Cell Biol.* **50**:550–557.

Locke, M., Krishnan, N., and McMahon, J. T., 1971, A routine method for obtaining high contrast without staining sections, *J. Cell Biol.* **50**:540–544.

Loftus, J. C., Choate, J., and Albrecht, R. M., 1984, Platelet activation and cytoskeletal reorganization: High voltage electron microscopic examination of intact and Triton-extracted whole mounts, *J. Cell Biol.* **98**:2019–2025.

Mannella, C. A., and Ratkowski, A. J., 1983, Adaptation of a HVEM for routine biological dark-field imaging, *Ultramicroscopy* **11**:21–34.

Massover, W. H., 1972, Ultra-high voltage electron microscopy (1–3 MeV) of biological macromolecules, in: *Proceedings of the 30th Annual Meeting of the Electron Microscopy Society of America*, pp. 182–183.

Massover, W. H., 1974, Effective resolution in biological thin sections: Experimental results with ultra-high voltage electron microscopy (1–3 MeV), in: *Proceedings of the 3rd International Congress on High Voltage Electron Microscopy, Oxford*, pp. 163–166.

Massover, W. H., Lacaze, J.-C., and Durrieu, L., 1973, The ultrastructure of ferritin macromolecules. I. Ultrahigh voltage electron microscopy (1–3 MeV), *J. Ultrastruct. Res.* **43**:460–475.

Mazurkiewicz, J. E., and Nakane, P. K., 1972, Light and electron microscopic localization of antigens in tissues embedded in polyethylene glycol with a peroxidase-labelled antibody method, *J. Histochem. Cytochem.* **20**:969–974.

Meek, G. A., 1976, *Practical Electron Microscopy for Biologists*, 2nd ed., pp. 365–375, Wiley, New York.

Nierzwicki-Bauer, S. A., Balkwill, D. L., and Stevens, S. E., Jr., 1983, Three-dimensional ultrastructure of a unicellular cyanobacterium, *J. Cell Biol.* **97**:713–722.

Nierzwicki-Bauer, S. A., Balkwill, D. L., and Stevens, S. E., Jr., 1984, The use of high-voltage electron microscopy and semi-thick sections for examination of cyanobacterial thylakoid membrane arrangements, *J. Microsc. (Oxford)* **133**:55–60.

Palay, S. L., and Chan-Palay, V., 1973, High voltage electron microscopy of the central nervous system in Golgi preparations, *J. Microsc. (Oxford)* **97**:41–47.

Parsons, D. F., and Ratkowski, A. J., 1980, A survey of high voltage electron microscope operations, *Ultramicroscopy* **5**:209–213.

Paulin, J. J., 1975, The chondriome of selected trypanosomatids: A three-dimensional study based on serial thick sections and high voltage electron microscopy, *J. Cell Biol.* **66**:404–413.

Pesacreta, T. C., and Parthasarathy, M. V., 1984, Improved staining of microfilament bundles in plant cells for high voltage electron microscopy, *J. Microsc. (Oxford)* **133**:73–77.

Porter, K. R., and Tucker, J. B., 1981, The ground substance of the living cell, *Sci. Am.* **244**(3):41–51.

Poux, N., Favard, P., and Carasso, N., 1974, Étude en microscopie électronique haute tension de l'appareil vacuolaire dans les cellules méristématiques de racines de concombre, *J. Microsc. (Paris)* **21**:173–180.

Rambourg, A., Marraud, A., and Chretiem, M., 1973, Tri-dimensional structure of the forming face of the Golgi apparatus as seen in the high voltage electron microscope after osmium impregnation of the small nerve cells in the semilunar ganglion of the trigeminal nerve, *J. Microsc. (Oxford)* **97**:49–57.

Rambourg, A., Clermont, Y., and Marraud, A., 1974, Three dimensional structure of the osmium impregnated Golgi apparatus as seen in the high voltage electron microscope, *Am. J. Anat.* **140**:27–45.

Research Resources Information Center, 1983, Biotechnology Resources: A Research Resources Directory, Division of Research Resources, National Institutes of Health, NIH Publication 83–1430.

Reynolds, E. S., 1963, The use of lead citrate at high pH as an electron-opaque stain in electron microscopy, *J. Cell Biol.* **17**:208–212.

Rieder, C. L., and Bowser, S. S., 1983, Factors which influence the light microscopic visualization of biological material in sections prepared for electron microscopy, *J. Microsc. (Oxford)* **132**:71–80.

Rieder, C. L., and Bowser, S. S., 1985, Correlative immunofluorescence and electron microscopy on the same section of Epon-embedded material, *J. Histochem. Cytochem.* **33**:165–171.

Rieder, C. L., Rupp, G., and Bowser, S. S., 1985, Electron microscopy of semithick sections: Advantages for biomedical research, *J. Electron Microsc. Tech.* **2**:11–28.

Spurr, A. R., 1969, A low-viscosity epoxy resin embedding medium for electron microscopy, *J. Ultrastruct. Res.* **26**:31–43.

Tivol, W. F., Ratkowski, A. J., and Parsons, D. F., 1983, EDX and EELS in the high-voltage electron microscope: Localization of elements in thick specimens, *Neurotoxicology* **4**:161–163.

Todd, W. J., Wray, G. P., and Hitchcock, P. J., 1984, Arrangement of pili in colonies of *Neisseria gonorrhoeae, J. Bacteriol.* **159**:312–320.

Turner, J. N., See, C. W., Ratkowski, A. J., Chang, B. B., and Parsons, D. F., 1981, Design and operation of a differentially pumped environmental chamber for the HVEM, *Ultramicroscopy* **6**:267–280.

Venable, J. H., and Coggeshall, R., 1965, A simplified lead citrate stain for use in electron microscopy, *J. Cell Biol.* **25**:407–408.

Weibull, C., 1974, Studies on thick sections of microorganisms using electron microscopes working at accelerating voltages from 60 to 1000 kV, *J. Ultrastruct. Res.* **47:**106–114.

Wolosewick, J. J., 1980, The application of polyethylene glycol to electron microscopy, *J. Cell Biol.* **86:**675–681.

Wolosewick, J. J., and Porter, K. R., 1976, Stereo high-voltage electron microscopy of whole cells of the human diploid line W 1-38, *Am. J. Anat.* **147:**303–324.

Wolosewick, J. J., and Porter, K. R., 1979, Microtrabecular lattice of the cytoplasmic ground substance: Artifact or reality?, *J. Cell Biol.* **82:**114–139.

Chapter 12

Computer Analysis of Ordered Microbiological Objects

Murray Stewart

Medical Research Council
Laboratory of Molecular Biology
Cambridge CB2 2QH, England

1. INTRODUCTION

Electron microscopy has yielded a wealth of information on the structure of microorganisms and their constituents and in many instances simple inspection of micrographs has given considerable insight into the structures being investigated. However, to obtain high-resolution information from electron micrographs of microbiological material, the simple, subjective methods generally used to interpret images are often inadequate. Objective methods of image analysis are needed to extract as much reliable information as possible, for example, about the position of individual protein subunits within an assembly or to give an indication of subunit size and shape.

Unlike metallurgical and ceramic specimens, where high resolution is more easily attained, biological specimens have a number of inherent drawbacks. They generally have very low intrinsic contrast, and so are usually viewed embedded in a heavy atom negative stain, such as uranyl acetate or sodium phosphotungstate. Although negative staining does improve the contrast of the specimen, high-resolution features are usually still faint and tend to be obscured by the background that arises from irregularities in the carbon film on which the specimen is supported in the microscope and by the granularity of the stain. Furthermore, both the stain and the protein are degraded by the electron beam used to form the image. To circumvent this, exposures are usually kept low in high-resolution studies and this, of course, makes the image even fainter and its contrast lower. The image forming process in the microscope also contributes a fine background granularity (phase granules) that further obscures fine detail. Finally, the large depth of focus of the electron microscope means that information from all levels of the object is superimposed. This is quite unlike the case in a light microscope where, because of the small depth of focus, it is possible to focus at different levels of the specimen in turn. The superimposition of information from different levels often produces a confused image, which, even though it contains quite high-resolution data, cannot be interpreted directly. Furthermore, the electron microscope image gives only a two-dimensional representation of the object and often the shape and disposition of subunits can only be fully appreciated in a three-dimensional representation.

Image processing techniques have been devised to overcome many of the difficulties associated with high-resolution images of biological specimens, particularly those deriving from poor contrast and superimposition problems. These techniques are almost indispensible when working at resolutions of between 1 and 3 nm and when reconstructing three-dimensional images at the molecular level. Ways of assessing the reliability of images and of averaging information from different specimens are important features of these methods. They are especially powerful when applied to periodically repeating structures, but can, in principle, be extended to any asymmetric particle. However, because of difficulties of data collection at present, their application is generally limited to regular objects such as virus particles, enzyme and membrane protein crystals, and bacterial surface layers. Analysis generally proceeds by first using optical diffraction to select the best images, which are then digitized to enable data processing by computer. Fourier transforms are then calculated, which enable the data to be analyzed and manipulated more easily. Finally, images are then reconstructed from the processed Fourier data by inverse transformation.

This chapter presents an overview of the principles of these methods and illustrates their application on a number of microbiological objects. A detailed and comprehensive treatment of the general methods is given in Misell (1978) whereas in-depth reviews of early work in the field are given by Amos (1974) and Crowther and Klug (1975).

2. DIFFRACTION PATTERNS AND FOURIER TRANSFORMS

2.1. Optical Diffraction

Optical diffraction is usually employed to assess images and to select areas for further processing. Because it is a relatively simple and rapid technique and requires little specialized apparatus, a very large number of micrographs can be examined and assessed for resolution, focus, astigmatism, and distortion. This enables the best images to be selected objectively and is a much more reliable measure of micrograph quality than simple visual inspection.

Optical diffraction patterns are formed in the back focal plane of a lens and analogous patterns are formed by the lenses of the electron microscope. These patterns are not usually observed in most optical devices because the focal length of the lenses is usually too short. This produces a very small pattern because the size of the pattern is proportional to the focal length of the lens. However, patterns are easily observed if lenses of focal length of the order of 1 m are employed. Figure 1 illustrates a simple laboratory apparatus for the production of optical diffraction patterns and consists of a laser with expanding telescope, a long (0.5–1 m)-focal-length lens, and a viewing screen. The telescope expands the laser beam to about 5-cm diameter so that a micrograph can be illuminated by parallel light. The lens then produces the diffraction pattern at its back focal plane where the viewing screen is located. To minimize the risk of eye damage, this screen should be made of frosted glass and a small (about 1-mm diameter) piece of opaque material should be placed at the position of the direct beam. It is often convenient to use an SLR camera back as the viewing screen, as the patterns can then be easily recorded photographically as required.

The optical diffraction pattern is related to the original image, but not in a simple way. It is, however, a useful way to represent the data present in the micrograph, as resolution, defocus, astigmatism, and preservation of the object are usually much more easily analyzed in the diffraction pattern format. In this context it is helpful to think of the process of image formation in terms of the diffraction theory formulated by Abbe. The key to this formulation is to express the image in terms of spatial frequencies. An image is usually thought of as being a pattern of areas of light and darkness, or, in other words, a spatial distribution of image intensity. In everyday experience we do not come across the idea of the frequency spectrum of an image, but in many ways this concept is similar to the way sound is commonly thought of in terms of its frequency spectrum. Thus, one can think of a complex sound as being made up of a number of waves of different frequencies added together. In the case of sound, the waves are a measure of displacement as a function of position. In an image, a similar wave concept is

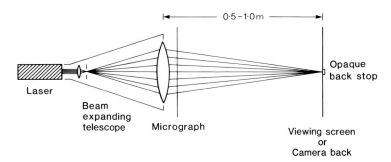

FIGURE 1. Diagrammatic representation of a simple system for generating optical diffraction patterns from electron micrographs. A telescope attached to a laser is used to illuminate an area up to 5 cm in diameter on the micrograph and to focus the diffraction pattern on a screen.

employed, except that the waves now measure the variation of optical density with position. Just as it is possible to build up a complex sound by superimposition of a number of sound waves of different frequencies, so it is possible to build up a complex image by the superimposition of optical density waves of different frequencies. Because the optical density varies with position, these are called spatial frequencies. In an optical diffraction pattern, the different spatial frequencies contribute at different positions and, at least for the small angles of scattering observed, the angle of scattering is proportional to the spatial frequency in the object. This gives rise to a very important difference between the diffraction pattern and the original image. Because the diffraction pattern represents the image in terms of spatial frequencies, distances in the diffraction pattern are the reciprocal of the corresponding distances in the image. Thus, fine spatial frequencies, corresponding to high-resolution data, will scatter to high angles whereas coarser frequencies will be found closer to the origin of the diffraction pattern.

2.2. Computed Diffraction Patterns (Fourier Transforms)

The frequency distribution of an image can also be generated by computer using Fourier transformation. This relies on the fact that it is possible to express any intensity distribution as a sum of waves. Each wave in the sum will have a particular amplitude and frequency. It will also have a phase term indicating the position of its origin relative to the other waves. Since Fourier transformation enables any density pattern to be broken down into a sum of waves of different frequencies, it allows a spatial frequency distribution to be generated a similar way as optically. The advantage of the computed Fourier transform is that it contains both amplitude and phase information whereas the phase information is lost when recording the optical diffraction pattern on film.

To produce a Fourier transform by computer, it is first necessary to express the image in numerical form. This is done by a digitizer that scans the image in a raster pattern and records its optical density at regular intervals. Typically, an image would be expressed as a matrix of density values with 512×512 elements, which represents over 250,000 numbers, although areas up to 4192×4192 picture elements are sometimes employed. Typical raster spacings are 10 to 25 μm on the film, which is near the resolution of most photographic films.

When processing micrographs, the optical and computer methods tend to complement one another. Because they can be obtained quickly and at negligible cost, optical transforms are used to assess image quality and so enable only the best examples to be selected for subsequent computer processing. However, the Fourier data derived from computer processing can be more thoroughly analyzed and are more easily manipulated and averaged than optical data. Consequently, although they are more expensive and time-consuming, computer methods are generally used for the actual processing of the images.

2.3. Fourier Transforms of Regular Objects

Regular periodic objects, such as crystals, have the useful property that their diffraction pattern is concentrated into a number of spots (Figure 2). This comes about because the only waves that can contribute to a periodic object are those with frequencies that are integral multiples of the repeat frequency of the object. This is because the pattern has to be the same in each repeat in the object and is analogous to the fundamental and overtone frequencies of a note played on a musical instrument. Regardless of the type of instrument on which the note is played, the fundamental and overtone frequencies are always the same. What is different is the amplitude of the different frequencies in the series and it is this that enables us to distinguish the type of instrument. A similar relationship applies for the frequency spectrum of a periodic object. Here too we find a fundamental frequency (which corresponds to the period of the object) and a series of overtones: objects with the same periodicity will have the same fundamental and overtone frequencies but, if the structure of the repeating unit is different, this will be reflected in a different distribution of intensities in the series. The Fourier transform of objects lacking a regular periodic structure is, by comparison, featureless with intensity distributed over the entire frequency spectrum (Figure 2). Electron micrographs of regular objects contain both signal and noise and so can be thought of as a sum of the two. Similarly, the Fourier transform of a composite image is the sum of the transforms of the two and so has the sharp spots deriving from the lattice superimposed on the continuous background due to the noise (Figure 2). As illustrated in Figure 3, as the amount of noise increases (producing a decrease in the signal-to-noise ratio), it becomes progressively more difficult to make out the features of the regular object. High-resolution features are lost first, but eventually, when the signal-to-noise ratio is very low, one cannot even make out the coarse features of the underlying lattice, although spots are still present in the Fourier transform.

2.4. Image Reconstruction

Images can be reconstructed from their transforms by inverse Fourier transformation. This entails synthesis of the image by adding together all the waves from which it is constituted. The process is analogous to the case in sound where, if the frequency spectrum is known, the original sound can be reconstituted. An example is an electronic synthesizer. In the optical case the image is produced by recombining the waves corresponding to different spatial frequencies in the image plane of the lens and it is often useful to view the process in terms of a simple optical system in which the diffraction pattern or optical Fourier transform is an intermediate between the object and the final image. It is therefore intuitively obvious that the object, Fourier transform, and image are intimately related. Furthermore, if a mask were to be placed over part of the diffraction

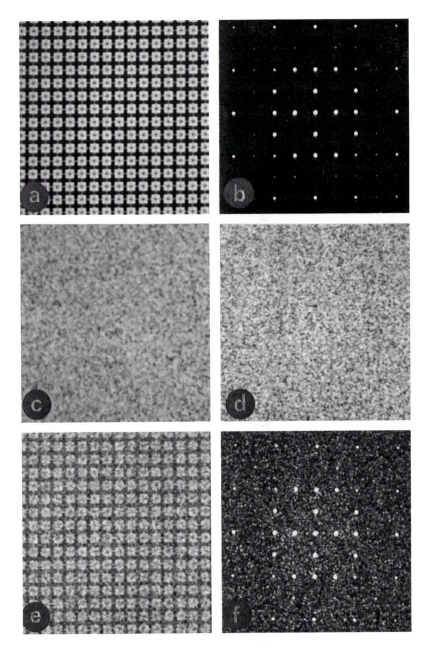

FIGURE 2. Fourier transform patterns from different sorts of objects. A regular object (a) gives rise to a transform (b) with sharp spots, whereas a random object (c) gives rise to a pattern (d) in which the intensity is evenly spread. A normal electron micrograph of a regular object can be thought of as the sum of a regular and random image (e). The transform (f) of this composite image is likewise the

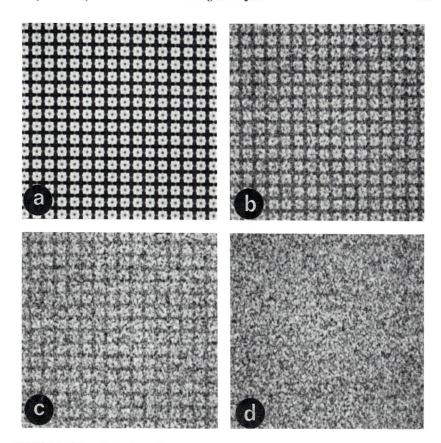

FIGURE 3. Effect of signal-to-noise ratio on the appearance of a regular array. The four images shown here all have the same regular object but with different amounts of random noise added. In (a) the signal-to-noise ratio is 1:1. The structure of the repeating unit can be clearly made out and fine details discerned. When the signal-to-noise ratio is 1:10 (b), the fine details of the object become obscured, although the gross features of the array, such as the general position of subunits and the hole in the center of each subunit, can still be made out. When the signal-to-noise ratio is decreased to 1:20 (c), only the general positions of the subunits on the square array can be easily seen, and when the signal-to-noise ratio is reduced still further to 1:100 (d), not even the subunit positions are visible. Electron micrographs of negatively stained material generally resemble (b) and (c) whereas low-dose images, particularly of unstained material, resemble (d). Filtering aims to increase the signal-to-noise ratio and so make the image more like (a) in which fine detail can be made out.

sum of the two diffraction patterns [note that the pattern has been printed so that the noise is less intense than in (d)]. Because the data from the regular object are concentrated into spots in the transform, they can be easily separated from the noise in this format and so a "filtered" image reconstructed in which the noise content is substantially reduced. In all cases, the origin is at the center of the transform.

pattern, it is clear that this would alter the final image produced by the optical system. This is a very important concept because it means that it is possible to reconstruct an image in which only some of the original data are present. Thus, there is the potential to remove unwanted elements in the original image, such as, for example, the granularity of the supporting film and stain, which obscure the fine details of the image.

The reconstruction of images from their Fourier transforms is a key element in the computer processing of electron micrographs of regular arrays and the general procedure of Fourier transformation—manipulation of data—and inverse transformation will be a recurring theme throughout. Although it may at first seem cumbersome to transform the data in this way only to retransform them back into an image later, it is extremely powerful because a large range of manipulations are far more easily carried out on the Fourier transform than on the original data. In this respect the technique is analogous to logarithms: after expressing numbers as exponents, one can carry out complex tasks (such as, for example, taking a fifth root) simply and then regenerate the answer by reversing the transformation and taking inverse logarithms.

Although the output from the computer processing is in numerical form, it can be easily represented in a visual format. This can be done, for example, by producing a contour plot of image density or by reversing the digitizing step and translating the numbers back into optical density. A photographic negative can be produced by scanning a film with a fine beam of light and varying the intensity in accord with the image density required. Alternatively, the numbers can be used to generate a video signal for display on a TV screen.

3. IMAGE PROCESSING

3.1. Filtering

One of the most effective methods for processing electron microscope images is to filter out the contributions to the image made by the background associated with support film and stain granularity. This technique is analogous to the filters often provided on hi-fi equipment, which enable unwanted or annoying sounds, such as those from scratches on records or the hiss from cassette tapes, to be suppressed. These contributions are usually referred to as "noise" to distinguish them from the portion of the output we are interested in, referred to as the "signal." In hi-fi equipment, filtering is performed by taking advantage of the fact that often the noise occurs mainly at a different frequency from that of the signal and so can be filtered out by suppressing a particular portion of the frequency spectrum. Scratches and hiss, for example, result primarily from high frequencies and can be decreased substantially by a high-frequency filter. In the process, a small amount of signal is also lost, but the overall effect is usually well worth the effort because the proportion of noise in the output is substantially reduced. In other words, filtering has increased the signal-to-noise ratio. The important feature of this filtering is that it relies on the separation of the signal from the noise. An analogous filtering of images can be performed for periodic objects. Because the Fourier transform of a periodic object is concentrated into a relatively small number of spots whereas the transform of the background noise is spread over the entire frequency spectrum, Fourier transformation allows the periodic signal to be separated effectively from most of the random noise. In filtering, we take advantage of this property to reconstruct the image using only the parts of the transform corresponding to the lattice frequencies. These are the lattice points. In the optical case, this is done by placing an opaque mask in the back focal plane of the objective lens. The mask has a series of holes cut in it at the lattice points and so only light of the desired frequencies can pass through it and form the filtered image. The noise is spread over all other frequencies, so most will be removed from the image, with only a small fraction passing through the holes. In the computer method, all of the transform, except at the points corresponding to the lattice, is simply set to zero. A second (inverse) Fourier transformation step then produces the filtered image. Again the noise is removed effectively from the final image. It is vital that the lattice points be correctly identified and so it is important to inspect both optical diffraction patterns and computed Fourier transforms carefully to ensure that the spots selected lie strictly on a lattice and that some spots have not been inadvertently omitted from the lattice.

Although they give essentially the same result, in practice computer filtering methods of image processing have a number of advantages over optical ones. It can, for example, be difficult to make and align optical masks because very small holes have to be exactly located (often to within a few tens of micrometers) and the optical apparatus then has to be aligned exactly to give the correct result. In addition to not having these technical difficulties, the computer methods allow information from a number of images to be easily averaged or compared and allow further symmetry-related averaging. The different symmetries that can be found in two-dimensional lattices are discussed in detail by Henry and Lonsdale (1969). Although symmetry-related averaging can be a powerful method for increasing the signal-to-noise ratio, some care is needed in correctly identifying the symmetry group present and it is always useful to discuss the result with an experienced crystallographer. Computer methods also enable other data, such as electron diffraction, to be incorporated into the final reconstruction and for corrections to be made for the effect of microscopy aberrations and defocus on the image. Finally, because the result is expressed in numerical form, it is more easily analyzed by quantitative methods and can be manipulated and displayed in different ways to enable the features of interest to be most easily made out.

3.1.1. Relationship between Filtering and Superimposition

An alternate method of enhancing the signal-to-noise ratio is to superimpose a number of different images of the repeating unit or unit cell of the array. This is most easily achieved by making a composite print of the micrograph in which a large number of exposures are made with the paper being translated by the lattice repeat distance between exposures. This technique of translational superimposition was first described by Markham *et al.* (1963), but it is very laborious compared to filtering and it is often difficult to determine the translation distance objectively and accurately. It can be shown (e.g., Fraser and Millward, 1970) that filtering and translational averaging are essentially equivalent and that filtering, because it represents a convolution of the image with its lattice, amounts to averaging the data present in the unit cells of the image. However, although filtering and translational superimposition are broadly similar techniques, filtering has two distinct practical advantages: first, it is easier to include a large number of unit cells in the final image; and, second, one is less likely to make a mistake in determining lattice parameters. The degree of averaging in filtering is determined by the size of the "holes" around each lattice point in the Fourier transform: the smaller the holes, the more unit cells are included in the average (see Fraser and Millward, 1970). This way of looking at filtering also gives an insight into why it is effective because it shows that it can be considered as an averaging method. The degree of averaging is greatest with small holes and so the increase in signal-to-noise ratio is also greatest under these circumstances.

The spot size in the transform can thus become a limitation and broadly the diameter of the spots decreases as more unit cells contribute to the Fourier transform. However, imperfections in the crystalline lattice can increase the size of the spots and also decrease their intensity, particularly at higher resolutions. It is for this reason that selection of the best areas by optical diffraction is so important, as then averaging can take place over the largest number of repeats. In cases where there is substantial disorder, some improvement can often be obtained by correcting for lattice distortions (e.g., see Crowther and Sleytr, 1977; Saxton and Baumeister, 1982).

3.1.2. Examples of Filtering

An appreciation of how filtering can be used in studies of protein structure can be gained by looking at a number of examples.

3.1.2a. One-Dimensional Object: Tropomyosin Paracrystals. The first example is a simple problem using the muscle protein tropomyosin, a rod-shaped molecule about 40 nm long. In the presence of magnesium ions, tropomyosin forms cigar-shaped paracrystals in which the molecules line up parallel to the long axis of the paracrystal. The amino acid sequence of tropomyosin is known and it contains a single cysteine residue. It became necessary to know the location of this cysteine residue in the paracrystals; this was done by labeling the residue with mercury and then examining the paracrystals by electron microscopy (Stewart, 1975). Because the mercury was rapidly vaporized in the electron beam, it was necessary to record micrographs with low exposures (about one-tenth that usually employed) and so it was almost impossible to make out any pattern in the images (Figure 4). Computer processing demonstrated the location of the mercury convincingly, by producing one-dimensional filtered images of the density variation along the paracrystal (Figure 5). There is an extra peak present in the material stained with mercury, which can be seen much more clearly when the signal due to the underlying paracrystal material is subtracted (Figure 5). There is an extra peak present in the material stained with mercury, which can be seen much more clearly when the signal due to the underlying paracrystal material is subtracted (Figure 5d). Furthermore, superimposition of the traces before and after mercury staining (Figure 5c) shows that the changes are localized to the region of the peak and that the patterns correspond well over the remainder. This example shows not only how powerful this technique can be in extracting a periodic signal from a seemingly featureless image, but also how,

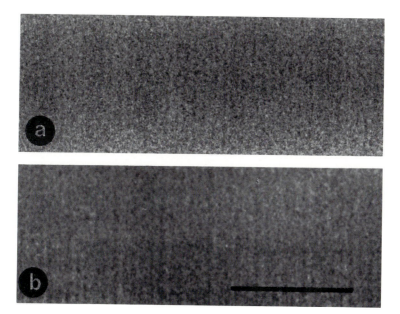

FIGURE 4. Electron micrographs of unstained magnesium paracrystals of the muscle protein tropomyosin. The paracrystal in (b) has been labeled with methyl mercury on its cysteine residues whereas that in (a) is unlabeled. Bar = 100 nm. Reproduced, with permission, from Stewart (1975).

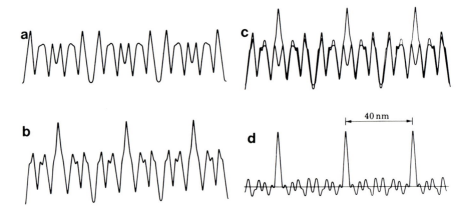

FIGURE 5. Filtered axially-projected density traces from electron micrographs of unstained tropomyosin magnesium paracrystals. (a) Unlabeled; (b) labeled with methyl mercury. Trace (c) shows (a) and (b) superimposed; (d) shows (a) subtracted from (b). Reproduced, with permission, from Stewart (1975).

by manipulating the numerical results obtained from the filtering step (in this case, by subtracting one image from the other), features of special interest can be accentuated. Thus, computer filtering methods are powerful in detecting small differences between objects.

3.1.2b. Two-Dimensional Arrays: Bacterial Regular Surface (RS) Layers.

Computer image processing has proved valuable in the study of the RS layers found on the outermost surfaces of many bacteria of both gram-positive and gram-negative genera (Beveridge, 1981; Sleytr and Messner, 1983). These layers are thought to protect the underlying layers of the bacterial surface (the cell wall and, in the case of gram-negative bacteria, the outer membrane as well) from attack by hostile elements of their environment such as parasitic bacteria, phage particles, and lytic enzymes. They may also help shield the bacterium from heavy metals. Some of these RS layers are very highly ordered and make ideal specimens for high-resolution structural investigations based on computer processing of electron micrographs.

An interesting illustration of the power of filtering techniques is presented by the RS layer frbm the gram-positive bacterium *Sporosarcina ureae* (Stewart and Beveridge, 1980). Electron micrographs of negatively stained fragments of this layer show two different types of pattern, because the layer binds tightly to the carbon support film and so the negative stain only outlines the upper side of the specimen (Figure 6a,b). One pattern is quite distinct and studies using sectioned material demonstrated that this pattern corresponds to the inner (cytoplasmic) side of the layer. The second pattern corresponds to the outer surface of the layer and appears to be completely featureless. However, optical diffraction

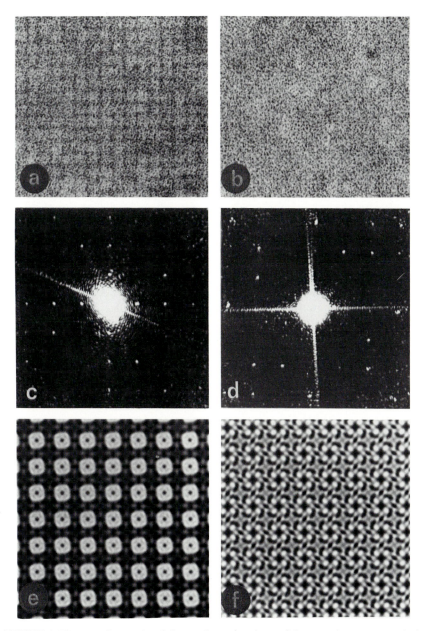

FIGURE 6. Electron micrographs of the regular surface layer of *Sporosarcina ureae* negatively stained with phosphotungstate and uranyl acetate. Two different types of pattern (a, b), deriving from different sides of the layer, were seen and which gave strong optical diffraction patterns (c, d). Computer-filtered images of the two types of pattern (e, f) with the protein white and the stain dark, as in the original negatively stained images. The distribution of protein is clearly very different in the two images, but there is a striking similarity in the distribution of dark areas. This indicates that there are holes or pores running through the layer and which are seen as dark areas in the patterns from each side of the layer. Reproduced, with permission, from Stewart and Beveridge (1980).

shows clearly that there is a regular structure present (Figure 6c,d). After filtering and rotational averaging based on the fourfold symmetry of both objects, the patterns present are quite easily appreciated (Figure 6e,f). These two patterns seem unrelated, but close inspection reveals a remarkable correspondence between the dark areas, which correspond to regions of high stain density. The fact that they correspond so well on either side suggests strongly that there are holes or gaps that can fill with stain between the protein units in the layer. On the basis of the filtered patterns, it would appear that these gaps in the layer probably have a limiting diameter of about 2 nm. This example also illustrates how signal-to-noise ratio can be increased by averaging both within and between micrographs. Averaging between micrographs is straightforward and the increase in signal-to-noise ratio is proportional to the square root of the number of images, as in any averaging process. Averaging within the images relies on the symmetry of the layer, which in the case of the two *Sporasarcina* images was p4 (see Henry and Lonsdale, 1969). This means the repeating motif has a fourfold rotation axis and a further twofold improvement in signal-to-noise ratio was obtained by fourfold rotational averaging about this axis (Steward and Beveridge, 1980).

3.1.2c. Other Examples. Other examples of objects analyzed by filtering are the RS layers of *Micrococcus radiodurans* (Baumeister *et al.*, 1981), *Bacillus polymyxa* (Burley and Murray, 1983), and *Bacillus brevis* (Aebi *et al.*, 1973; Lepault and Pitt, 1984); the porin in the *Escherichia coli* outer membrane (Steven *et al.*, 1977); and crystals of elongation factor EF-Tu from *E. coli* (Cremers *et al.*, 1981).

3.1.3. Some Simple Pitfalls

3.1.3a. Failure to Index Properly. When reconstructing a filtered image, it is most important that all the spots due to the array in question be included in the synthesis. If some spots are omitted (except those beyond the resolution of the reconstruction), then an artifactual result will be produced. This is quite a common problem and arises if the lattice chosen does not include all the spots deriving from the layer; in other words, the indexing scheme chosen is incorrect. This usually arises by failing to see weak spots that occur at double the frequency (i.e., at half the spacing in the Fourier transform) of the strong spots in the pattern. These weak spots indicate that the lattice spacing in the original array is double that of the most obvious repeat. This usually comes about by there being some slight difference between successive subunits. If the image is reconstructed without including these weak spots, then the difference between successive subunits will disappear and an artifactual result, in which all subunits appear identical, will be produced.

An example of the sort of artifact that could be produced is given by the sheath of the archaebacterium *Methanospirillum hungatei* (Stewart *et al.*, 1985).

FIGURE 7. Electron micrograph of a sheath of *Methanospirillum hungatei* negatively stained with uranyl acetate. The structure has a fine repeat at about 2.8 nm both vertically and horizontally together with irregular axial stripes at multiples of 2.8 nm. The optical diffraction pattern (inset) shows strong spots at about 2.8 nm vertically and horizontally. However, there is also a weak series of spots vertically at a spacing of 5.7 nm (arrows) indicating that the true repeat in the vertical direction is actually twice the apparent one. This means that alternate subunits in this direction must be slightly different.

Electron micrographs of this material (Figure 7) show a fine underlying array made up of subunits that appear to be arranged on a 2.8 × 2.8-nm tetragonal lattice. However, optical diffraction (Figure 7 inset) shows a faint repeat at a spacing of 5.7 nm in one direction, which indicates that alternate subunits in this direction are different in some way. A reconstructed image using all spots to a resolution of 2 nm (Figure 8a) shows this difference and suggests that it may arise from the orientation of alternate subunits being different. However, if the

FIGURE 8. Reconstructed images of the *M. hungatei* sheath (a) properly and (b) with incorrect indexing in which the weak peaks in the Fourier transform at a spacing of 5.7 nm were omitted. In (a), successive subunits in a vertical direction are clearly different, whereas in the incorrectly reconstructed image (b), this difference is lost.

weak spots at 5.7 nm are omitted and only the spots at 2.8 nm are included in the reconstruction (Figure 8b), then this difference is lost and the subunits all appear (incorrectly) to be identical.

Other examples of how artifactual results may be produced by incorrectly indexing a pattern are the polyoma virus "hexamer" tubes, which actually contain paired pentamer subunits instead of the hexamers originally proposed (Baker *et al.*, 1983), and the arrangement of subunits in bacteriophage T4 polyheads (Moody, 1967).

3.1.3b. Introducing Too High Symmetry into the Pattern. Artifactual results can also be introduced if too high symmetry is assigned to the pattern. If, for example, the repeating unit contains four lobes, it is possible that it may have fourfold rotational symmetry. However, if it actually has two type A subunits and two type B, this difference will be lost if the structure is fourfold rotationally averaged and instead it will appear that all four subunits are identical. Because this sort of processing is, in effect, a self-fulfilling prophecy, it is impossible to detect the artifact by inspecting the filtered image. It is therefore vital to only use symmetry-related averaging when the particular symmetry element has been established unequivocally by detailed analysis of the Fourier transform, employing the restraints appropriate to each symmetry group listed in Henry and Lonsdale (1969). Certainly if there is any doubt about whether a symmetry element should be included or not, it is much safer to omit it and to assume lower symmetry.

3.2. Rotational Filtering

It is possible to filter objects having rotational symmetry in an analogous manner to that employed for crystals having translational symmetry (Crowther and Amos, 1971). This can be done by a rotational superimposition analogous to the translational method, but it is sometimes difficult to determine rotational symmetries correctly and one is unable to verify the result objectively. A more quantitative method involves expressing the object in terms of a sum of rotational harmonics. This is equivalent to representing the object on a polar coordinate grid and then transforming over Θ and R instead of X and Y in the Cartesian coordinate system. One can then build up a rotational power spectrum that gives a measure of the strength of the different rotational frequencies from which the image is constituted. The fundamental frequency can then be objectively determined and the image reconstructed using only the fundamental rotational harmonic and its overtones.

Figure 9 shows the application of this method to the stacked disk form of tobacco mosaic virus (Crowther and Amos, 1971). It is difficult to make out the individual capsomere units in the electron micrograph or to decide on the rotational symmetry of the disk. However, the rotational power spectrum shows a very clear and strong peak at a rotational frequency of 17, which establishes the symmetry. The reconstructed image shows the individual capsomere subunits clearly. Other microbiological specimens analyzed in this way include the base plate of bacteriophage T4 and its structural changes on activation (Crowther *et al.*, 1977); the structure of the head–tail connector of bacteriophage φ29 (Carrascosa *et al.*, 1982); and the structure of the gene 20 product of bacteriophage T4 (Driedonks *et al.*, 1981).

3.3. High-Resolution Work (beyond 2 nm)

3.3.1. Problems

Two major obstacles to obtaining high-resolution electron micrographs of microbiological material are the damage produced in the specimen by electron irradiation and the changes introduced into the image as a result of the microscope imaging system. Radiation damage is a very severe problem at resolutions below about 2nm and results both from destruction of the biological material itself and from migration of the negative stain (Unwin, 1974). Generally, radiation damage is more severe with biological material than with negative stain and so becomes a major problem when using unstained specimens preserved either in glucose derivatives (Unwin and Henderson, 1975) or in vitreous ice (Dubochet *et al.*, 1982; Lepault *et al.*, 1983; Stewart and Vigers, 1986). In order to prevent these processes, or at least reduce them to an acceptable level, it is necessary to

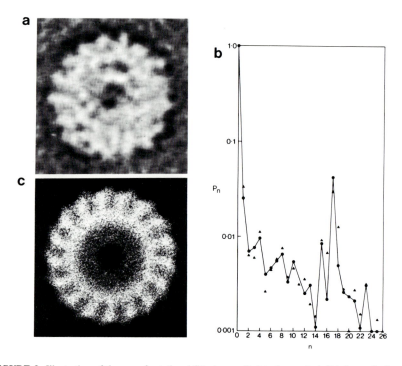

FIGURE 9. Illustration of the use of rotational filtering applied to the stacked disk form of tobacco mosaic virus protein. The original micrograph (a) does not enable an unequivocal decision to be made about the rotational symmetry, or the shape and arrangement of subunits. However, (b) the rotational power spectrum (analogous to the Fourier transform of a planear array) shows a clear peak at 17, indicating that the particle has 17-fold rotational symmetry. The reconstructed image (c) enables subunit shape and orientation to be seen. Reproduced, with permission, from Crowther and Amos (1971), copyright by Academic Press (London).

reduce the electron dose to about $\frac{1}{100}$ that normally employed to record micrographs. However, when micrographs are recorded using such low electron doses, the image is very faint and has lower contrast and much higher noise than normal exposures. Prints of these micrographs are usually almost featureless, although optical diffraction shows that they do contain high-resolution information. Consequently, image processing can be used to increase the signal-to-noise ratio.

A second problem at high resolution comes from the manner of image formation in the electron microscope. Most of the contrast in thin biological objects is phase contrast and so will not be present in a perfectly focused image. This phase contrast does, however, become manifest when the microscope is defocused and is the major source of the increased contrast seen as one moves slightly below true focus. The production of phase-contrast images in the electron microscope is not as straightforward as the production of phase-contrast images in the light microscope. In the light microscope, a retarding plate is used to introduce a phase shift between the central (unscattered) beam and the scattered beams that form the image. This produces the phase image by creating an interference between the scattered and direct beams. Suitable retarding plates cannot easily be introduced into electron microscopes and so the phase difference between scattered and unscattered beams is instead produced by placing the objective lens slightly out of focus. This has the result of creating a phase image because some of the scattered beams will be out of phase with the unscattered beam and these will form an image by interference. However, the phase difference introduced by this method relies on the path length differences between scattered and unscattered beams and this will only be correct for beams scattered at particular angles. Since the scattering angle is proportional to the spatial frequency, it follows that only some spatial frequencies will contribute correctly to the phase-contrast image in the electron microscope; others will have the wrong phase. By defocusing to about 0.5 μm, most frequencies in the range 10–2 nm will contribute to the image, but higher-resolution information will not be correctly represented (see Erickson and Klug, 1971). More importantly, the amplitudes will be modulated somewhat by the process of image formation. Fortunately, it is easy to represent the effect of image formation in the Fourier transform of the object: it can be considered as multiplying the Fourier transform of the object by a "phase contrast transfer function" such as that illustrated in Figure 10. The effect of this is to change the amplitudes of most spatial frequencies quite substantially, but the phases are either correct or changed by 180°. Thus, the Fourier transform of a high-resolution image contains phase information that is either correct or is easily corrected, but amplitude information that has been substantially altered. In principle, the amplitudes can be corrected, but are usually more conveniently obtained by electron diffraction.

3.3.2. Use of Low-Dose and Electron Diffraction

One can overcome most of the effects of radiation damage by taking low-dose micrographs and these can then be used to produce reliable phase information about the object's Fourier transform, whereas the problem of the modulation of the Fourier transform amplitude data can be avoided by recording the diffraction pattern of the object directly in the microscope using it in electron diffraction mode. The film image of the electron diffraction pattern does not contain the phases, but these can be obtained by computer Fourier transformation of the image. Thus, one can obtain reliable amplitude data from electron diffraction patterns of the object and reliable phases from computer processing of images, which are usually recorded consecutively from the same area. Fourier inversion of these data can be used to reconstruct a high-resolution image of the object (Unwin and Henderson, 1975).

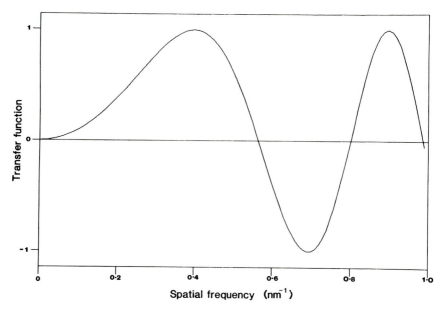

FIGURE 10. Typical electron microscope contrast transfer function for the degree of defocus generally employed to record images of biological objects. The Fourier transform is multiplied by this function and clearly this results in the transform, and so the resultant image, being somewhat changed. The contrast transfer function is near 1 for spatial frequencies corresponding to about 2 nm and so in this region, which corresponds to the fine detail usually easily seen in negatively stained material, the Fourier transform (and so the final image) is not greatly altered. However, at higher spatial frequencies the transfer function decreases rapidly and then becomes negative, which causes a considerable change to the Fourier transform and so to the final image. Note, however, that while the transfer function is positive, only the amplitudes of the transform are altered and the phases are correct. Moreover, even when the transfer function is negative, the phases are shifted by 180° and so can be easily corrected provided one knows the points at which the transfer function changes sign. These are easily obtained by optical diffraction of high-dose images recorded immediately after the working images (see Unwin and Henderson, 1975).

3.3.3. Examples of High-Resolution Studies

3.3.3a. The Surface Layer of *Aquaspirillum putridiconchylium*. An example of the application of this combination of low-dose imaging and electron diffraction is the analysis of the RS layer of the gram-negative bacterium *A. putridiconchylium* (Stewart *et al.*, 1980). Electron micrographs of negatively stained preparations of this layer show a distinctive pattern of parallel double striations (Figure 11). However, the fine structure of the layer and particularly the subunit shape cannot be made out from the micrograph. The computer-filtered image at about 3-nm resolution obtained from high-dose images (Figure 12a) shows that the striations are made up of hourglass-shaped units, with those in one line of the striation being staggered relative to those in the adjacent line. The units in one pair of striations are connected to adjacent pairs by linkers so that the structure resembles a ladder with the linkers as the rungs. Because it is possible to obtain large fragments of this layer, the resolution of this study was extended to below 1.5 nm by recording micrographs with a low (about $\frac{1}{15}$ normal) exposure and combining the results obtained by computer image processing with electron diffraction data. This is more demanding technically and also more laborious than more conventional methods, but it gives a better representation of the structure because radiation damage to the specimen is reduced and the effects of the microscope phase contrast transfer function are minimized. As illustrated

FIGURE 11. Electron micrograph of a negatively stained sample of the surface layer of *Aquaspirillum putridiconchylium*. The image shows prominent ribs, but little detail about the structure and arrangement of the subunits. Inset is an optical diffraction pattern showing that the layer is highly ordered and that structural information is present to high resolution. See Stewart *et al.* (1980).

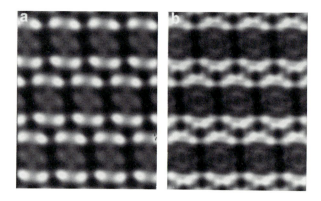

FIGURE 12. Reconstructed images from negatively stained specimens of *A. putridiconchylium*. Simple filtering (a) to a resolution of about 2.5 nm increases the signal-to-noise ratio sufficiently to enable the subunits to be made out and their arrangement deduced. However, (b) by combining electron diffraction data with phases from low-dose images, the resolution can be extended to about 1.5 nm and the contribution of the phase contrast transfer function to be eliminated. This enables the fine structure of the subunit to be more easily seen.

in Figure 12b, there is a wealth of fine detail present when compared to Figure 12a, although the basic structure of striations and linkers remains the same.

A probable relationship between the structure of RS layers and their function was suggested by the results obtained by image processing (Stewart *et al.,* 1980). The RS layer probably protects the underlying layers of the wall by preventing hostile elements of the environment from reaching them, in a manner analogous to the protective function of virus capsids. Unlike a virus, a bacterium has an active metabolism and must obtain nutrients from and expel waste into the environment. This would be prevented by a continuous and impermeable RS layer and it is in this respect that the gaps are believed to be important. They would enable passage of small nutrient and waste molecules while still maintaining a barrier against larger particles, such as enzymes, parasitic bacteria, and phage particles. Of course, this is unlikely to be the only role of the RS layer, but it does serve to illustrate how insights into the function of biological systems can be obtained by image processing.

3.3.3b. Other Examples. Other objects investigated to high resolution using a combination of electron diffraction and computer image processing of low-dose micrographs include the structure of the purple membrane of *Halobacterium halobium* (Unwin and Henderson, 1975; Dumont *et al.,* 1981) and the surface layer of *Sulfolobus Acidocaldarius* (Deatherage *et al.,* 1983; Taylor *et al.,* 1982).

3.4. Superimposition

Because the depth of focus in an electron microscope is very large compared to the thickness of specimen, all levels of the object are in focus simultaneously and so information from all levels is superimposed in the image. Thus, unlike light microscopy, one cannot focus up and down through different layers of the specimen. The superimposition of information from different levels of the object can produce confusing images. This is particularly true if different levels contain regular arrays, which can produce moiré interference patterns and so give a quite misleading impression of the underlying structure. This commonly happens with sheetlike structures such as bacterial RS layers, membrane crystals, or tubular structures, where a moiré pattern forms between the top and bottom of the tube. A special case of this arises in helical structures (Section 4.1).

3.4.1. Rotational Moiré Patterns

Rotational moiré patterns can often be easily decomposed into their constituent layers by image processing, because it often proves possible to separate the different components in the Fourier transform. Thus, if one thinks of the transform of the composite image as representing the sum of the transforms of its constituents, then the problem becomes very similar to that considered earlier in filtering. Figure 13 illustrates the general method employed and shows the production of the composite image by the superimposition of two regular arrays. Each individual image is clear, but the pattern becomes obscured when they are superimposed. However, it is quite easy to identify the spots from each lattice in the transform and so an image can be reconstructed from only a single layer.

A powerful example of the resolution of moiré patterns in this way came from the detailed analysis of bacteriophage T4 polyheads. These objects exist as long tubes and interference between the top and bottom of the tubes produces a very confused image. However, when the image of a single side of the tube was reconstructed, the hexagonal nature of the subunit became clear. Moreover, analysis of early and late tubes indicated that changes in their morphology could be correlated with removal of part of the capsid protein by proteolysis (DeRosier and Klug, 1972; Yanagida et al., 1972; Laemmli et al., 1976; Steven et al., 1976). This sort of study was extended using antibodies to locate material within the unit cell more accurately (Kistler et al., 1978).

3.4.2. Translational Moiré Patterns

A more complicated case arises when the layers are related by a translation rather than a rotation. In this instance the spots from the two layers overlap and so the two cannot be easily separated. However, if it is possible to produce

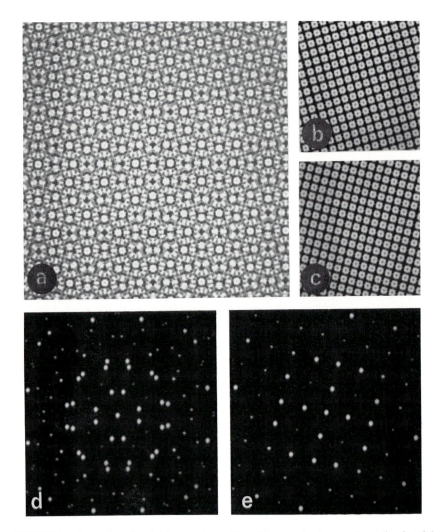

FIGURE 13. Generation of moiré by superimposition of two regular images. A rotational moiré pattern (a) is produced by superimposing two regular arrays (b, c) that have been rotated relative to one another. Although the structure of the subunits (which is the same as in Figures 2 and 3) can be clearly made out in a single layer, the moiré pattern produced by their superimposition is very confused. Image processing can be employed to extract the single layers from the moiré by identifying in the transform the spots that correspond to a single layer and reconstructing the image from these alone. The transform of the composite layer (d) shows spots from both layers as can be seen by comparing it with the transform from a single layer (e). It is well worth doing this by hand to convince yourself that the composite transform is made up in this way [the second single layer transform will be the mirror image of that shown in (e)]. Because the spots from the two layers can be separated in the transform, the two layers can be separated in exactly the same way that signal was separated from noise in filtering (except that the second layer becomes the unwanted feature of the image). This enables the individual layers to be reconstructed from the composite.

fragments containing a single layer, least-squares methods can be used to analyze the Fourier transforms of both single and composite layers to see if it is possible to explain the composite layer on the basis of superimposition alone. This was done, for example, with the complex RS layer of *Aquaspirillum serpens* MW5 (Stewart and Murray, 1982). This layer normally appears as a series of ribs on the surface of the bacterium. But, as illustrated in Figure 14, its appearance can be accounted for by superimposition of two hexagonal layers similar to those found in other *A. serpens* strains.

3.4.3. Objects Having Helical Symmetry

Helical objects have superimposition problems similar to those seen with overlying sheets or tubes, in that an interference of moiré pattern forms between the patterns on the top and bottom of the helix. Generally the signals from the top and bottom of a helical object can be separated in a way similar to that described for tubes by exploiting the fact that the signals from each can be easily separated in Fourier space. This is illustrated in Figure 15 where the composite pattern from both sides of the helix is confused and shows little easily interpreted information whereas the pattern from either the top or the bottom shows clearly

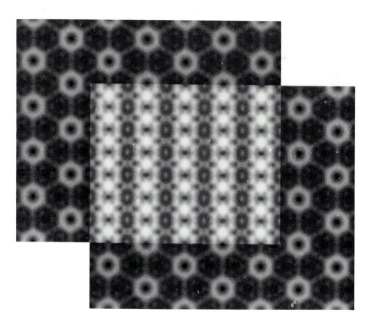

FIGURE 14. Illustration of the production of a translational moiré pattern from two superimposed hexagonal layers as found in the regular surface layer of *Aquaspirillum serpens* MW5. Reproduced, with permission, from Stewart and Murray (1982).

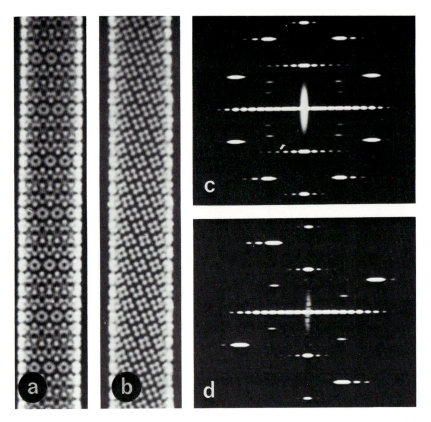

FIGURE 15. Moiré pattern produced by superimposition of the top and bottom of a helical particle. One side (b) shows the subunit structure and position clearly, but this becomes confused when top and bottom are superimposed (a). In an analogous manner to the superimposition of planar arrays, the one-sided view can be reconstructed if the contributions from the top and bottom of the particle can be separated in the Fourier transform. The transform of the whole particle is shown in (c) and consists of a regular array of intensity arranged on layer lines instead of the simple spots seen with a planar array. However, it is still possible to identify the intensity due to one side of the particle, as shown in (d), and thereby to reconstruct a one-sided view such as (b) from the composite two-sided view (a).

the arrangement of the individual subunits and also their general shape. The diffraction pattern from a helix is not quite as simple to analyze as that from a plane sheet, because the data are concentrated into lines perpendicular to the helical axis. These are called layer lines and can be described mathematically in terms of sums of Bessel functions (see Klug *et al.*, 1958). They tend to form a lattice similar to that seen for sheets, but some care is needed in indexing these patterns and it is always best to discuss the problem with someone with experience in helical diffraction theory. Fraser and MacRae (1973) give a good general introduction to the subject.

The production of one-sided images has been used extensively with helical viruses and bacteriophage tails, such as T4 (Klug and DeRosier, 1966), and for other objects such as bacterial flagella (O'Brien and Bennett, 1972) and flagellar microtubules (Amos and Klug, 1974).

4. THREE-DIMENSIONAL RECONSTRUCTION

4.1. Helical Objects

The Fourier transforms of helical objects have a rather useful property—the intensity on a layer line does not change as the helix is rotated about its axis. In other words, the amplitude of the transform does not change with azimuth for a given radius and axial height. The phase of the transform does change under these circumstances, but in a predictable way (Klug et al., 1958; see Fraser and MacRae, 1973, or Misell, 1978, for simplified treatments). This means that it is possible to deduce the entire three-dimensional transform of a helical object from a two-dimensional transform. This comes about as a consequence of the projection theorem (discussed in detail by DeRosier and Klug, 1968). Because one can deduce the three-dimensional Fourier transform from the transform of a two-dimensional projection (which corresponds to an electron micrograph), the three-dimensional density distribution of the object can then be reconstructed by Fourier inversion (DeRosier and Moore, 1970). The process can be understood in image space by realizing that a helical particle produces a number of views of its subunit from different angles and so one can obtain a three-dimensional picture of the object by combining the data from the different independent views. Difficulties can arise if there are terms deriving from Bessel functions of more than one order present on a layer line, but usually this does not occur until quite high resolution and, in many instances, can be overcome by analyzing images with a number of different azimuthal orientations.

Figure 16 illustrates the application of this method to the structure of the bacteriophage T4 tail sheath and shows the structural rearrangement that takes place on activation. Other examples of the application of this method to microbiological objects are the structure of flagellar hooks (Wagenknecht et al., 1981); helical protein aggregates, such as that of E. coli elongation factor EF-Tu (Cremers et al., 1981); and the structure of helical virus particles, such as tobacco mosaic virus (Unwin and Klug, 1974) and the tail of bacteriophage Mu (Admiral and Mellema, 1976).

4.2. Three-Dimensional Reconstruction of Two-Dimensional Crystals

Methods have also been devised for obtaining three-dimensional models from two-dimensional crystalline arrays by combining data from a number of tilt

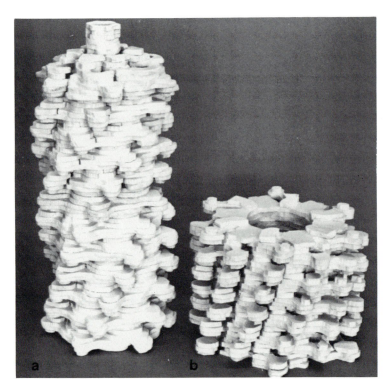

FIGURE 16. Three-dimensional reconstruction of the tail from bacteriophage T4 shown before (a) and after (b) activation. Production of three-dimensional images such as these is possible with helical objects because it is possible to deduce the three-dimensional Fourier transform from a two-dimensional transform, which one can obtain from the normal two-dimensional electron micrograph. Reproduced, with permission, from Amos and Klug (1975), copyright by Academic Press (London).

series. This methodology has been comprehensively reviewed by Amos *et al.* (1982) and is based on the reconstruction of the purple membrane of *Halobacterium halobium* by Henderson and Unwin (1975). Other microbiological objects examined in this way include the surface layers of *S. acidocaldarius* (Deatherage *et al.*, 1983; Taylor *et al.*, 1982) and *Chlamydia trachomatis* (Chang *et al.*, 1982).

The computational and methodological problems associated with this method are formidable as it is necessary to merge data from a large number of tilt series and to compensate for focal changes across the micrograph. The resolution is limited in a direction perpendicular to the plane of the crystal because these lie in an "excluded cone" of the Fourier transform that cannot be accessed because it is generally not possible to tilt specimens past about 60°. With particularly favorable objects, such as the purple membrane, it is possible to obtain quite spectacular results and, for example, determine the arrangement of units of secondary structure such as α-helices (Henderson and Unwin, 1975). However, with many objects the result may not be so impressive and, if only data to a resolution of about 3nm are available, the new information obtained may not seem large when compared to the very considerable effort required.

5. CONCLUSIONS

Computer image processing can often yield spectacular results when applied to electron micrographs of microbiological material, but is currently known to give reliable information only for objects in which subunits are arranged in a regular manner. Methods are being developed for isolated objects and arrays in which the order is only poorly preserved, but it is too soon to assess their practical value. All methods are most effective when making very good micrographs even better, and not, as is often hoped, improving poor-quality micrographs. The most important ingredient in any study involving computer image processing is to have the best possible micrographs in the first place. With good data, computer image processing can enable fine detail, which is often masked by noise in the image, to be seen and can also enable the generation of three-dimensional models of the objects. These methods are most powerful, however, in determining differences between similar objects or in determining the position and orientation of subunits rather than their detailed shape. Although it may impress the credulous, obtaining images that are simply pretty seldom leads to significant advances in the understanding of the microbiological world and so, before undertaking any study using computer image processing, it is vital to define the problem that one wishes to solve. In many instances, these questions can be answered by careful inspection of the original micrographs or, alternatively, might be addressed by a simple technique such as optical diffraction. It is also important to assess the degree of resolution that will be required to answer a question and to ascertain whether this can be (or has been) obtained by electron microscopy. In this context, it is worth remembering that good information on subunit position, orientation, and overall shape can usually be obtained at resolutions of 2 to 3 nm and that at resolutions higher than these, great care is needed in interpreting images because of positive staining effects. Unstained material is usually very sensitive to radiation damage and this may well prove a limitation. It is also noteworthy that only limited information is obtained on most microbiological objects by increasing resolution until one is able to detect secondary structure reliably at about 0.5- to 0.7-nm resolution. Thus, if one cannot obtain this sort of resolution, it may well not be worth proceeding past a resolution of 2 to 3 nm.

Finally, it cannot be stressed too strongly that these methods require consid-

erable computing power and some mathematical knowledge. Computer programs are available to do the processing described in this chapter, but individual
microbiologists might do well to consider collaborating with established groups,
at least in the initial stages of a study. It is very easy to produce an artifactual
result by, for example, indexing a diffraction pattern incorrectly or assuming the
wrong symmetry. Many assumptions in image processing become self-fulfilling
prophecies and are very difficult to detect by analyzing the final result. Like most
other techniques in electron microscopy, computer image processing is not without its problems and dangers, but the information obtained is often well worth the
effort involved.

ACKNOWLEDGMENTS. I am most grateful to Linda Amos, Tom Ceska, and
Aaron Klug for their helpful comments and criticisms.

6. REFERENCES

Admiral, G., and Mellema, J. E., 1976, The structure of the contractile sheath of bacteriophage Mu,
    J. Ultrastruct. Res. **56**:48–64.
Aebi, U., Smith, P. R., Dubochet, J., Henry, C., and Kellenberger, E., 1973, A study of the
    structure of the T-layer of *Bacillus brevis, J. Supramol. Struct.* **1**:498–522.
Amos, L. A., 1974, Image analysis of macromolecular structures, *J. Microsc. (Oxford)* **100**:143–
    152.
Amos, L. A., and Klug, A., 1974, The arrangement of subunits in flagellar microtubules, *J. Cell Sci.*
    14:523–549.
Amos, L. A., and Klug, A., 1975, Three-dimensional image reconstruction of the contractile tail of
    T-4 bacteriophage, *J. Mol. Biol.* **99**:51–73.
Amos, L. A., Henderson, R., and Unwin, P. N. T., 1982, Three-dimensional structure determina-
    tion by electron microscopy of two-dimensional crystals, *Prog. Biophys. Mol. Biol.* **39**:183–
    231.
Baker, T. S., Casper, D. L. D., and Murakami, W. T., 1983, Polyoma virus 'hexamer' tubes consist
    of paired pentamers, *Nature (London)* **303**:446–448.
Baumeister, W., Kübler, O., and Zingsheim, H. P., 1981, The structure of the cell envelope of
    Micrococcus radiodurans revealed by metal shadowing and decoration, *J. Ultrastruct. Res.*
    75:60–71.
Beveridge, T. J., 1981, Ultrastructure, chemistry and function of the bacterial cell wall, *Int. Rev.
    Cytol.* **72**:229–317.
Burley, S. K., and Murray, R. G. E., 1983, Structure of the regular surface layer of *Bacillus
    polymyxa, Can. J. Microbiol.* **29**:775–780.
Carrascosa, J. L., Vinuela, E., Garcia, N., and Santisteban, A., 1982, Structure of the head–tail
    connector of bacteriophage φ29, *J. Mol. Biol.* **154**:311–324.
Chang, J.-J., Leonard, K., Talmon, A., Pitt, T., Zhang, Y.-X., and Li-hua, Z., 1982, Structural
    studies of the outer envelope of *Chlamydia trachomatis* by electron microscopy, *J. Mol. Biol.*
    161:579–590.
Cremers, A. F. M., Sam, A. P., Bosch, L., and Mellema, J. E., 1981, Characterisation of regular
    polymeric products of elongation factor EF-Tu from *Escherichia coli, J. Mol. Biol.* **153**:477–
    486.

Crowther, R. A., and Amos, L. A., 1971, Harmonic analysis of electron microscope images with rotational symmetry, *J. Mol. Biol.* **60**:123–130.

Crowther, R. A., and Klug, A., 1975, Structural analysis of macromolecular assemblies by image reconstruction from electron micrographs, *Annu. Rev. Biochem.* **44**:161–182.

Crowther, R. A., and Sleytr, U. B., 1977, An analysis of the fine structure of the surface layers from two strains of *Clostridia*, including correction for distorted images, *J. Ultrastruct. Res.* **58**:41–49.

Crowther, R. A., Lenk, E. V., Kikuchi, Y., and King, J., 1977, Molecular reorganisation in the hexagon to star transition of the baseplate of bacteriophage T4, *J. Mol. Biol.* **116**:489–523.

Deatherage, J. F., Taylor, K. A., and Amos, L. A., 1983, Three-dimensional arrangement of the cell wall protein of *Sulfolobus acidocaldarius, J. Mol. Biol.* **167**:823–852.

DeRosier, D. J., and Klug, A., 1968, Reconstruction of three-dimensional structures from electron micrographs, *Nature (London)* **217**:130–134.

DeRosier, D. J., and Klug, A., 1972, Structure of tubular variants of the head of bacteriophage T4 (polyheads). I. Arrangement of the subunits in some classes of polyhead, *J. Mol. Biol.* **65**:469–488.

DeRosier, D. J., and Moore, P. B., 1970, Reconstruction of three-dimensional images from electron micrographs of structures having helical symmetry, *J. Mol. Biol.* **52**:355–369.

Driedonks, R. A., Engel, A., ten Heggeler, B., and van Driel, R., 1981, Gene 20 product of bacteriophage T4: Its purification and structure, *J. Mol. Biol.* **152**:641–662.

Dubochet, J., Lepault, J., Freeman, R., Berriman, J. A., and Homo, J.-C., 1982, Electron microscopy of frozen water and acqueous solutions, *J. Microscopy* **128**:219–237.

Dumont, M. E., Wiggins, W. E., and Hayward, S. C., 1981, Location of platinum binding sites on bacteriorhodopsin by electron diffraction, *Proc. Natl. Acad. Sci. USA* **78**:2947–2952.

Erickson, H. A., and Klug, A., 1971, Measurement and compensation of defocusing and aberrations by Fourier processing of electron micrographs, *Philos. Trans. R. Soc. London Ser. B* **261**:105–118.

Fraser, R. D. B., and MacRae, T. P., 1973, *Conformation in Fibrous Proteins and Related Synthetic Polypeptides*, Academic Press, New York.

Fraser, R. D. B., and Millward, G. R., 1970, Image averaging by optical filtering, *J. Ultrastruct. Res.* **31**:203–211.

Henderson, R., and Unwin, P. N. T., 1975, Three-dimensional model of purple membrane obtained by electron microscopy, *Nature (London)* **257**:28–32.

Henry, N. F. M., and Lonsdale, K. (eds.), 1969, *International Tables for X-ray Crystallography,* Vol. I, Kynoch Press, Birmingham, England.

Kistler, J., Aebi, U., Onorato, L., ten Heggeler, B., and Showe, M. K., 1978, Structural changes during the transformation of bacteriophage T4 polyheads—Characterisation of the initial and final stages by freeze-drying and shadowing of F_{ab}-labelled preparations, *J. Mol. Biol.* **126**:571–589.

Klug, A., and DeRosier, D. J., 1966, Optical filtering of electron micrographs: Reconstruction of one-sided images, *Nature (London)* **212**:29–32.

Klug, A., Crick, F. H. C., and Wykoff, H. W., 1958, Diffraction by helical structures, *Acta Crystallogr.* **11**:199–213.

Laemmli, U., Amos, L. A., and Klug, A., 1976, Correlation between structural transformation and cleavage of the major head protein of T4 bacteriophage, *Cell* **7**:191–203.

Lepault, J., and Pitt, T., 1984, Projected structure of unstained, frozen-hydrated T-layer from *Bacillus brevis, EMBO J.* **3**:101–105.

Lepault, J., Booy, F. P., and Dubochet, J., 1983, Electron microscopy of frozen biological specimens, *J. Microsc. (Oxford)* **129**:89–102.

Markahm, R., Frey, S., and Hills, G., 1963, Methods for the enhancement of detail and accentuation of structure in electron micrographs, *Virology* **20**:88–102.

Misell, D. L., 1978, Image analysis, enhancement and interpretation, in: *Practical Methods in Electron Microscopy*, Vol. 7 (A. M. Glauert, ed.), North-Holland, Amsterdam.

Moody, M. F., 1967, Structure of the sheath of bacteriophage T4. I. Structure of the contracted sheath and polysheath, *J. Mol. Biol.* **25:**167–200.

O'Brien, E. J., and Bennett, P. M., 1972, Structure of a straight flagella from a mutant *Salmonella*, *J. Mol. Biol.* **70:**133–152.

Saxton, O. W., and Baumeister, W., 1982, The correlation averaging of a regularly arranged bacterial cell envelope, *J. Microsc. (Oxford)* **127:**127–138.

Sleytr, U. B., and Messner, P., 1983, Crystalline surface layers on bacteria, *Annu. Rev. Microbiol.* **37:**311–339.

Steven, A. C., Couture, E., Aebi, U., and Showe, M. K., 1976, Structure of T4 polyheads. II. A pathway of polyhead transformations as a model for T4 capsid maturation, *J. Mol. Biol.* **106:**187–221.

Steven, A. C., ten Heggeler, B., Müller, R., Kistler, J., and Rosenbruch, J. P., 1977, The ultrastructure of a periodic protein layer on the outer membrane of *Escherichia coli*, *J. Cell Biol.* **72:**292–301.

Stewart, M., 1975, Location of the troponin binding site on tropomyosin, *Proc. R. Soc. London Ser. B* **190:**257–266.

Stewart, M., and Beveridge, T. J., 1980, Structure of the regular surface layer of *Sporosarcina ureae*, *J. Bacteriol.* **142:**302–309.

Stewart, M., and Murray, R. G. E., 1982, Structure of the regular surface layer of *Aquaspirillum serpens* MW5, *J. Bacteriol.* **150:**348–357.

Stewart, M., and Vigers, G., 1986, Electron microscopy of frozen hydrated biological material, *Nature (London)* **319:**631–636.

Stewart, M., Beveridge, T. J., and Murray, R. G. E., 1980, Structure of the regular surface layer of *Spirillum putridiconchylium*, *J. Mol. Biol.* **137:**1–8.

Stewart, M., Beveridge, T. J., and Sprott, G. D., 1985, Crystalline order to high resolution in the sheath of *Methanospirillum hungatei*, *J. Mol. Biol.* **183:**509–515.

Taylor, K., Deatherage, J. F., and Amos, L. A., 1982, Structure of the S-layer of *Sulfolobus acidocaldarius*, *Nature (London)* **299:**840–842.

Unwin, P. N. T., 1974, Electron microscopy of the stacked disc aggregate of tobacco mosaic virus. II. The influence of electron irradiation on stain distribution, *J. Mol. Biol.* **87:**657–670.

Unwin, P. N. T., and Henderson, R., 1975, Molecular structure determination by electron microscopy of unstained crystalline specimens, *J. Mol. Biol.* **94:**425–440.

Unwin, P. N. T., and Klug, A., 1974, Electron microscopy of the stacked disc aggregate of tobacco mosaic virus protein. I. Three-dimensional reconstruction, *J. Mol. Biol.* **87:**641–656.

Wagenknecht, T., DeRosier, D., Shapiro, L., and Weissborn, A., 1981, Three-dimensional reconstruction of the flagellar hook from *Caulobacter crescentus*, *J. Mol. Biol.* **151:**439–465.

Yanagida, M., DeRosier, D. J., and Klug, A., 1972, The structure of the tubular variants of the head of bacteriophage T4 (polyheads). II. Structural transition from a hexamer to a 6+1 morphological unit, *J. Mol. Biol.* **65:**489–499.

Chapter 13

Digitizing and Quantitation

Edwin S. Boatman

Department of Environmental Health
School of Public Health and Community Medicine
University of Washington
Seattle, Washington 98195

1. INTRODUCTION

The aim of quantitative morphology as specifically restricted here to the electron microscopy of microorganisms, is to estimate in an as unbiased manner as possible, the volume, area, number, and size of cells or components of cells derived from the analysis of two-dimensional images of sectioned material by the use of relatively simple to moderately complex techniques and equipment.

Only recently have microbiologists turned their thoughts to the possibility of describing the morphological characteristics of the organisms they use in quantitative terms other than by estimates of viability and cell number. Cell size, if mentioned at all, is often qualitatively determined and at best of doubtful value. This chapter is an attempt to illustrate ways in which quantitative appraisals of cell morphology can be undertaken.

2. HISTORICAL BACKGROUND

The science of quantitation is based firmly on the theorems of geometrical probability, the calculus, and statistical analysis. Although slow to start, morphometry and stereology (the distinction between the two will be defined later) received considerable impetus from the field of geology due to the work of the French geologist M. A. Delesse (1847). At that time, the volumetric analysis of rock into its mineral constituents (e.g., particularly gold and silver) required prolonged separation techniques and chemical analyses. Delesse determined that the volume fraction of different minerals contained within a sample of rock could be estimated from their areal profiles on polished rock surfaces. This was, in essence, a method for obtaining information about three-dimensional structures (in turn, a structure consists of objects) based upon quantitative analyses of their two-dimensional surfaces or transections. Somewhat later, a similar principle was introduced into geology (1856) and metallurgy (1865) by Sorby (cited by Aherne and Dunnill, 1982).

In subsequent years, other relatively simple techniques were devised such as lineal analysis (Rosiwal, 1898), and the use of linear intercepts and point-counting (Glagolev, 1933, 1934). In this particular point-counting method, the specimen is moved systematically in discrete steps under a fixed point situated in the microscope eyepiece. An equivalent but reverse procedure was adopted by Chalkley (1943) whereby the specimen field is held stationary while an array of points are projected over the area of the field. A new field is selected and the point "hits" on tissue components counted. In both situations, the idea is to count the number of times a point (or a number of points) overlays or "hits" a specific component of the sample. The greater the number of "hits" made on a specific component, the greater is the volume occupied by that component and thus the larger the volume fraction or volume density of that component in relation to the reference volume and to the sample as a whole. Or, to put it another way, the areal fraction of a component in a representative two-dimensional section of an object is directly proportional to its volume fraction in the original three-dimensional object. The beauty of this simple strategem is that volume fractions of a number of components (e.g., different minerals, or for biologists, mitochondria, Golgi apparatus, cytoplasmic inclusions, and so on) can be estimated simultaneously. It was only a matter of time before variations of this technique employing point-counting eyepiece graticules (Haug, 1955; Hennig, 1956), and transparent plastic overlays transcribed by lines, points, or grid patterns were devised (Dunnill, 1962; Weibel and Bolender, 1973; Weibel, 1979). Currently, estimates of area, volume fraction, cell number, and other parameters can be obtained by the use of microprocessor-assisted digitizing tablets or semi-automatic image analyzers. Time efficiency, however, is as important a criterion to be considered in this work as accuracy. We have now reached the stage where, as elegantly stated by Williams (1977), "The effects of experimental treatments on cell structures and the time course of morphological effects can now be precisely charted on the basis of cell or organelle volume, membrane area, organelle number or size." Although by use of present-day morphometric techniques observer subjectivity has been greatly reduced, it is important to realize that not every form of image observed under a microscope lends itself to morphometric analysis.

Stereology and Morphometry Literature

While profusion of literature exists relating to the analysis of cultures of eukaryotic cells, normal and diseased tissues derived from a variety of organs of human, animal, and fish origin, publications concerning the morphometric analysis of microorganisms are few in number. This is probably because the questions that can be answered by morphometric analyses of bacteria are more limited in number than those that can be asked of eukaryotic cells and also because the morphometric component of a published work may be small and consequently not merit inclusion in the title of the article; identification and retrieval is therefore less likely. At the conclusion of this chapter a list of books and papers for further reading is presented.

3. DEFINITIONS AND SYMBOLS

Morphometry and stereology are two words that in the recent past were used somewhat interchangeably. Current usage defines morphometry as the measurement of structures and stereology as comprising a series of mathematical methods used to obtain three-dimensional parameters of a structure based on two-dimensional measurements of flat images. Since an array of measurements can be made encompassing length, area, volume, and so on, in both fractional and absolute values, a collection of symbols is necessary; Table I is a partial list appropriate for the methods and applications to be described in this chapter.

4. DIGITIZING TABLETS, IMAGE ANALYZERS, AND POINT-COUNTING

Methods for the morphometric analysis of cells and tissues range from the relatively simple to the complex: planimetry, lineal analysis, point-counting, and microprocessor-assisted digitizing tablets (e.g., MOP, manual optical picture analyzer; ZIDAS, Zeiss interactive digital analysis system) are in the former category and semiautomatic or automatic image analyzers in the latter.

With random fields of biological images observed in two dimensions and photographed, planimetry, lineal analysis, and point-counting methods (for the latter, using test grids of various designs) give average values and the variations within a single field are compensated for by the potentially large number of fields (pictures of random profiles in section) available. The number of points or intersections required in the design of the test grids to keep the limits of error within a given range (say ±10%) can be estimated and, consequently, also the time necessary for each manual analysis to be completed.

Digitizers of varying complexity (e.g., Leitz ASM; Reichert/Kontron M.O.P.; Nikon Microplan II; Zeiss Videoplan) have been in use for a number of years. A recent and reasonably priced edition is the ZIDAS, which was the device used to make the measurements on profiles of bacteria and HeLa cells described in this chapter. The basic design consists of a flat board with an active electromagnetic measuring surface of about 28 × 27 cm and an active height of 2.0 cm. The surface is activated from a grid of magnetized steel wires embedded just below the

Table I
List of Symbols and Related Dimensions[a]

P	Number of points	—
P_T	Number of test points	—
P_i	Number of points that hit a structure	—
P_P	Point fraction (P_i/P_T)	—
P_L	Number of intersections per unit length of test line	No/cm
P_A	Number of points per test area	No/cm^2
P_V	Number of points per unit volume	No/cm^3
L_L	Lineal fraction; length of intercept per unit length of test line	cm/cm
L_A	Length of lineal elements per unit test area	cm/cm^2
L_V	Length of lineal elements per unit test volume	cm/cm^3
A_T	Test area	cm^2
A_A	Area fraction	cm^2/cm^2
S_V	Surface area per unit volume	cm^2/cm^3
V_T	Test volume	cm^3
V_V	Volume fraction[b]	cm^3/cm^3
N_V[c]	Number of particles per unit volume	No/cm^3
N_A	Number of profiles per unit test area	No/cm^2
$\bar{L}$	Mean linear intercept length L_L/N_L	cm
L_T	Test line length	cm
$\bar{A}$	Mean profile area	cm^2
$\bar{S}$	Mean surface area	cm^2
$\bar{V}$	Mean volume	cm^3
$\bar{D}$ (or $\bar{H}$)	Mean particle diameter	cm
d	Mean diameter of a population of profiles	cm
$\bar{V}_C$	Mean volume of an individual cell or organelle ($\bar{V}_{org}$)	cm^3

[a]The term intercept means that portion of a test line falling within the boundary of an area profile, whereas intersection means the point where a test line crosses a profile boundary.

[b]Variously termed by different authors as volume density, volume fraction, or volume proportion. Density (Weibel, 1979) implies the quantity per unit volume, unit area, or unit length; indicating respectively volume density (V_V), surface density (S_V), and length density (L_V).

[c]N_V determinations are straightforward only for objects of simple shape or of a readily calculable or known mean volume.

surface of the tablet. The magnetostrictive pulses generated along the wires are intercepted by the tracing movements of a hand-held cursor with cross hairs or a pointed stylus. A microprocessor in the electronics of the tablet establishes the position of the cursor at any moment. The spatial resolution of profiles can be discrete or continuous solely by depressing a button on the cursor for the appropriate period of time; in this way, the operator can select and measure any portions of the image desired. Images may take the form of a microscope projection onto the measuring surface, a drawing or tracing, or a photograph. In all cases, the magnification of the image and its clarity should be such as to permit rapid tracing of the desired components or contours. Tracing speeds with a cursor or stylus can approach 100 mm/sec.

As with the use of any tracing device, certain precautions are necessary. Tracing of contours should be done as carefully as possible; even then, depending on the size and complexity of the profile traced, average deviations from the mean of about 5% are common. Very small areas of 5 to 20 mm² usually have deviations that are considerably larger and in order to reduce this error higher magnifications of a projected image or photograph are necessary. In this respect, it is a good idea to undertake pilot tracings of profiles of standard dimensions and areas and the accuracy and reproducibility of the measurements obtained tabulated for future reference. Some test data will be presented later.

Current digitizing tablets have a varied menufield and are able to calculate about 16 measurement functions including area, perimeter, equivalent circle diameter, minimum diameter, maximum diameter, distance, count per unit area, some statistical analysis, histograms, and some stereology (volume and surface densities); data are transferred to a printer. However, the work involved in the measurement of large numbers of profiles may be tedious and somewhat time-consuming.

Computer-assisted semiautomatic or automatic image analyzers used for the analysis of photographic images obtained by transmission electron microscopy (TEM) have to deal with structures discernible (detectable) from one another in terms of a scale of densities from white to gray to black. The level of detection desired and background homogeneity (shading) is set by the operator. There are a number of supplementary manipulations by which the operator can modify the image by erosion, dilation, cover, deletion, and others, thus improving the accuracy by which an item of interest can be analyzed. If the image is an appropriate one for analysis, data can be very quickly obtained. The uses of automatic image analyzers are beyond the scope of this chapter and will only be briefly mentioned further.

In the morphometric analysis of electron micrographs, the magnification of the final image is critical to both labor and results irrespective of whether the method used is simple or complex. Objects or areas for analysis must be easily discernible by eye or both time efficiency and accuracy will be compromised. At progressively higher magnifications the contours of membranes (e.g., irregular contours of sectioned gram-negative bacterial cell wall and/or cytoplasmic membrane) are seen in greater detail, whereas at lower magnifications small irregularities in shape and direction cannot be accurately followed. Thus, estimates of perimeter and surface areas obtained by tracing devices as well as by point-counting may progressively increase as the magnification is raised (Paumgartner et al., 1980).

Over the last 3 to 4 years, evaluations have been made of the relative efficiency of planar image analysis comparing point-counting at different densities of test points with digitizing analyzer systems and with automatic image analysis. In a study by Mathieu et al. (1980), the precision in estimating the volume fraction (V_V) of a specific organelle profile from ten pictures by counting

64 total points (P_T) per picture was comparable in accuracy to the digitizing optical analyzer but about 11 times faster. As the overall error inherent in a picture sample is mainly determined by the variation between sample units (individual pictures), it is more efficient to evaluate more pictures less precisely than to achieve very high precision on fewer samples. These findings were confirmed in a study by Gundersen et al. (1981) using a more complex sample consisting of normal and pathologically altered cells. Acquisition of data by point-counting was up to 3 times faster than by use of a digitizing tablet. The resolution of the digitizing tracing operation is, however, superior to point-counting but as mentioned above, the degree of error of the final estimate rests more on the sampling efficiency than on the measuring precision. Biological variation between individuals or animals can be quite large and in terms of coefficient of variation is rarely below 10%; efforts to increase the estimate in a single animal or sample below 10% are of limited value or of no benefit (Gundersen and Osterby, 1980). However, overall, much of the variation in stereological estimates comes from the methods of sampling and analysis. Similar difficulties may arise on comparing the morphology of the same strain of bacteria grown on markedly different culture media, conditions of culture, or different methods of fixation and processing. A further study of Mathieu et al. (1981) entailed comparisons between point-counting, digital analysis and automatic image analysis systems using both synthetic and biological test materials. Point-counting grids with square lattices containing from 4 to 400 points were used. Initialization time and setting time were both longer for digital and automatic analysis than for point-counting. Measuring time, however, was shortest for automatic image analysis and for point-counting with a 100-point grid. These two techniques were also the most efficient for reducing the relative standard error to the lowest expected level. Automatic image analyzers work extremely well when presented with discrete objects of high contrast, particularly synthetic materials, dispersed cell suspensions, and certain biological tissues. Profiles of complex tissue in ultrathin sections, wherein cell components may only differ slightly from one another in contrast, and although easily discernible by the sensory system of the eye, constitute a difficult task in differentiation for the automatic image analyzer.

The overall costs of the respective systems vary from one to three orders of magnitude. The choice is largely determined by the nature of the material to be evaluated, the number of stereological parameters to be measured, and the estimated long-term use for the equipment.

5. METHODS OF MEASUREMENT

Under this heading a description of each method is given. In a subsequent portion some results will be given using these methods.

5.1. Measurement of Cell Perimeter from Micrographs: Cartographer's Wheel

Perimeters of well-defined objects at large magnifications can be estimated on nonglossy micrographs most simply by means of a cartographer's wheel (Figure 1A). The magnification needs to be large enough so that the manipulation of the tracing wheel is easily managed. Distances may be read off the dial in inches or centimeters and converted to the actual value in micrometers by the appropriate conversion and magnification factors. Perimeters of objects may also be estimated by use of a digitizing tablet; some examples will be given later.

5.2. Measurement of Profile Areas from Micrographs

5.2.1. Planimetry

A compensating polar planimeter is a manually operated tracing device that will record areas directly in either square inches or square centimeters. It consists

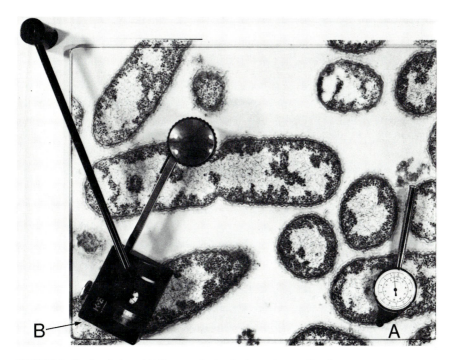

FIGURE 1. Tracing devices. (A) Cartographer's wheel to estimate cell perimeter and (B) planimeter to estimate cell area; shown is an electron micrograph of *Aeromonas salmonicida* enlarged 73,500× for tracing.

of a double-pivoted movable arm attached to a recording unit. The tracing arm has, at its terminal end, a magnifying lens at the center of which is scribed a small circle; the circle is aligned with, and moved over, the profile being traced (Figure 1B). The direction of the trace is always clockwise and motion is concluded precisely at the starting point or point of origin. It is important to maintain the angle of the arm within an arc of between 15° and 165°. The distance traveled is recorded by a rotary vernier scale one unit of which is equivalent to 0.01 in². Initially, a number of tracings should be made of a particular profile to determine intra- and, if necessary, interobserver error. Accuracy is about ±0.1%. It is useful to start out by tracing profiles of known areas drawn on paper to check recordings and calculations and to determine the smallest area that can be accurately traced, usually about ≥ 20 cm² or 5 cm diameter. When initially tracing new profile areas on micrographs, it is worthwhile to estimate the time taken to complete one or more micrographs.

5.2.2. Digitizer

After the initializing information is entered into a program, profiles of images may be traced with cursor or stylus in any direction desired. The tablet and cursor components of a digitizer are shown in Figure 2.

5.2.3. Lineal Analysis

This is a simple procedure but not often used. A plastic sheet on which are drawn a series of equally spaced parallel lines is superimposed on each micrograph in a random orientation. The spacing between lines is such that an adequate number of lines lie over the profile whose area (A_A) is to be measured. The length of all lines that intercept on the feature of interest (A) is measured and the total length of the intercepts (L_A) is expressed as a fraction of the total line length (L_T) on the micrographs such that

$$A_A = L_A/L_T \tag{1}$$

5.2.4. Point-Counting

A clear plastic sheet transcribed with an array of points and linear probes is superimposed over each micrograph in turn and counts tallied of the number of points (P_i) over the object whose area fraction (A_A) is to be determined. This tally of points is divided by the total number of points (P_T) over all the micrographs such that

$$A_A = P_i/P_T \tag{2}$$

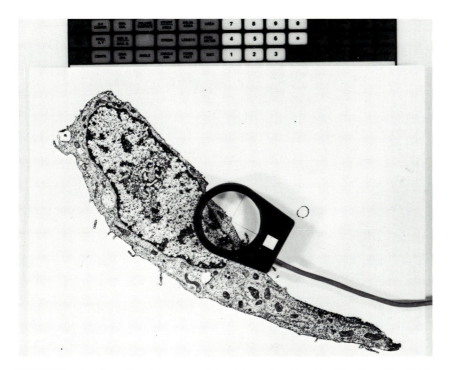

FIGURE 2. Digitizing tablet with cursor positioned for tracing the cell profile of a sectioned HeLa cell, 15,000×.

The magnification of the micrograph should be such that all items to be evaluated are clearly visible and that the number and arrangement of superimposed points permits about one point "hit"/small cytoplasmic or tissue component. The superimposed array of points can consist of a regularly spaced array, a coherent form of multipurpose test grid composed of short lines with terminal points (Figure 3), or a square lattice array of coarse and fine points (Figure 4), the coarse for large areas (whole cell profiles) and the fine for counting "hits" on small areas (e.g., profiles of mitochondria, cytoplasmic granules). Use of a large square lattice avoids counting an unnecessarily large number of point "hits" over tissue compartments covering large areas. Within limits, the more points superimposed, the greater is the accuracy. The number of points necessary to keep the limits within a certain range (say ±10%) can be estimated by the formula of Weibel (1979) where the relative standard error (RSE) is given by

$$RSE = 0.6745 \sqrt{\frac{1 - V_V}{V_V \times P_T}} \qquad (3)$$

where P_T is the total number of points on the reference area and V_V is the volume density of the tissue compartment.

To reduce to a minimum the effort and time involved, the construction of cumulative mean plots is recommended. Such plots determine where, at a given level of error, the counting of more total points has little effect in reducing the desired level of error (see Williams, 1977; Weibel, 1979; Aherne and Dunnill, 1982).

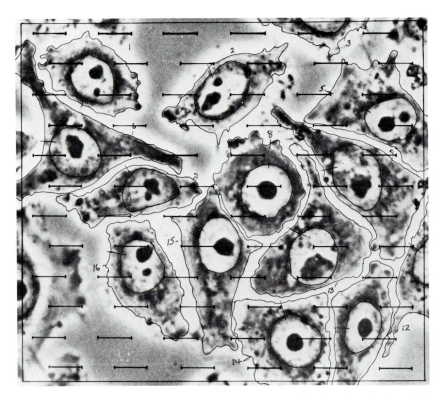

FIGURE 3. Phase-contrast micrograph of unfixed HeLa cells (with cell and nuclear boundaries outlined in ink) superimposed by a multipurpose test grid of 72 short lines with a point at each end (P_T, 144) suitable for point- and intersection-counting.

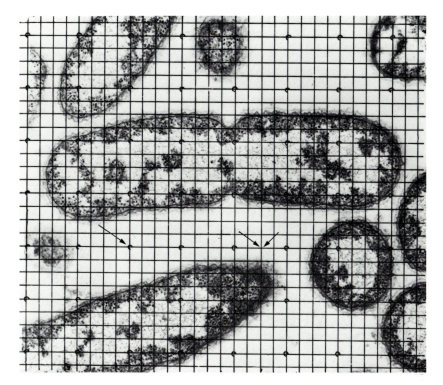

FIGURE 4. Electron micrograph of a sectioned bacterium overlaid by a square grid system consisting of coarse points (arrow) and fine points (double arrow). For point- and intersection-counting.

5.3. Measurement of Surface Density

Surface density (S_V) is an estimate of the amount of surface per volume of tissue. S_V can be estimated by:
(1) the superimposition of lines that completely traverse the micrograph and the counting of the intersections of these lines on the items of interest:

$$S_V = 2I/L_T \qquad (4)$$

where I = the number of intersections per compartment, L_T = total test line length; units of length depend upon the final magnification of the image.
(2) Superimposition of an array of equally spaced short lines with a point at each end:

$$S_V = 4I/PL \qquad (5)$$

where P = the number of points on the item of interest, L = the length of one line, I = number of intersections.

(3) Superimposition of an array of equally spaced vertical and horizontal lines forming a grid of squares the dimensions of which are dictated by the size of the item(s) of interest:

$$S_V = 2I/L_T \qquad (6)$$

5.4. Measurement of Numerical Density

Numerical density (N_V) is the number of component bodies in a certain unit of volume. Assumptions have to be made on the shape and the size distributions of the bodies and, if possible, on the thickness of the section. Usually the thickness of the section can be ignored if the size (caliper diameter) of the body is much larger than the section thickness.

$$N_V = N_A/\bar{H} \qquad (7)$$

where N_A is the average number of sectioned profiles per area of section and $\bar{H}$ is the mean caliper diameter. Mean caliper (tangent) diameter $(\bar{H})$ for spherical particles of radius R is $\bar{H} = 2R$; for a particle of any other shape, however, the tangent diameter will depend on the orientation of the particle with respect to the sectioning plane. For non-spherical particles, one will have to be satisfied with an approximation by use of a solid (e.g., ellipsoid, cylinder, polyhedron) which describes the shape of the particle as well as possible (see Weibel, 1979, p. 49; Elias and Hyde, 1983, p. 57). If section thickness is considered, then:

$$N_V = N_A/(\bar{H} + t) \qquad (8)$$

where t is the section thickness. If assumptions are made on shape and size, these values are incorporated into the formula of Weibel and Gomez (1962):

$$N_V = \frac{K}{\beta} \frac{(N_A)^{3/2}}{(P_p)^{1/2}} \qquad (9)$$

where K is the size distribution coefficient, B is the shape coefficient, N_A is the mean number of profiles per sectioned area of tissue, and P_P is the mean point fraction of sectioned profiles (which, in fact, constitutes an estimate of V_V).

5.5. Material for Morphometric Analysis

The requirements to be discussed here are confined to the use of ultrathin sections as a source material for the quantitative analysis of electron micrographs.

Morphometric measurements are usually made on the surfaces of cells or tissues that have been sectioned and stained. The fact that sections used in TEM have finite thickness gives rise to two main problems. Apart from the well-known influence section thickness has on resolution, it has, in addition, an effect—the "Holmes effect" (Holmes, 1921)—on the visualization of the "true" image surface depending upon the size, shape, and orientation of the object relative to the section. Section thickness also has a detectable effect on the contrast of the visualized image. Membrane profiles in thin section are best defined when the membrane segment is oriented vertical to the electron beam, which, for spherical particles larger than the thickness of the section, occurs when they are sectioned through their maximum diameter. This occurrence is well demonstrated in Figure 5, which shows profiles of three consecutive sections through the membranes of poly-β-hydroxybutyrate granules in the cytoplasm of the photosynthetic organism *Rhodospirillum rubrum*. Figure 6 demonstrates this sectional change in granule membrane appearance diagrammatically. Furthermore, depending on the position of a granule in relation to the plane of section and on the contrast, surface areas and diameters of sectioned granules may be overestimated morphometrically or, if contrast is poor, particularly in the cap region these values may be underestimated. For example, in Figure 7, only granule 3 is sectioned close to its equator; granule 1 is cut at a polar cap and is likely to be barely detectable, while granules 2 and 4 are sectioned close to a cap and will be detectable but with reduced contrast. Profiles of granules 1 and 2 would underestimate the true granule diameter. Note that the average diameter of a chromatophore is close to that of the section thickness and the images of the two adjacent chromatophores (arrow) would appear as one (Boatman, 1964).

Many investigators consider the effects of plane of section and poor contrast to cancel each other out, but under some circumstances a correction factor for the "Holmes effect" may be necessary if the section is thicker than $\frac{2}{11}$ the diameter of the particle being measured. The correction factor (K) takes the form:

$$K = (1 + 3t)/2\bar{D} \qquad (10)$$

where t is the section thickness and $\bar{D}$ is the mean particle diameter. Estimation of section thickness by comparing interference colors is subject to large errors, so if it is necessary to make this estimation it can be made by use of an interference microscope or scanning interferometer (Williams, 1977) or, better still, by use of the minimal fold method (Elias and Hyde, 1983, p. 60).

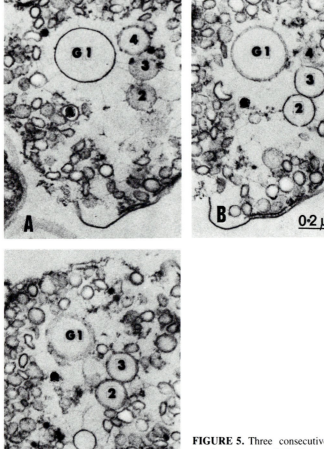

FIGURE 5. Three consecutive sections through a lysed spheroplast of *Rhodospirillum rubrum* showing profiles of poly-β-hydroxybutyrate granules G1 to 4. Note changes in membrane appearance (reduced sharpness and contrast) as level of section changes from A to C (48,000×).

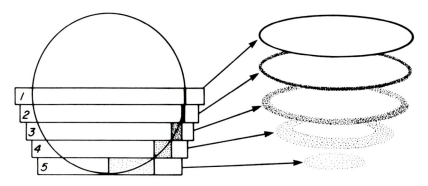

FIGURE 6. Diagrammatic representation of granule changes shown in Figure 5. Assuming the granules to be spherical, the best resolution of the granule membrane is to be seen at the equatorial region (section 1) and thereafter diminishes progressively toward the cap (section 5).

Having broached the subject, beginners are nevertheless advised to first become thoroughly conversant with the methods of morphometric analysis, collect some data, and then worry about whether or not a correction factor for the "Holmes effect" should be used for their particular material and requirements after a review of the literature on the subject. Errors due to the "Holmes effect" are in the range of 12 to 20% for objects with a mean diameter approaching the thickness of the section. (for further discussion, see Williams, 1977; Weibel, 1979; Aherne and Dunnill, 1982; Collan *et al.*, 1983) and are largest for small

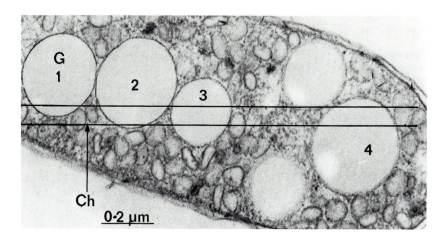

FIGURE 7. Electron micrograph of a section profile of *Rhodospirillum rubrum* cytoplasm showing β-hydroxybutyrate granules (G) and chromatophores (Ch). The two parallel lines represent a hypothetical section (600 Å thick) through the interior of the cell. (See text for details.)

structures such as ribosomes. For most work, the aim should be to obtain sections as thin as possible and possessing high contrast.

5.6. Selection of Samples for Analysis

Unquestionably, the most critical aspect to the undertaking of morphometric analysis of biological material by ultrathin sectioning and transmission electron microscopy (TEM) is the collection of samples. How shall the material be sampled? What constitutes an adequate sample? This presupposes that the source of the material and all details associated with its prior treatment (e.g., details of bacterial culture or tissue culture) are precisely known.

The dimensions of the problem of adequate sampling are clearly related to the nature of the material. For example: (1) The parenchyma of the human lung contains about 300×10^6 alveoli, each of about 300 μm in diameter with a shape approximating that of a dodecahedron in an organ that when distended (by instillation of fixative) reaches a volume of 5 liters or so. The size of a block of tissue for TEM may be 1.5 mm^3 and when sectioned may show 50 or so alveolar profiles per grid, each 60 nm thick and each representing 1/5000th of a whole unsectioned alveolus. In what way can one relate the contents of this single block of tissue to the whole organ? Furthermore, what if pathological changes are present in focal and widely separated areas? (2) A suspension of tissue culture cells poses a similar problem but of a different magnitude. Here we may have a population of 10^8 cells of generally spherical shape each with a diameter of 30 μm; when embedded in a 1.5-mm^3 resin block and sectioned we may observe profiles of several hundred cells per electron miscropy (EM) grid, each representing $\frac{1}{500}$ of the whole cell. In one plane of section some cells will show a profile of a nucleus, others may not. An advantage is that the suspension of cells can be observed directly by light microscopy and their size, shape, viability (trypan blue dye exclusion test), and aberrant cells, if any, determined prior to electron microscopy.

Bacterial cultures are handled in the same way as suspensions of eukaryotic cells. When sectioned, the numbers of profiles per field may amount to several hundred and a single bacterial profile may represent $\frac{1}{17}$ of the whole cell. Appropriate light microscopy will indicate the uniformity of the population and any divergence in shape and size as the tendency to form filaments.

The next critical step is to determine the procedure to be used for sampling. With all material it is necessary to know whether the structures to be observed are randomly distributed in space (i.e., isotropic) or have a preferred orientation (i.e., anisotropic). If, in a series of random sections cut from a random sample of specimen blocks, the profiles of the structures look very similar from one section to another, a generous assumption is that the structures are (for biological tissue) isotropic. All three examples quoted above are normally found to be isotropic. Examples of tissue structures that would be anisotropic are striated muscle, kidney cortex as opposed to medulla, and also neurological tissue. Morphometric analyses of isotropic oriented structures should give values from section to section that vary only by the experimental error of the analyses. Anisotropic structures will give values that vary markedly depending on the plane of section relative to the orientation of the structure. When dealing with suspensions of cells, a major concern is whether preparative procedures involving filtration or centrifugation modify the normal distribution of the cell population giving rise to layering of the cells according to their surface charge, size, or density. If not intended, the layering can be overcome by washing the cells from the surface of a filter or by resuspension of a centrifuged pellet. Here again, light microscopy should be used to determine the overall distribution of the concentrated cells.

We are now ready to discuss methods for sampling tissues, again using our three examples of lung, eukaryotic cells, and bacterial cultures.

The lung, as distinct from other organs, is usually fixed in the distended state at a precise pressure of 20 or 25 cm H_2O before sampling is attempted. Lung volume (V_L) is estimated by water displacement, where V_L = displaced water level volume − original water level volume. Knowledge of the volume of an organ is necessary if conversion of data to absolute values is desired. Having estimated the volume of the fixed lung, individual lobes are removed and sliced transversely into slices of equal thickness. The slices are laid out in sequence from apical to basal. Depending on the size of the lung, every slice or alternate slice is subsampled by use of a transparent overlay ruled into a grid of squares; each square is numbered and blocks of tissue (e.g., 2 × 1 × 0.5 cm) are randomly selected by use of random number tables or by a method known as area-weighted sampling (Miles, 1978). For whole lung values, the number of blocks from each lobe should be proportional to the lobe volume. A portion of each tissue block is processed for light microscopy and a portion is cut into smaller blocks for postfixation and embedding for TEM. From each batch of small blocks, one to three blocks are chosen at random and processed. Each block is sectioned and one section only per block is photographed to produce a set number of electron micrograph negatives. The tissue section must be photographed without bias, most often by taking a picture at each top left-hand corner of a succession of grid squares even if the field contains only a small profile of tissue. Fields should not be rejected on the basis of appearance alone. It is always worthwhile to keep track of the origin of the tissue sections photographed; e.g., right or left lung, upper, middle, or lower lobe, slice number, block number. This sequence of sampling steps from organ to sections to micrographs is typical of a systematic stratified random sampling procedure. In nonrandom sampling, the lung or organ is sampled at given intervals laterally and longitudinally throughout an organ. Both procedures give a good representative sample providing the organ does not possess structures of natural periodicity or is affected by areas of disease. Finally, it is necessary to decide how many sites are to be sampled per organ. This depends on the number of animals per group, the number of groups per experiment, the questions being asked, and the level of accuracy desired (Shay, 1975).

For eukaryotic cells or bacterial cells in suspensibn, purely random sampling is sufficient providing, as stated earlier, the cells or bacteria remain homogeneous in distribution when concentrated by centrifugation or filtration following fixation. Aliquots of cell deposits from replica cultures are processed and embedded and two or three blocks selected at random from each culture, trimmed, and sectioned. A 1.0-μm section is cut to demonstrate the presence of organisms and then an ultrathin section. Unlike tissues, profiles of sectioned cells or organisms may not be uniformly disposed over the area of the section. For morphometry, it is best to select a grid square at random at a low magnification, increase the magnification to 7000 to 20,000×, and photograph whatever cell profiles are present in that particular area. Other grid squares are randomly selected until about 20–30 negatives have been collected. It is important that no serial sections are photographed.

If mixtures of different cell types (e.g., white blood cells, peritoneal cells, or lavaged cells from lung) are to be analyzed, it may be necessary in the absence of specific markers to identify the desired cell type by nuclear profile or by certain cytoplasmic constituents (e.g., granules, lipid droplets, lamellar bodies, cilia) and photograph only the selected cells. A nuclear-biased sample (specifically for spherical cells) is necessary if cell number (N_A), mean cell volume (V_c), or nuclear/cytoplasmic volume ratios are to be estimated. For these kinds of estimations, a final print magnification of 30,000× is desirable. For calculations of surface densities of bacterial cell wall and cytoplasmic membrane or inner and outer mitochondrial membranes of eukaryotic cells, a print magnification of around 80,000× is necessary.

Variation from sample site to sample site should be insignificant statistically. When working with a familiar tissue, the investigator should be able to judge whether the chosen samples are representative of the whole. With unfamiliar tissue, it may be necessary to assess the mean values of the samples by analysis of variance aiming for a SD appropriate for the experiment (Aherne and Dunnill, 1982).

5.7. Processing Samples for Microscopy

The need for very careful processing of material for electron microscopy prior to morphometric analysis cannot be too highly stressed and is critical both to the accuracy of a particular analysis and to intrasample and intersample comparisons (test versus control). As stressed by Weibel (1979), the standardization of procedures is most important "lest your morphometric data reveal more about artefacts than about real experimental changes." Likewise, for the publication of work, it is important to include all details of the methods used and particularly any steps taken to control or modify changes in shape and size due to processing.

With respect to processing bacterial cultures, it is obligatory to know the point on the growth curve at which a culture was harvested and also the viability of the population. For suspensions of eukaryotic cells, assessment of viability by use of the trypan blue dye exclusion test is useful. As a prelude to electron microscopy, observations by light microscopy including the taking of photographs is advisable, particularly for the record and to document before and after fixation comparisons.

Procedures for the preparation of specimens for TEM involve the accepted sequence of fixation, dehydration, and embedding in resin. Other chapters in this book give in explicit detail recipes for processing bacterial and other cells. Here it is necessary only to specify certain precautions of which to be aware when processing specimens. Clearly there are no set standards for what an optimally "fixed" bacterium or cell should look like in either two dimensions or three. Whatever system of primary and secondary fixatives and buffers, etc. is used, the results should be reproducible. The tonicity or osmolarity of the solutions used should be determined and adjusted, avoiding the use of concentrations whose tonicities diverge markedly from *in vivo* or *in vitro* growth conditions. Alternatively, it may be necessary to use processing regimens that are known to preserve cellular components in a certain manner without regard to osmolarity; e.g., the preservation of bacterial nucleoplasm in the fibrillar form, by the Ryter–Kellenberger (Kellenberger *et al.*, 1958) method of fixation (which, incidentally, produces a fixative/buffer osmolarity of about 340 mOsm), or the detection of material in the periplasmic space by low-temperature embedding (Hobot *et al.*, 1984). For animal tissues, many buffer/fixative solutions are prepared to give osmolarity levels of 330 to 360 mOsm/kg H_2O approaching that of blood plasma.

The widely varying ecology of different microbial species is reason for special care when fixation for electron microscopy is being considered. Species of mycoplasmas are going to be more susceptible to changes in osmolarity than most Gram-positive and Gram-negative bacteria. Yeasts are probably quite resistant compared to tissue culture cell lines. If specimen shrinkage or swelling is pronounced (and this will have to be deduced from measurements made before and after processing)—on the order of >10% linear, about 30% by volume—then dimensional correction factors are required to be incorporated into the morphometric data analyses and calculations. It is now believed that most swelling or shrinkage is caused by the molar strength of the buffer employed rather than (within limits) by the concentration of the fixative. Changes due to dehydration through alcohols and the final polymerization in resin are not considered to be large (i.e., they are on the order of 5% linear). For additional discussion of this point, see Chapter 4. The difficulties associated with the processing of microorganisms for the determination of cell volumes by electronic particle sizing, scanning electron microscopy (SEM) and TEM have been documented by Montesinos, *et al.*, 1983).

5.8. Producing the Micrographs

This topic concerns the collection of EM negatives, their photographic enlargement, and the final photographic prints. The overriding factors here are (1) knowledge of the precise magnification at which the negatives were obtained, (2) the precise magnification of the enlarged and printed image, and (3) the contrast of the image. The subject matter to be discussed presupposes that the electron microscope to be used for obtaining negatives for morphometric analysis is in the best possible working order. That is, lens currents and high voltage are stable, astigmatism is corrected, and distortion is minimal.

5.8.1. Magnification at the Electron Microscope

For morphometry, it is most important to prepare an EM negative of a carbon grating replica or silicon monoxide grating replica of say 28,800 lines/inch (0.882 μm/line) or 54,864 lines/inch (0.463 μm/line), respectively, at each magnification step at which a series of negatives are to be photographed. The procedure is as follows.

Magnification = Find the distance in millimeters (between the first and last line) and divide by the number of spaces between the first and last line = magnified width of 1 space

$$\text{Instrument magnification} = \frac{\text{magnified width of 1 space}}{0.882 \ \mu\text{m (actual width of 1 space)}}$$

Example:

$$\frac{36 \text{ mm}}{10 \text{ spaces}} = \frac{36{,}000 \ \mu\text{m}}{10} = 3600 \ \mu\text{m} = \text{width of 1 space}$$

$$\text{Magnification} = \frac{3600 \ \mu\text{m}}{0.882 \ \mu\text{m}} = 4081\text{X}$$

Thus, if the digital readout at this magnification step of the TEM is say 4.0 ($\times1000$), the actual magnification at that time is $4081\times$.

5.8.2. Photographic Enlargers

For maximum resolution in printing EM negatives, a point-light source enlarger with a double-condenser lens system is advisable. This enlarging system provides the best method for reproducing the image and the fine detail in the negative. The current applied to the enlarger light source should be controlled by use of a constant-voltage regulator. The image portrayed onto the enlarger easel should be focused with an appropriate focusing magnifier of about 20×. The degree of negative enlargement must be determined and recorded on printing paper so that the magnification can be determined *after* the paper has been processed and dried. This can be most easily accomplished by using a transparent scale located at one end of the negative holder. The image of this scale is projected vertically and horizontally onto a sheet of photographic paper and processed along with prints of the actual work. After completion of processing, the scale is measured and the magnification setting of the enlarger computed and the total magnification of all prints recorded. It can be easily demonstrated that changes in the overall dimensions of photographic paper occur following drying in a print drying machine. In a simple experiment using two sizes of Kodabromide single-weight paper (8 × 10 and 11 × 14 inches; No. 1 paper), the *linear stretching* of the papers, placed with the shortest dimension at right angles to the direction of travel of the apron on the print dryer, were 2.5% linear to the direction of travel and 0.5% at right angles to the direction of travel. This is, however, not a problem with resin-coated material.

5.8.3. Photographic Contrast

The final contrast of the printed image can be improved within limits by judicious selection of the appropriate grade of photographic paper (e.g., Kodabromide paper No. 1, 2, 3, 4, 5, and 6). For morphometric analysis, the items to be analyzed must be extremely well delineated and, if need be, at the expense of other components in the image. Best results are obtained when following the manufacturer's recommendations for use of photographic papers and processing solutions.

5.9. Practical Examples of Methods: Testing the Accuracy of the Tracing Devices

The particular type of compensating polar planimeter used here has a reputed accuracy of 0.1% on areas greater than 6425 mm^2 (89-mm diameter). From Table II, it can be seen that at circle diameters of 25.0 mm (491 mm^2) and below, the level of recording accuracy as measured by the planimeter drops substantially. At these relatively small tracing areas, however, the performance of the digitizer is superior to that of the planimeter, giving reasonable accuracy at a diameter of 5 mm (Table II). Repetitive measurements would improve the accuracy a little but the time factor required for measuring profiles of a discrete number of objects many times over would be prohibitive. It is better to increase the magnification of the object of interest and make one trace per image on a larger number. Because the digitizing system can display up to six different measuring functions following a single trace, it is clearly the more useful of the

Table II
Planimeter and Digitizer Tracings of Drawn Circles of Various Diameters[a]

Actual circle area (mm²)	Area (mm²), planimeter	% error[b]	Area (mm²), digitizer		Actual circle diameter (mm)	Equivalent circle diameter (mm), digitizer	% error[c]
8107	8103	−0.05	8105		101.6	101.4	−0.2
4656	4671	+0.32	4666		77.0	77.0	0
2027	2026	−0.05	2030		50.8	50.8	0
491	477	−2.85	497		25.0	25.0	0
314	323	+2.87	313		20.0	19.8	−1.0
177	193	+9.04	182		15.0	15.2	+1.3
79	64	−18.99	80		10.0	10.0	0
50	52	+4.00	49		8.0	7.9	−1.2
28	19	−32.14	27		6.0	5.9	−1.7
20	13	−35.00	18	−10.0	5.0	4.9	−2.0
7	6	−14.29	6	−14.3	3.0	2.8	−6.7

[a]Each value derived from a single trace.
[b]Percent error is difference between actual area and planimeter area.
[c]Error is difference between actual diameter and digitized diameter.

two tracing devices. Comparisions of area and perimeter measurements made on an electron micrograph of a sectioned HeLa cell after single and repetitive traces with a planimeter, digitizer, and a cartographer's wheel are given in Table III. The comparisons between methods are quite good even at a single trace. It was found that the cursor of the digitizer could be moved at a faster rate than the arm of the planimeter. While a cartographer's wheel is a simple device, it is troublesome to use, and although its tracing error appears small this is due to the rather gross scale where, within the smallest division of 1.0 cm, the end point can only be approximated.

Determination of an optimal range of tracing speeds for the digitizer cursor by measurement of a straight line 150 mm in length indicated speeds from 7.5 mm/sec up to the maximum of 100 mm/sec to be appropriate. At slow speeds averaging 3.25 mm/sec, length was overestimated by about 2.3%. The relationship between tracing speed and associated error depends largely on the algorithms used to process the data (Cornelisse and van den Berg, 1984).

5.10. HeLa Cells

5.10.1. Cell Shrinkage

Before processing cells for electron microscopy, it is always advisable to acquire dimensions of unfixed cells for estimates of shrinkage factors.

Table III
Comparisons of Area and Perimeter from an Electron Micrograph (8700 ×) of a Sectioned HeLa Cell Measured by Different Methods

	Area (μm²)		Perimeter (μm)	
Measurements	Planimeter	Digitizer	Cartographer's wheel	Digitizer
1	89.15	92.13	50.57	48.98
10	89.59 ± 0.50[a]	91.64 ± 0.71	50.50 ± 0.30	50.07 ± 0.59
Time (min)	12	10	15	8

[a]Mean ± 1 SD

Table IV shows mean values of cell and nuclear areas, perimeters and equivalent circle diameters for HeLa cell monolayers grown on coverslips, photographed under phase-contrast, and measured by planimetry and digitizer at 2100× magnification. Duplicate monolayers were fixed *in situ* by the addition of 2.0% glutaraldehyde for 2 hr at 21°C (osmolarity of medium with fixative was 310 mOsm/kg H_2O). In terms of area estimates, planimeter and digitizer values were identical. Linear shrinkage after fixation was 7.4% and it is notable that

Table IV
Area, Perimeter, and Equivalent Circle Diameter Estimates Obtained from Phase-Contrast Micrographs (2100 ×) of Unfixed and Fixed HeLa Cells

	Area (μm²)		Perimeter (μm)		Circle diameter (μm)	
Method	Cell	Nucleus	Cell	Nucleus	Cell	Nucleus
Planimeter, unfixed cells	606.3 ± 156.65[a]	151.6 ± 28.26	—	—	—	—
	25.98 ± 1.05[b]					
Digitizer, unfixed cells	606.2 ± 153.14	151.7 ± 27.45	114.9 ± 19.23	42.6 ± 3.93	27.6 ± 3.41	13.8 ± 1.24
	25.02 ± 1.04[b]					
Digitizer, fixed cells	520.2 ± 107.08	147.7 ± 20.36	170.3 ± 19.65	42.5 ± 3.16	25.5[c] ± 2.65	13.6[c] ± 0.97
	28.40 ± 1.01[b]					

[a]Mean ± 1 SD
[b]Percentage of cell area occupied by nucleus.
[c]Linear shrinkage for fixed cell and nucleus (equivalent circle diameters) were 7.4 and 1.3%, respectively.

nuclei appeared to shrink less than cell cytoplasm. Also, the average volume density of the nuclear component with reference to the cell was about 0.25. That is, the nucleus occupied 25% of the area of the cell.

5.10.2. Lineal Analysis and Digitizer

A test grid for estimating cell and nuclear profile area and volume density of nuclei by lineal analysis is shown in Figure 8. The grid consists of 25 parallel lines each 180 mm in length (the equated length of the line depends on the magnification of the image to be analyzed), the total length of lines (L_T) is 4500 mm, and the distance between lines is 8.0 mm. The sectioned HeLa cell superim-

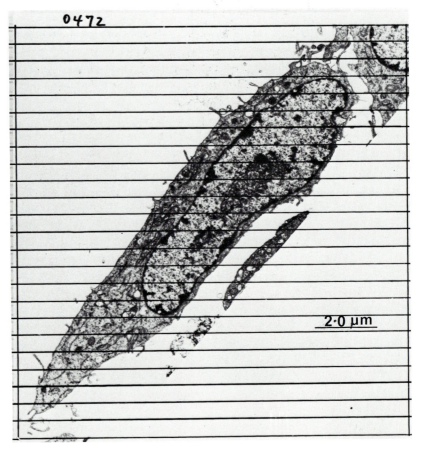

FIGURE 8. Line grid for lineal analysis superimposed over an electron micrograph of a HeLa cell. Intersection counts were made at 8700×.

posed by the grid was analyzed at $8700\times$; thus, L_T was 517.24 μm and the grid area 428 μm^2. The length of each line intercepting the cell boundary (i.e., lying within the cell) was measured and together gave a total length (L_{Ac}) of 107 μm. Since from equation (1)

$$A_{Ac} = V_{Vc} = \frac{L_{Ac}}{L_T}; \qquad \frac{107.0}{517.2} = 0.21$$

thus, this cell profile occupied 21% of the total area of the grid giving a cell area of 88.54 μm^2. The same procedure was used to tally lines intercepting the nuclear boundary (L_{An}) but using the cell as the reference area, i.e.,

$$\frac{L_{An}}{L_{Ac}} = \frac{53.2}{107.0} = 0.50$$

thus, the volume density of nucleus in relation to the cell was 0.50 or 50%. With a cell area of 89.88 μm^2 the nuclear area becomes $89.88 \times 0.50 = 44.94$ μm^2. The same components were measured by the digitizer and gave 92.79 μm^2 and 47.19 μm^2, respectively, with a volume density of 0.51, in good agreement with lineal analysis. The reason why the cell area here differs substantially from that obtained from fixed whole cells (Table IV) is that in the sectioned cell population, the image plane photographed did not encompass the entire area of the cell. This means that with sectioned material, a large number (e.g., several hundred) of cells have to be photographed and analyzed.

5.10.3. Point-Counting and Digitizer

The volume density, size, and number of cytoplasmic components (e.g., mitochondria, Golgi, lysosomal bodies) present in cells can be determined by point- and intersection-counting using a square lattice grid.

For example, Figure 9 is an electron micrograph of a portion of a HeLa cell at $20,000\times$ overlaid by a grid consisting of 25 vertical and 25 horizontal lines producing 64 coarse points and 576 fine points. Point-counts on mitochondria (P_m) and on cytoplasm (P_{cyt}) were tallied and the volume density of mitochondria (V_{vm}) in reference to cytoplasm determined:

$$V_{vm} = \frac{P_m}{P_{cyt} + P_m} = \frac{40}{290 + 40} = 0.12 \text{ or } 12.0\%$$

V_{vcyt} in reference to the whole grid (total grid points) was:

$$V_{vcyt} = \frac{P_{cyt}}{P_T} = \frac{330}{576} = 0.57 \text{ or } 57\%$$

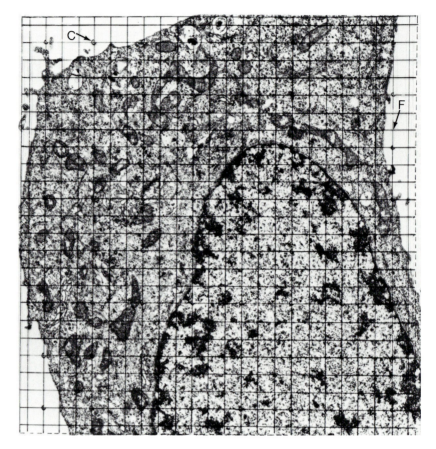

FIGURE 9. Portion of a HeLa cell enlarged 20,000× and overlaid by a square grid lattice for estimating volume density of cytoplasmic components. Grid contains both coarse (C) and fine (F) points.

Each vertical line is 240 mm in length and each horizontal line 242 mm; at 20,000× these translate to 12 μm and 12.1 μm, respectively, giving a total grid area of 145.2 μm². Cell cytoplasm occupied 57% of the grid area, i.e., (145.2 × 57)/100 = 82.76 μm²; and mitochondria 12.12% of the cytoplasm, i.e., (82.76 × 12.12)/100 = 10.0 μm². It is a simple matter to count the number of mitochondrial profiles present (which was 36) in that portion of cytoplasm under the grid and divide the area occupied by the mitochondria by this number to get the average area occupied by one mitochondrion and finally the average diameter; where $1.128\sqrt{\text{area}}$ = diameter:

$$A_m = \frac{10.0 \ \mu m^2}{36} = 0.28 \ \mu m^2; \qquad 1.128\sqrt{0.28} = 0.59\text{-}\mu m \ \text{diameter}$$

The same electron micrograph was then evaluated using the digitizing tablet. Results were in excellent agreement, being $V_{vcyt} = 54\%$, $A_m = 0.29 \ \mu m^2$, and average equivalent circle diameter $= 0.57 \ \mu m$. If the total number of mitochondria N_v is required, then it is necessary to approximate the shape of the cell and estimate its volume.

There are two problems relating to the counting of mitochondria. First, if the mitochondria in question are not generally oval in shape but tend to be long and irregular (see Figure 9), some may reenter the section face and be counted twice. Second, an estimate of mitochondrial size (volume and surface area) by conversion to diameter is valid only if the mitochondria are randomly disposed in the sectioned cell or tissue and they are spherical or approximate prolate or oblate spheroids in shape.

To estimate the probable total number of mitochondria in a cell, it is necessary to approximate the shape of the cell (actually several hundred cells) and estimate its volume. The equivalent circle diameter of the cell shown in Figure 9 (only a part of the whole cell is shown) is 15.32 μm, which gives a sphere volume of 1882.6 μm^3; the nuclear diameter is 9.72 μm, which gives a sphere volume of 480.8 μm^3; the volume of cytoplasm is therefore 1882.6 $-$ 480.8 $=$ 1401.842 μm^3. We have estimated that mitochondria occupy 12.12% of this volume; therefore, the volume occupied by mitochondria $=$ 169.9 μm^3. If the volume of one mitochondrion is 0.11 μm^3, then the number of mitochondria distributed in the cytoplasm is about 1502. Cell volumes can be determined from nuclear measurements; one needs to know the mean nuclear volume (by digitizing, this was estimated to be 480.8 μm^3) and the volume fraction of the cell occupied by the nucleus, which, in the present example, is 0.25. By use of the formula cell volume (μm^3) $=$ mean nuclear volume (μm^3)/volume fraction, nucleus:

$$\text{Cell volume} = \frac{480.8}{0.25} = 1923.2 \ \mu m^3$$

which is very close to that derived by use of the equivalent circle diameter of the cell.

Mean nuclear volume may be derived by the formula mean nuclear volume (μm^3) $= \frac{4\pi}{3} \times \left(\frac{\bar{D}}{2}\right)^3$, where $\bar{D}$ is the mean diameter of the nucleus. By digitizing, D was found to be 9.72 μm, and:

$$\text{Mean nuclear volume } (\mu^3) = \frac{12.566}{3} \times \left(\frac{9.72}{2}\right)^3 = 480.8 \ \mu m^3$$

It is necessary to note that perhaps the only way to accurately estimate the average nuclear volume or mitochondrial volume (of which the above calculations constitute an exercise) is by serial section reconstructions. The number of mitochondria (or other organelles that are spheres or regular convex objects can be estimated by use of the formula of Weibel and Gomez (1962):

$$N_V = \frac{K}{\beta} \times \frac{(\bar{N}_A)^{3/2}}{(\bar{V}_V)^{1/2}}$$

where K is a constant that depends on the size distribution of the objects and may be given an arbitrary value between 1.02 and 1.1; B is the shape coefficient; $\bar{N}_A$ is the mean number of objects per test area; and $\bar{V}_V$ is the volume fraction occupied by objects. B can be estimated by $B = V/a_3^2$, where V is the object's volume and a is the object's area.

5.11. Bacterial Cultures

Aeromonas salmonicida was grown in brain heart infusion broth at 18°C for 24 hr (late log phase). Cultures were centrifuged at 1000 rpm for 15 min and resuspended in Ryter–Kellenberger's fixative and fixed at 21°C for 4 hr after which cultures were dehydrated in ascending concentrations of ethanol and embedded in Epon 812. Random sections were cut and negatives taken at 21,000× and enlarged photographically.

5.11.1 Digitizer

A test was done to determine to what degree a reading of cell perimeter is increased as the magnification at which the image is traced is increased. A TEM negative of a sectioned *A. salmonicida* cell was taken at 21,000× and micrographs prepared at magnifications of 47,250×, 73,500× and 141,750× (Figure 10). Perimeters of cell wall and cytoplasmic membrane were traced by use of the digitizer cursor. Measurements were made after a single trace on one day and after five repetitive tracings on the next day. The mean value of the five tracings was only 0.8% higher than after a single trace. Perimeter values for the cell wall at 141,750× were 3.9% higher than values at 73,500× and 6.4% higher than at 47,250×. The latter percentage was significantly different at $p < 0.01$. For the cytoplasmic membrane, which was less irregular in contour, the difference in perimeter between the highest and lowest magnification was only 0.8%. It is concluded that an appropriate magnification for digitizing images of sectioned bacteria with irregular contours would be about 75,000×. At these high magnifications and to include an adequate number of cells, it is best to use 11 × 14-inch printing paper.

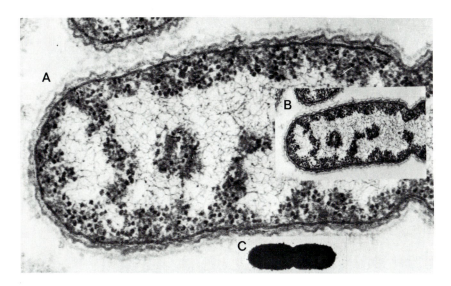

FIGURE 10. Composite photograph of a portion of a dividing cell of *Aeromonas salmonicida* at a high (A, 141,750×) and a low level (B, 47,250×) of enlargement used to determine the optimal magnification for tracing cell wall and cytoplasmic membrane contours on a digitizing tablet. Unsectioned bacterium (C) represents a typical profile of a dividing cell (17,500×).

Random sections of the rod-shaped bacterium *A. salmonicida* present profiles in the longitudinal plane, oblique plane, and in cross section. Counts of the percentage of profiles in each plane in four random fields at 18,200× (186 profiles counted) were 18, 39, and 44%, respectively. Profiles of cell wall, cytoplasmic membrane, and area occupied by nuclear material were traced by a digitizer cursor to estimate cell area, perimeter, and equivalent circle diameter. The results are shown in Table V. Under the stated conditions of growth, fixation, and processing, the nuclear material occupies about 33.5% of the cytoplasm, the periplasmic space 13.5%, and the cytoplasm proper 53%. From Table V it can be seen that the equivalent circle diameter of the average rod-shaped cell is 0.980 μm, which, by use of simple mensuration formulas for spheres, gives a volume of 0.493 μm^3, a surface area of 3.017 μm^2, and a surface-to-volume ratio (S/V) of 6.123. The average rod-shaped cell by direct measurement is 1.14 × 0.82 μm; assuming this cell to be cylindrical with hemispherical ends, the surface area, volume, and S/V become 2.937 μm^2, 0.458 μm^3, and 6.413, respectively. The combined equivalent circle diameter of all sectioned bacterial profiles measured (38) was 0.858 ± 0.24 μm. It is concluded that the average equivalent circle diameter of a population of single cells as recorded by the digitizing tablet gives a good estimate of cell volume, surface area, and S/V.

Table V
Area, Perimeter, and Equivalent Circle Diameter Measurements of Profiles of *Aeromonas salmonicida* from Electron Micrographs Traced by a Digitizer at 73,500×

Plane of section	Item	Area (μm^2)	Perimeter (μm)	Equivalent circle diameter (μm)
Longitudinal (single cells)	CW[a]	0.76 ± 0.07[b]	3.36 ± 0.28	0.98 ± 0.05
	CM[c]	0.68 ± 0.06	3.23 ± 0.25	0.93 ± 0.04
	DNA	0.29 ± 0.05	—	—
Cross section	CW	0.37 ± 0.08	2.10 ± 0.22	0.68 ± 0.07
	CM	0.31 ± 0.07	1.91 ± 0.21	0.63 ± 0.07
Oblique section	CW	0.65 ± 0.27	3.00 ± 0.79	0.89 ± 0.19
	CM	0.56 ± 0.24	2.76 ± 0.76	0.83 ± 0.18
	DNA	0.19 ± 0.08	—	—

[a]CW, cell wall.
[b]Mean ± 1 SD
[c]CM, cytoplasmic membrane.

5.11.2. Point-Counting and Digitizer

The same series of electron micrographs that were analyzed by the digitizer (Table V) were evaluated by use of a square lattice grid of 576 fine points and 50 lines as shown in Figure 9. The length of one line is 241 mm and the total length is 50 × 241 divided by the magnification of the micrographs (i.e., 73,500×) = 163.9 μm. Point "hits" were counted on the cell wall–periplasmic space–cytoplasmic membrane complex (wc), on cytoplasm (cyt), and on DNA (d). The results in terms of volume density (V_V), with the cell as a reference and compared with the digitizer values, were:

$$V_{Vwc} = \frac{P_{wc}}{P_T} = \frac{142}{921} = 0.154 = 15.4\% \qquad \overset{\text{Digitizer}}{13.5\%}$$

$$V_{Vd} = \frac{P_d}{P_T} = \frac{325}{921} = 0.353 = 35.3\% \qquad 33.5\%$$

$$V_{Vcyt} = \frac{P_{cyt}}{P_T} = \frac{454}{921} = 0.493 = 49.3\% \qquad 53.0\%$$

Surface density for cell profiles (S_{Vc}) and volume density with the whole grid as a reference (V_{Vc}) were determined by intersection-counts and point-counts, respectively:

$$S_{VC} = \frac{2I}{L_T} = \frac{2 \times 151}{163.9} = 1.84 \qquad \begin{array}{c} \text{Digitizer} \\ \hline 6.12 \end{array}$$

$$V_{VC} = \frac{P_C}{P_T} = \frac{181}{576} = 0.31 \qquad S/V = \frac{1.84}{0.31} = 5.94$$

It is concluded that a satisfactory agreement was obtained between values recorded by the digitizer and by point-counting analysis. In each case, a series of 11 × 14-inch photographic prints were analyzed and a total of 38 bacterial profiles per case examined.

5.12. Statistical Analysis

Since the value of morphometric analyses is based on statistics and probability theory, it is wise for the investigator unfamiliar with morphometric/stereological methods to consult a statistician and arrive at an appropriate procedure before samples are collected. As Aherne and Dunnill (1982) aptly pointed out, the work required for completion of a task involving morphometry will vary inversely as the square of the error. "A task which takes two hours to complete with an error of 5% will take fifty hours to complete with an error of 1%!" For most work, an error of 5% is adequate. An estimate of the relative standard error (RSE) can be readily obtained by the use of the Weibel and Gomez formula (equation 3). Another appropriate formulation (Hally, 1964) is as follows:

$$\text{RSE} = \frac{\sqrt{1 - V_V}}{\sqrt{n}}$$

where V_v is the volume density fraction and n is the number of points that must fall on the region whose V_v is being estimated. If an RSE of 5% is desired, then:

$$5\% = 0.05 = \frac{\sqrt{1 - V_V}}{\sqrt{n}} \quad \text{and} \quad \sqrt{n} = \frac{\sqrt{1 - V_V}}{0.05}$$

Taking the previously calculated volume fraction of bacterial cytoplasm of 0.493, n becomes:

$$\sqrt{n} = \frac{\sqrt{1 - 0.493}}{0.05} = \frac{0.7120}{0.05} = 14.24$$

$$n = 14.24^2 = 203 \text{ points}$$

Thus, at least 203 points must fall over cytoplasm and since cytoplasm is 49% of the bacterial cell, the total number of points that must be superimposed over the photographs of sectioned bacteria is $203 \frac{100}{49} = 416$. In this particular analysis, 921 points were used giving an RSE of 2.4%. The volume density estimates for V_{vwc} (cell wall complex) and V_{vd} (DNA) gave RSEs of 8% and 4%, respectively. Because the volume density of V_{vwc} was small (15.4%), the total number of points superimposed must be increased about three times to obtain an RSE of 5%.

Data that are normally distributed (Gaussian) can be evaluated by the determination of standard deviation, variance, and standard error. Data distribution that is other than normal requires additional procedures. Tests of significance employing some form of Student's t test are necessary. In most cases, the use of elementary statistics is all that is required. Again, when in doubt consult a statistician. It is recognized however, that not all statisticians are adequately acquainted with morphometric/stereological procedures.

5.13. Utility of Current Methods

Stereology has been widely used to explore the architecture of human and animal lungs. Estimates have been obtained of alveolar number and diameter, the width of the air–blood barrier, the surface area of the lung, and the diameters, numbers, and volumes of various cells and their cytoplasmic components. The kidney and brain have also received much attention. In the former, the number, volume, surface area, and mean diameter of glomeruli have been calculated in adult and neonatal specimens. In the brain, motor neurons in the anterior horn of the spinal cord have been counted, as have Purkinje cells in the cerebellar cortex, to name a few. Stereology is being used to investigate many changes in cell morphology following exposure to a wide variety of gaseous and particulate materials, and, where subjective assessments of cell injury were previously made, it is now possible to describe changes in numerical terms. In fact, rarely are these kinds of investigations now done without recourse to the use of morphometry and stereology. As stated earlier, these techniques have not been notably employed to study microorganisms. Much can be done in this area with respect to effects on organisms following exposure to various fixatives, freeze-substitution, antibiotics, irradiation, and so on; agents that affect cell division, cause swelling or shrinkage, and so on. The best approach is to determine what questions are being asked and to decide whether the determinations desired would be substantially strengthened by the use of the stereological approach. This is particularly important if morphological comparisons between tissues over time, or between different cell populations, or microorganisms are envisaged.

Finally, it is important to recall that the constancy of dimensional parameters relies heavily on the methods used to prepare the sample, and artifacts may arise in the most cursory manner if vigilance is allowed to lapse.

6. REFERENCES

Aherne, M. A., and Dunnill, M. S., 1982, *Morphometry,* Arnold, London.
Boatman, E. S., 1964, Observations on the fine structure of spheroplasts of *Rhodospirillum rubrum,* *J. Cell Biol.* **20:**297–311.
Chalkley, H. W., 1943, Method for the quantitative morphologic analysis of tissue, *J. Natl. Cancer Inst.* **4:**47–53.
Collan, Y., Oja, E., and Whimster, W. F., 1983, Mathematical background to stereology and morphometry for diagnostic pathologists, *Acta Stereol.* **2:**214–238.
Cornelisse, J. T. W. A., and van den Berg, T. J. T. P., 1984, Profile boundary length can be overestimated by as much as 41% when using a digitizer tablet, *J. Microsc. (Oxford)* **136:**341–344.
Delesse, M. A., 1847, Procédé mécanique pour determiner la composition des roches. *C. R. Acad. Sci. (Paris)* **25:**544.
Dunnill, M. S., 1962, Quantitative methods in the study of pulmonary pathology, *Thorax* **17:**320–328.
Glagolev, A. A., 1933, On the geometrical methods of quantitative mineralogic analysis of rocks, *Trans. Inst. Econ. Min. Moscow* **59:**1.
Glagolev, A. A., 1934, Quantitative analysis with the microscope by the point method, *Eng. Min. J.* **135:**399–402.
Gundersen, H. J. G., and Osterby, R., 1980, Sampling efficiency and biological variation in stereology, *Mikroskopie* **37:**143–148.
Gundersen, H. J. G., Boysen, M., and Reith, A., 1981, Comparison of semiautomatic digitizer-tablet and simple point counting performance in morphometry, *Virchows Arch. B* **37:**317–325.
Hally, A. D., 1964, A counting method for measuring the volumes of tissue components in microscopical sections, *Q. J. Microsc. Sci.* **105:**503.
Haug, H., 1955, Die Treffermethode, ein Verfahren zur quantitativen analyse in histologischen Schnitt, *Z. Anat. Entwicklungsgesch.* **118:**302–312.
Hennig, A., 1956, Bestimmung der Oberflache beliebig geformter Korper mit besonderer Anwendung auf Korpenhaufen im mikroskopischen Bereish, *Mikroskopie* **11:**1–20.
Hobot, J. A., Carlemalm, E., Villiger, W., and Kellenberger, E., 1984, Periplasmic gel: New concept resulting from the reinvestigation of bacterial cell envelope ultrastructure by new methods, *J. Bacteriol.* **160:**143–152.
Holmes, A. H., 1921, *Petrographic Methods and Calculations,* Murby, London.
Kellenberger, E., Ryter, A., and Sechaud, J., 1958, Electron microscope study of DNA-containing plasms. II, *J. Biophys. Biochem. Cytol.* **4:**671–678.
Mathieu, O., Hoppeler, H., and Weibel, E. R., 1980, Evaluation of tracing device as compared to standard point-counting, *Mikroskopie* **37:**413–414.
Mathieu, O., Cruz-Orive, L. M., Hoppeler, H., and Weibel, E. R., 1981, Measuring error and sampling variation in stereology: Comparison of the efficiency of various methods for planar image analysis, *J. Microsc. (Oxford)* **121:**75–88.
Miles, R. E., 1978, The sampling, by quadrats, of planar aggregates, *J. Microsc. (Oxford)* **113:**257–267.
Montesinos, E., Esteve, I., and Guerrero, R., 1983, Comparison between direct methods for determination of microbial cell volume: Electron microscopy and electronic particle sizing. *Appl. Environ. Microbiol.* **45:**1651–1658.
Paumgartner, D., Losa, G., and Weibel, E. R., 1980, Resolution effect on the stereological estimation of surface and volume and its interpretation in terms of fractal dimensions. *J. Microsc.* **121:**51–63 (Oxford).

Rosiwal, A., 1898, Uber geometrische Gesteinsanalysen Verhandel, K. K. *Geol. Reichsanst Wien*
    5–6, 143.
Shay, J., 1975, Economy of effort in electron microscope morphometry. *Am. J. Path.* **81**:503–512.
Sorby, H. C., 1856, On slaty cleavage, as exhibited in the Devonian limestones in Devonshire,
    Philos. Mag. **11**:20.
Sorby, H. C., 1865, On the microscopical structure of meteoric iron, *Proc. R. Soc.* **13**:333.
Weibel, E. R. (ed.), 1979, *Stereological Methods: Practical Methods for Biological Morphometry,*
    Vol. 1, Academic Press, New York.
Weibel, E. R., and Bolender, R. P., 1973, Stereological techniques for electron microscopic mor-
    phometry, in: *Principles and Techniques of Electron Microscopy: Biological Applications,* Vol.
    3 (M. A. Hayat, ed.), pp. 237–296, Van Nostrand–Reinhold, Princeton, New Jersey.
Weibel, E. R., and Gomez, D. M., 1962, A principle for counting tissue structures on random
    sections, *J. Appl. Physiol.* **17**:343–348.
Williams, M. A., 1977, Quantitative methods in biology, in: *Practical Methods in Electron Micros-
    copy,* Vol. 6 (A. M. Glauert, ed .), pp. 1–84, Elsevier/North-Holland, Amsterdam.

7. SUGGESTED READING

The foregoing survey of methods applicable to morphometric and stereological analysis of profiles of sectioned material is of necessity brief.
It is equally necessary for the beginner to acquire a greater depth of understanding on how the various formulas have been derived and in what manner they
are to be used, along with construction of the appropriate grid systems for
analysis. The following books and papers cover in great detail all aspects of the
art of stereology and should be consulted accordingly.

Aherne, W. A., 1967, Methods of counting discrete tissue components in microscopical sections, *J.
    R. Microsc. Soc.* **87**:493.
Aherne, W. A., and Dunnill, M. S., 1982, *Morphometry,* Arnold, London.
Bolender, R. P., 1978, Correlation of morphometry and stereology with biochemical analysis of cell
    fractions, *Int. Rev. Cytol.* **55**:247–289.
De Hoff, R. T., and Rhines, F. N., 1968, *Quantitative Microscopy,* McGraw–Hill, New York.
Romppanen, T., and Collan, Y., 1983, Practical guidelines for a morphometric study, *Acta Stereol.*
    2:274–297.
Underwood, E. E., 1970, *Quantitative Stereology,* Addison–Wesley, Reading, Massachusetts.
Weibel, E. R. (ed.), 1979, *Stereological Methods: Practical Methods for Biological Morphometry,*
    Vol. 1, Academic Press, New York.
Weibel, E. R. 1980, *Stereological Methods: Theoretical Foundations,* Vol. 2, Academic Press, New
    York.
Weibel, E. R., and Bolender, R. P., 1973, Stereological techniques for electron microscopic mor-
    phometry, in: *Principles and Techniques of Electron Microscopy: Biological Applications,* Vol.
    3 (M. A. Hayat, ed.), pp. 237–296, Van Nostrand–Reinhold, Princeton, New Jersey.
Weibel, E. R., and Elias, H. (eds.), 1967, *Quantitative Methods in Morphology,* Springer-Verlag,
    Berlin.
Weibel, E. R., Kistler, G. S., and Scherle, W. F., 1966, Practical stereological methods for
    morphometric cytology, *J. Cell Biol.* **30**:23–38.
Williams, M. A., 1977, Quantitative methods in biology, in: *Practical Methods in Electron Micros-
    copy,* Vol. 6 (A. M. Glauert, ed.), pp. 1–84, Elsevier/North-Holland, Amsterdam.

Chapter 14

Localization of Carbohydrate-Containing Molecules

Gregory W. Erdos

Department of Microbiology and Cell Science
University of Florida
Gainesville, Florida 32611

1. INTRODUCTION

A variety of carbohydrate-containing molecules can be found at both extracellular and intracellular locations in microorganisms. These can be subdivided into several broad categories based on their general composition and would include the polysaccharides (both homo- and heteropolymers), glycoproteins, glycolipids, and proteoglycans. These macromolecules, especially those exposed at the cell surface, have taken on considerable importance in the study of self–nonself recognition, cell attachment, adhesion, and pathogenesis, and receptor-mediated uptake. Thus, *in situ* localization of these molecules has become an important adjunct to biochemical studies of their structure. Although histochemical methods provide precise localization of carbohydrates, they have, until recently, given us only very basic information about the specific structure of the molecule in question. Most investigations have been limited to the detection of anionic groups exposed at the cell surface or to the presence of hexose sugars. Both of these techniques are based on light microscopic histochemistry. With the discovery of more and more lectins, some of which have very specific binding characteristics, more precise information can be gained about the saccharide structure of a molecule in question. Increasingly sophisticated technology in the use of lectins can provide information about sugars at the cell surface as well as in the cell interior. Most recent is the application of carbohydrate-specific polyclonal and monoclonal antibodies coupled to electron-dense markers for the ultrastructural localization of specific polysaccharides and glycoconjugates.

2. CATIONIC REAGENTS

These reagents are used to demonstrate the presence of acid mucosubstances exposed on the surface of most cells so far investigated. This surface coat, which can be chemically quite variable, has been termed the glycocalyx (Bennett, 1963). Demonstration of this cell coat is thought to occur by means of the electrostatic interaction of negatively charged carboxyl or sulfate groups of the glycocalyx with electron-dense, positively charged ions or colloids.

2.1. Ruthenium Red

Like many of the reagents used in electron microscopy, staining with ruthenium red (RR) finds its origins in light microscopic histology. This dye historically has been used as a semispecific stain for plant pectins. Although pectins stain intensely with RR, there is little if any increase in electron density when it is used alone. Its use for electron microscopy was introduced by Luft (1964) and considered in more detail by him in later works (Luft, 1971a,b). The method of Luft (1971a), where RR is used in conjunction with osmium tetroxide, is the one most commonly employed. Its first application in bacteria was by Pate and Ordal (1967) in their study of the surface layers of *Chondrococcus columnaris*. Since then, it has been employed frequently for the study of bacterial capsules and extracellular slime material. In his original work, Luft was able to demonstrate the presence of the cell coat on a variety of animal tissues and bacteria. It is used most often as an *en bloc* stain and, because of its poor diffusibility, is limited to extracellular binding sites. Its binding seems to be dependent on exposed carboxyl groups. The chemical specificity of RR is questionable, since it has been shown to react with pure lipids (Luft, 1971a) and with certain intracellular lipids (Vidic, 1973). Considering that RR is a strong oxidant, it may be acting as much as a fixative as a stain. Further, one must consider that in conjunction with osmium, RuO_4 may be formed, which could then interact with protein, glycogen, polar lipids, and other oligosaccharides (Hayat, 1981). So if one's intent is to demonstrate the presence of a glycocalyx at the cell surface with little concern about the chemical nature of its components, RR is an effective reagent in most cases. An example of such an application is shown in Figure 1.

For most purposes the following protocol can be followed:

- Primary fixative:
  One part aqueous glutaraldehyde (4–8%)
  One part 0.2 M cacodylate buffer (pH 7.3)
  One part RR stock solution (0.15% in water)
- Secondary fixative:
  One part aqueous OsO_4 (5%)

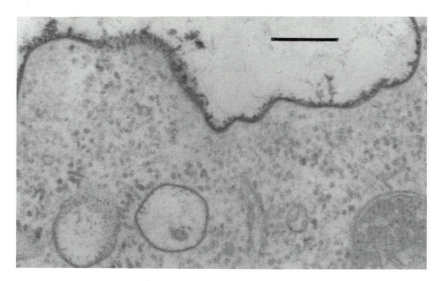

FIGURE 1. Demonstration of the glycocalyx of the amoeba of the cellular slime mold *Dictyostelium discoideum* using ruthenium red. Poststained with lead citrate. Bar = 0.25 μm.

One part 0.2 M cacodylate buffer (pH 7.3)
One part RR stock solution (0.15% in water)

Each fixative should be mixed just prior to use. Cells should be fixed for 1 hr in the primary fixative followed by buffer wash. The cells are then postfixed in the secondary fixative for 1–3 hr, washed, dehydrated, and embedded.

RR has also been used as a poststain for ultrathin sections and works best when the cells have been embedded in a methacrylate-based resin. For details, one should refer to Gustafson and Pihl (1967).

2.2. Iron-Containing Stains

Since it is the 20% iron content of cationized ferritin that is responsible for its electron density, I include cationized ferritin in this group. The first use of this reagent seems to be by Nachmias and Marshall (1961), who methylated native ferritin in order to give it a positive charge. Their work involved the demonstration of acidic groups associated with the cell cost of amoebae. The technique apparently was "rediscovered" in the work of Danon *et al.* (1972), who used a different method for imparting the positive charge to the ferritin by converting the carboxyl groups into tertiary amino groups. Cationized ferritin is available from several commercial sources. It appears that the binding of this reagent is entirely electrostatic in nature and can be considered specific for sulfate and carboxyl groups of glycosaminoglycans and acidic glycoconjugates. Since it can be used at physiological pHs, one can apply it to living cells as well as to prefixed cells prior to embedding. Cationized ferritin has also been applied as a poststain on ultrathin sections of both epoxy- and methacrylate-embedded material, the latter being preferred (Thiéry and Ovtracht, 1979). By first incubating sections with Alcian blue at pH 0.5, these workers were able to effectively block sulfate groups. Subsequent treatment of the sections with cationized ferritin at pH 2.5 labeled only the carboxyl groups. This marker is particularly useful because of its size (10-nm diameter) and its distinctive appearance, especially when the bismuth intensification procedure is used (Ainsworth and Karnovsky, 1972). Moreover, it tends not to obscure the underlying ultrastructure.

Other iron-containing reagents have been used to localize acidic sites both at the cell surface and at intracellular locations when performed on thin sections. In either case they must be used at low pH (1.2–2.4) (Wetzel *et al.*, 1966). Two of these are dialyzed colloidal iron, which detects both carboxyl and sulfate groups, and high-iron diamine, which detects sulfates only. These, as well as cationized ferritin, have been investigated intensely in the laboratory of Samuel Spicer (reviewed by Spicer and Schulte, 1982) in their studies of acid mucosubstances in mammalian tissue. These methods have not been applied to microbial systems.

The dialyzed iron reagent is based on that of Rinehart and Abul-Haj (1951), who developed it for light microscopy. The method was adapted by Hardin and Spicer (1971) for use as both a pre- and postembedding reagent for electron microscopy. A stock solution of dialyzed iron is prepared by dissolving 75 g of anhydrous ferric chloride in 250 ml of distilled water. To this is added 100 ml of glycerin. Then, with constant stirring, 55 ml of concentrated ammonium hydroxide (28%) is slowly added over a 45-min period. This solution is dialyzed for 3 days against distilled water, two changes per day. The dialysis bag should be as full as possible to prevent dilution of the reagent. This stock solution can be stored at 4°C in the dark for several months. Just prior to use, 1 part of glacial acetic acid is added to 4 parts of the iron solution so that the final pH is between 1.8 and 2.0.

For *en bloc* labeling, cells first should be fixed with an appropriate aldehyde. Some methods also include an osmium postfixation at this point but it is probably best avoided until later. After washing, the material is treated with the dialyzed iron for 3 hr or more. The material then may be postfixed in osmium, dehydrated, and embedded in the plastic of choice. The major drawback of this technique is that it often results in poor morphological preservation, presumably due to the long exposure of the tissue to the low pH environment of the iron-containing reagent.

This problem of poor preservation can be overcome by using an iron colloid that maintains its positive charge at physiological pHs. One such reagent is the cacodylate iron colloid (Seno *et al.*, 1983) that is applied before any fixation. This reagent is prepared by slowly adding 1 part of a 0.1 M $FeCl_3$ solution to 9 parts of boiling distilled water. The water should be stirred continuously during this addition and always kept boiling. On cooling to room temperature, 1 part of the resulting iron colloid is added to 5 parts of 0.1 M sodium cacodylate buffer (pH 7.3). The resulting solution can be adjusted to the appropriate pH with HCl or NaHCO$_3$. After application of this reagent, the material is then fixed in aldehyde and postfixed in osmium prior to embedding. Although this iron reagent is stable at room temperature, the authors recommend that it be prepared freshly before each use. The cacodylate iron colloid produces particles from 3 to 5 nm in diameter and effectively labels negatively charged groups over a wide pH range (1.8–7.6).

The dialyzed iron colloid reagent discussed above also can be used on thin sections for the demonstration of both cell surface and intracellular mucosubstances. Thomopoulos et al. (1983a) have investigated the effect of fixation and embedding media on the staining of thin sections. In general, they conclude that osmium should be omitted from the fixation protocol and that nonepoxy resins give heavier labeling than epoxy resins. They recommend the use of a styrene–Vestopal W mixture for nonosmicated material and the use of a styrene–methacrylate mixture for briefly osmicated (5 min) material. They did not test any of the various methacrylate resins alone.

The general technique is to mount thin sections on inert grids (Formvar coated for the styrene–methacrylate) and immerse them in water for 10 min. This is followed by immersion in the dialyzed iron reagent for 2 hr (longer times did not increase the level of labeling). The grids are washed four times for 1 min each and observed without further staining.

The high-iron diamine reagent can be used to label sites that are negatively charged by virtue of their sulfate groups. The use of this reagent for ultrastructural studies was introduced by Spicer et al. (1978). They were able to demonstrate the specificity of this reagent by prior methylation, which removes the sulfate esters. The reagent is prepared by simultaneously adding 120 mg of N,N-dimethyl-m-phenylenediamine (HCl)$_2$ and 20 mg of N,N-dimethyl-p-phenylenediamine HCl to 50 ml of distilled water. To this solution is added 1.4 ml of a 40% (w/v) solution) of ferric chloride. Specimens previously fixed in glutaraldehyde are stained for 24 hr at room temperature. After washing, the tissue is postfixed in osmium and embedded in an epoxy resin of choice. The staining can be enhanced by further treatment of the sections with periodic acid–thiocarbohydrazide–silver proteinate (Thiéry, 1967). By comparing results from dialyzed iron-treated cells with those stained with high-iron diamine, one can determine whether the negative charge of the carbohydrate is due solely to carboxyl groups or to sulfate groups or both. That is, those positive for dialyzed iron but negative for high-iron diamine would contain only carboxyls and, conversely, those negative for dialyzed iron but positive for high-iron diamine would have only sulfates. A double positive would indicate the presence of both.

2.3. Alcian Blue

This stain has been used as a specific indicator of acid mucosubstances for light microscopy (Pearse, 1968). It binds with polyanions in tissue via sulfate and carboxyl groups. However, at low pH (0.5–1.0), it is specific for sulfate esters (Jones and Reid, 1973) and can thus be used to differentiate between sulfomucins and sialomucins. It has been applied in the study of bacterial exopolysaccharides with uneven results (Progulske and Holt, 1980). These workers found that different strains of the same species did not label, even though they were able to demonstrate the presence of acid mucopolysaccharides by other methods. Similar problems have been encountered when Alcian blue was used on mammalian tissues (Behnke and Zelander, 1970). In both cases, specimens were fixed in glutaraldehyde containing 1–3% Alcian blue, followed by osmium postfixation.

The difficulties in using Alcian blue have been overcome partly by using it in combination with other reagents. In a study of the cell coats of a number of amoeboid organisms, Dykstra and Aldrich (1978) found that by using 0.5% Alcian blue in the primary fixative and 0.05% RR in the secondary fixative, they could demonstrate a cell coat where none could be shown with either stain alone. They also were able to enhance the staining in cases where the results were marginal using a single reagent. Geyer (1971) and Geyer et al. (1971) were able to intensify the contrast imparted by the Alcian blue by impregnating the sections with silver sulfide. This was shown both for specimens that had been stained en bloc and for sections of glutaraldehyde-fixed material that were first stained with Alcian blue.

Alcian blue is a family of dyes and the one most commonly used is 8G or 8GX having a molecular weight of 1300. Stocks of these dyes can be found in many histology laboratories. The manufacturers have ceased production of the Alcian blues but one may substitute Astra blue 6GLL (Scott, 1980). For a discussion of the specificity and mode of action of these dyes, one should consult Scott (1972, 1980).

2.4. Other Cations

Another reagent that is thought to interact with mucosubstances is colloidal thorium (Revel, 1964), which is applied to thin sections of methacrylate-embedded material. It reacts with both sulfate and carboxyl groups and, although specific, the technique results in a heavy, coarse deposition over the reactive sites, effectively obscuring the ultrastructure. Thorium has been applied similarly to bacteria by Woo et al. (1979). A number of reagents that may belong to this group have been used for ultrastructural localizations but their specificity and/or the nature of their reactivity have not been well established. These include phosphotungstic acid (Pease, 1970), lanthanum (Shea, 1971), cetylpyridinium chloride–ferric thiocyanate (Courtoy et al., 1974), and tannic acid–metal salts (Sannes et al., 1978).

Recently, a new cationic reagent has been synthesized that has certain advantages over others previously used (Ghinea and Simionescu, 1985). These workers have made both positively and negatively charged hemeundecapeptides as derivatives of microperoxidase M-11. The markers can be used on unfixed material, in vivo and in vitro, and because of their small size (ca. 2 nm) they have access to cellular sites that might not be available to other reagents because of steric hindrances. The bound reagent is visualized via the diaminobenzidine (DAB) reaction. (The DAB reaction is discussed in Chapter 15.)

3. PAS-BASED REACTIONS

The periodic acid–Schiff (PAS) technique has long been used for the detection of polysaccharides in both plant and animal light microscopic histology. The chemical specificity of this reaction is well documented (Pearse, 1968). The reaction is dependent on the presence of aldehyde groups that have been engendered by the oxidation of free hydroxyl and/or amino groups on adjacent carbons of the sugar. This oxidation is caused by the periodic acid. Generally, the reaction is visualized for electron microscopy by the reduction of silver ions to elemental silver at these sites. Since the reaction detects aldehydes and not sugars per se, one must be aware of any other aldehydes that might be present in the specimen, either those occurring naturally or those introduced during processing. One must carefully observe unoxidized controls to determine how much of the labeling is due to preexisting aldehydes. Possible sources of nonspecific label can include free aldehydes due to fixation with glutaraldehyde or acrolein (but not formaldehyde). Osmium may introduce reactive keto groups and is best avoided whenever possible. If osmium tetroxide is used, it should be removed by first treating the sections with H_2O_2. Ultraviolet irradiation can also introduce aldehydes that will result in nonspecificity of PAS-based reactions. Therefore, one must avoid embedding in a resin that is polymerized by UV such as Lowicryl K4M. This does not rule out the use of other acrylic resins such as LR White. Nonspecific deposition of silver can result from chelation, but this is easily overcome by treating the sections with sodium thiosulfate, which would leave only reduced silver. Another problem that may be encountered is that catecholamines, free sulfhydryl groups, and disulfide bonds can reduce silver ions.

If one determines, from the unoxidized control that has been treated with the aldehyde-sensitive reagent, that nonspecific silver deposition will interfere with accurate interpretation of the results, several control treatments can be applied to the sections prior to the periodic acid oxidation step. One can mask or block preexisting aldehydes by treatment with sodium borohydride (Craig, 1974), dimedone, iodoacetate, or bisulfite (Pickett-Heaps, 1967). To eliminate reaction with sulfhydryl groups and disulfide bonds, one can apply reducing and alkylating agents prior to oxidation (Lewis and Knight, 1977).

3.1. Silver Methenamine

Silver methanamine was first introduced as an aldehyde-detecting reagent by Gomori (1946, 1952). It was later used to detect periodate-induced aldehyde groups for the electron microscopic localization of carbohydrates (Rambourg and Leblond, 1967) and it is this basic technique that is used most often. (Methenamine has also been referred to as hexamine, hexamethylenetetramine.) Thin sections are cut and mounted on inert grids (or, alternatively, free-floating sections may be used). The sections are oxidized with 1% periodic acid for 20 min at room temperature. (Longer or shorter time may be appropriate depending on the nature of the specimen and the embedding resin used.) After two short washes in distilled water, the sections are exposed to the silver methanamine solution in the dark at 60°C for 1 hr or until the sections start to turn brown. The reaction should take place in a tightly sealed moist chamber. The silver methenamine solution should be prepared freshly by adding 2 ml of 5% silver nitrate dropwise to 18 ml of 3% methenamine. To this is added 2 ml of 2% sodium borate. The solution should have a pH of 8–9. This should be verified by pH paper and not with a pH meter. Adjustment can be made with 4 N NaOH. It is also advisable to centrifuge the reagent prior to use to remove any undissolved material. The sections are then washed in distilled water and floated on a 5% solution of sodium thiosulfate for 5 min and washed in water again. The sections may be viewed without further staining unless additional contrasting is desired.

Silver deposition usually corresponds to the areas found to be PAS positive at the light microscope level. Glycogen does not stain. This "fault" in the procedure can be an advantage where large amounts of stained glycogen might obscure the positive label of other cell constituents. Starch grains in plants are generally eroded away, leaving holes in the sections. A low level of nonspecific label may occur over the cytoplasm. Hernandez et al. (1968) and Rambourg et al. (1969) inserted two additional steps in the procedure to help overcome this. Prior to the incubation with silver methenamine, the sections were treated for 5 min with 10% chromic acid followed by 1 min with 1% sodium bisufite. In addition to increasing specificity, these extra steps result in the deposition of silver over glycogen. The silver methenamine technique is most useful for medium- to low-resolution microscopy due to the fact that the silver grains are about 10 nm in diameter and tend to aggregate.

3.2. Silver Proteinate

The use of the osmiophilic reagents, thiocarbohydrazide (TCH) and thiosemicarbazide (TSC), for the detection of aldehydes was developed by Hanker et *al.* (1964) and Seligman *et al.* (1965). Periodate-induced aldehydes condense with the hydrazine groups, which in turn reduce osmium tetroxide, yielding an electron-dense product at the site of the reaction. This technique was modified by Thiéry (1967), who substituted silver proteinate for osmium tetroxide, resulting in the deposition of metallic silver at the site of the reaction. The reaction product is of much finer grain than that from the methenamine technique and there is less background deposition, in addition, both glycogen and starch are detected. Using a variety of plant and animal specimens, Courtoy and Simar (1974) compared the methenamine technique and the silver proteinate technique and found the latter to be more specific for carbohydrates. They also emphasized the need for proper controls in both methods. Although osmium postfixation has been used in many of the applications of this technique (e.g., Courtoy and Simar, 1974), Thomopoulos *et al.* (1983b) found it to be detrimental especially when the material was embedded in epoxy resins. Others also have warned against the use of osmium when carbohydrate cytochemistry is to be investigated (Rambourg, 1971; Lewis and Knight, 1977; Simson, 1977).

The silver proteinate technique is similar to that for silver methenamine. Again, free sections or those mounted on inert grids can be used. After periodic acid oxidation and washing, the sections are floated either on 1.0% TSC in 10% acetic acid or on 0.2% TCH in 20% acetic acid at room temperature for 30–45 min. (These solutions are stable for several days when stored at 4°C.) This is followed by two rapid washes in 10% acetic acid, two 15-min washes in 10% acetic, 5 min in 5% acetic, 5 min in 1% acetic, and two 5-min water washes. The sections then are reacted with 1% silver proteinate for 30 min at room temperature in the dark. (The solution is prepared by sprinkling silver proteinate over water in a shallow container and allowing it to stand undisturbed for 20 min; then stir gently. It may be stored at 4°C for a week or more in a lightproof container.) Finally, the sections are washed several times in distilled water before viewing. A comparison of the silver methenamine and silver proteinate techniques is given in Figure 2.

4. LECTINS

More specific information about the nature of complex carbohydrates may be obtained by using lectins as probes. Lectins are a group of carbohydrate-binding proteins (usually glycoproteins) that occur naturally in all groups of organisms that have been examined. They are either divalent or multivalent, which makes them agglutinins. They are distinguished further by the fact that they are of nonimmune origin and are nonenzymatic in function. Little is known about the endogenous role of lectins, but because of their carbohydrate-binding specificity, they have found wide application in the study of the structure and function of complex carbohydrates (Roth, 1978). When complexed to an electron-dense marker,

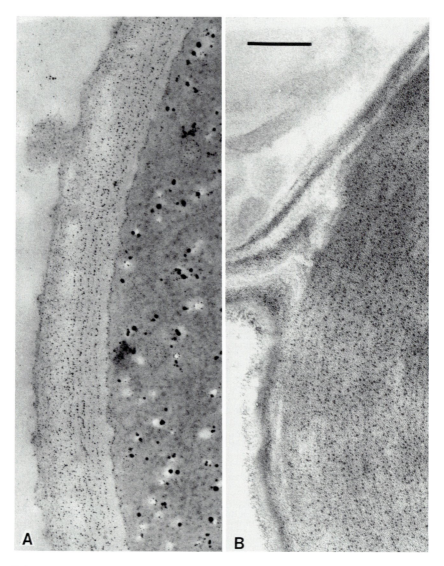

FIGURE 2. (A) Application of the silver methenamine technique for the localization of wall poly-saccharides in an encysted myxamoeba of the plasmodial slime mold *Didymium iridis*. Note the tendency of the silver grains to form aggregates leading to an uneven deposition. Micrograph by Thomas Raub. (B) Demonstration of cell wall polysaccharides using the silver proteinate technique on the fungus *Endogone pisiformis*. Silver deposition using this technique is more particulate with less tendency to aggregate. Micrograph by Jack Gibson. Bar = 0.25 μm for both panels.

lectins provide the means to localize specific sugar residues at the ultrastructural level. In some cases one can even determine whether the sugar is terminal or internal by choosing the appropriate lectin. Currently, a large number of lectins with different sugar-binding characteristics have been purified from both animal and plant sources and are available from several commercial vendors. Also available are lectins attached to a variety of electron-opaque markers, putting this technique easily within the reach of most workers.

Concanavalin A (Con A) was the first lectin to be used as an ultrastructural cytochemical reagent for the labeling of the cell surface coat in mammalian cells (Avrameas, 1970; Bernhard and Avrameas, 1971; Nicolson and Singer, 1971) owing to its affinity for α-D-mannoside and α-D-glucoside. This affinity thus makes it a generic reagent for the identification of glycoproteins since it will bind to their mannose backbone. Early applications to microorganisms included the localization of yeast wall mannans using mercury (Horisberger *et al.*, 1971) or colloidal gold (Bauer *et al.*, 1972) as the electron-dense marker. Shortly thereafter, the Con A–gold complex was used to localize glycosylated teichoic acid in the wall of *Streptococcus faecalis* (Garland, 1973). A number of other lectins also are known to interact with microorganisms and have been or can be used for ultrastructural localization (Pistole, 1981). These include wheat germ agglutinin (WGA); the lectins from pea, soybean, peanut, potato, *Helix pomatia, Lotus tetragonolobus;* phytohemagglutinin, limulin, and *Ricinus communis* agglutinin (RCA) I and II.

Lectins can be used in a variety of ways for the localization of sugar residues. They can be applied prior to fixation, after aldehyde fixation, or after thin sectioning. One has the option of using a direct method, where the electron-dense marker is complexed to the lectin, or an indirect method, where the marker is complexed to a secondary reagent that has affinity for the lectin. These secondary reagents can include natural glycoproteins (Geoghegan and Ackerman, 1977), synthetic glycoproteins (Kataoka and Tavassoli, 1984) or polysaccharides (Horisberger and Rosset, 1977) with appropriate sugar-binding sites, glycosylated markers (Kieda *et al.*, 1977; Schrevel *et al.*, 1979), antilectin antibodies (Horisberger and Tacchini-Vonlanthen, 1983; Briggman and Widnell, 1983), or the use of biotinylated lectins in conjunction with avidin-bound markers (Bayer *et al.*, 1976). The technique is useful not only in transmission electron microscopy but also in scanning electron microscopy (Horisberger, 1979) where lectin-binding sites over large areas of cell surface can be studied, provided the appropriate particulate marker is used (Molday and Moher, 1980). Unlabeled Con A has also been used for the stabilization of bacterial surface structures prior to their processing for transmission electron microscopy (Birdsell *et al.*, 1975).

Several possibilities exist for the type of electron-dense marker that one can use (Horisberger, 1984a). One of these is an enzymatic marker, such as horseradish peroxidase (HRP), that can be detected via the osmiophilic reaction using DAB. HRP can be covalently coupled to a lectin (Avrameas, 1969) or, in the case of Con A, lentil lectin, and pea lectin, merely applied after the lectin since it will bind by virtue of its mannosyl residues (Avrameas, 1970). WGA may be similarly visualized using glucose oxidase without any coupling agent (François and Mongiat, 1977). It is also possible to glycosylate HRP with an appropriate sugar so that it will bind to any unoccupied sites on the lectin being used. However, the reaction products formed by these enzymes are dense and amorphous. There is also the problem of diffusion of the reaction product away from the original site of binding and the possibility that endogenous oxidases present in the cell could react. For these reasons, this technique is impossible to quantify. Enzymatic markers have the advantage that they are less likely to cause steric hindrances to binding, that they can penetrate to levels of the cell surface complex close to the plasma membrane, and that they are visualized easily at low magnification.

A number of particulate markers can be used to visualize lectin-binding sites. Ferritin, the first to be so used, was coupled to Con A (Nicolson and Singer, 1971) and has since been used with other lectins. Because of its size and distinctive appearance, it is a particularly attractive marker for high-resolution work but less practical for low-magnification studies (Figure 3). Several markers have been used in conjunction with Con A, and other lectins with similar binding characteristics, because they bear the appropriate sugar residues and have distinctive morphologies. These include hemocyanin (Smith and Revel, 1972), iron–dextran (Martin and Spicer, 1974), and iron–mannan (Roth and Franz, 1975). These and other saccharide affinity markers have the potential of displacing the lectin from the cell by competing for the sugar-binding sites. Furthermore, once the lectin has bound to its primary target, only a small number of binding sites may remain available to the marker (Horisberger, 1984b). This would lead to a misleadingly low level of labeling. In addition, appropriate controls must be used to ensure that an endogenous lectin is not binding the affinity marker, or the lectin probe itself, which may be a glycoprotein.

A very effective method is to use a lectin (or a secondary reagent) that has been adsorbed to colloidal gold. Gold colloids are easily prepared in the laboratory in a range of sizes from 3 to 150 nm in diameter and a variety of macromolecules can be adsorbed to them. (This topic is discussed in Chapter 17 and will not be repeated here.) For details of preparing lectin–gold conjugates, one should consult Horisberger and Rosset (1977) and Geoghegan and Ackerman (1977). Gold markers are an attractive alternative to other types of markers since they are electron-opaque, have discrete borders, and are uniform in size, making them easily recognized in the electron microscope. In addition, they may be used in either direct or indirect procedures and as a preembedding (Geoghegan and Ackerman, 1977) or a postembedding label (Roth, 1983). They are particularly useful in postembedding procedures since other markers can obscure the associated structures or are difficult to resolve from the background ultrastructure. When simultaneous labeling with more than one lectin (or a lectin in conjunction

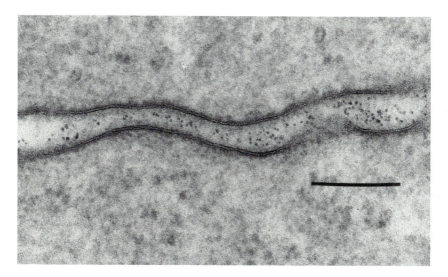

FIGURE 3. The use of ferritin-conjugated Con A to localize glucans and/or mannans at the cell surface of two adjacent cells of *Physarum polycephalum*. In this case the labeled lectin was applied prior to fixation and embedding. The section was poststained with $BaMnO_4$. Bar = 0.2 μm. Micrograph by Henry Aldrich.

with an antibody) is desired, gold particles can be used together with other markers (Roth and Binder, 1978) or with gold particles of different sizes (Horisberger and Vonlanthen, 1979). Quantification of labeling is easier when using colloidal gold rather than other markers since the former are discrete particles distinct from cellular components. The particles can be counted and their distribution per unit of linear distance or per unit area calculated.

For preembedding labeling of cell surfaces, cells may be fixed minimally in aldehyde or not fixed depending on the needs of the study. After washing, the cells are exposed to the labeled lectin for 15–60 min followed by postfixation (or DAB reaction). Unfixed cells can be processed by a standard fixation sequence appropriate for good morphological visualization. The cells are then embedded by any standard protocol. Appropriate controls include (1) preincubation of the cells with unlabeled lectin before application of the labeled lectin, and (2) the inclusion of an excess of the inhibitory sugar hapten with the labeled lectin. For indirect procedures, the fixed cells are incubated with the lectin at a concentration of 200 μg/ml for 15–60 min, followed by application of the labeled secondary reagent for a similar length of time. The cells are then postfixed and embedded. Three controls are used for indirect labeling: (1) omission of the lectin from the schedule, (2) inclusion of the inhibitory sugar with the lectin, (3) application of unlabeled secondary reagent prior to the labeled secondary reagent. Thorough washing is needed between each step. Some workers take the added precaution of treating aldehyde-fixed cells with ammonium chloride, lysine, ethanolamine, or glycine to block any remaining reactive aldehydes before using the lectin, thereby reducing the possibility of nonspecific binding. In a comparative study on the effects of various fixatives on the binding of Con A to lymphocytes, Renau-Piqueras *et al.* (1981) found that 1% glutaraldehyde gave the lowest level of labeling and 0.7% OsO_4/0.8% formaldehyde gave the highest level of labeling of the fixatives surveyed. Unfortunately, they did not compare these to unfixed controls to determine if there was enhancement or inhibition of binding due to these treatments. Very few of these types of studies have been done, making it difficult to generalize about optimal fixation for lectin binding.

Both direct and indirect methods may be applied to thin sections for the localization of intracellular lectin receptors as well as cell surface receptors. This has been accomplished most often using colloidal gold as the electron-dense marker. Roth (1983) has done a comparative study from which he concluded that the highest levels of specific labeling were found using an indirect procedure on tissue that had been fixed in 1% glutaraldehyde only and embedded at low temperature in Lowicryl K4M. He also observed that postfixation in osmium and/or embedding in Epon reduced the labeling by about 90%. To the contrary, Horisberger *et al.* (1978) presented quite good results using material embedded in Spurr's (Spurr, 1969) epoxy mixture; they localized the yeast wall galactomannan using RCA I and chitin using WGA. In another study, chitin was localized by a direct method using WGA on yeast cells that had been postfixed in osmium and embedded in Epon (Roberts *et al.*, 1983) (Figure 4). Since systematic comparisons using different kinds of specimens have not been done, it is difficult to recommend a given procedure.

The basic procedure for the labeling of thin sections with lectins is as follows. Sections mounted on inert grids are treated briefly with 1% bovine serum albumin in order to occupy any nonspecific protein-binding sites on the sections. The grids are blotted, but not to dryness, and floated on either gold-labeled lectin or unlabeled lectin for 1 hr. The grids are washed, and observed with or without poststaining in the case of the direct method, or reacted with the labeled secondary reagent in the case of the indirect method, followed by washing and poststaining. During the procedure it is important that the grids are never allowed to become completely dry. Controls similar to those used for whole cell labeling also need to be applied to this procedure. Etching of the sections with a strong oxidizing agent before the lectin treatment is to be avoided.

A recent technique that may find utility beyond the original subject involves the use of both a cationic reagent and a lectin for the localization of soluble glycoprotein allergens associated with pollen grains (Grote and Fromme, 1984). Fixation was accomplished by including in the aldehyde fixative 0.5% cetylpyridinium chloride, which apparently precipitated the carbohydrates that in turn could be labeled using Con A. Without this addition to the primary fixative, the

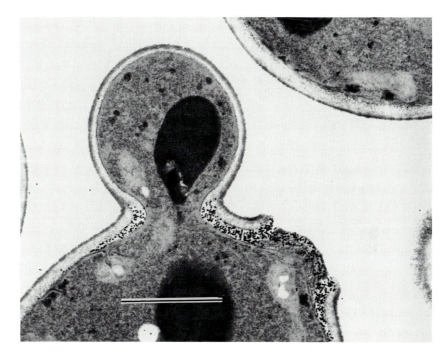

FIGURE 4. Localization of chitin in the cell wall of *Saccharomyces cerevisiae* using wheat germ agglutinin coupled to 17-nm gold particles. The chitin is restricted to the area of bud formation. Bar = 1.0 μm. Micrograph by Blair Bowers.

outer layer was lost in the aqueous fixative. The possibility that similar material is being lost from the surfaces of other organisms should be considered.

5. GLYCOSIDIC ENZYMES

Extraction of substrates with glycosidic enzymes was introduced by Monneron and Bernhard (1966) as a technique for the localization of polysaccharides in Epon-embedded tissue. The technique has been used in two ways. First, one can examine the section for evidence of the loss of material. Alternatively, one can observe whether or not a carbohydrate-specific staining reaction has been reduced or possibly enhanced as a direct result of treating the specimen with an enzyme. The technique has been used with various fixation and embedding regimes and there seems to be no best choice. Enzymatic extraction also can be used prior to embedding if adequate access to the substrate is provided. These techniques have been applied almost exclusively to higher organisms (Lewis and Knight, 1977; Spicer and Schulte, 1982).

A more recent and promising technique takes advantage of the binding

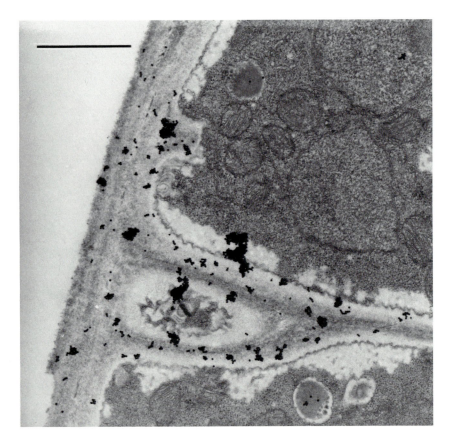

FIGURE 5. Chitin localization in a thin section of the fungus *Wangiella dermatiditis* using chitinase coupled to colloidal gold. The labeling is specific for the cell wall. This micrograph also illustrates one of the problems that may be encountered with gold-labeled probes, i.e., aggregation of the probe. Bar = 1.0 μm. Micrograph by James Harris.

affinity of an enzyme for its substrate. The technique was developed by Bendayan (1981) for the localization of nucleic acids in thin sections. The procedure uses enzymes, bound to colloidal gold as the electron-dense marker. Subsequently, other enzymes besides nucleases have been used successfully for the localization of specific substrates. These include chitin (Harris and Szaniszlo, 1983) in the fungus *Wangiella* (Figure 5), xylans in the cell wall of wood (Vian *et al.*, 1983), elastin, glycogen, and collagen in mammalian tissue (Bendayan, 1984), and chitin in the plant parasitic fungus *Fusarium oxysporum* in tomato tissue (Chamberland *et al.*, 1985). Again, various fixations and embedding media have been used and it appears that the best method must be determined empirically for each substrate. The accuracy of the method depends mostly on the purity and specificity of the enzyme used. Even minor contaminants could lead to erroneous conclusions. This approach to the cytochemistry of complex carbohydrates holds the promise of supplying very specific information about the nature and location of a substance in question.

6. CARBOHYDRATE-SPECIFIC ANTIBODIES

Immunoelectron microscopy has become a valuable tool in the study of microorganisms and is considered in some detail in Chapter 17. Here I will briefly discuss the use of antibodies raised against complex carbohydrates for the preservation and localization of the antigens. Antibodies directed against purified capsular polysaccharides of bacteria have been used to stabilize the capsule so that it could withstand the rigors of preparation for electron microscopy (Bayer and Thurow, 1977). Capsules that were not so stabilized often either collapsed or were lost entirely during processing. Once stabilized, the capsule is seen in what is presumed to be its natural configuration. The capsule is available then for further probing with carbohydrate-specific reagents such as RR (Whitfield *et al.*, 1984).

As early as 1973 (Linssen *et al.*, 1973), labeled antibodies against surface polysaccharides were used to localize capsular material both by pre- and postembedding techniques. The method has been used with eukaryotic microbes as well as for the identification of cell wall components in yeast (e.g., Horisberger and Rosset, 1977). With the advent of monoclonal antibodies, even more specific

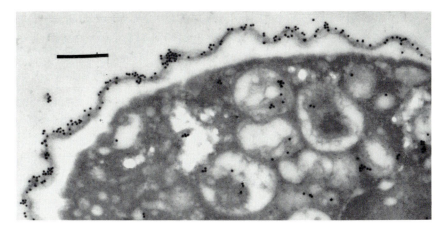

FIGURE 6. The use of carbohydrate-specific monoclonal antibodies on thin sections using an indirect method. The primary antibody is specific for the spore coat glycoprotein of *Dictyostelium discoideum*. Gold-labeled goat anti-mouse antibody was used as the secondary label. Bar = 0.5 μm.

probes are available. Currently, we are using monoclonal antibodies in the study of developmentally regulated glycoproteins in *Dictyostelium* and have been able to identify several families of WGA-binding glycoproteins based on common carbohydrate epitopes (West and Erdos, 1986). These are easily localized to the plasma membrane, internal compartments, or the extracellular matrix on thin sections using an indirect, colloidal gold method (Figure 6). One can test the carbohydrate specificity of the antibody on thin sections. Prior treatment of the sections with 20mM NaIO$_4$ at pH6 will abolish antibody binding. This can be used in conjunction with other evidence to establish the carbohydrate nature of the epitope being investigated.

7. REFERENCES

Ainsworth, S. K., and Karnovsky, M. J., 1972, An ultrastructural staining method for enhancing the size and electron opacity of ferritin in thin sections, *J. Histochem. Cytochem.* **20:**225–229.

Avrameas, S., 1969, Coupling of enzymes to proteins with glutaraldehyde: Use of conjugates for the detection of antigens and antibodies, *Immunochemistry* **6:**43–52.

Avrameas, S., 1970, Emploi de la concanavaline-A pour l'isolement, la detection et le resume des glycoproteines a glucides extra ou endocellulaire, *C.R. Acad. Sci.* **270:**2205–2208.

Bauer, H., Horisberger, M., Bush, D. D., and Sigarlakie, E., 1972, Mannan as the major component of the bud scars of *Saccharomyces cerevisiae, Arch. Microbiol.* **85:**202–208.

Bayer, E. A., Wilchek, M., and Skutelsky, E., 1976, Affinity cytochemistry: The localization of lectin and antibody receptors on erythrocytes via the avidin–biotin complex, *FEBS Lett.* **68:**240–244.

Bayer, M. E., and Thurow, H., 1977, Polysaccharide capsule of *Escherichia coli:* Microscope study of its size, structure and its sites of synthesis, *J. Bacteriol.* **130:**911–936.

Behnke, O., and Zelander, T., 1970, Preservation of intercellular substances by the cationic dye Alcian blue in preparative procedures for electron microscopy, *J. Ultrastruct. Res.* **31:**424–438.

Bendayan, M., 1981, Ultrastructural localization of nucleic acids by the use of enzyme–gold complexes, *J. Histochem. Cytochem.* **29:**531–541.

Bendayan, M., 1984, Enzyme–gold electron microscopic cytochemistry: A new affinity approach for the ultrastructural localization of macromolecules, *J. Electron Microsc. Tech.* **1:**349–372.

Bennett, H. S., 1963, Morphological aspects of extracellular polysaccharides, *J. Histochem. Cytochem.* **11:**14–23.

Bernhard, W., and Avrameas, S., 1971, Ultrastructural visualization of cellular carbohydrate components by means of concanavalin A, *Exp. Cell Res.* **64:**232–236.

Birdsell, D. C., Doyle, R. J., and Morgenstern, M., 1975, Organization of teichoic acid in the cell wall of *Bacillus subtilis, J. Bacteriol.* **121:**726–734.

Briggman, J. V., and Widnell, C. C., 1983, A comparison of direct and indirect techniques using ferritin-conjugated ligands for the localization of concanavalin A binding sites on isolated hepatocyte plasma membranes, *J. Histochem. Cytochem.* **31:**579–590.

Chamberland, H., Charest, P. M., Ouellette, G. B., and Pauze, F. J., 1985, Chitinase gold complex to localize chitin ultrastructurally in tomato root cells infected by *Fusarium oxysporum* f. sp. *radicis-lycopersici* compared with chitin-specific gold-conjugated lectin, *Histochem. J.* **17:**313–322.

Courtoy, R., and Simar, L. J., 1974, Importance of controls for the demonstration of carbohydrates in electron microscopy with the silver methenamine or the thiocarbohydrazide–silver proteinate methods, *J. Microsc. (Oxford)* **100:**199–211.

Courtoy, R., Boniver, J., and Simar, L. J., 1974, A cetylpyridinium chloride (CPC) and ferric thiocyanate (FeTh) method for polyanion demonstration on thin sections for electron microscopy, *Histochemistry* **42:**133–139.

Craig, A. S., 1974, Sodium borohydride as an aldehyde blocking reagent for electron microscope histochemistry, *Histochemistry* **42:**141–144.

Danon, D., Goldstein, L., Marikovsy, Y., and Skutelsky, E., 1972, Use of cationized ferritin as a label of negative charges on cell surfaces, *J. Ultrastruct. Res.* **38:**500–510.

Dykstra, M. J., and Aldrich, H. C., 1978, Successful demonstration of an elusive cell coat in amebae, *J. Protozool.* **25:**38–41.

François, D., and Mongiat, F., 1977, An ultrastructural study of wheat germ agglutinin binding sites using glucose oxidase as a marker, *J. Ultrastruct. Res.* **59:**119–125.

Garland, J. M., 1973, Preparation and performance of gold-labelled concanavalin A for the location of specifically reactive sites in walls of *S. faecalis* 8191, in: *Electron Microscopy and Cytochemistry* (E. Wisse, W. T. Daems, I. Molenaar, and P. van Duijn, eds.), pp. 303–307, North-Holland, Amsterdam.

Geoghegan, W. D., and Ackerman, G. A., 1977, Adsorption of horseradish peroxidase, ovomucoid and anti-immunoglobulin to colloidal gold for the indirect detection of concanavalin A, wheat germ agglutinin and goat anti-human immunoglobulin G on cell surfaces at the electron microscope level: A new method, theory and application, *J. Histochem. Cytochem.* **25:**1187–1200.

Geyer, G., 1971, New histochemical techniques for the demonstration of carboxyl groups in mucosubstances, *Histochem. J.* **3:**241–250.

Geyer, G., Helmke, U., and Christner, A., 1971, Ultrahistochemical demonstration of Alcian blue stained mucosubstances by the sulfide–silver reaction, *Acta Histochem.* **46:**244–249.

Ghinea, N., and Simionescu, N., 1985, Anionized and cationized hemeundecapeptides as probes for cell surface charge and permeability studies: Differentiated labeling of endothelial plasmalemmal vesicles, *J. Cell Biol.* **100:**606–612.

Gomori, G., 1946, A new histochemical test for glycogen and mucin, *Am. J. Clin. Pathol. Tech. Sect.* **10:**177–179.

Gomori, G., 1952, *Microscopic Histochemistry: Principles and Practice,* p. 57, University of Chicago Press, Chicago.

Grote, M., and Fromme, H. G., 1984, Electron microscopic localization of concanavalin A receptor sites in pollen surface material after fixation with glutaraldehyde–cetylpyridinium chloride, *J. Histochem. Cytochem.* **32:**869–871.

Gustafson, G. T., and Pihl, E., 1967, Staining of mast cell acid glycosaminoglycans in ultrathin section with ruthenium red, *Nature (London)* **216:**697–698.

Hanker, J. S., Seaman, A. R., Weiss, L. P., Ueno, H., Bergmann, R. A., and Seligman, A. M., 1964, Osmiophilic reagents: New cytochemical principals for light and electron microscopy, *Science* **146:**1039–1043.

Hardin, J. H., and Spicer, S. S., 1971, Ultrastructural localization of dialyzed iron-reactive mucosubstance in rabbit heterophils, basophils and eosinophils, *J. Cell Biol.* **48:**368–386.

Harris, J. L., and Szaniszlo, P. J., 1983, Electron microscopic demonstration of chitin in cell walls of *Wangiella dermatiditis* by chitinase–colloidal gold, *Abstr. Annu. Meet. Am. Soc. Microbiol.* **1983:**144.

Hayat, M. A. (ed.), 1981, *Principles and Techniques of Electron Microscopy: Biological Applications,* Vol. 1, 2nd ed., University Park Press, Baltimore.

Hernandez, W., Rambourg, A., and Leblond, C. P., 1968, Periodic acid–chromic acid–methenamine silver technique for glycoprotein detection in the electron microscope, *J. Histochem. Cytochem.* **16:**507.

Horisberger, M., 1979, Evaluation of colloidal gold as a cytochemical marker for transmission and scanning electron microscopy, *Biol. Cell.* **36:**253–258.

Horisberger, M., 1984a, Electron-opaque markers: A review, in: *Immunolabeling for Electron Microscopy* (J. M. Polak and I. M. Varndell, eds.), pp. 17–26, Elsevier, Amsterdam.

Horisberger, M., 1984b, Lectin cytochemistry, in: *Immunolabeling for Electron Microscopy* (J. M. Polak and I. M. Varndell, eds.), pp. 249–258, Elsevier, Amsterdam.

Horisberger, M., and Rosset, J., 1977, Colloidal gold, a useful marker for transmission and scanning electron microscopy, *J. Histochem. Cytochem.* **25**:295–305.

Horisberger, M., and Tacchini-Vonlanthen, M., 1983, Stability and steric hindrance of lectin-labelled gold markers in transmission and scanning electron microscopy, in: *Lectins,* Vol. 3 (T. C. Bog-Hansen and G. A. Spengler, eds.), pp. 189–197, de Gruyter, Berlin.

Horisberger, M., and Vonlanthen, M., 1979, Multiple marking of cell surface receptors by gold granules: Simultaneous localization of three lectin receptors on human erythrocytes, *J. Microsc. (Oxford)* **115**:97–102.

Horisberger, M., Bauer, H., and Bush, D. A., 1971, Mercury labelled concanavalin A as a marker in electron microscopy—Localization of mannan in yeast cell walls, *FEBS Lett.* **18**:311–314.

Horisberger, M., Vonlanthen, M., and Rosset, J., 1978, Localization of galactomannan and wheat germ agglutinin receptors in *Schizosaccharomyces pombe, Arch. Microbiol.* **119**:107–111.

Jones, R., and Reid, L., 1973, The effect of pH on Alcian blue staining of epithelial acid glycoproteins. I. Sialomucins and sulfomucins (singly or in simple combinations), *Histochem. J.* **5**:9–18.

Kataoka, M., and Tavassoli, M., 1984, Synthetic neoglycoproteins: A class of reagents for the detection of sugar recognizing substances, *J. Histochem. Cytochem.* **32**:1091–1098.

Kieda, C., Delmotte, F., and Monsigny, M., 1977, Preparation and properties of glycosylated cytochemical markers, *FEBS Lett.* **76**:257–261.

Lewis, P. R., and Knight, D. P., 1977, Staining methods for sectioned material, in: *Practical Methods for Electron Microscopy* (A. M. Glauert, ed.), pp. 77–136, Elsevier/North-Holland, Amsterdam.

Linssen, W. H., Huis in 't Veld, J. H. J., Poort, C., Slot, J. W., and Geuze, J. J., 1973, Immuno-electron microscope study of two types of streptococcal carbohydrate antigens: A comparison of two different incubation techniques, in: *Electron Microscopy and Cytochemistry* (E. Wisse, W. T. Daems, I. Molenaar, and P. van Duijn, eds.), pp. 193–196, North-Holland, Amsterdam.

Luft, J. H., 1964, Electron microscopy of cell extraneous coats as revealed by ruthenium red staining, *J. Cell Biol.* **23**:54a.

Luft, J. H., 1971a, Ruthenium red and violet. I. Chemistry, purification, methods of use for electron microscopy and mechanism of action, *Anat. Rec.* **171**:347–368.

Luft, J. H., 1971b, Ruthenium red and violet. II. Fine structural localization in animal tissues, *Anat. Rec.* **171**:369–416.

Martin, B. J., and Spicer, S. S., 1974, Concanavalin A–iron dextran technique for staining cell surface mucosubstances, *J. Histochem. Cytochem.* **22**:206–209.

Molday, R., and Moher, P., 1980. A review of cell surface markers and labelling techniques for scanning electron microscopy, *Histochem. J.* **12**:273–315.

Monneron, A., and Bernhard, W., 1966, Action de certaines enzymes sur des tissus inclus en Epon, *J. Microsc. (Paris)* **5**:697–714.

Nachmias, V. T., and Marshall, J. M., Jr., 1961, Protein uptake by pinocytosis in amoebae: Studies on ferritin and methylated ferritin, in: *Biological Structure and Function,* Vol. 2 (T. Goodwin and O. Lindberg, eds.), pp. 605–619, Academic Press, New York.

Nicolson, G. L., and Singer, S. J., 1971, Ferritin-conjugated plant agglutinins as specific saccharide stains for electron microscopy, *Proc. Natl. Acad. Sci. USA* **68**:942–945.

Pate, J. L., and Ordal, E. J., 1967, The fine structure of *Chondrococcus columnaris.* III. The surface layers of *Chondrococcus columnaris, J. Cell Biol.* **35**:37–50.

Pearse, A. G. E., 1968, *Histochemistry, Theoretical and Applied,* Vol. 1, 3rd ed., Churchill, London.

Pease, D. C., 1970, Phosphotungstic acid as a specific electron stain for complex carbohydrates, *J. Histochem. Cytochem.* **18**:455–458.

Pickett-Heaps, J. D., 1967, Preliminary attempts at ultrastructural polysaccharide localization in root tip cells, *J. Histochem. Cytochem.* **15**:442–455.

Pistole, T. G., 1981, Interaction of bacteria and fungi with lectins and lectin-like substances, *Annu. Rev. Microbiol.* **35**:85–112.

Progulske, A., and Holt, S. C., 1980, Transmission–scanning electron microscopic observations of selected *Eikenella corodens* strains, *J. Bacteriol.* **143**:1003–1018.

Rambourg, A., 1971, Morphological and histochemical aspects of glycoproteins at the surface of animal cells, *Int. Rev. Cytol.* **31**:57–114.

Rambourg, A., and Leblond, C. P., 1967, Electron microscopic observations on the carbohydrate-rich cell coat present at the surface of cells in the rat, *J. Cell Biol.* **32**:27–53.

Rambourg, A., Hernandez, W., and Leblond, C. P., 1969, Detection of complex carbohydrates in the Golgi apparatus of rat cells, *J. Cell Biol.* **40**:395–414.

Renau-Piqueras, J., Knecht, E., and Hernandez-Yago, J., 1981, Effects of different fixative solutions on labeling of concanavalin-A receptor sites in human T-lymphocytes, *Histochemistry* **71**:559–565.

Revel, J.-P., 1964, A stain for ultrastructural localization of acid mucopolysaccharides, *J. Microsc. (Paris)* **3**:535–544.

Rinehart, J. F., and Abul-Haj, S. K., 1951, An improved method for histochemical demonstration of acid mucopolysaccharides in tissues, *Arch. Pathol.* **52**:189–194.

Roberts, R. L., Bowers, B., Slater, M. L., and Cabib, E., 1983, Chitin synthesis and localization in cell division cycle mutants of *Saccaromyces cerevisiae, Mol. Cell. Biol.* **3**:922–930.

Roth, J., 1978, The lectins: Molecular probes in cell biology and membrane research, *Exp. Pathol.* Suppl. **3**:5–186.

Roth, J., 1983, Application of lectin–gold complexes for electron microscopic localization of glycoconjugates on thin sections, *J. Histochem. Cytochem.* **31**:987–999.

Roth, J., and Binder, M., 1978, Colloidal gold, ferritin and peroxidase as markers for electron microscopic double labelling lectin techniques, *J. Histochem. Cytochem.* **26**:163–169.

Roth, J., and Franz, H., 1975, Ultrastructural detection of lectin receptors by cytochemical affinity reaction using mannan–iron complex, *Histochemistry* **41**:365–368.

Sannes, P. L., Katsuyama, T., and Spicer, S. S., 1978, Tannic acid–metal salts sequences for light and electron microscopic localization of complex carbohydrates, *J. Histochem. Cytochem.* **26**:55–61.

Schrevel, J., Kieda, C., Caigneaux, E., Gros, D., Delmotte, F., and Monsigny, M., 1979, Visualization of cell surface carbohydrates by a general two-step lectin technique: Lectins and glycosylated cytochemical markers, *Biol. Cell.* **36**:259–266.

Scott, J. E., 1972, Histochemistry of Alcian blue. III. The molecular biological basis of staining by Alcian blue 8GX and analogous phthalocyanins, *Histochemie* **32**:191–212.

Scott, J. E., 1980, The molecular biology of histochemical staining by cationic phthalocyanin dyes: The design of replacements of Alcian blue, *J. Microsc. (Oxford)* **119**:373–381.

Seligman, A. M., Hanker, J. S., Wasserkrug, H., DiMochowski, H., and Katzoff, L., 1965, Histochemical demonstration of some oxidized macromolecules with thiocarbohydrazide (TCH) or thiosemicarbazide (TSC) and osmium tetroxide, *J. Histochem. Cytochem.* **13**:629–639.

Seno, S., Tsujii, T., Ono, T., and Ukita, S., 1983, Cationic cacodylate iron colloid for the detection of anionic sites on the cell surface and the histochemical stain of acid mucopolysaccharides, *Histochemistry* **78**:27–31.

Shea, S. M., 1971, Lanthanum staining of the surface coat of cells. Its enhancement by the use of fixatives containing Alcian blue or cetylpyridinium chloride, *J. Cell Biol.* **51**:611–620.

Simson, J. A. V., 1977, The influence of fixation on the carbohydrate cytochemistry of rat salivary gland secretory granules, *Histochem. J.* **9**:645–657.

Smith, S. B., and Revel, J.-P., 1972, Mapping of concanavalin A binding sites on the surface of several cell types, *Dev. Biol.* **27**:434–441.

Spicer, S., and Schulte, B., 1982, Ultrastructural methods for localizing complex carbohydrates, *Hum. Pathol.* **13**:343–354.

Spicer, S. S., Hardin, J. H., and Setser, M. E., 1978, Ultrastructural visualization of sulfated complex carbohydrates in blood and epithelial cell with the high iron diamine procedure, *Histochem. J.* **10**:435–452.

Spurr, A. R., 1969, A low viscosity resin embedding medium for electron microscopy, *J. Ultrastruct. Res.* **26**:31–43.

Thiéry, J. P., 1967, Mise en evidence des polysaccharides sur coup fine en microscopie electronique, *J. Microsc. (Paris)* **6**:987–1018.

Thiéry, J. P., and Ovtracht, L., 1979, Differential characterization of carboxyl and sulfate groups in thin sections for electron microscopy, *Biol. Cell.* **36**:281–288.

Thomopoulos, G. N., Schulte, B. A., and Spicer, S. S., 1983a, The influence of embedding medium and fixation on the post-embedment ultrastructural demonstration of complex carbohydrates. II. Dialyzed iron staining, *Histochemistry* **79**:417–431.

Thomopoulos, G. N., Schulte, B. A., and Spicer, S. S., 1983b, The influence of embedding media and fixation on the post-embedment ultrastructural demonstration of complex carbohydrates. I. Morphology and periodic acid–thiocarbohydrazide–silver proteinate staining of *vicinal* diols, *Histochem. J.,* **15**:763–784.

Vian, B., Brillouet, J.-M., and Satiat-Jeunemaitre, B., 1983, Ultrastructural visualization of xylans in cell walls of hardwood by means of xylanase–gold complex, *Biol. Cell.* **49**:179–182.

Vidic, B., 1973, Structure and cytochemistry of the acinar cell in the rat maxillary gland, *Am. J. Anat.* **137**:103–117.

West, C. M., and Erdos, G. W., 1986, Glycoconjugates from several serologically-defined families are secreted and accumulate in the matrix of *Dictyostelium discoideum, Differentiation* (submitted).

Wetzel, M. G., Wetzel, B. K., and Spicer, S. S., 1966, Ultrastructural localization of acid mucosubstances in the mouse colon with iron containing stains, *J. Cell Biol.* **30**:299–315.

Whitfield, C., Vimr, E. R., Costerton, J. W., and Troy, F. A., 1984, Protein synthesis is required for in vivo activation of polysialic acid capsule synthesis in *Escherichia coli* K1, *J. Bacteriol.* **159**:321–328.

Woo, D. D. L., Holt, S. C., and Leadbetter, E. R., 1979, Ultrastructure of *Bacteroides* species: *Bacteroides asaccharlyticus, Bacteroides fragilis, Bacteroides melaninogenicus* subspecies *melaninogenicus,* and *B. melaninogenicus* subspecies *intermedius, J. Infect. Dis.* **139**:534–546.

Chapter 15

Cytochemical Techniques for the Subcellular Localization of Enzymes in Microorganisms

Martha J. Powell

Department of Botany
Miami University
Oxford, Ohio 45056

1. INTRODUCTION

1.1. Strategies for Localization of Enzymes

Methods to localize intracellular sites of enzymes with electron microscopy grew directly from light microscopic histochemistry. Techniques, such as Gomori's (1952) for acid phosphatase activity, which had as end products heavy metal ions with sufficient mass to scatter electrons, were directly applicable for electron microscopy if noncoagulative fixatives were used. Introduction of catalytic osmiophilic polymer generation (Hanker *et al.*, 1972a) made substrates previously valuable only for light microscopy useful for electron microscopy. In the most widely used strategies for electron microscopic localization of enzymes in microorganisms, enzymes are not viewed directly, but reactions with end products impart electron density to the sites of enzyme activity. There are numerous strategies for the cytochemical localization of enzymes, and this chapter will emphasize three of these: (1) ion capture and precipitation of products; (2) ferricyanide reduction and product amplification; and (3) oxidative polymerization of diaminobenzidine. Thus, this discussion is not exhaustive of all techniques used for microorganisms but is illustrative of rationales used in attempting to identify sites of enzyme activities.

Other approaches to enzyme localization are beyond the scope of this chapter, but are important correlates. For example, some techniques visualize enzymes directly. Immunocytochemical and autoradiographic methods are based on the antigenic property of enzymes or the affinity of enzymes for radioactively labeled substrates (Murry *et al.*, 1984; Sentandreu *et al.*, 1984; Sternberger, 1973; Veser *et al.*, 1981; Wientjes *et al.*, 1980). Development of gold-labeled protein A complexes has improved resolution of reaction deposits and made possible automatic quantitation of reaction product density using computer-aided image analyzers (Beier and Fahimi, 1985; Litwin *et al.*, 1984; Yokota *et al.*, 1983). Mirroring cytochemical methods for enzyme localization, gold-labeled enzymes have been used to localize specific substrates such as chitin (Danscher and Rytter-Nörgaard, 1983; Horisberger and Vonlanthen, 1977).

1.2. General Considerations

The ability to localize an enzyme cytochemically with electron microscopy depends, in part, on maintaining enzyme activity and immobilization during tissue preparation. Similarly, the cytochemical reaction product must be relatively insoluble and stable during subsequent preparative stages. Van Duijn (1974) discussed the requirements and rationales of substrate transport into cells, kinetics of substrate turnover, diffusion of products and capturing agents, and necessity for concentrated accumulation of visible end products. For proper interpretation of any cytochemical procedure, several factors should be considered:

1. Does fixation alter the level of enzyme activity? For many enzymes, activity is preserved best with (para)formaldehyde and is inhibited or inactivated with glutaraldehyde (Hand, 1975). This is problematic because formaldehyde alone does not give the best tissue preservation. As a consequence, a mixture of formaldehyde and low concentrations of glutaraldehyde is frequently used (Gruber and Frederick, 1977). Long incubation in glutaraldehyde, on the other hand, does not alter activity of some enzymes, such as aryl sulfatase (Hanker *et al.*, 1975). With other enzymes such as catalase, prior fixation in glutaraldehyde enhances the density of reaction product (Roels *et al.*, 1975). Addition of substrate during fixation protects the active site of some enzymes (Van Duijn, 1974). To preserve enzyme activity, fixations and washes prior to incubation in the reaction medium should be at 4°C. The influence of fixation on enzyme activity can be determined easily with spectrophotometric assays (Trelease, 1975; Trelease *et al.*, 1974).

2. Is the precision of enzyme localization influenced by altering concentrations of chemicals or composition of the reaction medium? Just as with spectrophotometric assays of enzyme activity, enzyme cytochemistry generally follows standardized procedures. Although the "cookbook" approach does not take into consideration differences in optimal pH, temperature, or K_m of enzymes in different organisms, without enzyme purification and characterization alterations in reaction media could provide discordant results. For example, in the ferricyanide reduction method, decreasing the concentration of capturing ion

or increasing the concentration of electron acceptor or chelating agent increases diffusion of reaction product (Shnitka and Talibi, 1971). Gruber and Frederick (1977) note that copper cannot capture ferrocyanide before it diffuses from the site of enzyme activity if the oxidation of the substrate is too rapid. With the Gomori (1952) method for acid phosphatase, high concentrations of the capturing ion, lead, can inhibit enzyme activity (Hanker et al., 1972b).

The condition of the reaction medium is important. For example, in some techniques, different enzymes are visualized at different pHs although the composition of the reaction medium is the same (Section 4; Powell, 1977). In addition, it is important that the reactants do not become depleted during the incubation. Thus, at least a 10 : 1 ratio of reaction medium to tissue is recommended. For some techniques, the reaction medium should be replaced during the incubation (Powell, 1977; Shnitka and Talibi, 1971). Final pH of the reaction medium should be monitored to ensure the stability of pH during the reaction.

3. What happens if cells are impermeable to reactants? Typically, fixations in aldehydes make membranes more permeable to chemicals, but, because of the instability of some enzymes to aldehydes, it is necessary to incubate cells in the reaction medium before fixation (Beezley et al., 1976). With some systems this presents no problem as with peroxidase (Roels et al., 1975), but in others cells are gently sonicated to increase membrane permeability (Beezley et al., 1976). Dimethyl sulfoxide is also used to improve penetration of reactants into cells (Shnitka and Talibi, 1971). Simultaneous cellular penetration of reactants is particularly important in substrate capture and ferricyanide reduction strategies. For example, if a substrate enters the cell before the ion to capture the enzyme product, the product can diffuse from the site of enzyme activity before it is precipitated. Thus, nonspecific cytoplasmic staining or electron-dense deposits around organelles occur. Because of difficulty with penetration of reactants into cells, some cells contain no reaction product whereas other cells in the same fixation have reaction product (Gruber and Frederick, 1977). When penetration of reactants is a problem, one can expect to find more reaction product at the tissue surface than at its core.

4. How can product stability and immobilization be increased? Generally, prolonged incubation in reaction medium, in buffers, and in dehydration solutions increases solubility and diffusion of reaction product. Thus, minimal times for subsequent preparative stages and rapid dehydration, as with 2-methoxyethanol (Wick and Hepler, 1980), are recommended.

5. Are the electron-dense deposits necessarily enzyme specific? Because there can be naturally occurring electron-dense deposits or because solutions used in tissue preparation can be contaminated with heavy metal ions, the enzyme-associated nature of deposits should be verified. One of the most direct ways to evaluate the nature of deposits is the use of adequate controls. A detailed scheme for use of controls is explained in the diaminobenzidine reaction for catalase (Section 4). In general, correlation of the following controls can be used:

A. To inactivate enzymes, tissue is heated to 90°C for 5 min prior to incubation in the standard reaction medium. If deposits similar to those found in active tissue incubated in the same medium are not found in the heat-treated tissue, then enzymes are possibly involved in formation of deposits.

B. Exclusion of the substrate or capturing ion from the reaction medium should eliminate enzyme-associated deposits.

C. Addition of specific enzyme inhibitors to the reaction medium should prevent deposits if enzymes are responsible for their production. Because the substrate could reach enzymes before the inhibitor, it is necessary to incubate tissue in buffered solutions of inhibitors prior to incubation in the reaction medium.

Another way to evaluate the validity of cytochemical results is to use more than one strategy for enzyme localization. For example, acid phosphatase and glycolate oxidase can be visualized with both the ion capture technique (Section 2) and the ferricyanide reduction method (Section 3).

X-ray microanalysis is an important tool in determining the ionic nature of electron-dense deposits when enzymes are visualized by ion capture of products or by ferricyanide reduction. This is particularly significant because deposits may be composed of a mixture of ions (Van Steveninck, 1979).

6. What happens when results are inconsistent with standard cytochemical techniques? The thickness of sections makes considerable difference in density of reaction products. Sometimes thin sections (gray) do not contain enough reaction product to be conspicuous. If this is the case, then section thickness should be increased. It is important that when comparisons are made, sections are of comparable thicknesses. The organism or tissue used makes a difference in sites of cytochemical enzyme localization. As shown with the acid phosphatase technique (Section 2.1), an enzyme may be found in different organelles in different organisms. Differences in sites of enzyme localization may also result from differences in developmental stage because an enzyme may be active in one stage and not in another. The influence of developmental stage on subcellular localization of peroxidase is illustrated in Section 4. On the other hand, some techniques, such as the copper capture of glucose for cellulase (Bal, 1974), have had limited use because of difficulty in consistent localization of reaction products. In these situations, modification of techniques is needed to increase their utility.

2. ION CAPTURE OF ENZYME PRODUCT

2.1. Lead Capture of Phosphate

The ultrastructural localization of phosphatases is based on the strategy that heavy metal ions, such as lead, can capture and precipitate *in situ* phosphate ions cleaved from substrates by phosphatases. Although these are simple techniques to carry out, sources of artifacts limit their precision. Endogenous phosphates, diffusion of phosphate products prior to capture, and requirement for adequate nucleating phosphates are sources of artifactual localization of enzyme activities. Staining of nucleoplasm and chromatin, commonly cited artifacts, is prevented by increasing the lead concentration to 5.4 mM (Pfeifer *et al.*, 1973). The pitfall of this alteration is that high concentrations of lead are inhibitory to acid phosphatase activity. Because precipitates of phosphates in the Gomori reaction are large, reaction product sometimes covers up cellular detail. Essner (1973) and Goldfischer *et al.* (1964) outline the history of these techniques.

Despite these limitations, the lead capture method for phosphatases is readily interpreted because the deposits are conspicuous. Figure 1 demonstrates precise localization of acid phosphatase reaction product in cyst vesicles of the fungus *Phytophthora palmivora*. Reaction product (lead phosphate) appears as large clumps that do not completely fill the matrix of the vesicles. In the sporangium of this fungus, reaction product is so dense in dictyosomes that it fills cisternae and appears homogeneous (Figure 2). Using the same technique with another fungus, *Entophlyctis variabilis,* deposition of reaction product is more generally localized (Figure 4). Finely granular reaction product is in the lumen of all endomembranes and the nuclear envelope as well as in vacuoles (Figure 4). The differences in sites of localization of reaction product for acid phosphatase in these two fungi demonstrate that the organism used influences where enzymes are localized. With these two examples there is no detectable diffusion of reaction product because incubation times in reaction media were short and incubations in solution for processing the tissues were minimal. Figure 6 shows diffuse reaction product over all cell contents as a result of longer incubations. Figures 3 and 5 illustrate the necessity of controls for proper interpretation of results. When the substrate is omitted from the reaction medium, electron-dense deposits are still found in microvesicles around dictyosomes of sporangia of *P. palmivora* (Figure 3) and in vacuoles in zoospores of the fungus *Coelomomyces punctatus* (Figure 5). These deposits, therefore, are the results of intrinsic free phosphates.

2.1.1. Acid and Alkaline Phosphatases

Despite questions of reliability (Washitani and Sato, 1976), the lead capture method for acid (Barka and Anderson, 1962; Gomori, 1952) and alkaline (Hugon and Borgers, 1966) phosphatases are widely used for microorganisms (Aliaga and Ellzey, 1984; Armentrout *et al.*, 1976; Costerton and Marcks, 1977; Doonan and Jensen, 1980; DuBois *et al.*, 1984; Esteve, 1970; Gezelius, 1971; Hänssler *et al.*, 1975; Holt and Beveridge, 1982; Kazama, 1973; Maxwell *et al.*, 1978; Meyer *et al.*, 1976; Noguchi, 1976; Powell, 1977; Rosing, 1984; Rudzinska, 1972, 1974; Vorisek, 1977). Essner (1973) has provided an extensive review of variations in these techniques. The techniques that I have found most reliable among fungi are described below.

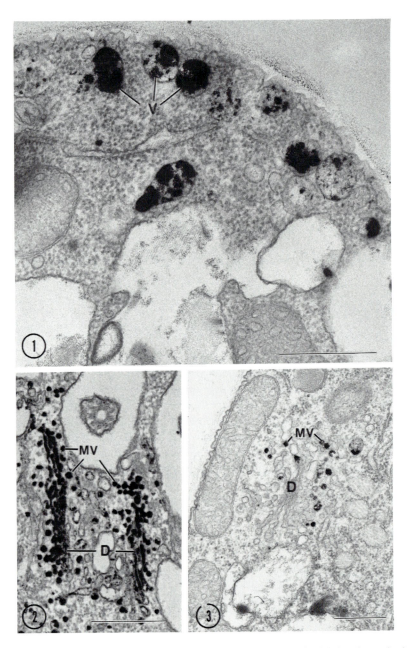

FIGURES 1–3. Lead capture of phosphate technique for localization of acid phosphatase in the fungus *Phytophthora palmivora* (M. J. Powell and C. E. Bracker). (1) Reaction product is precisely localized in cyst vesicles (V). (2) In the sporangium during zoospore formation, reaction product fills dictyosomes (D) and microvesicles (MV), such that the product appears homogeneous. (3) In the minus substrate control, reaction product is absent from dictyosomal cisternal membranes (D) but is still present in the lumen of microvesicles (MV). Thus, orthophosphate occurs naturally in the microvesicles, and the deposition of reaction product in this site is not necessarily due to acid phosphatase activity. Bars = 0.5 μm.

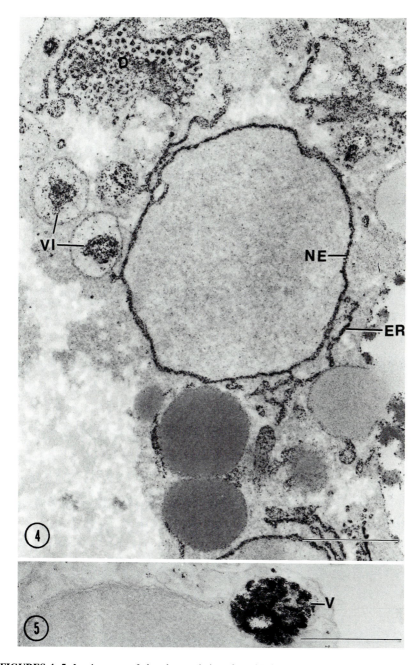

FIGURES 4, 5. Lead capture of phosphate technique for acid phosphatase localization in the fungi *Entophlyctis variabilis* and *Coelomomyces punctatus*. (4) In sporangium of *E. variabilis*, reaction product is widely distributed in endomembranes including the endoplasmic reticulum (ER), dictyosomes (D), and nuclear envelope (NE) as well as on the vacuolar inclusion (VI). (5) In minus substrate control, reaction product is in the vacuole (V), indicating free phosphate occurs in the matrix of the vacuole of zoospores of *C. punctatus*. Bars = 0.5 μm.

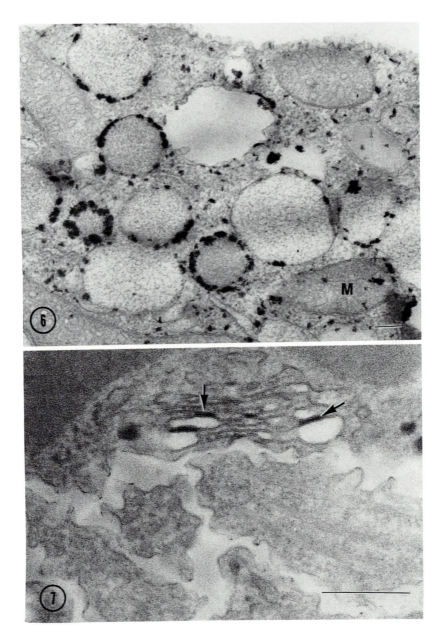

FIGURES 6, 7. Lead capture method for acid phosphatase in *P. palmivora* and thiamine pyrophosphatase in *E. variabilis*, respectively. (6) Random cytoplasmic and mitochondrial (M) distribution of reaction product, indicating diffusion of product during incubation in the medium for acid phosphatase. (7) Thiamine pyrophosphatase reaction product (arrows) on membranes of dictyosomes and forming vesicles during zoosporogenesis. Bars = 0.25 μm.

Tissues are fixed in a mixture of 2.0% paraformaldehyde–1.0% glutaraldehyde in 0.1 M sodium cacodylate buffer for 15–60 min. Fixation in 2.5% glutaraldehyde in 0.1 M sodium cacodylate buffer sometimes gives satisfactory results. Phosphate buffers should not be used in this technique, as in all other phosphate capture techniques, because they introduce an additional source of phosphate. Tissues are washed in 0.1 M sodium cacodylate buffer (pH 7.0) and then rinsed several times in reaction medium buffer.

For localization of acid phosphatase, the Barka and Anderson (1962) modification of Gomori's (1952) method gives the least random background deposits. The medium is made by adding the following reagents sequentially with constant stirring:

0.5 M Tris–maleate buffer, pH 5.0	10 ml
Double-distilled water	10 ml
1.25% Na β-glycerophosphate, pH 5.0	10 ml
0.2% lead nitrate (added dropwise with continuous stirring)	20 ml

Alkaline phosphatase (Hugon and Borgers, 1966) is localized with a similar medium except the pH is higher and the lead nitrate concentration is greater than that used for acid phosphatase. The medium (Essner, 1973) is made by adding the following reagents sequentially with constant stirring:

Na β-glycerophosphate	25 mg
Distilled water	6.7 ml
0.2 M Tris–maleate buffer, pH 9.0	2.0 ml
1% lead nitrate (added dropwise)	1.3 ml

The pH of each medium should be adjusted after final mixing and the solution then filtered and used immediately. Cells are incubated in one of these media for 15–60 min at 37°C. Possible enzyme inhibitors for acid phosphatase include sodium fluoride and for alkaline phosphatase, levamisole (Aliaga and Ellzey, 1984).

After incubation, tissues are washed in buffer, postfixed in buffered osmium tetroxide, and processed for electron microscopy.

2.1.2. Adenosine Triphosphatase

Adenosine triphosphatase (Wachstein and Meisel, 1957) is commonly found in the plasma membrane of protists (Meirelles and DeSouza, 1984; Price and Whitecross, 1983). Caution in functional interpretations of this reaction must be exercised. The actual nature of the enzyme visualized is not clear, but it does not appear to be transport ATPase. In the reaction medium, Mg^{2+} functions as an activator of the ATPase, and the enzyme is not inhibited by ouabain (Essner,

1973). In addition, transport ATPase is inhibited by the concentration of lead used in the Wachstein and Meisel (1957) medium. Thus, the cytochemically localized enzyme can best be considered Mg^{2+}-activated ATPase; other techniques are available for Na^+,K^+-ATPase (Meirelles and DeSouza, 1984). Several investigators (Hall *et al.*, 1980; Van Steveninck, 1979) have explored the validity of this cytochemical technique.

After fixation in buffered formaldehyde, tissues are incubated at 37°C for 20 min in the following medium (Essner, 1973) in which the reagents are added sequentially with constant stirring:

Sodium adenosine 5'-triphosphate	25 mg
Distilled water	22 ml
0.2 M Tris–maleate buffer, pH 7.2	20 ml
1.2% magnesium sulfate (anhydrous)	5 ml
2.0% lead nitrate (added dropwise)	3 ml

The medium is adjusted to pH 7.2, filtered, and used immediately.

2.1.3. Thiamine Pyrophosphatase

Thiamine pyrophosphatase (Novikoff and Goldfischer, 1961) is of interest because it serves as a marker enzyme for the Golgi apparatus. In some fungi it can be found in dictyosomes and associated membranes (Powell, 1979a) and in Golgi equivalents in other fungi lacking stacked cisternae characteristic of the typical Golgi apparatus configuration (Dargent *et al.*, 1982; Feeney and Triemer, 1979; Powell *et al.*, 1981). Reaction product can occur as patches in cisternae or can completely fill the cisternal lumen (Dargent *et al.*, 1982; Powell, 1979a). In some organisms, reaction product is localized only on membranes of dictyosomes and forming vesicles (Figure 7).

Cells are fixed at 4°C in a mixture of 1.0% glutaraldehyde–3% formaldehyde buffered with 0.05 M cacodylate buffer (pH 7.2) for 15–60 min. Cells are rinsed in fixative buffer and then washed repeatedly in reaction medium buffer, 0.04 M Tris–maleate (pH 7.2). The reaction medium, freshly prepared, consists of the following reagents added sequentially with constant stirring:

Thiamine pyrophosphate	25 mg
Distilled water	7 ml
0.2 M Tris–maleate buffer, pH 7.2	10 ml
25 mM manganese chloride	5 ml
1% lead nitrate (added dropwise)	3 ml

The medium settles for 10 min and is then filtered through Whatman No. 1 filter paper before use. Cells are incubated in the reaction medium for 30–60 min at

37°C. The medium should be replaced with fresh medium 30 min after incubation is initiated. After buffer washes, cells are postfixed in 1.0% osmium tetroxide in 0.05 M cacodylate buffer (pH 7.2) and prepared for electron microscopy.

2.2. Cerium Capture of Hydrogen Peroxide

Another example of the product capture approach is the cerous ion capture of hydrogen peroxide produced by oxidases. The technique is based on the deposition of cerium perhydroxide. For some enzymes such as glycolate oxidase, this technique results in less background precipitation than alternative techniques for the same enzyme (Thomas and Trelease, 1981).

2.2.1. NADH Oxidase

In mitochondria and vacuoles of yeast (Borgers et al., 1977), hydrogen peroxide is a product of NADH oxidase activity (Briggs et al., 1975). Cerous ions produce finely granular electron-dense precipitates in the presence of peroxides. Cells are fixed in 2.0% glutaraldehyde in 0.1 M cacodylate buffer (pH 7.3) and repeatedly washed in buffer. In the standard technique (Briggs et al., 1975), cells are not prefixed before incubation in the reaction medium. To inhibit cellular catalase activity, which would remove hydrogen peroxide produced by the oxidase, cells are preincubated for 10 min at 37°C in 0.1 M Tris–maleate (pH 7.5) containing the catalase inhibitor 1 mM 3-amino-1,2,4-triazole. Reagents added sequentially to the reaction medium are:

0.2 M Tris–maleate buffer, pH 7.5	5 ml
0.1 M 3-amino-1,2,4-triazole	1 ml
10 mM CeCl$_3$	1 ml
7.1 mM NADH	1 ml
Double-distilled water	1 ml

The medium is filtered through 0.45-μm filters before use. Cells are incubated in reaction medium for 20 min at 37°C and then are washed in pH 6.0 buffer to remove any cerium hydroxide precipitates formed incidentally during incubation. As a control, 0.01% catalase is added to the media to remove hydrogen peroxides and eliminate oxidase-induced precipitates (Briggs et al., 1975).

Cells are postfixed in osmium tetroxide and processed for electron microscopy.

2.2.2. Other Enzymes

In yeasts, Veenhuis et al. (1976) have used the cerium capture of hydrogen peroxide strategy to visualize methanol oxidase and other oxidases. Thomas and

Trelease (1981) have found that this technique for glycolate oxidase is superior to the ferricyanide reduction technique used for the same enzyme. Visualization of these oxidases requires prolonged incubations, but diffusion of reaction product is not a problem.

2.3. Copper Capture of Glucose: Cellulase

Heterotrophic microorganisms secrete a battery of enzymes important in substrate catabolism and decay, as well as in morphogenic responses. Cytochemical detection of organelles involved in transport of these extracellular enzymes and of sites of secretion is needed to elucidate the mechanisms of action. Bal (1974) reported that the important extracellular enzyme, cellulase, could be cytochemically visualized with a modification of the Benedict reaction. Cellulase is an enzyme complex (Chapman *et al.*, 1983), and the cytochemical reaction is based on activity of the $\beta1,4$-glucanase, the endoglucanase portion of the enzyme, with glucose as the end product. In Bal's (1974) technique, glucose reduces cupric salts to heavy metal precipitates of cuprous oxide.

Published micrographs representing this technique are scarce for microorganisms (Chapman *et al.*, 1983; Nolan and Bal, 1974) where reaction products consist of dense rod-shaped deposits. Artifactual staining is possible with this technique because endogenous glucose or other reducing sugars could be present in tissues. The coincidence of immunocytochemical localization of cellulase and cytochemically deposited cuprous oxide in a cellulase-secreting fungus, however, supports the power of this technique. Because of the importance of this method, modifications in the technique making it more widely applicable are desirable.

After fixation of tissues in Karnovsky's (1965) paraformaldehyde–glutaraldehyde in phosphate buffer (pH 7.2) at 0°C for 60 min, tissues are washed repeatedly in buffer and soaked overnight in the cold. Incubation in the substrate medium, 0.02% carboxymethylcellulose in 10 mM phosphate buffer (pH 6.0), is for 5–10 min at 25°C. The reaction medium is Benedict's solution (Eichhorn *et al.*, 1981), which is composed of two parts:

- Solution A = 17.3 g sodium citrate ($Na_3C_6H_5O_7\cdot11H_2O$); 9 g sodium carbonate (anhydrous) dissolved in 50 ml 60°C water; filtered and adjusted to 85 ml aqueous.
- Solution B = 1.73 g cupric sulfate ($CuSO_4\cdot5H_2O$) in 50 ml water.

Solution B is slowly added to solution A with constant stirring. Cells are incubated in the reaction medium at 70–100°C for 5–10 min, washed in water, postfixed for 60 min in 1% osmium tetroxide in phosphate buffer (pH 7.2), and processed for electron microscopy.

3. FERRICYANIDE REDUCTION

Because hydrolases and oxidases catalyze electron liberation from substrates, these enzymes can be cytochemically localized with modifications of Shnitka and Talibi's (1971) technique of ferricyanide reduction. In the reaction, potassium ferricyanide functions as the electron acceptor and phenazine methosulfate is the electron carrier. The enzymatic oxidation of the substrate is simultaneously coupled with the reduction of ferricyanide to ferrocyanide. Copper captures ferrocyanide and forms the electron-opaque precipitate, cupric ferrocyanide (also called Hatchett's brown). Potassium sodium tartrate is in the reaction medium as a chelating agent for copper, and dimethyl sulfoxide increases permeability of cells to the reactants. A problem with this approach is that cupric ferrocyanide is not highly contrasted in sections, and longer incubations, which would impart more contrast, increase diffusion of reaction product. Hanker *et al.* (1972a,b) intensified the contrast of the copper deposits by bridging them to osmium tetroxide with thiocarbohydrazide, resulting in deposits called osmium black. An alternate method of product amplification uses copper's ability to catalyze the polymerization of 3,3'-diaminobenzidine (DAB), which is subsequently osmicated (Hanker *et al.*, 1972a).

3.1. L-α-Hydroxy Acid Oxidases

The ferricyanide reduction technique was first used to visualize L-α-hydroxy acid oxidase activity (Hand, 1975; Shnitka and Talibi, 1971). These enzymes oxidize a variety of hydroxy acids to keto acids. As initially used, tissues were incubated in the reaction medium prior to fixation because of the enzymes' sensitivity to aldehydes. Incubation is for 30–40 min at 37°C in the following medium, which is made just before use, adding the reagents sequentially with constant stirring:

0.1 M Sorenson's phosphate buffer, pH 7.2	6.5 ml
40 mM potassium sodium tartrate	0.5 ml
30 mM copper sulfate	1.0 ml
Double-distilled water	1.0 ml
5 mM potassium ferricyanide	1.0 ml
Substrate (L-α-hydroxy acids)	37.0 mg
Phenazine methosulfate	5.0 mg
Dimethyl sulfoxide	1.0 ml

After incubation in the reaction medium, cells are washed in buffer, and fixed in a mixture of formaldehyde and glutaraldehyde for 10 min at 4°C.

Gruber and Frederick (1977) used a variation of this technique on algae to visualize the organellar site of the L-α-hydroxy acid oxidase, glycolate oxidase. Cells are fixed for a short period in a low concentration of glutaraldehyde, i.e., 1% glutaraldehyde buffered with 50 mM potassium phosphate (pH 7.2) for 20 min in the dark. Cells are then preincubated in a ferricyanide solution: 50 mM potassium phosphate (pH 7.2), 5 mM potassium sodium tartrate, 2 mM potassium ferricyanide followed by buffer washes. The reaction medium contains:

0.2 M potassium phosphate, pH 7.2	2.5 ml
25 mM potassium sodium tartrate	2.0 ml
25 mM copper sulfate	2.0 ml
50 mM potassium ferricyanide	0.4 ml
0.5 M sodium glycolate or 0.5 M L-lactate	0.4 ml
Distilled water	2.7 ml
Flavin mononucleotide	2.0 mg
Phenazine methosulfate	5.0 mg

Cells are incubated at 25°C for 30 min in the dark. After repeated buffer washes, tissues are postfixed at room temperature for 6 min in 2.0% osmium tetroxide and processed for electron microscopy.

3.2. Acid Phosphatase

In the method of Hanker et al. (1972b), acid phosphatase cleaves off a thiol group from the substrate 2-naphthylthiolphosphate (DDNTP), which in turn reduces ferricyanide to ferrocyanide. Copper ions precipitate ferrocyanide, and these deposits are amplified with thiocarbohydrazide or DAB, producing osmium black (Hanker et al., 1972a). In contrast to the lead capture method for acid phosphatase, this method does not have the disadvantages of lead inhibition of enzyme activity and the nonenzymatic deposition of lead on free cytoplasmic phosphates. In addition, the deposits are very finely granular, not obscuring cellular details, and there is little random cytoplasmic precipitation (Figure 8). In a fungus, products were precisely located in a cisterna involved in producing an autophagic vacuole around a mitochondrion (Figure 8).

Fixation in formaldehyde–glutaraldehyde mixtures is recommended for maintenance of enzyme activity and cell fine structure. If enzyme activity is lost, then glutaraldehyde should be omitted from the initial fixation. The incubation medium is prepared by adding the following reagents sequentially with constant stirring:

Didicyclohexylammonium 2-naphthylthiolphosphate (DDNTP)	5 mg
0.06 N sodium acetate, pH 5.6	7.9 ml
0.1 N acetic acid	0.25 ml
0.1 M sodium citrate	0.60 ml

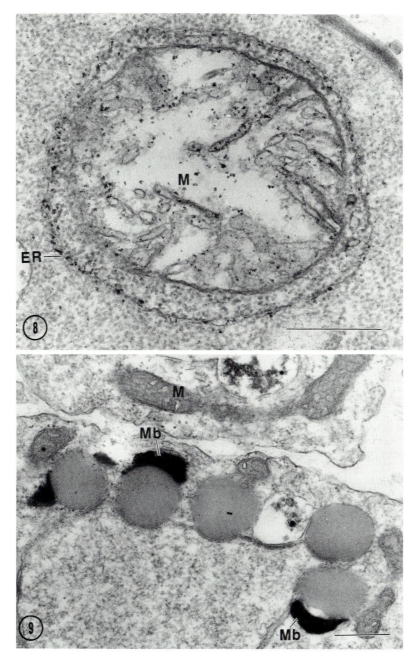

FIGURES 8, 9. The ferricyanide reduction and product amplification with DAB method for acid phosphatase in *Gilbertella persicaria,* and the DAB method for catalase in *C. punctatus,* respectively. (8) Fine-grained reaction product is localized precisely in the cisterna (ER) and mitochondrion (M) during formation of an autophagic vacuole (M. J. Powell, C. E. Bracker, and D. Sternshein). (9) In the pH 9.2 standard DAB reaction medium, homogeneously distributed reaction product fills the matrix of microbodies (Mb) of the zoospore. There is no reaction product in mitochondria (M). Bars = 0.5 μm.

30 mM copper sulfate (added dropwise)   1.25 ml
5 mM potassium ferricyanide (added dropwise)   1.25 ml

The reaction medium is filtered before use, and cells are incubated for 15–20 min at room temperature. Following repeated washes in water, the cells are incubated in a solution of 5 mg DAB in 10 ml 0.05 M acetate buffer (pH 5.6) for 30 min at room temperature to amplify deposits. An alternative method for copper product intensification is to incubate cells in 0.5% thiocarbohydrazide for 5 min. Cells are then washed in water, layered on a coverslip and exposed for 20 min. to vapors from 2% osmium tetroxide heated to 55°C. It is important that the osmication step be carried out under a well-vented hood. Cells are washed in distilled water and processed for electron microscopy.

3.3. Aryl Sulfatase

Hanker *et al.* (1975) use the strategy that aryl sulfatase will cleave phenols from substrates that in turn reduce ferricyanide. The reaction medium is essentially the same as that prepared for acid phosphatase activity (Section 3.2; Hanker *et al.*, 1972b) except that the pH is 5.2–5.6 and the substrate is 20 mg of 4-nitro-1,2-benzenediol mono(hydrogen sulfate) dissolved in 7.9 ml of 0.06 N sodium acetate (trihydrate). Brown and Romanovicz (1976) used this technique to demonstrate the role of aryl sulfatase in Golgi-mediated scale formation in algae. This method is more precise than the widely used one for light microscopy of fungi (Reiss, 1974). In the latter method, aryl sulfatase catalyzes the hydrolysis of phenol sulfates into phenol and sulfate. The sulfate is reacted with ammonium sulfide, producing lead sulfite. Although the lead deposits would be visible with electron microscopy, this technique has not been sufficiently refined for electron microscopy.

3.4. Other Enzymes

A number of other enzymes have been successfully localized using the strategy of production of Hatchett's brown. These include:

1. Esterase (Hanker *et al.*, 1972b) where the reaction medium is made as for acid phosphatase activity (Section 3.2) except that 2.5 mg of 2-thiolactoxybenzanilide dissolved in 0.1 ml of acetone is used as the substrate.
2. Malate synthase (Burke and Trelease, 1975; Trelease, 1975; Trelease *et al.*, 1974) is useful for localizing enzyme activity in glyoxysomes of fungal zoospores (Powell, 1976).
3. Glycolate dehydrogenase (Beezley *et al.*, 1976) is useful for localizing enzyme activity in mitochondria of green algae.

4. DIAMINOBENZIDINE OXIDATION

The DAB cytochemical procedure (Novikoff and Goldfischer, 1969) is widely used in studies of microorganisms to localize peroxidatic activity of heme-containing enzymes (Dorward and Powell, 1980; Gierczak et al., 1982; Hänssler et al., 1981; Henry, 1975; Hirai, 1974; Huang et al., 1983; Kawamoto et al., 1977; Menzel, 1979; Morrison, 1977; Oakley and Dodge, 1974; Osumi et al., 1975; Philippi et al., 1975; Powell, 1976, 1977, 1978, 1979a,b, 1981; Pueschel, 1980; Silverberg, 1975; Stempen and Evans, 1982; Stevens et al., 1977; Todd and Vigil, 1972; Ton-That et al., 1983; Van Dijken et al., 1975; Veenhuis et al., 1979; Williams and Stewart, 1976). Since several different heme-containing enzymes or proteins oxidize DAB, conditions of the reaction and controls must be analyzed for proper interpretations of results. Reaction products appear as homogeneously electron-dense deposits of osmium black and tend to fill compartments where enzymes are localized (Figures 9–11). At high reaction pH catalase is localized (Figure 9), whereas at lower pH peroxidase and cytochrome c are visualized (Figures 10 and 11). As an example of the influence of developmental stage on localization of enzymes, Figure 10 reveals that peroxidase activity is absent from the endomembranes of a developing sporangium of a zoosporic fungus, and Figure 11 demonstrates its appearance in the sporangial nuclear envelopes and endoplasmic reticulum during zoospore formation. In chloroplast-containing organisms, polyphenol oxidase (Henry, 1975) can oxidize DAB. Sodium cyanide and sodium azide are both used as nonspecific inhibitors of heme enzymes, and aminotriazole is an inhibitor of catalase activity.

4.1. Catalase

An example of DAB reaction product due to catalase activity is illustrated in Figure 9. Fixation prior to incubation in the reaction medium enhances deposition of reaction product (Roels et al., 1975). Hänssler et al. (1981) found that in fungi, use of sodium cacodylate buffer rather than phosphate buffer resulted in localization of reaction product in microbodies of organisms that had not previously been stained with DAB. Because cacodylate buffer contains arsenic, these results suggest that these microbody catalases are stimulated by arsenic.

Cells are fixed in 2.5% glutaraldehyde in 0.05 M sodium cacodylate buffer (pH 7.2) at 4°C. Tissues are washed in fixative buffer and then in repeated changes of incubation medium buffer. The reaction medium is prepared at twice the concentration used to facilitate dilution when inhibitors are used. This 2× complete medium is:

0.1 M 2-amino-2-methyl-1,3-propanediol buffer (AMP), pH 9.2	24.5 ml
6% hydrogen peroxide	0.5 ml
3,3'-diaminobenzidine-HCl (DAB)	0.1 g

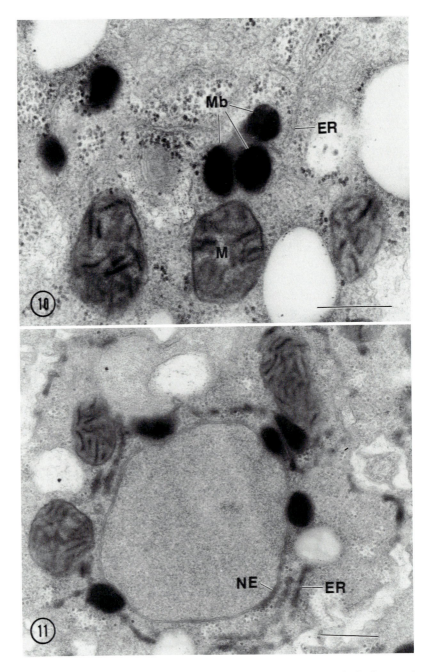

FIGURES 10, 11. DAB reaction at pH 7.0 in the fungus *E. variabilis.* (10) In the maturing sporangium, reaction product is in microbodies (Mb) due to peroxidase and in mitochondria (M) due to cytochrome *c.* (11) As the cytoplasm is cleaved into zoospores, reaction product due to peroxidase also appears in the nuclear envelope (NE) and endoplasmic reticulum (ER). Bars = 0.05 μm.

The final pH of the medium is adjusted to 9.2 prior to use. For incubations in the standard medium, the medium is diluted 1 : 1 with double-distilled water and cells are incubated at 37°C for 30 min. The ratio of tissue to reaction medium should be 1 : 10. Cells are then washed in reaction medium buffer and then fixative buffer before postfixation in buffered 1.0% osmium tetroxide.

As controls, inhibitors can be added to the reaction medium. Prior to incubation in the reaction medium with inhibitors, however, cells should be preincubated in inhibitor solutions. The following inhibitor solutions should be used for preincubation of tissue:

a. 0.1 M sodium azide in 0.06% peroxide and buffered with 0.05 M AMP
b. 0.1 M sodium cyanide in 0.06% peroxide and buffered with 0.05 M AMP
c. 0.02 M 3-amino-1,2,4-triazole in 0.06% peroxide and buffered with 0.05 M AMP

Cells are preincubated for 45 min at room temperature and then incubated at 37°C for 30 min in incubation medium that has been diluted 1 : 1 with double-strength solutions of inhibitors (i.e., 0.2 M sodium azide, 0.2 M sodium cyanide, or 0.04 M 3-amino-1,2,4-triazole). These tissues are processed as above with osmium tetroxide. All tissues are then prepared for electron microscopy. If the reaction product is mediated by catalase activity, then reaction product deposits should be reduced in tissues treated with any of the three inhibitors. A complete scheme for use of controls is given in Figure 12.

4.2. Peroxidase

Whereas catalase is visualized at alkaline pHs, 9.2–10.0, peroxidases are active at pHs closer to neutrality, pH 7.0–8.5. Thus, peroxidase is localized using the same reaction medium as for catalase, but with a more acid pH and buffer. Roels et al. (1975) reported that peroxidase can be distinguished by varying fixation, reaction pH, temperature, and substrate concentration. Aldehyde fixation before incubation in reaction medium is required for visualization of catalase, but peroxidase can be localized with no prior fixation. In contrast to high incubation temperature and peroxide concentration for catalase, peroxidase can be visualized with lower peroxide concentrations and at room temperature. In some organisms, catalase and peroxidase activities are found in the same organelle. For example, in an aquatic fungus (Powell, 1977), microbodies contained reaction product in the DAB reaction at the pH for catalase (pH 9.2; Figure 9) and for peroxidase (pH 7.0; Figure 10).

4.3. Cytochrome c Oxidase

In addition to peroxidase, mitochondrial staining is common in the DAB reaction at pH 7.0–8.5, activity that has been interpreted as cytochrome oxidase

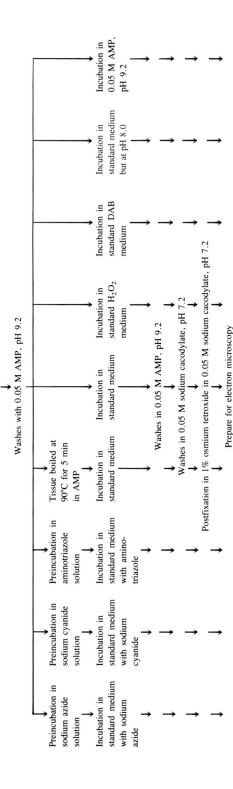

FIGURE 12. Scheme for use of controls in the DAB reaction for catalase activity at pH 9.2.

or cytochrome peroxidase. Roels (1974) makes the case that what is visualized is actually cytochrome c, an indicator of cytochrome oxidase activity. In Figures 10 and 11, peroxidase activity in microbodies and endomembranes is visualized simultaneously with cytochrome c activity in mitochondria where reaction product fills the intracristal spaces of mitochondria. As expected, based on differences of pH requirements for enzyme activity, no mitochondrial staining occurs at the higher pH, 9.2 (Figure 9).

5. CONCLUSIONS

In general, cytochemical techniques used to localize enzymes in plants and animals are applicable, with few or no modifications, for microorganisms. The techniques I have discussed are those most widely used, but other diverse approaches are available and should be employed for microorganisms. Correlations between developmental, physiological, and enzyme cytochemical studies are necessary avenues for expanding our understanding of the functional morphology of microorganisms.

ACKNOWLEDGMENTS. Some of the examples illustrated are from research supported by National Science Foundation Grants DEB 8010913 and DEB 8207181 to the author and by a grant from the Brown-Hazen Fund of Research Corporation to Dr. Charles E. Bracker.

6. REFERENCES

Aliaga, G. R., and Ellzey, J. T., 1984, Ultrastructural localization of acid phosphatase and alkaline phosphatase within oogonia of *Achlya recurva, Mycologia* **76**:85–98.

Armentrout, V. N., Hänssler, G., and Maxwell, D. P., 1976, Acid phosphatase localization in the fungus *Whetzelinia sclerotiorum, Arch. Microbiol.* **107**:7–14.

Bal, A. K., 1974, Cellulase, in: *Electron Microscopy of Enzymes: Principles and Methods,* Vol. 3 (M. A. Hayat, ed.), pp. 68–76, Van Nostrand–Reinhold, Princeton, New Jersey.

Barka, T., and Anderson, P., 1962, Histochemical methods for acid phosphatase using hexazonium pararosanilin as coupler, *J. Histochem. Cytochem.* **10**:741–753.

Beezley, B. B., Gruber, P. J., and Frederick, S. E., 1976, Cytochemical localization of glycolate dehydrogenase in mitochondria of *Chlamydomonas, Plant Physiol.* **58**:315–319.

Beier, K., and Fahimi, H. D., 1985, Automatic determination of labeling density in protein A–gold immunocytochemical preparations using an image analyzer, *Histochemistry* **82**:99–100.

Borgers, M., DeNollin, S., Thorné, F., and Van Belle, H., 1977, Cytochemical localization of NADH oxidase in *Candida albicans, J. Histochem. Cytochem.* **25**:193–199.

Briggs, R. T., Brath, D. B., Karnovsky, M. L., and Karnovsky, M. J., 1975, Localization of NADH oxidase on the surface of human polymorphonuclear leukocytes by a new cytochemical method, *J. Cell Biol.* **67**:566–589.

Brown, R. M., and Romanovicz, D. W., 1976, Biogenesis and structure of Golgi-derived cellulosic scales in *Pleurochrysis.* I. Role of the endomembrane system in scale assembly and exocytosis, *Appl. Polym. Symp.* **28**:537–585.

Burke, J. J., and Trelease, R. N., 1975, Cytochemical demonstration of malate synthase and glycolate oxidase in microbodies of cucumber cotyledons, *Plant Physiol.* **56**:710–717.

Chapman, C. M., Lowenberg, J. R., Schaller, M. J., and Piechura, J. E., 1983, Ultrastructural localization of cellulase in *Trichoderma reesei* using immunocytochemistry and enzyme cytochemistry, *J. Histochem. Cytochem.* **31**:1363–1366.

Costerton, J. W., and Marcks, I., 1977, Localization of enzymes in procaryotic cells, in: *Electron Microscopy of Enzymes: Principles and Methods*, Vol. 5 (M. A. Hayat, ed.), pp. 98–134, Van Nostrand–Reinhold, Princeton, New Jersey.

Danscher, G., and Rytter-Nörgaard, J. O., 1983, Light microscopic visualization of colloidal gold on resin-embedded tissue, *J. Histochem. Cytochem.* **31**:1394–1398.

Dargent, R., Touze-Soulet, J. M., Rami, J., and Montant, C., 1982, Cytochemical characterization of Golgi apparatus in some filamentous fungi, *Exp. Mycol.* **6**:101–114.

Doonan, B. B., and Jensen, T. E., 1980, Ultrastructural localization of alkaline phosphatase in the cyanobacteria *Coccochloris peniocytis* and *Anabaena cylindrica*, *Protoplasma* **102**:189–197.

Dorward, D. W., and Powell, M. J., 1980, Microbodies in *Monoblepharella* sp., *Mycologia* **72**:549–557.

DuBois, J. D., Roberts, K. R., and Kapustka, L. A., 1984, Polyphosphate body and acid phosphatase localization in *Nostoc* sp., *Can. J. Microbiol.* **30**:8–15.

Eichhorn, S. E., Perry, J. W., and Evert, R. F., 1981, *Preparation Guide for Laboratory Topics in Botany*, p. 65, Worth Publishers, New York.

Essner, E., 1973, Phosphatases, in: *Electron Microscopy of Enzymes: Principles and Methods*, Vol. 1 (M. A. Hayat, ed.), pp. 44–76, Van Nostrand–Reinhold, Princeton, New Jersey.

Esteve, J. C., 1970, Distribution of acid phosphatase in *Paramecium caudatum:* Its relations with the process of digestion, *J. Protozool.* **17**:24–35.

Feeney, D. M., and Triemer, R. E., 1979, Cytochemical localization of Golgi marker enzymes in *Allomyces*, *Exp. Mycol.* **3**:157–163.

Gezelius, K., 1971, Acid phosphatase localization in myxamoebae of *Dictyostelium discoideum*, *Arch. Mikrobiol.* **75**:327–337.

Gierczak, J. S., Stevens, F. J., Pankratz, H. S., and Uffen, R. L., 1982, Cytochemical localization and measurement of aerobic 3,3'-diaminobenzidine oxidation reactions in photosynthetically grown *Rhodospirillum rubrum*, *J. Histochem. Cytochem.* **30**:901–907.

Goldfischer, S., Essner, E., and Novikoff, A. B., 1964, The localization of phosphatase activities at the level of ultrastructure, *J. Histochem. Cytochem.* **12**:72–95.

Gomori, G., 1952, *Microscopic Histochemistry: Principles and Practice*, University of Chicago Press, Chicago.

Gruber, P. J., and Frederick, S. E., 1977, Cytochemical localization of glycolate oxidase in microbodies of *Klebsormidium*, *Planta* **135**:45–49.

Hall, J. L., Browning, A. J., and Harvey, D. M. R., 1980, The validity of the lead precipitation technique for the localization of ATPase activity in plant cells, *Protoplasma* **104**:193–200.

Hand, A. R., 1975, Ultrastructural localization of L-α-hydroxy acid oxidase in rat liver peroxisomes, *Histochemistry* **41**:195–206.

Hanker, J. S., Anderson, W. A., and Bloom, F. E., 1972a, Osmiophilic polymer generation catalysis by transition metal compounds in ultrastructural cytochemistry, *Science* **175**:991–993.

Hanker, J. S., Yates, P. E., Clapp, D. H., and Anderson, W. A., 1972b, New methods for the demonstration of lysosomal hydrolases by the formation of osmium blacks, *Histochemie* **30**:201–214.

Hanker, J. S., Thornburg, L. P., Yates, P. E., and Romanovicz, D. K., 1975, The demonstration of arylsulfatases with 4-nitro-1,2-benzenediol mono(hydrogen sulfate) by the formation of osmium blacks at the sites of copper capture, *Histochemistry* **41**:207–225.

Hänssler, G., Maxwell, D. P., and Maxwell, M. D., 1975, Demonstration of acid phosphatase-containing vacuoles in hyphal tip cells of *Sclerotium rolfsii*, *J. Bacteriol.* **124**:997–1006.

Hänssler, G., Muhlenbacker, D., and Reisener, J. J., 1981, Cytochemical localization of microbodies in *Puccinia graminis* var. *tritici*, *Exp. Mycol.* **5:**209–216.

Henry, E. W., 1975, Polyphenyl oxidase activity in thylakoids and membrane-bound granular components of *Nicotiana tabacum* chloroplasts, *J. Microsc. (Paris)* **22:**109–166.

Hirai, K. I., 1974, Distribution of peroxidase in *Tetrahymena pyriformis* mitochondria, *J. Histochem. Cytochem.* **22:**189–202.

Holt, S. C., and Beveridge, T. J. (eds.), 1982, *Electron Microscopy: Its Development and Application to Microbiology, Can. J. Microbiol.* **28:**1–718.

Horisberger, M., and Vonlanthen, M., 1977, Location of mannan and chitin on thin sections of budding yeasts with gold markers, *Arch. Microbiol.* **115:**1–7.

Huang, A. H. C., Trelease, R. N., and Moore, T. S., Jr., 1983, *Plant Peroxisomes*, Academic Press, New York.

Hugon, J., and Borgers, M., 1966, A direct lead method for electron microscope visualization of alkaline phosphatase activity, *J. Histochem. Cytochem.* **14:**429–431.

Karnovsky, M. J., 1965, A formaldehyde–glutaraldehyde fixative of high osmolality for use in electron microscopy, *J. Cell Biol.* **27:**137a.

Kawamoto, S., Tanaka, A., Yamamura, M., Teranishi, Y., Fukui, S., and Osumi, M., 1977, Microbody of n-alkane-grown yeast: Enzyme localization in the isolated microbody, *Arch. Microbiol.* **112:**1–8.

Kazama, F., 1973, Ultrastructure of *Thraustochytrium* sp. zoospores. III. Cytolysomes and acid phosphatase distribution, *Arch. Mikrobiol.* **89:**95–104.

Litwin, J. A., Tokota, S., and Fahimi, H. D., 1984, Light microscopic immunocytochemical demonstration of peroxisomal enzymes in Epon sections, *Histochemistry* **81:**15–22.

Maxwell, D. P., Hänssler, G., and Maxwell, M. D., 1978, Ultrastructural localization of acid phosphatase in *Pythium paroecandrum, Protoplasma* **94:**73–82.

Meirelles, M. N. L., and DeSouza, W., 1984, Localization of Mg^{2+}-activated ATPase in the plasma membrane of *Trypanosoma cruzi, J. Protozool.* **31:**135–140.

Menzel, D., 1979, Accumulation of peroxidase in the cap rays of *Acetabularia* during the development of gametangia, *J. Histochem. Cytochem.* **27:**1003–1010.

Meyer, R., Parish, R. W., and Hohl, H. R., 1976, Hyphal tip growth in *Phytophthora:* Gradient distribution and ultrahistochemistry of enzymes, *Arch. Microbiol.* **110:**215–224.

Morrison, P. J., 1977, Gametangial development in *Allomyces macrogynus*. II. Evidence against mitochondrial involvement in sexual differentiation, *Arch. Microbiol.* **113:**173–179.

Murry, M. A., Hallenbeck, P. C., and Benemann, J. R., 1984, Immunochemical evidence that nitrogenase is restricted to heterocysts in *Anabaena cylindrica, Arch. Microbiol.* **137:**194–199.

Noguchi, T., 1976, Phosphatase activities and osmium reduction in cell organelles of *Micrasterias americana, Protoplasma* **87:**163–178.

Nolan, R. A., and Bal, A. K., 1974, Cellulase localization in hyphae of *Achlya ambisexualis, J. Bacteriol.* **117:**840–843.

Novikoff, A. B., and Goldfischer, S., 1961, Nucleoside diphosphatase activity in the Golgi apparatus and its usefulness for cytological studies, *Proc. Natl. Acad. Sci. USA* **47:**802–810.

Novikoff, A. B., and Goldfischer, S., 1969, Visualization of peroxisomes (microbodies) and mitochondria with diaminobenzidine, *J. Histochem. Cytochem.* **17:**675–680.

Oakley, B. R., and Dodge, J. D., 1974, The ultrastructure and cytochemistry of microbodies in *Porphyridium, Protoplasma* **80:**233–244.

Osumi, M., Imaizumi, F., Imai, M., Sato, H., and Yamaguchi, H., 1975, Isolation and characterization of microbodies from *Candida tropicalis* PK233 cells grown on normal alkanes, *J. Gen. Appl. Microbiol.* **21:**375–387.

Pfeifer, V., Poelmann, E., and Witschel, H., 1973, Kinetics of the accumulation of lead phosphate in acid phosphatase studies, in: *Electron Microscopy and Cytochemistry* (E. Wisse, W. T. Daems, I. Moelnaar, and P. Van Duijn, eds.), pp. 25–28, Elsevier, Amsterdam.

Philippi, M. L., Parish, R. W., and Hohl, H. R., 1975, Histochemical and biochemical evidence for the presence of microbodies in *Phytophthora palmivora, Arch. Microbiol.* **103:**127–132.

Powell, M. J., 1976, Ultrastructure and isolation of glyoxysomes (microbodies) in zoospores of the fungus *Entophlyctis, Protoplasma* **89:**1–27.

Powell, M. J., 1977, Ultrastructural cytochemistry of the diaminobenzidine reaction in the aquatic fungus *Entophlyctis variabilis, Arch. Microbiol.* **114:**123–136.

Powell, M. J., 1978, Phylogenetic implications of the microbody lipid globule complex in zoosporic fungi, *BioSystems* **10:**167–180.

Powell, M. J., 1979a, The structure of microbodies and their associations with other organelles in zoosporangia of *Entophlyctis variabilis, Protoplasma* **98:**177–198.

Powell, M. J., 1979b, What are the chromidia of *Polyphagus euglenae?, Am. J. Bot.* **66:**1173–1180.

Powell, M. J., 1981, Ultrastructure of *Polyphagus euglenae* zoospores, *Can. J. Bot.* **59:**2049–2061.

Powell, M. J., Bracker, C. E., and Sternshein, D. J., 1981, Formation of chlamydospores in *Gilbertella persicaria, Can. J. Bot.* **59:**908–928.

Price, G. D., and Whitecross, M. I., 1983, Cytochemical localization of ATPase activity on the plasmalemma of *Chara corallina, Protoplasma* **116:**65–74.

Pueschel, C. M., 1980, Evidence for two classes of microbodies in meiocytes of the red alga *Palmaria palmata, Protoplasma* **104:**273–282.

Reiss, J., 1974, Cytochemical detection of hydrolases in fungus cells. III. Aryl sulfatase, *J. Histochem. Cytochem.* **22:**183–188.

Roels, F., 1974, Cytochrome c and cytochrome oxidase in diaminobenzidine staining of mitochondria, *J. Histochem. Cytochem.* **22:**442–446.

Roels, F., Wisse, E., DePrest, B., and van der Meulen, J., 1975, Cytochemical discrimination between catalases and peroxidases using diaminobenzidine, *Histochemistry* **41:**281–312.

Rosing, W. C., 1984, Ultracytochemical localization of acid phosphatase within deliquescing asci of *Chaetomium brasiliense, Mycologia* **76:**67–73.

Rudzinska, M. A., 1972, Ultrastructural localization of acid phosphatase in feeding *Tokophrya infusionum, J. Protozool.* **19:**618–629.

Rudzinska, M. A., 1974, Ultrastructural localization of acid phosphatase in starved *Tokophrya infusionum, J. Protozool.* **21:**721–728.

Sentandreu, R., Martinez-Ramon, A., and Ruiz-Herrera, J., 1984, Localization of chitin synthase in *Mucor rouxii* by an autoradiographic method, *J. Gen. Microbiol.* **130:**1193–1199.

Shnitka, T. K., and Talibi, G. G., 1971, Cytochemical localization of ferricyanide reduction of α hydroxy acid oxidase activity in peroxisomes of rat kidney, *Histochemie* **27:**137–158.

Silverberg, B. A., 1975, 3,3'-Diaminobenzidine (DAB) ultrastructural cytochemistry of microbodies in *Chlorogonium elongatum, Protoplasma* **85:**373–376.

Stempen, H., and Evans, R. C., 1982, Behavior of the inner wall layer of the germinating *Fuligo septica* spore: Evidence of peroxidase activity, *Mycologia* **74:**26–35.

Sternberger, L. A., 1973, Enzyme immunocytochemistry, in: *Electron Microscopy of Enzymes: Principles and Methods,* Vol. 1 (M. A. Hayat, ed.), pp. 150–191, Van Nostrand–Reinhold, Princeton, New Jersey.

Stevens, F. J., Pankratz, H. S., and Uffen, R. L., 1977, Demonstration of two 3,3'-diaminobenzidine oxidation reactions associated with photosynthetic membranes in anaerobic light-grown *Rhodospirillum rubrum, J. Histochem. Cytochem.* **25:**1264–1268.

Thomas, J., and Trelease, R. N., 1981, Cytochemical localization of glycolate oxidase in microbodies (glyoxysomes and peroxisomes) of higher plant tissues with the $CeCl_3$ technique, *Protoplasma* **108:**39–53.

Todd, M. M., and Vigil, E. L., 1972, Cytochemical localization of peroxidase activity in *Saccharomyces cerevisiae, J. Histochem. Cytochem.* **20:**344–349.

Ton-That, T. C., Michea-Hamzehpour, M., and Turian, G., 1983, Ultrastructural demonstration of loss and recovery of cytochrome oxidase activity during and after heat induction of microcycle conidiation in *Neurospora crassa*, *Protoplasma* **116**:149–154.

Trelease, R. N., 1975, Malate synthase, in: *Electron Microscopy of Enzymes: Principles and Methods*, Vol. 4 (M. A. Hayat, ed.), pp. 157–176, Van Nostrand–Reinhold, Princeton, New Jersey.

Trelease, R. N., Becker, W. M., and Burke, J. J., 1974, Cytochemical localization of malate synthase in glyoxysomes, *J. Cell Biol.* **60**:483–495.

Van Dijken, J. P., Veenhuis, M., Vermeulen, C. A., and Harder, W., 1975, Cytochemical localization of catalase activity in methanol-grown *Hansenula polymorpha*, *Arch. Microbiol.* **105**:261–267.

Van Duijn, P., 1974, Fundamental aspects of enzyme cytochemistry, in: *Electron Microscopy and Cytochemistry* (E. Wisse, W. T. Daems, I. Moelnaar, and P. Van Duijn, eds.), pp. 3–23, Elsevier, Amsterdam.

Van Steveninck, R. F. M., 1979, The verification of cytochemical tests for ATP-ase activity in plant cells using X-ray microanalysis, *Protoplasma* **99**:211–220.

Veenhuis, M., Van Dijken, J. P., and Harder, W., 1976, Cytochemical studies on the localization of methanol oxidase and other oxidases in: peroxisomes of methanol-grown *Hansenula polymorpha*, *Arch. Microbiol.* **111**:123–135.

Veenhuis, M., Keizer, I., and Harder, W., 1979, Characterization of peroxisomes in glucose-grown *Hansenula polymorpha* and their development after the transfer of cells into methanol-containing media, *Arch. Microbiol.* **120**:167–175.

Veser, J., Martin, R., and Thomas, H., 1981, Immunocytochemical demonstration of catechol methyltransferase in *Candida tropicalis*, *J. Gen. Microbiol.* **126**:97–101.

Vorisek, J., 1977, Electron-cytochemical demonstration of acid phosphatase in saprophytic *Claviceps purpurea*, *Arch. Microbiol.* **111**:289–295.

Wachstein, M., and Meisel, E., 1957, Histochemistry of hepatic phosphatases at physiological pH with special reference to the demonstration of bile canaliculi, *Am. J. Clin. Pathol.* **27**:13–23.

Washitani, I., and Sato, S., 1976, On the reliability of the lead salt precipitation method of phosphatase localization in plant cells, *Protoplasma* **89**:157–170.

Wick, S. M., and Hepler, P. K., 1980, Localization of Ca^{2+}-containing antimonate precipitates during mitosis, *J. Cell Biol.* **86**:500–513.

Wientjes, F. B., Riet, J. V., and Nanninga, N., 1980, Immunoferritin labeling of respiratory nitrate reductase in membrane vesicles of *Bacillus licheniformis* and *Klebsiella aerogenea*, *Arch. Microbiol.* **127**:39–46.

Williams, P. G., and Stewart, P. R., 1976, The intramitochondrial location of cytochrome c peroxidase in the wild-type and petite *Saccharomyces cerevisiae*, *Arch. Microbiol.* **107**:63–70.

Yokota, S., Deiman, W., Hashimoto, T., and Fahimi, H. D., 1983, Immunocytochemical localization of two peroxisomal enzymes of lipid α-oxidation in specific granules of rat eosinophils, *Histochemistry* **78**:425–433.

Chapter 16

Localization of Nucleic Acids in Sections

H. C. Aldrich

Department of Microbiology and Cell Science
University of Florida
Gainesville, Florida 32611

and

Sylvia E. Coleman

Research Service
Veterans Administration Medical Center
Gainesville, Florida 32602

1. INTRODUCTION

Occasionally, structures of unknown composition are observed inside the nucleus or cytoplasm of eukaryotic microbes or in the cytoplasm of prokaryotes. If the structure is suspected to contain DNA or RNA, there is a varied arsenal of techniques that will aid in identification. Nucleic acids can be detected by staining methods that utilize heavy metals, by labeling, and by extraction. We will review the most useful of these techniques, emphasizing procedures that can be used on ultrathin sections.

2. "SELECTIVE" STAINS

There is no electron-dense stain truly specific for either DNA or RNA. However, under controlled conditions, some compounds stain the two molecular species in a manner usually called "selective." This means that one or the other is stained preferentially, but that results must be interpreted with caution. In this section we will review the best of the methods of this type and then proceed to more specific procedures.

2.1. Uranyl Compounds

The uranyl ion exhibits a definite, if not specific, affinity for nucleic acids (Gautier, 1976). It was observed by Kellenberger et al. (1958) that DNA fibrils in the nucleoid of bacteria were stabilized by uranyl acetate after fixation by osmium tetroxide, and the interaction of uranyl acetate with DNA was observed by Valentine in the same year. Although after the usual osmium fixation, neither uranyl acetate nor lead hydroxide will preferentially stain chromatin, Marinozzi and Gautier (1962) demonstrated that after fixation with formaldehyde, chromatin is stained intensely, and the nucleolus less intensely, by uranyl acetate but not by lead hydroxide. On the other hand, if fixation by formaldehyde is followed by postfixation with either osmium or uranyl acetate, the chromatin will stain with lead hydroxide while retaining its affinity for uranyl acetate. These authors found that the best conditions for preferential staining of deoxyribonucleoproteins on thin sections were the use of formaldehyde fixation with postosmication followed by staining with 1% uranyl acetate at pH 5.

Uranyl acetate is now widely used as a contrasting agent in combination with lead citrate or as a component of the EDTA regressive method for ribonucleoproteins (Moyne, 1980). The properties of uranyl acetate that make it useful for augmenting contrast on sections are preferential binding to nucleic acids by interaction with phosphate groups, the possibility of binding to proteins, and the increased contrast of chromatin.

Derksen and Meekes (1984) modified conventional uranyl acetate–lead staining procedures to selectively stain nucleic acid structures in glutaraldehyde-fixed, Epon-embedded material. The method recommended is:

1. Stain sections in 2% uranyl acetate at pH 1.8 for 4 min.
2. Poststain with 0.2% lead citrate (Venable and Coggeshall, 1965) for 10 to 30 sec.
3. Wash in 0.1% NH_4OH.

Longer staining times or washing in water alone, rather than in ammonium hydroxide, resulted in a fine precipitate over membranes and the nucleolus. No selective staining occurred in OsO_4-fixed material with or without glutaraldehyde fixation or in material embedded in glycol methacrylate. Structures known to contain DNA and RNA stained well, such as nucleoli, chromosomes, ribonucleoprotein in puffs and nucleoplasm, and ribosomes. Structures such as membranes, nuclear envelopes, and basement membranes remained unstained.

The authors felt that the selectivity of this modified uranyl acetate–lead citrate protocol was probably due to the fact that carboxyl groups of proteins are not charged at low pH while the PO_4 groups of the nucleic acids remain charged and capable of binding uranyl acetate.

2.2. Bismuth

Albersheim and Killias (1963) introduced bismuth as an electron stain. Bismuth complexes and precipitates with inorganic phosphate, DNA, and RNA. In osmium-fixed *Allium cepa* (onion), although there was a background stain from the osmium fixation, the increased contrast of the structures containing nucleic acid was unmistakable. Brown and Locke (1978) evaluated the mechanism of the reaction and concluded that part of the staining described by Albersheim and Killias was due to bismuth binding to reduced osmium.

2.3. Sodium Tungstate

A third protocol for the rapid detection of nucleic acids involves staining with sodium tungstate. Swift and Adams (1966) used sodium tungstate at pH 5.6 (adjusted with HCl) and obtained a relatively specific staining of nucleic acids. In 1977, Stockert reevaluated the stain using various plant and animal tissues. Structures containing RNA and DNA were stained selectively, but structures containing glycoprotein were stained as well. Earlier, Marinozzi (1968) had pointed out that acidified solutions of sodium tungstate at pH 1 to 4.5 increased the contrast of glycoproteins and polysaccharides, similar to the action of acidified phosphotungstic acid, which stains glycoproteins under the same conditions.

The technique for sodium tungstate staining is as follows:

1. Fix tissue in glutaraldehyde.
2. Embed in Epon.
3. Float thin section on 10% sodium tungstate in distilled water, pH 5.5 (adjusted with 1 N HCl), for 1 to 2 hr at room temperature.
4. Rinse in distilled water.

3. "FEULGEN"-TYPE STAINS FOR DNA

The Feulgen stain for light microscopy involves controlled HCl hydrolysis of sections, followed by staining with leuco-basic fuchsin. DNA stains magenta, while RNA is unstained. Parallel procedures have been developed for electron microscopy, using heavy metals instead of the fuchsin to visualize the DNA.

3.1. Silver Stains

The use of metallic silver at the electron microscope level was investigated by Bretschneider in 1949, applying a variant of the Feulgen reaction using ammoniacal silver nitrate to identify chromosomes in whole bull sperm. Using formaldehyde-fixed tissue, Churg *et al.* (1958) obtained an unspecific nuclear stain. The blocks were oxidized with periodic acid and impregnated with silver methenamine. Marinozzi (1961) stained thin sections of osmicated tissue embedded in methacrylate with ammoniacal silver but could not observe nuclear features. Izard and Bernhard (1962) adapted the light microscope method of Estable and Sotelo (1951) and obtained a heavy staining that was dense over the nucleoli. Thiéry (1966) showed that the contrast of nucleoproteins was enhanced when silver methenamine or silver proteinate was used on aldehyde-fixed thin sections.

A technique remarkably preferential for the nucleolus was devised by Risueno *et al.* (1973). The contrast of the silver particles permits identification of the nucleolus and evaluation of structural differences, although the size of the particles interferes with high-magnification observation. The technique is as follows (Risueno *et al.*, 1973; Moreno Diaz de la Espina and Risueno, 1976):

1. Immerse small blocks overnight at 4°C in a fixative containing 10% formaldehyde and 1% hydroquinone. To prepare formaldehyde by depolymerization of paraformaldehyde: boil 50 ml of distilled water, remove from heat, and slowly add 4 g of paraformaldehyde under hood. Stir the suspension (it will be milky), add 4 drops of 1 N NaOH, and stir until dissolved. Cool in crushed ice, fill to 50 ml with distilled water, and mix with 50 ml of 0.2 M phosphate buffer. Store at 4°C and use on same day.
2. Rinse in distilled water three times for 30 min each.
3. Impregnate with 2% $AgNO_3$ in water at 70°C. Leave for 4 hr in the dark.
4. Rinse in distilled water three times for 30 min each.
5. Postfix for 1 to 2 hr in the formaldehyde–hydroquinone fixative at room temperature.
6. Embed in Epon.

3.2. Osmium Ammine

The Feulgen-like reaction with osmium ammine-SO_2 (commercially available from Polysciences) is one of the best techniques for the specific staining of DNA. The result of applying this stain to the eukaryotic nucleus of the cellular slime mold *Dictyostelium discoideum* is shown in Figure 1. The reaction product in this procedure reveals the presence of apurinic acid. The technique was used by Cogliati and Gautier in 1973; Moyne (1980) modified the method as follows:

1. Fix tissue in glutaraldehyde using almost any embedding medium or ultrathin frozen sections. Thin sections of plastic-embedded material may be handled with Marinozzi rings (Marinozzi, 1964) and placed on

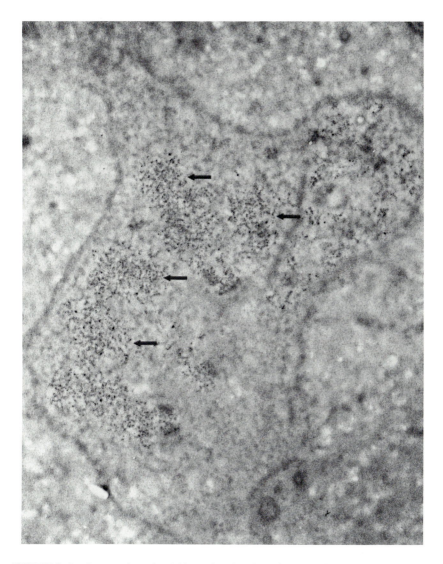

FIGURE 1. Osmium ammine stain of thin-sectioned nucleus of *Dictyostelium discoideum*. Fixed in glutaraldehyde, embedded in glycol methacrylate after "etching" with ethanol, hydrolysis for 30 min with 5 N HCl at 20°C and then treated 4.5 hr at 40°C with 0.4% osmium ammine-SO$_2$. Arrows indicate positively stained chromatin. ×21,500. Courtesy of Dr. Urs-Peter Roos, University of Zürich.

gold grids. If frozen sections are used, the gold grids must be Formvar–carbon coated.

2. Hydrolyze in 5 N HCl for 25 min at room temperature. Rinse with distilled water.

3. Float sections on osmium ammine-SO_2 for 1 hr in a closed vessel at 37°C. Prepare a 1 to 0.1% suspension of osmium ammine in distilled water, and bubble SO_2 through it under a fume hood for 30 min. Wait until the darker part of the solution sediments and occupies the lower half of the mixture. Draw out the middle region of the suspension with a Pasteur pipette to use for staining.

   Use room temperature for frozen sections. Rinse thoroughly with distilled water.

4. Immerse grids rapidly in 0.1 N H_2SO_4 to eliminate precipitates, using caution since the reaction product will also dissolve.

4. DISTINGUISHING BETWEEN DNA AND RNA

Two procedures in this category have found consistent application in biological electron microscopy. Although they are not strictly considered specific stains, they work dependably and are recommended for the neophyte.

4.1. Diaminobenzidine (Selective for DNA)

The oxidized form of 3,3′-diaminobenzidine (DAB) is highly osmiophilic, and posttreatment with osmium tetroxide renders it quite visible. The use of oxidized DAB as a stain that binds to the phosphate groups of both nucleic acids was first reported by Roels and Goldfischer (1971, 1972) who used oxidized DAB at pH 2.8–3.0. In this technique, the tissue was fixed with formaldehyde and embedded in glycol methacrylate. Sections were treated with an oxidized solution of DAB containing $FeCl_3$, with thick sections being observed immediately and thin sections osmicated and treated with $Pb(OH)_2$. This method increases contrast for all nucleoproteins but especially for deoxyribonucleoproteins.

Anteunis et al. (1973) modified the technique using fibroblast monolayers and obtained a preferential stain for DNA by fixing the cells in formaldehyde, incubating in an oxidized DAB solution, postfixing with OsO_4 followed by uranyl acetate and embedding in Epon. The chromatin was much more contrasted than the nucleoli. It was proposed that the differences in contrast were due to the double-stranded structure of chromatin DNA, which would exhibit more free phosphate groups than nucleolar RNA.

4.2. Bernhard's Regressive Stain (EDTA)

Introduced by Bernhard (1968, 1969), the EDTA regressive stain has been widely used for preferentially staining ribonucleoproteins. The basic mechanism appears to be that uranyl ions bound to deoxyribonucleoproteins are removed more easily by a chelating agent than are ions bound to RNP. Other chelating agents can be used, but EDTA gives better results. The method is defined as "regressive" by analogy to classical histology, since the EDTA bleaches chromatin and leaves stain in structures containing RNP. The bleaching depends on the length of exposure to EDTA, and thus it is necessary to determine the optimum treatment time for each different type of specimen. If the time of bleaching is too short, the chromatin will not be adequately bleached, and if too long, the nucleolus and ribosomes also lose contrast.

The EDTA regressive reaction is carried out on aldehyde-fixed material from which thin sections or ultrathin frozen sections are prepared, stained with uranyl acetate, treated with EDTA, and poststained with lead citrate. The techniques vary as follows:

Epon/Araldite Embedment

1. Cut silver-to-gold sections of glutaraldehyde (only)-fixed material; no osmium or uranyl fixation. Pick up on Formvar-coated copper grids.
2. Poststain for 1 min with 5% aqueous uranyl acetate at room temperature.
3. Wash with distilled water using a squirt bottle.
4. Float on *freshly made* 0.2 M EDTA (0.7 g in 10 ml water, brought to pH 7.0 with NaOH) for 60–75 min.
5. Wash with distilled water using a squirt bottle.
6. Float for 1 min on standard lead citrate at room temperature.
7. Wash and view.

Some Epon/Araldite-embedded eukaryotic material requires even longer treatment. Corn chromatin resisted destaining in our hands until we increased the time on EDTA to 90 min.

Lowicryl K4M-Embedded Material

1. Cut silver-to-gold sections; collect on Formvar grids.
2. Stain for 1 min on 0.5% aqueous uranyl acetate at room temperature.
3. Wash with distilled water using a squirt bottle.
4. Float on 0.2 M EDTA, pH 7.0 (see above), for 1 min at room temperature.
5. Wash with squirt bottle.

6. Float on lead citrate for 1 min at room temperature.

7. Wash with squirt bottle and view.

Ultrathin Frozen Sections

Mount on Formvar-coated copper grids and "stabilize" for 10 sec with 5% aqueous uranyl acetate to reduce disintegration of the tissue. Dry overnight and oxidize sections by floating for 90 sec on 2% H_2O_2. Treat with 0.2 M EDTA (pH 7) for only 15 to 20 sec., or replace with citrate, which is a weaker chelating agent (Puvion and Bernhard, 1975). Stain the ultrathin frozen sections for 4 min on a mixture of 2.5% uranyl acetate in 0.1 M citric acid (pH 5.5). Poststain with lead citrate for 1 min.

The overnight drying is an indispensable step and the treatment time limits are very critical. However, time is not as crucial if EDTA is replaced with citrate.

Staining and EDTA treatment times vary with embedding medium and tissue. We have found that water-soluble embedding media such as Lowicryl K4M require only very brief treatments, as shown in Figures 2 and 3. These cells were stained with 0.5% aqueous uranyl acetate for 1 min, destained on 0.2 M EDTA (pH 7.0) for 2 min, and stained with lead citrate for 1 min. The eukaryotic nucleus shown in Figure 4 was embedded in Epon Araldite, stained with 0.5% aqueous uranyl acetate for 1 min, destained with 0.2 M EDTA for 40 min, and stained with lead citrate for 1 min.

The regressive EDTA technique produces a low-contrast image in which DNA is electron-light and RNP electron-dense. The contrast of RNP varies from one specimen to the next but can be controlled by varying the treatment time. The procedure was considered by Bernhard to be preferential but not strictly specific for structures that are rich in RNA. The EDTA does not extract DNA, since after EDTA treatment chromatin can be restained with uranyl. In spite of its nonspecificity, the technique permits the identification of nuclear structures more easily than most other conventional stains, and it can be applied to many different types of material.

Certain cautions should be observed in interpreting the results of the EDTA regressive stain, however. It has been established that some structures that contain no RNP (e.g., desmosomes and glycogen) are still contrasted by this technique, due to their affinity for lead citrate. Some protein inclusions in nuclei that occur in certain pathological conditions, and which contain no chromatin, are bleached by the EDTA regressive method at a more rapid rate than is chromatin (Recher et al., 1973; Puvion et al., 1976). It was also recognized by Bachellerie et al. (1975) that the biochemical procedures used in isolating chromatin can modify EDTA staining patterns.

In making a critical evaluation of Bernhard's EDTA regressive staining

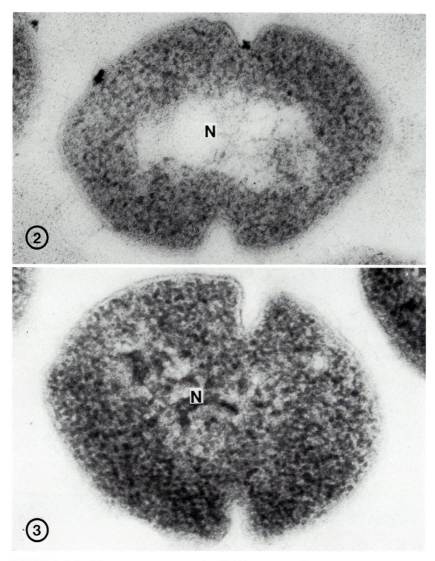

FIGURE 2. Cell of *Streptococcus sanguis* after EDTA treatment. Fixation: glutaraldehyde only. Embedment: Lowicryl K4M. Staining times: 0.5% uranyl acetate, 1 min; EDTA, 2 min; lead citrate, 1 min. Cytoplasmic ribosomes are stained. Nucleoid (N) has been destained. ×122,000.

FIGURE 3. Control cell of *S. sanguis*. Treatment as for the cell in Figure 2 except that EDTA treatment was omitted. Nucleoid (N) is stained. ×122,000.

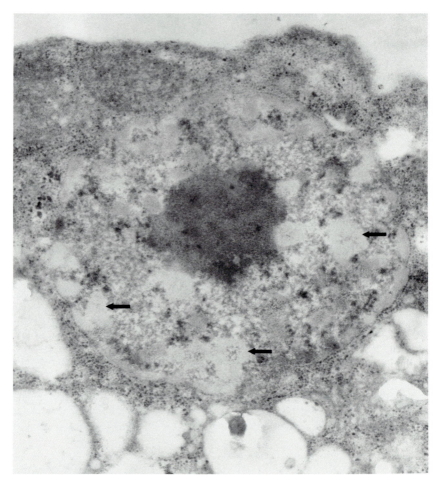

FIGURE 4. Eukaryotic nucleus of myxamoeba of the myxomycete *Physarum polycephalum* after EDTA treatment. Fixation: glutaraldehyde only. Embedment: Epon/Araldite. Staining times: 0.5% uranyl acetate, 1 min; EDTA, 40 min; lead citrate, 1 min. Central nucleolus and cytoplasmic ribosomes are stained; chromatin (arrows) is not. ×22,000.

technique, Pearson and Davies (1982) defined the following conditions under which the procedure retains its specificity:

1. The sections should be stained for only a short period with uranyl before treatment with EDTA.
2. Sections stained only in lead should be compared as controls with sections stained by Bernhard's technique so that any specificity of lead for

subcellular components is not confused with a positive indication of the presence of RNA.

5. GOLD-LABELED ANTIBODY STAINING
OF SPECIFIC TYPES OF RNA

In a very interesting recent study, Fakan *et al.* (1984) succeeded in localizing particular types of RNP within the nucleus of mouse and rat liver cells. Some specimens were embedded in Lowicryl K4M, while others were cryoultrathin sectioned for comparison. K4M sections were comparably labeled and better contrasted. Three antibodies were used: (1) Total human immunoglobulin from humans suffering from mixed connective tissue disease; this bound specifically to small nuclear ribonucleic acids (snRNA); (2) Mouse monoclonal IgG from a hybridoma derived from an autoimmune mouse; this also was specific for snRNA; (3) Mouse monoclonal IgM specific for heterogeneous nuclear RNP (hnRNP). Protein A–gold was used as the marker after prelabeling the thin sections with the antibody.

The anti-snRNA antibodies bound to interchromatin granules and perichromatin fibrils, while the anti-hnRNP antibody bound to perichromatin fibrils and perichromatin granules. Both types of antibodies labeled a type of coiled body characteristic of mouse nucleoli. The success of these experiments and the availability of antibodies specific for separate species of RNA augur well for the future of nuclear cytochemistry in eukaryotic cells. Perhaps we shall soon begin to identify the functions of some of the enigmatic structures present in the interphase eukaryotic nucleus.

6. GOLD-LABELED DNase AND RNase

The enzyme–gold postembedding method for the ultrastructural localization of macromolecules is based on the affinity of an enzyme for its substrate. This technique was developed by Bendayan (1981a,b, 1982) for detecting nucleic acids and has now been extended to other substrates (Bendayan, 1984a). Enzyme–gold labeling has also been performed on osmium-fixed tissues (Bendayan and Puvion, 1984). The technique provides high resolution and specificity since it is based on enzyme–substrate interactions and uses small electron-dense particles of gold. Specific nucleases are tagged with colloidal gold particles and the complex applied to thin sections of tissue, thus permitting the direct ultrastructural localization of their corresponding substrates (Bendayan, 1981a,b).

Due to its versatility, colloidal gold, which was introduced by Faulk and Taylor in 1971, seems to be the electron-dense marker of choice (Horisberger, 1981; Roth, 1983). It is a negatively charged hydrophobic sol that binds to macromolecules by noncovalent electrostatic adsorption which is dependent upon pH and concentration.

The technique for preparation of colloidal gold is given in Chapter 17 (see also Roth, 1982, 1983). Highly purified enzymes may be obtained commercially, and the procedure for enzyme–gold labeling is as follows (Bendayan, 1984a,b).

6.1. Preparation of Enzyme–Gold Complex

1. To determine the minimum amount of protein for full stabilization of the colloidal sol (Horisberger, 1981), a constant volume of colloidal gold (1 ml) is added to 0.1 ml of serial dilutions of protein. After 10 min, 0.1 ml of 10% NaCl is added and flocculation judged visually (a change from wine-red to a purple-blue color indicates the occurrence of flocculation). The lowest protein amount that prevents the color change from red to blue after addition of NaCl is considered to be sufficiently stabilizing (Roth, 1983). Alternatively, flocculation may be estimated spectrophotometrically (max 510–550). The minimum amount of protein needed to stabilize the colloidal gold depends upon the size of the particles.

2. To prepare each enzyme–gold complex, 0.1 to 0.5 mg of protein is dissolved in 0.2 ml of double-distilled water in a centrifuge tube and mixed with 10 ml of the colloidal gold suspension. The optimum pH for the binding is determined by the isoelectric point of the protein.

3. Adsorption of the protein to the surface of the colloidal particles occurs upon mixing. To further stabilize the coloidal gold, two drops of 1% polyethylene glycol (M_r 20,000) is added.

4. The complex is centrifuged for 30 min at 4°C at 25,000 rpm (Beckman ultracentrifuge, Ti-50 rotor) to remove excess unbound enzyme. At the completion of centrifugation, three phases are obtained: clear supernatant containing free enzyme; dark-red sediment at the bottom containing enzyme–gold complex; and a black spot near the bottom that represents metallic gold that was not stabilized by the protein.

5. The supernatant is carefully aspirated as completely as possible and discarded since it contains free enzyme that will compete with enzyme–gold complex during labeling.

6. The enzyme–gold complex is recovered and resuspended in 1.5 ml of 0.01 M phosphate-buffered saline (NaH_2PO_4/Na_2HPO_4 in 0.14 M NaCl) containing 0.02% polyethylene glycol. The metallic gold that has not been stabilized by the protein will remain at the bottom of the tube.

7. At the moment of recovery, the pH of the buffer should be adjusted to the value of the optimum enzyme activity, and tissue sections should be incubated at the same pH.

8. The stock solution of enzyme–gold complex should be kept at 4°C and

can be diluted further at the time of incubation. The enzyme–gold complexes have to be used within 10 days during which time they are enzymatically active.

6.2. Preparation of Tissue

The enzyme–gold labeling is performed on thin sections of tissue as a postembedding technique. Fixation may be with either glutaraldehyde or paraformaldehyde, with postfixation with osmium tetroxide if appropriate. The procedure is that of Bendayan (1984a):

1. 1% glutaraldehyde or 4% paraformaldehyde in 0.1 M phosphate buffer (pH 7.2) for 2 hr at room temperature.
2. 1% osmium tetroxide postfixation (where appropriate) for 1 hr at 4°C.

Embedding may be performed in various media including Epon, Araldite, Spurr's, glycol methacrylate, Lowicryl K4M or in some cases with tissues processed by cryoultramicrotomy. The thin sections are mounted on nickel grids with carbon-coated Parlodion films.

Osmium, if present, must be removed from sections before exposing them to the enzyme–gold complex. Float grids section side down on a drop of saturated aqueous sodium metaperiodate for 1 hr at room temperature. Rinse in distilled water (Bendayan and Puvion, 1984).

6.3. Cytochemical Labeling

1. Sections on nickel grids are floated face down on a drop of 0.01 M phosphate buffered-saline (PBS) adjusted to the optimum pH of the enzyme being used.
2. Grids are transferred onto a drop of the enzyme–gold complex and incubated 30 min at room temperature (this may vary, i.e., ultrathin frozen sections may require only 5 min).
3. The grids are washed thoroughly with a jet of PBS, rinsed with distilled water, and dried.
4. Stain with uranyl acetate and lead citrate before examination.

6.4. Controls for Enzyme–Gold Labeling

The choice of a control depends on the particular system, but a number are recommended by Bendayan (1984a):

1. Incubation of tissue sections with an enzyme–gold complex to which the specific substrate (about 1 mg/ml) has been added previously
2. Incubation with an enzyme-gold complex whose enzymatic activity has been abolished (i.e., heat treated)

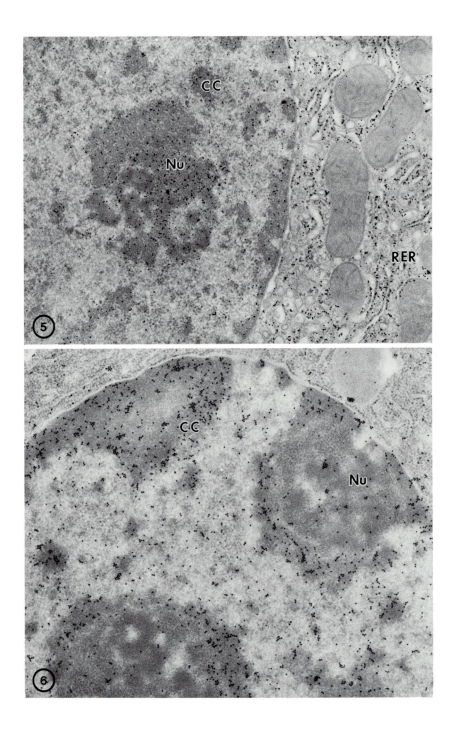

3. Incubation with an enzyme–gold complex that has been absorbed with the specific antibody against the enzyme
4. Incubation with enzyme–gold complexes adjusted to different pH values or incubated at 4°C
5. Incubation with a known inhibitor of the enzyme used to prepare the enzyme–gold complex
6. Incubation with a nonenzymatic protein–gold complex (i.e., protein A–gold, albumin–gold, or polyethylene glycol-stabilized gold particles)
7. Removal of the substrate from the tissue prior to incubation with the enzyme–gold complex

The results of the enzyme–gold technique are illustrated in Figures 5 and 6. Figure 5 is tissue incubated with RNase–gold complex. The nucleolus appears intensely labeled by gold particles. Label also appears over the heterochromatin, nuclear envelope, and RER. Figure 6 shows a section incubated with DNase–gold. Label is concentrated over the euchromatin, with some particles over the heterochromatin and nucleolus but few over the RER.

7. EXTRACTION METHODS

Nucleic acids may be identified by comparing untreated sections or tissue with sections or tissue from which material has been selectively removed by acid hydrolysis or by enzymes. Enzymatic extraction has been more frequently used in recent years, but we will review both types of procedures.

7.1. Mild Acid Hydrolysis

As discussed in Section 3, mild acid hydrolysis liberates certain aldehyde groups from the purine bases of DNA (Feulgen and Rossenbeck, 1924). These groups are able to react with Schiff's reagent and also to reduce silver from solutions. Procedures have been described using hydrochloric, perchloric, and trichloroacetic acid. The most frequently used method is HCl hydrolysis.

One method of hydrolysis with HCl uses immersion of tissue blocks in 1 N HCl at 60°C for 10 to 15 min. At this high temperature the time is critical, and it also varies with the specimen. A second method uses 5 N HCl on tissue blocks at

←——

FIGURE 5. RNase–gold treatment of rat liver fixed with glutaraldehyde and osmium. Labeling by the gold particles is intense over the rough endoplasmic reticulum (RER) and granular component of the nucleolus (Nu). A few particles are present over the condensed chromatin (CC). ×19,000. Courtesy of Dr. Moise Bendayan, University of Montreal.

FIGURE 6. DNase–gold treatment of rat liver fixed with glutaraldehyde and osmium. The condensed chromatin (CC) at the periphery of the nucleus and around the nucleolus (Nu) is intensely labeled with gold particles. ×25,000. Courtesy of Dr. Moise Bendayan, University of Montreal.

20°C (Deitch *et al.*, 1968). Here the time of exposure is less critical and depurination without loss of DNA occurs over a time range of 15 to 75 min.

The hydrolysis of thin sections is usually carried out (Gautier, 1976) by floating them on Marinozzi rings in 5 N HCl for 15 to 45 min at room temperature. The same procedure can also be used to treat slides holding thick-thin sections of the same material to provide a control for light or fluorescence microscopy.

Perchloric acid is also used as a hydrolyzing agent, but changes in temperature modify its characteristics. At temperatures of 60 to 90°C, rapid depurination occurs with a relatively rapid extraction of DNA. At 37 to 40°C, there is progressive extraction of both DNA and RNA.

Hydrolysis with perchloric acid is most useful at the lower temperature of 4°C where DNA is slowly depurinized without being extracted. At the same time, 80–90% of the RNA is extracted (Aldridge and Watson, 1963). The procedure may be carried out successfully on both blocks of tissue and sections.

Trichloroacetic acid hydrolysis has not been systematically studied, although it has been used either before or after nuclear treatment to extract nucleic acids and by Peters (1966a,b) to extract viral DNA. Citric acid hydrolysis appears to give identical results to HCl.

7.2. DNase and RNase Digestion of Sections

Removal of material from ultrathin sections with purified enzymes is considered diagnostic for identification, when images are compared with undigested adjacent sections. For example, DNase I will remove DNA from glycol methacrylate sections. Even if the enzyme is pure, however, negative results do not necessarily indicate that DNA is not present; this digestion may sometimes fail inexplicably. Preembedding digestion of tissue or pellets sometimes works better than digestion from sections. Solutions of DNase I are used at 0.05 to 0.1% concentrations in distilled water (pH 6.8) at 37°C usually with the addition of 3 mM Mg^{2+}. Digestion time is 1 to 4 hr.

Pancreatic RNase will remove RNA from glycol methacrylate-embedded blocks or frozen sections. RNase is often contaminated with other enzymes, but these may be eliminated by heating for 10 min at 80°C (ribonuclease survives this treatment). Blocks or sections are digested in 0.1 to 1% solutions in distilled water (pH 6.8) at 37°C for 1 to 24 hr (Moyne, 1980).

8. *IN SITU* HYBRIDIZATION

In situ hybridization was developed at the light microscope level in 1969 (Pardue and Gall, 1969; John *et al.*, 1969), and until recently, autoradiographic detection and radioactive probes were the only effective electron microscopic methods for *in situ* hybridization (Hutchison, 1984). These methods had limited resolution and were technically difficult to perform. A new procedure was reported by Manning *et al.* (1975), based on the binding of polymer spheres covalently coupled with the protein avidin to biotin-coupled nucleic acid probes hybridized to polytene chromosomes. This system depends upon the strong interaction between biotin, which is a vitamin from the B series, and avidin, a protein from egg white (Boorsma, 1983).

The procedure was further refined by Langer *et al.* (1981), who synthesized a uridine nucleotide derivative that has biotin on an alkylamine linker arm extending from the pyrimidine base. This procedure makes it possible to rapidly label DNA with biotin by nick translation, or cRNA by using the ribonucleotide derivative and RNA polymerase.

An EM-level hybridization procedure was developed by Hutchison *et al.* (1982) using mouse satellite DNA and mouse L929 whole mount chromosomes from tissue culture as a test system. This procedure produced specific labeling of C-band-positive heterochromatic regions, principally centromeres. On the average, labeling is ten times that of background binding, and the method is rapid and possesses the potential to permit ultrastructural localization of DNA sequences in chromosomes and in chromatin.

Complete details of the procedure are given in Hutchison *et al.* (1982). In brief, they are as follows (Hutchison, 1984):

1. Whole mount chromosome preparations are prepared using a modified spreading procedure (Miller and Beatty, 1969; Bakken and Hamkalo, 1978) on gold EM grids.
2. Chromosomes on grids are fixed with 0.1 to 0.5% glutaraldehyde, washed, rinsed in 0.1% Photoflo solution, and dried on filter paper.
3. DNA samples to be used are denatured in $2 \times$ SSC (0.3 M NaCl, 0.03 M Na citrate, pH 7) and adjusted carefully to pH 12 with NaOH.
4. The hybridization solution includes 0.6 M NaCl, 10% dextran sulfate, 50% formamide, and the nick-translated biotin-labeled probe DNA.
5. After hybridization, the grids are washed and incubated with primary antibiotin serum, rinsed, and incubated with colloidal gold tagged with the secondary antibody.
6. Following further washes to remove unbound colloidal gold, the grids are rinsed in Photoflo, dried, stained, and examined in the electron microscope.

Both 20-nm and 5-nm colloidal gold particles have been used in these studies, with 5-nm particles giving a greatly increased labeling intensity over satellite DNA-containing regions and a fourfold improvement in resolution over the use of 20-nm particles (Hutchison, 1984).

This author also reports that a number of the reagents used for *in situ* hybridization are now available commercially. The biotinylated nucleotide atom linker arm and goat antibiotin antibody are available from Enzo Biochemicals, New York, and many companies supply secondary antibodies. Avidin and streptavidin, as well as combinations of these with coupled antibodies or enzymatic complexes are available from several companies. Gold particles of 20- and 5-nm sizes with or without antibody tags may be purchased from Janssen Pharmaceutica Life Sciences in Beerse, Belgium.

9. GAUTIER'S HAPTA TECHNIQUE

The HAPTA (hydrochloric acid–phosphotungstic acid) technique was developed by Gautier (1968), who observed that after mild hydrolysis of aldehyde-fixed thin sections it was possible to use phosphotungstic acid (PTA), under certain conditions of dissociation, to increase the contrast of all nuclear constituents, except for those containing DNA and/or histones. Thus, the HAPTA reaction is a "nonstaining" technique that permits detection of constituents rich in DNA in a reproducible manner. The steps in the procedure are as follows.

1. Tissues are fixed with one or several aldehydes.
2. After dehydration, the tissues are embedded in Epon or Vestopal by the usual methods.
3. Sections are mounted on gold grids and hydrolyzed with 5 N HCl for 45 min at 20°C.
4. Sections are rinsed in distilled water, immersed in 1% PTA solution in 80% ethanol, pH 3.5, for 2 min at 20°C and again rinsed in distilled water.

Gautier (1976) points out that the embedding material is important, since the results with Araldite are not consistent and are poor with glycol methacrylate. The type of hydrolysis is not critical, but it is an important step in rendering the majority of biological constituents stainable by alcoholic PTA, whereas those containing DNA and histones are not stained. The HAPTA technique does not extract a significant amount of DNA or histones, since sections treated with uranyl acetate after the HAPTA reaction yield chromatin that stains as well as or better than other nuclear constituents.

The HAPTA reaction is useful in identification of eukaryotic chromatin, since it stains all cellular components *except* DNA and histones.

10. SUMMARY AND PERSPECTIVES

Clearly, the ideal stain procedure for DNA and/or RNA does not yet exist. Major steps that have improved our abilities in this area over the past several years include definition of conditions under which uranyl stains nucleic acids (Derksen and Meekes, 1984), utilization of purified enzymes to remove DNA and RNA from water-soluble embedding resins (Moyne, 1980), and critical evaluation of the EDTA regressive stain (Pearson and Davies, 1982). Recent developments in which colloidal gold-labeled enzymes are used to localize specific substrates (Bendayan, 1984a), plus proof that antibodies specific for subclasses of RNA (Fakan *et al.*, 1984) can identify these subclasses in thin sections, are very encouraging. As monoclonal antibodies are produced, screened, and used with colloidal gold labels, we can look forward to great advances in our understanding of nuclear structure and function in the near future.

11. REFERENCES

Albersheim, P., and Killias, U., 1963, The use of bismuth as an electron stain for nucleic acids. *J. Cell Biol.* **17:**93–103.

Aldridge, W. G., and Watson, M. L., 1963, Perchloric acid extraction as a histochemical technique, *J. Histochem. Cytochem.* **11:**773–781.

Anteunis, A., Pouchelet, M., Robineaux, R., and Vial, M., 1973, Ultrastructure des acides nucleiques et en particulier de l'ADN nucleolaire des cellules L 929, apres coloration par la 3-3′ diaminobenzidine (DAB) oxydee, *C. R. Acad. Sci. Ser. D* **277:**1169–1171.

Bachellerie, J.-P., Puvion, E., and Zalta, J.-P., 1975, Ultrastructural organization and biochemical characterization of chromatin–RNA–protein complexes isolated from mammalian cell nuclei, *Eur. J. Biochem.* **58:**327–337.

Bakken, A. H., and Hamkalo, B. A., 1978, Techniques for visualization of genetic material, in: *Principles and Techniques of Electron Microscopy* Vol. 9 (M. A. Hayat, ed.), pp. 84–106, Van Nostrand–Reinhold, Princeton, New Jersey.

Bendayan, M., 1981a, Ultrastructural localization of nucleic acids by the use of enzyme–gold complexes, *J. Histochem. Cytochem.* **29:**531–541.

Bendayan, M., 1981b, Electron microscopical localization of nucleic acids by means of nuclease–gold complexes, *Histochem. J.* **13:**699–710.

Bendayan, M., 1982, Double immunocytochemical labeling applying the protein A–gold technique, *J. Histochem. Cytochem.* **30:**81–85.

Bendayan, M., 1984a, Enzyme–gold electron microscopic cytochemistry: A new affinity approach for the ultrastructural localization of macromolecules, *J. Electron Microsc. Tech.* **1:**349–372.

Bendayan, M., 1984b, Protein A–gold electron microscopic immunocytochemistry: Methods, applications and limitations, *J. Electron Microsc. Tech.* **1:**243–270.

Bendayan, M., and Puvion, E., 1984, Ultrastructural localization of nucleic acids through several cytochemical techniques on osmium-fixed tissues: Comparative evaluation of the different labelings, *J. Histochem. Cytochem.* **32:**1185–1191.

Bernhard, W., 1968, Une methode de coloration regressive a l'usage de la microscopie electronique, *C. R. Acad. Sci. Ser. D* **267:**2170–2173.

Bernhard, W., 1969, A new staining procedure for electron microscopical cytology, *J. Ultrastruct. Res.* **27:**250–265.

Boorsma, D. M., 1983, Conjugation methods and biotin–avidin systems, in: *Techniques in Immunocytochemistry*, Vol. 2 (G. R. Bullock and P. Petrusz, eds.), pp. 155–174. Academic Press, New York.

Bretschneider, L. H., 1949, An electron-microscopical study of bull sperm. III, *Proc. K. Akad. Wet.* **52:**301–309.

Brown, G. L., and Locke, M., 1978, Nucleoprotein localization by bismuth staining, *Tissue Cell* **10**:365–388.

Churg, J., Mautner, W., and Grishman, E., 1958, Silver impregnation for electron microscopy, *J. Biophys. Biochem. Cytol.* **4**:841–842.

Cogliati, R., and Gautier, A., 1973, Mise en evidence de l'ADN et des polysaccharides a l'aide d'un nouveau reactif "de type Schiff," *C. R. Acad. Sci. Ser. D* **276**:3041–3044.

Deitch, A. D., Wagner, D., and Richart, R. M., 1968, Conditions influencing the intensity of the Feulgen reaction, *J. Histochem. Cytochem.* **16**:371–379.

Derksen, J., and Meekes, H., 1984, Selective staining of nucleic acid containing structures by uranyl acetate–lead citrate, *Micron Microsc. Acta* **15**:55–58.

Estable, C., and Sotelo, J., 1951, Una nueva estructura celular: el nucleolonema, *Inst. Inv. Cienc. Biol. Publ.* **1**:105–126.

Fakan, S., Leser, G., and Martin, T. E., 1984, Ultrastructural distribution of nuclear ribonucleoproteins as visualized by immunocytochemistry on thin sections, *J. Cell Biol.* **98**:358–363.

Faulk, W. P., and Taylor, G. M., 1971, An immunocolloid method for the electron microscope, *Immunochemistry* **8**:1081–1083.

Feulgen, R., and Rossenbeck, H., 1924, Mikroskopisch-chemischer Nachweis einer Nucleinsaure vom Typus des Thymonucleinsaure und die darauf beruhende elektive Farbung von Zellkernen in mikroskopischen Praparaten, *Hoppe Seyler's Z. Physiol. Chem.* **135**:203–248.

Gautier, A., 1968, Mise en evidence, sur coupes, du complexe DNA-histones par la technique "HAPTA," in: *Electron Microscopy*, Vol. II (D. S. Bocciarelli, ed.), pp. 81–82, Tipografia poliglotta vaticana, Rome.

Gautier, A., 1976, Ultrastructural localization of DNA in ultrathin tissue sections, *Int. Rev. Cytol.* **44**:113–191.

Horisberger, M., 1981, Colloidal gold: A cytochemical marker for light and fluorescent microscopy and for transmission and scanning electron microscopy, *Scanning Electron Microsc.* **II**:9–31.

Hutchison, N. J., 1984, Hybridisation histochemistry: *In situ* hybridisation at the electron microscope level, in: *Immunolabelling for Electron Microscopy* (J. M. Polak and I. M. Varndell, eds.), pp. 341–351, Elsevier, Amsterdam.

Hutchison, N. J., Langer-Safer, P., Ward, D. C., and Hamkalo, B. A., 1982, *In situ* hybridization at the electron microscope level: Hybrid detection by autoradiography and colloidal gold, *J. Cell Biol.* **95**:609–618.

Izard, J., and Bernhard, W., 1962, Analyse ultrastructurale de l'argentophilie du nucleole, *J. Microsc. (Paris)* **1**:421–434.

John, H. A., Birnstiel, M. L., and Jones, K. W., 1969, RNA–DNA hybrids at the cytological level, *Nature (London)* **223**:582–587.

Kellenberger, E., Ryter, A., and Sechaud, J., 1958, Electron microscope study of DNA-containing plasms. II. Vegetative and mature phage DNA as compared with normal bacterial nucleoids in different physiological states, *J. Biophys. Biochem. Cytol.* **4**:671–678.

Langer, P. R., Waldrop, A. A., and Ward, D. C., 1981, Enzymatic synthesis of biotin-labelled polynucleotides: Novel nucleic acid affinity probes, *Proc. Natl. Acad. Sci. USA* **78**:6633–6637.

Manning, J. E., Hershey, N. D., Broker, T. R., Pellegrini, M., Mitchell, H. K., and Davidson, N., 1975, A new method of *in situ* hybridization, *Chromosoma* **53**:107–117.

Marinozzi, V., 1961, Silver impregnation of ultrathin sections for electron microscopy, *J. Biophys. Biochem. Cytol.* **9**:121–133.

Marinozzi, V., 1964, Cytochimie ultrastructurale du nucleole RNA et proteines intranucleolaires, *J. Ultrastruct. Res.* **10**:433–456.

Marinozzi, V., 1968, Phosphotungstic acid (PTA) as a stain for polysaccharides and glycoproteins in electron microscopy, in: *Electron Microscopy*, Vol. II (D. S. Bocciarelli, ed.), pp. 55–56, Tipografia poliglotta vaticana, Rome.

Marinozzi, V., and Gautier, A., 1962, Fixations et colorations: Etude des affinites des composants nucleoproteiniques pour l'hydroxyde de plomb et l'acetate d'uranyle, *J. Ultrastruct. Res.* 7:436–451.

Miller, O. L., Jr., and Beatty, B. R., 1969, Visualization of nucleolar genes, *Science* **164**:955–957.

Moreno Diaz de la Espina, S., and Risueno, M. C., 1976, Effect of α-amanitine on the nucleolus of meristematic cells of *Allium cepa* in interphase and mitosis: An ultrastructural analysis, *Cytobiologie* **12**:175–188.

Moyne, G., 1980, Methods in ultrastructural cytochemistry of the cell nucleus, *Prog. Histochem. Cytochem.* **13**:1–72.

Pardue, M. L., and Gall, J. G., 1969, Molecular hybridization of radioactive DNA to the DNA of cytological preparations, *Proc. Natl. Acad. Sci. USA* **64**:600–604.

Pearson, E. C., and Davies, H. G., 1982, A critical evaluation of Bernhard's EDTA regressive staining technique for RNA, *J. Cell Sci.* **54**:207–240.

Peters, D., 1966a, Electron microscopic studies on the localization of deoxyribonucleic acid inside of DNA viruses, in: *Electron Microscopy,* Vol. II (R. Uyeda, ed.), pp. 195–196, Maruzen Co., Tokyo.

Peters, D., 1966b, Location of deoxyribonucleic acid inside of vaccinia virions, *J. Histochem. Cytochem.* **14**:759.

Puvion, E., and Bernhard, W., 1975, Ribonucleoprotein components in liver cell nuclei as visualized by cryoultramicrotomy, *J. Cell Biol.* **67**:200–214.

Puvion, E., Moyne, G., and Bernhard, W., 1976, Action of 3′deoxyadenosine (Cordycepin) on the nuclear ribonucleoproteins of isolated liver cells, *J. Microsc. Biol. Cell.* **25**:17–32.

Recher, L., Chan, H., and Sykes, J. A., 1973, Comparative study of microspherules and acridine orange reaction products, *J. Ultrastruct. Res.* **44**:347–354.

Risueno, M. C., Fernandez-Gomez, M. E., and Gimenez-Martin, G., 1973, Nucleoli under the electron microscope by silver impregnation, *Mikroskopie* **29**:292–298.

Roels, F., and Goldfischer, S., 1971, Staining of nucleic acids with 3,3′-diaminobenzidine (DAB), *J. Histochem. Cytochem.* **19**:713–714.

Roels, F., and Goldfischer, S., 1972, Nucleic acid staining with diaminobenzidine. II. GMA sections, in: *Int. Congr. Histochem. Cytochem.* Kyoto, pp. 211–212.

Roth, J., 1982, The protein A–gold (pAg) technique—A qualitative and quantitative approach for antigen localization on thin sections, in: *Techniques in Immunocytochemistry,* Vol. 1 (G. R. Bullock and P. Petrusz, eds.), pp. 107–133, Academic Press, New York.

Roth, J., 1983, The colloidal gold marker system for light and electron microscopic cytochemistry, in: *Techniques in Immunocytochemistry,* Vol. 2 (G. R. Bullock and P. Petrusz, eds.), pp. 217–284, Academic Press, New York.

Stockert, J. C., 1977, Sodium tungstate as a stain in electron microscopy, *Biol. Cell.* **29**:211–214.

Swift, H., and Adams, B. J., 1966, Nucleic acid cytochemistry of mitochondria and chloroplasts, *J. Histochem. Cytochem.* **14**:744.

Thiéry, J. P., 1966, Coloration sur coupe des nucleoproteines utilisable a la fois en microscopie photonique et en microscopie electronique, in: *Electron Microscopy,* Vol. II (R. Uyeda, ed.), pp. 73–74, Maruzen Co., Tokyo.

Valentine, R. C., 1958, Quantitative electron staining of virus particles, *J. R. Microsc. Soc.* **78**:26–29.

Venable, J. H., and Coggeshall, R., 1965, A simplified lead citrate stain for use in electron microscopy, *J. Cell Biol.* **25**:407–408.

Chapter 17

Colloidal Gold Labels for Immunocytochemical Analysis of Microbes

John Smit
Naval Biosciences Laboratory
School of Public Health
University of California
Berkeley, CA 94720

and

William J. Todd
Department of Veterinary Science
Louisiana Agricultural Experiment Station
Louisiana State University Agricultural Center
and
Department of Veterinary Microbiology and Parasitology
School of Veterinary Medicine
Louisiana State University
Baton Rouge, Louisiana 70803

1. INTRODUCTION

1.1. Purpose of This Chapter

In considering the spectrum of research involving biochemical and genetic analysis of microorganisms, it is apparent that understanding the location and spatial orientation of the components studied often accelerates problem solving. Research on membrane structure and function, cell growth and development processes, and mechanisms of microbial pathogenesis is particularly likely to benefit from a visual determination of the position and movements of cell constituents. The application of molecular genetics and monoclonal antibody methods to vaccine development requires a clear understanding of what components are exposed on the surface of microbial pathogens and how they must be presented to the host immune system to be exploited as vaccines. Easy and rapid methods to determine such positional or structural information are needed.

Antibodies have been the predominant basis of the direct visual examination methods used to address such concerns. Because bacteria and other microorganisms are small, the immunocytochemical tools developed for or adapted to the electron microscope are most effective.

For many years the methods available for immunocytology at the EM level have been fewer and more limited than those available to the light microscopist. The situation is rapidly improving due to advancements in methods, especially the use of colloidal gold as an EM labeling reagent.

The purpose of this chapter, then, is to describe the preparation and use of colloidal gold for EM immunocytochemistry (EM ICC), particularly as the use relates to bacterial research. To that end, this chapter will not serve as a review of research that has utilized colloidal gold. Several excellent reviews have recently been published that address the breadth of applications of colloidal gold labels and these should be consulted for additional information (Handley and Chien, 1983; De Mey, 1983a,b; De Mey and Moeremans, 1986; Horisberger, 1979, 1981a,b, 1983; Goodman et al., 1980). We provide a compendium of the types of labeling that can be done with colloidal gold and specific procedures to produce reagents to accomplish the labeling. A major theme is that the use of colloidal gold as a tool for biochemical and genetic experimentation is a relatively uncomplicated undertaking, usually easier than attempting to obtain the same information with non-EM techniques. The most straightforward procedures are emphasized, in the expectation that many who will apply colloidal gold labeling methods to their research problem do not use electron microscopy as a primary tool.

1.2. Historical and Theoretical Perspective of Colloidal Gold Sol Production

Although colloidal gold was used as a tracer in transmission electron microscopy in the 1960s and Faulk and Taylor (1971) reported immunocytochemical applications in 1971, it was not until about 1975 that an extensive body of literature on the use of colloidal gold in microscopy began to develop. Despite the resulting impression of colloidal gold being a modern reagent, gold colloids actually have a long history in chemistry. As pointed out by Horisberger (1981a), the 17th century alchemists were familiar with colloidal gold, and preparative methods were described by Faraday in 1857. Of interest to us, Faraday also determined that gold colloids could be stabilized against electrolyte-induced precipitation by first coating the particles with protein.

Despite the early start, detailed information about the precise reactions involved in producing gold colloids remains limited. An analysis of the chemistry of gold colloid formation was presented by Turkevich et al. (1955). Applying principles of reaction kinetics, they compared the effects of different reducing agents and experimental conditions on the synthesis of colloidal gold. They concluded that the important features of gold colloids, such as the particle size, size homogeneity, and stability, are determined by the nucleation process, the type of reducing agent, and the growth law.

The nucleation process is perhaps the least understood but most important event. Turkevich *et al.* proposed two models for nucleation. In one model, the reducing agents either directly cross-link gold ions, or in the reaction solution are first chemically altered to form reagents capable of cross-linking gold ions. For example, the reducing agent sodium citrate is believed to be converted, in a solution of tetrachloroauric acid molecules, to acetone dicarboxylic acid, which will then cross-link gold ions. This results in the binding together of a large number of gold and reducing agent molecules to form a macromolecular complex. When the particle size becomes just greater than that required for stability, a molecular rearrangement occurs resulting in formation of a metallic gold nucleus. The kinetics of this reaction follow those of autocatalytic events. The nuclei formed by Turkevich *et al.* were about 3 nm in diameter; this would represent the minimum stable particle size produced by reduction with sodium citrate.

The reducing agents used for the preparation of gold colloids also promote growth of the nuclei. Growth of the nuclei proceeds by reduction of the remaining tetrachloroauric acid molecules to form metallic gold, which is preferentially deposited on the surface of the nuclei. The formation of new nuclei apparently does not occur in competition with deposition of newly reduced gold onto the surfaces of existing particles. The nuclei present during the growth phase increase in size, but not in number. By regulating the extent of the growth phase, colloids of defined particle size can be generated.

These observations are of practical significance. For example, if more reducing agent is added to a constant amount of tetrachloroauric acid, a greater number of stable nuclei will be formed, and more colloidal particles of decreased average diameter will result (see Figure 3). Thus, within certain limits, the size of the particles produced can be controlled.

The second method of nucleation is exemplified by reduction with phosphorus. A solution of phosphorus in diethyl ether is mixed with tetrachloroauric acid in water. A fine emulsion of phosphorus is formed; gold ions are adsorbed on the surface on the phosphorus droplets and are reduced there to form colloid particles of sufficient size to exist as stable nuclei. This approach yields gold colloids reasonably uniform in particle size and significantly smaller than those produced by the sodium citrate method.

1.3. Characteristics of Colloidal Gold Sols with Respect to Their Use as EM Labels

Ideally, markers that can be used for EM ICC should have as many of the following characteristics as possible. They should:

1. Have an easily recognizable shape
2. Possess high electron density
3. Be capable of coupling to a wide variety of cytochemical reagents easily and with no loss of reactivity
4. Be usable for all EM ICC methods
5. Also be usable as a marker for light microscope immunocytochemistry
6. Be small enough to be used for high-resolution labeling (i.e., high degree of spatial localization of the labeled component) at the EM level
7. Be stable to electron exposure and to chemicals commonly used in EM procedures
8. Be inexpensive to prepare.

In addition, there should be more than one label available with such characteristics.

Colloidal gold particles come very close to fulfilling these criteria. The round, uniform electron-dense particles can be produced easily with routine laboratory equipment in sizes ranging from 3 to 1000 nm and because of the high atomic number of gold, are of higher contrast in the EM than the most intensely stained biological material. Under appropriate conditions, the gold particles will tightly bind to nearly any protein via strong ionic interactions with no need for cross-linking chemical reagents. (Table I lists proteins successfully bound to colloidal gold.) The great majority of macromolecules retain biological activity; beef catalase is the only macromolecule so far reported to be inactivated when bound to colloidal gold (Horisberger and Rosset, 1977). The marker can be used with all preparation techniques designed for transmission electron microscopy and scanning electron microscopy. Via color and response to plane-polarized light, colloidal gold can be used for light microscope ICC as well. The particles are very stable in the electron beam, and with few exceptions, couples of colloidal gold and labeling reagents are stable in the presence of the chemicals used in EM sample preparation methods. Finally, the presumption that labels made from gold must be costly to produce is erroneous. For example, the quantity of tetrachloroauric acid required to produce colloidal gold in an amount that can be coupled to 1 mg of protein A (i.e., enough for many labeling experiments) costs less than 15¢.

1.4. Review of Other Immunocytochemical Labels

The value of colloidal gold labels in EM ICC is best understood in comparison to the alternative labels that have been available in past years.

Ferritin has been the most widely used EM ICC label. It was introduced by Singer (Singer and Schick, 1961), and is still widely used. The small size of the label (approximately 5.5-nm iron core and 11-nm total diameter) (Hsu, 1962)

Table I
Combinations of Molecules in Which Colloidal Gold
Was Used as Marker

Molecule labeled	Reference
Antibodies	Faulk and Taylor (1971)
Asialoglycoprotein	Geuze et al. (1983)
Avidin	Tolson et al. (1981)
Cholera toxin	Montesano et al. (1982)
Collagenase	Bendayan (1982a)
Concanavalin A	Horisberger and Rosset (1977)
Dextran	Hicks and Molday (1984)
DNase	Bendayan and Puvion (1984)
Elastase	Bendayan (1982a)
Factor VIII	Furlan et al. (1981)
Fetuin	Roth et al. (1984)
Glucagon	Ackerman et al. (1983)
Helix pomatia lectin	Roth (1983b)
Horseradish peroxidase	Geoghegan and Ackerman (1977)
Insulin	Cunningham et al. (1984)
Lens culinaris lectin	Roth and Binder (1978)
Low-density lipoprotein	de Bruijin et al. (1980)
Protein A	Romano and Romano (1977)
Ricinus communis lectin	Roth and Binder (1978)
RNA polymerase	Mace et al. (1977)
RNase	Bendayan and Puvion (1984)
Soybean lectin	Roth (1983b)
Tetanus toxin	Schwab and Thoenen (1978)
Thrombin	Handley et al. (1982)
Ulex europeus lectin	Roth (1983b)
Wheat germ agglutinin	Horisberger and Rosset (1977)

permits high spatial resolution. The label has proved to be applicable to almost all types of EM ICC, including surface labeling of cells for subsequent whole mount or thin-section viewing, labeling of frozen thin-section material (Fournier-Lafleche et al., 1975), and as a surface marker for the freeze-etch replica technique.

Ferritin is not an ideal label. The iron core is often difficult to visualize because it is not especially electron-dense (Figure 1). In addition, the iron core consists of four subunits lying in a plane forming a square and the random orientation of that plane with respect to the beam renders a variety of shapes on micrographs. The result is a label that can be difficult to distinguish, especially with stained material. It is usually not suitable for the postembedding labeling methodology (the labeling of plastic ultrathin sections) because it strongly adheres to many embedding resins (Sternberger, 1967; Singer and McLean, 1963) (Figure 2). There are, however, some reports of successful labeling of ultrathin sections (Garaud *et al.*, 1982; Takamiya *et al.*, 1979).

The preparation of ferritin from horse spleens is laborious, and products purchased from commercial sources often require further purification (Breese and Hsu, 1971). In storage, ferritin may lose its electron-dense iron core and become apoferritin. Coupling of the label to immunospecific reagents is a laborious and inefficient process requiring bifunctional cross-linking reagents that may inactivate the label or cross-link ferritin molecules (Breese and Hsu, 1971), resulting in multiple ferritin molecules per labeling unit. The net effect is that while ferritin has proved to be a very useful and broadly applicable labeling tool, its use requires a certain level of commitment to the methodology.

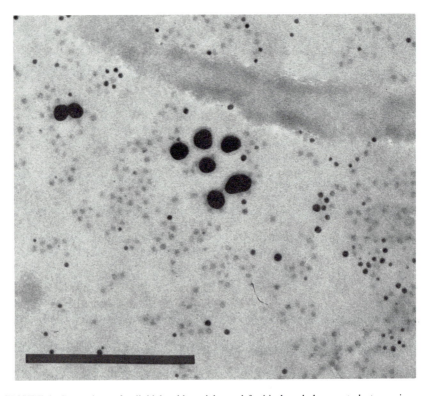

FIGURE 1. Comparison of colloidal gold particles and ferritin by whole-mount electron microscopy. Colloidal gold particles produced by phosphorus reduction (5- to 6-nm average diameter) and citrate reduction (15- to 20-nm average diameter) are seen with the lower contrast ferritin. Although the iron core of ferritin appears about the same size as the small colloidal gold particles, this unstained preparation does not reveal the proteinaceous shell, which increases the effective diameter of the particle to about twice that of the iron core. Bar = 0.2 μm.

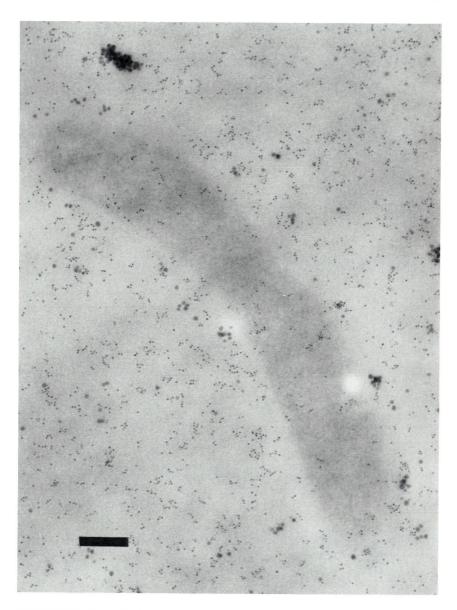

FIGURE 2. Ferritin–antibody postembedding labeling of ultrathin sections. This is an example of the difficulty in using ferritin to label plastic sections. Shown is a *Caulobacter crescentus* cell fixed with glutaraldehyde and embedded in Spurr's resin. The ferritin binds nonspecifically to the plastic matrix to such an extent that the presence of cell material reduces the labeling in the region of the cell. Antibody to *Caulobacter* flagellins was used for this micrograph but control antibody gave similar results. The larger, less-electron-dense particles are Ludox, a colloidal silica preparation that is used at an earlier step in the preparation of *Caulobacter* cells. Bar = 0.2 μm.

Sternberger (Sternberger *et al.*, 1970) introduced the unlabeled antibody peroxidase–antiperoxidase (PAP) method. In its simplest form, primary antibody and the label are connected by a secondary antibody. The label is composed of a complex of horseradish peroxidase and antibody to the peroxidase. Thus, for example, a rabbit-derived antibody to the structure of interest is connected to the PAP complex (commercially prepared from antibody raised in rabbits) with anti-rabbit antibody, often raised in a goat. The PAP complex is rendered electron-dense with the peroxidase-catalyzed reduction of compounds such as diamino-benzidine, which become insoluble when reduced and attract osmium tetroxide stain. The resulting labeling particles are of roughly uniform size and can some-times be recognized by a characteristic pentagonal shape. The PAP method is a lower-resolution label than ferritin, with a label size no smaller than about 20 nm. Moreover, because the marker complex is of relatively low contrast and of somewhat variable shape and size, debris from water or chemical sources can be mistaken for the label. Consequently, the method requires careful controls and scrupulous cleanliness. The low contrast of the label complex also affects the extent to which the labeled specimen can be stained to reveal morphological characters.

On the positive side, the PAP method can be used for postembedding labeling (Childs, 1982). Also, the method does not require cross-linking of the EM label with antibody or other reagents. Thus, much smaller amounts of antisera or lectin are required and there is no loss of activity due to chemical treatment.

Imposil, a commercially available iron dextran particle, was introduced for EM ICC by Dutton *et al.* (1979). It is a small particle with moderate electron density; its shape is distinguishable from ferritin in double-label studies. It also requires significant effort to attach antibodies or other proteins.

Other labeling reagents have also been developed. Keyhole limpet hemo-cyanin is used as a surface marker for SEM or freeze-etch replica studies (Gonda *et al.*, 1979), but is not electron-dense and thus is seldom used for ICC studies of thin-sectioned material or whole-mounted specimens. Labeling particles have also been constructed from dextran (Kent and Wilson, 1975), methacrylate (Re-mbaun and Dreyer, 1980), silica (Peters *et al.*, 1978), and other materials. All are similar to hemocyanin in that they suffer from (1) a lack of inherent electron density; and (2) difficulty in preparation of a specific label complex. In general, these latter particles are larger than ferritin, Imposil, or colloidal gold, and are at best marginally acceptable for work with bacteria.

As an overall evaluation, the methods necessary to prepare and use these (non-colloidal gold) particulate labels, and some of their inherent characteristics, make their use in EM ICC subject to more than a routine, readily acquired expertise. A high-resolution, high-contrast, easily recognized set of labels that can be quickly and efficiently coupled to immunospecific reagents was needed and has been provided by colloidal gold particles.

2. PRODUCTION OF GOLD COLLOIDS

2.1. General Considerations

All procedures for producing gold colloids are based on the reduction of tetrachloroauric acid (also called gold chloride or chloroauric acid) to elemental gold. The procedures vary both in the reductant and in reduction conditions. The result is the ability to produce colloidal gold particles in a wide variety of sizes.

Many agents will serve as reductants of tetrachloroauric acid. Elemental phosphorus, trisodium citrate, and sodium ascorbate are the three most commonly used to produce EM labels and these will be discussed in detail below. Such diverse agents as sodium borohydride, acetone, acetylene, ethanol, and tannic acid have also been used as reductants.

Once an elemental gold colloid is formed, it is stable for long periods of time by virtue of a strong negative surface charge that prevents coalescence and precipitation of gold particles. The particles are capable of interacting with more positively charged compounds. Small cations, such as sodium, disrupt the strong repellance of the gold particles and result in an irreversible precipitation of colloidal gold sols. Binding of proteins under appropriate conditions results in a stable colloidal gold preparation in the sense that colloidal gold is protected from the effects of small cations. This phenomenon is exploited in an assay to determine how much of a particular protein can be bound to the colloidal gold particles (see Section 3).

Several procedural details of colloidal gold sol preparation need to be addressed. Some of the chemicals used in colloidal gold preparation must be handled with some care. Tetrachloroauric acid is used in all the procedures and although only small quantities are needed, the chemical poses some hazard. It is a highly corrosive and hygroscopic solid; it will destroy the chromium plating on a microspatula and will cause persistent black spots on skin upon contact. Thus, latex gloves are necessary for weighing steps and flat wooden toothpicks work well in place of standard spatulas. However, this step need not be done frequently; a stock solution (generally 1%) is stable at room temperature indefinitely.

The smallest colloidal gold particles are produced by reduction of tetrachloroauric acid with elemental phosphorus. The phosphorus is used as a saturated solution in diethyl ether. Phosphorus spontaneously ignites when exposed to oxygen; diethyl ether boils at 35°C and ignites in a wide range of air/ether vapor ratios. The hazard is obvious but in practice, the problem can be significantly minimized. Solutions are prepared by cutting a piece of phosphorus (a waxy solid) and placing it in a glass container of ether. The solution is stored double-bottled, in a fume hood, in clear glass containers so that the ether level can be monitored. This solution does not require centrifugation before use and works reproducibly for at least 4 years. Many procedures specify the use of white phosphorus, which does not seem to be widely available. In our hands, yellow phosphorus (e.g., from Fisher Chemicals) works well. When the phosphorus solution is used, a copper sulfate solution (5%) should be nearby; it can be poured over a phosphorus spill to prevent spontaneous ignition.

Glassware used for the production of colloidal gold should be thoroughly cleaned and then dedicated to the procedure; siliconization is not necessary. High-quality water should be used. Particulates in water can serve as nucleating centers for colloidal gold reduction and thereby affect the size and shape of colloidal gold particles produced. Double-distilled water or water from systems such as Milli-Q (Millipore Corp.) are adequate.

Most methods for producing colloidal gold include a step in which the solutions are boiled for a period of time and most procedures specify refluxing. Refluxing is usually not necessary. Rather, the fluid level is marked before the boiling step and lost volume restored after heating.

pH adjustments of unstabilized gold colloids are often required and these should be measured with pH paper. Liquid-filled glass electrodes may release enough ions to cause some gold particle precipitation, which may ruin both the experiment and the electrode.

The following sections detail procedures for producing colloidal sol particles of various sizes, pointing out features of each. To provide a complete protocol for preparing an EM label, these procedures continue beyond colloid formation and include steps for coupling protein A to colloidal gold, resulting in a label that is adequate for most experiments (Section 2.2). Protein A is chosen as an example because it is the most common reagent coupled to colloidal gold. Section 3 is devoted to a discussion of general procedures for coupling proteins, and methods for coupling other commonly used reagents are detailed.

Most of the procedures have been adjusted to accommodate the researcher who does not do EM ICC daily. Thus, the procedures are designed to produce relatively small amounts of colloidal gold label from stock solutions using readily available equipment.

2.2. Reduction with Phosphorus—The Smallest Particles

This protocol, a slight modification of the procedure reported by Horisberger and Rosset (1977), produces particles of 5.5 to 6.0 nm in diameter. These particles provide adequate labeling resolution for nearly all studies on bacteria. Moreover, a higher labeling density is possible with this size label than for larger particles (see Section 4.1). The small particles are also generally more disperse, i.e., the particles tend not to clump after the proteins are coupled.

Procedure

A. To 24 ml of water in a 125-ml Erlenmeyer flask, add 300 μl of 1% tetrachloroauric acid and 400 μl of 0.2 N K_2CO_3. Mark the fluid level on the flask.

B. In a hood, mix 160 μl of diethyl ether and 40 μl of saturated phosphorus in diethyl ether and add to solution in A.

C. Shake for a few seconds and then let sit for 15 min at room temperature.

D. Heat solution to boiling on a hot plate; continue boiling 5 min. A deep red color will develop.

E. Cool to room temperature (an ice bath may be used if desired) and restore lost water volume.

F. Mix 20 ml of the colloidal gold solution quickly with 1 mg of protein A (0.5 ml of a 2 mg/ml stock in H_2O, stored frozen, works well). Let sit for 3 min.

G. Add 1 ml of 1% polyethylene glycol 20M (PEG 20M). This must be freshly prepared and filtered through a 0.45-μm filter just before use.

H. Centrifuge at approximately 100,000 g for 1½ hr. The supernatant is carefully aspirated away from the pellet soon after centrifugation; the pellet is somewhat loose and some supernatant will have to remain. The pellet is suspended in approximately 600 μl of 50 mM Tris (pH 7.5), 150 mM NaCl, 1 mg/ml PEG 20M. The preparation is left overnight, at 4°C, and centrifuged for 2 min in a microfuge before first use. To maintain sterility, sodium azide is added, or the colloid is filtered through a 0.45-μm filter, (See Section 3.5 for discussion of stability and long-term storage of the preparation.)

Roth (1982a) has developed a variation of this general method of reduction that results in 3-nm particles, the smallest thus far reported. Adapting that procedure to the protocol specified above, the following changes should be made:

1. Bring solution in A just to boiling.
2. Eliminate steps B through D.
3. Rapidly add 125 μl of the undiluted phosphorus in diethyl ether solution and continue boiling until an orange-red color is obtained.

Roth reports that although the 3-nm protein A–colloidal gold complex is physically stable, the protein A loses activity within 4 weeks. Protein A coupled to larger colloidal gold particles is stable for longer periods of time (see Section 3.5).

2.3. Reduction with Sodium Citrate—Intermediate to Large Particles

Frens (1973) established an effective procedure for producing gold colloid sols of uniform particle size. The average particle diameter can be adjusted from 12 to 100 nm by the amount of sodium citrate used (Figure 3).

Procedure

A. Heat 50 ml of 0.01% tetrachloroauric acid until steaming.

B. Add 1% sodium citrate rapidly with good initial mixing. The amount of

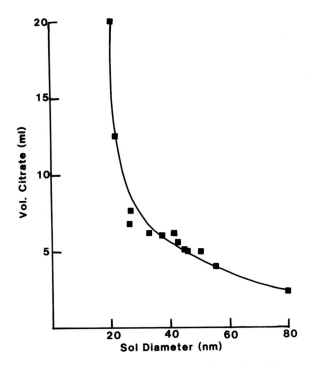

FIGURE 3. The mean diameter of colloidal gold particles produced by the reduction of tetrachlorauric acid with sodium citrate varies as a function of the amount of sodium citrate added to a constant amount of tetrachloroauric acid. Reproduced with permission from the *Journal of Microscopy (Oxford)* (Goodman *et al.*, 1981).

sodium citrate will vary from 0.2 to 1.5 ml, depending on the particle size desired. For example, 1.0 ml added will yield particles of about 17-nm average diameter.

C. Boil for 15–20 min; the solution will turn gray, then orange to red or violet depending on the particle size produced. Restore lost water volume.

D. To produce a protein A–colloidal gold complex, adjust the pH to approximately 7 with K_2CO_3 and continue with steps F through H of the phosphorus reduction procedure, except (1) use a ratio of 0.5 mg of protein A per 20 ml of gold colloid and (2) centrifuge at 60,000 *g*, 1 hr.

2.4. Reduction with Sodium Ascorbate—Broader Distribution of Particle Sizes

A more heterogeneous population of particle sizes results with ascorbate reduction. The particles can be used directly to make immunospecific labels, but more commonly this method provides starting material for size fractionation of colloidal gold particles (see Section 2.6). The procedure is adapted from Slot and Geuze (1981) (see also Horisberger *et al.*, 1978):

A. To 25 ml of distilled water chilled on ice, add 1 ml of 1% tetrachloroauric acid and 1 ml of 0.1 M K_2CO_3.
B. While stirring, quickly add 1 ml of 0.7% sodium ascorbate.
C. When the solution turns purple, adjust the volume to 100 ml with water and heat to near boiling for 20 min.
D. To produce a protein A–colloid gold complex, adjust the pH to approximately 6, continue with steps F–H of the phosphorus reduction procedure, except (1) use a ratio of 0.25 mg of protein A per 20 ml of gold colloid, and (2) centrifuge 50,000 *g*, 45 min.

2.5. Recent Methods

The methods described in the previous section are currently the most commonly used to prepare colloids of different particle diameters, but all have some deficiency. As discussed, phosphorus reduction has an element of hazard. The citrate reduction procedure produces particles that are larger than ideal for double-label experiments, and ascorbate reduction, without further steps, produces a wide distribution of particle sizes that is unsuitable for many experiments. Two newer methods have been reported that have not, as yet, received wide use but appear promising. The first (Muhlpfordt, 1982) uses tannic acid in combination with citrate and is reported to produce particles in the same size range as phosphorus reduction but without the hazard. The second method (De Mey and Moeremans, 1986), a modification of the citrate reduction method, produces 8-nm particles with a narrow size distribution, ideal for double-label experiments in combination with smaller particles. Both reports stress that the volumes and other conditions specified are critical to achieve the advertised results.

A. *Small particle sizes using tannic acid/sodium citrate reduction.* This method is reported to result in particles with an average diameter of 5.7 nm, with 95% of the particles between 3 and 8.4 nm.
  1. Bring 100 ml of 0.01% tetrachloroauric acid in a 500-ml Erlenmeyer flask to boil in 6.5 min with vigorous stirring.
  2. While the solution is boiling rapidly, add a mixture of 2 ml of 1% sodium citrate and 0.45 ml of freshly prepared 1% tannic acid (Merck).
  3. Boiling is continued for 5 min. The solution is cooled to room temperature; lost volume is restored. This solution may be stored at 4°C at this point.
  4. To produce a protein A–colloidal gold complex, adjust the pH to 7.2

with K_2CO_3 and continue with steps F through H in the phosphorus reduction procedure.

B. *8- to 10-nm particles—modified citrate reduction.* This method is essentially the reverse of the initial steps of the standard sodium citrate method, i.e., concentrated tetrachloroauric acid is added to a dilute sodium citrate solution.

1. Bring 250 ml of water to a boil.
2. Add 15 ml of freshly prepared 1% sodium citrate and continue boiling for several minutes.
3. Add 2.5 ml of 1% tetrachloroauric acid as quickly as possible to the boiling citrate solution.
4. Continue boiling for 15 min. Cool and restore lost volume.
5. To produce a protein A–colloidal gold complex, adjust the pH to approximately 7 with K_2CO_3 and continue with steps F through H of the phosphorus reduction procedure, except (1) use a ratio of 0.5 mg of protein A per 20 ml of gold colloid and (2) centrifuge at 60,000 *g*, 1 hr.

2.6. Sizing of Colloidal Gold Particles

Colloidal gold particles in the size ranges of 5 nm, 8–10 nm, or 12–100 nm can be made by the described procedures. Populations of gold particles of other size ranges can be obtained by size fractionation on a glycerol gradient of heterogeneous particle colloids produced by sodium ascorbate reduction (Section 2.4). The fractionation procedure is also useful to narrow the size range of particles used in multiple-labeling experiments to eliminate the possibility of overlapping size ranges. Experiments involving double or triple labeling can most reliably be accomplished by combining the procedures to produce particles of a defined size with the glycerol gradient method for final sizing of each population to ensure elimination of overlapping particle sizes.

Following is the method of Slot and Geuze (1981):

A. Stabilize the heterogeneous population of particles (Section 2.4) by coating them with a desired label (Section 3). It is not necessary to separate free label from bound label; removal of unbound label is also accomplished during this procedure.
B. Concentrate 30 ml of particles by centrifugation for 45 min at 50,000 *g*.
C. Aspirate the supernatant and overlay the loose portion of the pellet on a 10–30% glycerol gradient in phosphate-buffered saline (pH 7.2).
D. Centrifuge for 30 min at 20,000 rpm in an SW 41 Beckman rotor, or equivalent.
E. Collect the red-colored zone in successive 1.0-ml fractions. Fractions from the top half of the gradient contain smaller particles and are the

most uniform in size. The glycerol is usually removed by dialysis prior to use.

3. COUPLING PROTEINS TO THE COLLOIDAL GOLD SOLS

3.1. General Considerations

The mechanism by which proteins are bound to colloidal gold particles is not completely understood. The attraction of colloidal gold sols to proteins is very strong; by practical criteria, the coupled protein–colloidal gold complex does not dissociate during any of the manipulations (including long-term storage) involved with EM labeling. Binding is based on ionic interaction; the surface of the particle is apparently surrounded by negative charges formed by reorganization of water at the particle surface. The effective charge appears to be independent of pH. This simplifies the task of adapting the coupling procedure to a new protein; only the pH requirements of the untested protein need be addressed. For additional information and rationalization of the properties of colloidal gold and the factors involved in protein binding (including factors discussed below), the reader is referred to the cogent discussion given by Geoghegan and Ackerman (1977).

One important requirement for colloidal gold in preparation for binding to proteins is the maintenance of low ionic strength. Addition of cations to a colloidal gold solution neutralizes the strong repulsion of the colloidal gold particles; coalescence and flocculation results. However, some proteins are insoluble or unstable in low-ionic-strength solutions. Methods for dealing with some of these problem reagents have been developed (see Section 3.3.3).

Once a protein has been absorbed onto colloidal gold particles, the particles are protected from the effects of cations. This protection is the basis of the assay described below for titrating the amount of protein to be added to a colloidal gold preparation. Most of the protocols for producing labeled colloidal gold particles specify storage in a buffered saline solution. This may offer additional protection against the unwanted effects of any protein–colloidal gold dissociation; presumably, the high cation concentration causes precipitation of unprotected colloidal gold. Most procedures also include the addition of a high-molecular-weight PEG solution after protein coupling, an additional insurance that all binding sites are saturated. Horisberger (1979) demonstrated that PEG of 15,000–20,000 molecular weight (Union Carbide's Carbowax 20M) is dramatically more effective than polymers of larger or smaller size.

Experiments by several groups have shown that the pI of the protein must be considered when adapting the general coupling method to an untested protein (Geoghegan and Ackerman, 1977; Goodman et al., 1981). For most proteins, the best binding will occur at a pH somewhat above the pI of the protein.

Precipitation of the colloidal gold will result if too acidic a pH is chosen (apparently the protein then acts like a small cation and promotes coalescence), and poorer binding will result if the pH is set too high.

An inherent problem in coupling a polyclonal immunoglobulin to colloidal gold is that the polyclonal IgG preparations, by definition, contain a heterogeneous mixture of immunoglobulins with a range of pI values, such that at neutral pH some molecules bind well, while others will promote precipitation. Procedure modifications to reduce the problem are discussed in Section 3.3.

3.2. Titrating the Amount of Protein That Can Be Coupled

Prior to making a colloidal gold label complex with an untested protein, a titration should be done to determine the amount of protein that can be bound. The titration should be done at the pH dictated by the pI determination. Many also recommend that a titration be done for each new batch of colloidal gold or new source of protein because there is often some variation between batches. For general-purpose labels, this is probably not necessary once the amount of protein that will readily saturate all colloidal gold preparations has been established. Titration is appropriate for producing the highest labeling activity; too little protein will leave particles unlabeled, while too much protein may result in a residual amount of unlabeled protein in the preparation, which will compete for binding to the antigen. If an overabundance of probe protein is used, the colloidal gold particles may be coated with a thicker shell of the protein, thereby reducing the resolution somewhat, and increasing the likelihood that some of the probe protein may shed shortly after making the colloidal gold–probe protein complex.

In any case, a titration is clearly needed when evaluating an untested protein, or one that is expensive or difficult to prepare. The titration assay is based on the premise that unstabilized gold colloids will turn blue and precipitate upon addition of sodium chloride. This can be monitored visually or spectrophotometrically. The procedure presented here is similar to that reported by Roth and Binder (1978) and others (Slot and Geuze, 1981; Horisberger and Rosset, 1977).

A. Adjust the pH of the colloidal gold solution several tenths of a pH unit above the pI of the protein to be coupled. For polyclonal antibodies, adjust to pH 9.0; use 0.2 M K_2CO_3 to raise the pH and acetic, formic, hydrochloric, or phosphoric acid to lower the pH.

B. Mix 0.5 ml of pH-adjusted colloidal gold with 0.1 ml of serial twofold dilutions of the protein in buffer or salt solution of 5 mM or less. A starting stock solution of 0.5 to 1.0 mg/ml will generally be adequate. Wait 2–3 min.

C. Add 0.1 ml of 10% sodium chloride. Wait several minutes; if the amount of protein present is insufficient to protect the colloidal gold from salt-induced flocculation, the solution will turn from red to purple and finally blue and can be monitored spectrophotometrically at 520 nm.

D. Determine the minimum amount of protein necessary to stabilize (spectrophotometrically, the point above which the curve becomes asymptotic), increase the amount by 10–20%, and extrapolate to a preparative amount of colloidal gold sol.

3.3. Coupling Cytochemical Reagents

The great majority of colloidal gold label applications entail complexing the label with either protein A, antibody, or avidin. The coupling of these reagents to colloidal gold is addressed below, followed by a general discussion of other reagents, with particular regard to reagents that are problematical for the colloidal gold technique.

3.3.1. Protein A

In Section 2, complete protocols for making a labeling complex with this reagent were given, and a specific protocol is not repeated here. Most published studies with colloidal gold have used the label as a couple with protein A, and protein A–colloidal gold complexes are readily available commercially.

Several of the procedures in Section 2 specify a pH of about 7 for coupling, despite the fact that the pI of protein A is 5.6 and the general procedure (Section 3.3.4) would indicate an adjustment to approximately pH 6.0. However, many published accounts indicate coupling at about pH 7. It is likely that any pH between 6 and 7.5 is effective for protein A.

Advantages of protein A include high-resolution labeling due to its small size (42,000 molecular weight), bivalency for immunoglobulin (Goding, 1978), stability at low ionic strength, ready availability, and tendency to produce monodisperse labels with colloidal gold. One disadvantage in the use of protein A–colloidal gold couples has not been widely reported. The protein A–colloidal gold complex can be dissociated by treatment with osmium tetroxide solutions (Romano and Romano, 1977). For many procedures, this will have no consequence. However, surface labeling of bacteria, followed by a fixation procedure that includes treatment with osmium tetroxide, should not be attempted with protein A–colloidal gold labels. Immunoglobulins, avidin, or lectins do not seem to be dissociated by osmium tetroxide and should be used when postfixation of labeled material in osmium tetroxide is required.

3.3.2. Avidin

Avidin is a family of basic glycoproteins, approximate molecular weight 65,000, with high-affinity binding (K_d 10^{-15} M^{-1}) for biotin. Avidins are tetramers with four biotin-binding sites per molecule. They have isoelectric points of about pH 10 (Green, 1975). Avidins are produced by a variety of organisms. Those isolated from chicken eggs and *Streptomyces* are most commonly used.

Many characteristics of the avidin–biotin system are suitable for immunochemical assays. The high-affinity binding and multivalent features are most important. Avidin and biotin are available commercially, and biotin can also be purchased derivatized to react with protein or carbohydrates under mild conditions. Biotin can therefore be used to tag most molecules of biological interest and provide a marker for detection by avidin coupled to colloidal gold.

The biotin tag is small and has been covalently attached to antibodies, lectins, cells, subcellular organelles, hormones, or enzyme substrates such as nucleic acid bases, with few instances of interference in the steric-dependent characteristics of the labeled molecules (Bayer and Wilchek, 1978; Green, 1975; Morris and Saelinger, 1984). Biotin does not react with commonly used fixatives and so can be reacted with specimens prior to fixation, and later detected by the avidin–gold probe. An example of the avidin–biotin system and colloidal gold markers is shown in Figure 4.

Potential problems with the avidin–biotin system so far seem limited to occasional reports of nonspecific or unwanted binding of egg avidin (Korpela *et al.*, 1984), which is most likely due to the carbohydrate moiety of the avidin.

FIGURE 4. The survival of colloidal gold labels through the procedures of fixation in glutaraldehyde and osmium tetroxide and embedding in Spurr's resin, is demonstrated using the avidin–biotin indirect method. The capacity to produce N-acetylneuraminic acid was cloned into *E. coli*. Production was detected at polar locations during the early stages of induced expression. The location of this antigen was determined by first reacting the specimen with biotin-labeled anti-N-acetylneuramnic acid antibodies followed by incubation with avidin-coated colloidal gold. The bacteria and biotin-labeled antibodies used in this experiment were provided by Richard P. Silver at the National Center for Drugs and Biologics, Federal Drug Administration (FDA), Bethesda. Bar = 0.2 μm.

Avidin from *Streptomyces* (streptavidin) lacks carbohydrate and has been successfully used in cases where egg avidin has caused problems.

Procedure

The following procedure details the coupling of egg avidin and 15- to 20-nm
colloidal gold particles produced by citrate reduction. To couple avidin to particles of other size, a titration should be done to establish the appropriate amounts
to be added (Section 3.2).

A. Add 1.0 ml of 0.1 N K_2CO_3 to 25 ml of 15- to 20-nm-average-diameter
   colloidal gold sols. See Section 2.3 for preparation details. pH should
   be near 11.

B. Add 3.7 ml of avidin at 0.1 mg/ml (dialyzed to remove salts present in
   most commercial preparations).

C. After 5 min, add 1.5 ml of 0.1 N HCl while stirring. If aggregation
   occurs, treat 1-ml portions with 50 μl of 0.1 N HCl and then pool all
   portions.

D. Add 1.6 ml of 1% PEG 20M and adjust pH to 7.0–7.4.

E. The preparation may then be centrifuged and the pellet suspended in
   buffer similar to the procedure for protein A described in Section 2.3.

F. Alternatively, unbound avidin can be removed by ultrafiltration using
   an Amicon unit with an MX300 filter. In this case, buffers are added
   before the concentration step. The avoidance of centrifugation minimizes the formation of aggregates in avidin–colloidal gold couples and
   may be appropriate for other proteins. The method may not be suitable
   for smaller particles, which may pass through the filter, or large proteins, which may be retained.

3.3.3. Polyclonal Immunoglobulins

There are instances when indirect labeling, with colloidal gold coupled to
protein A, is inadvisable or inconvenient. For example, fragile structures such as
bacterial pili often require a direct label to reduce the number of manipulations
(see Section 5.2). Also, immunoglobulins from some animals, notably goat,
bind poorly to protein A, and preparation of colloidal gold couples with a
secondary antibody may be the best alternative to protein A.

Unfortunately, immunoglobulin has proven to be one of the more troublesome reagents to complex with colloidal gold. The problem lies in the formation of aggregates of colloidal gold particles and is at two levels. The first
problem appears after the centrifugation following mixing of antibody and colloidal gold. Often, much of the pellet will not resuspend, forming instead mac-
roscopic clumps of gold and protein that must be removed by low-speed centrifugation. The immediate effect is a significant loss of yield.

The second problem is that the label that does suspend after the first cen-trifugation step often consists of microaggregates, containing several colloidal gold particles. This can seriously reduce both the titer and the resolution of the label.

There are probably two reasons for aggregation. The first is that polyclonal IgG is a heterogeneous mix of protein molecules that share large regions of homology but also have significant portions of varying composition. Thus, the pI of an IgG preparation must be considered as a range. This causes problems; for example, a neutral pH may be too low for a fraction of the immunoglobulin population and these molecules assume the characteristics of single cations and promote flocculation. The second probable cause for aggregated labels is that IgG can self-aggregate, especially at low temperatures, and at concentrations greater than 1–2 mg/ml.

In principle, the effects of these two sources of aggregation could be mini-mized by (1) coupling at moderately high pH, such that the pI of nearly all the IgG species will be a lower value; (2) maintaining IgG solutions at approximately 1 mg/ml, and (3) centrifuging IgG preparations at high speed to remove microag-gregates. A buffer should be chosen that is effective at moderately high pH, such that low molarities can be used to avoid the problems of cationic strength. De Mey *et al.* (1981) have reported a procedure that accounts for these factors and an adaptation of that procedure is presented here. The use of PEG 20M as a secondary stabilizing agent, instead of bovine serum albumin (BSA), is the one significant variation from the method of De Mey *et al.* These authors argue that for their purposes (penetration of colloidal gold labels into eukaryotic tissue culture cells), BSA was superior to PEG 20 M. In general, however, BSA is less effective and a more problematical reagent (e.g., due to potential contamination with glycoproteins and immunoglobulins) than PEG 20M for secondary sta-bilization of immunoglobulin–colloidal gold couples.

Procedure

A. Dialyze IgG solutions (1 mg/ml) against 2 mM sodium borate (pH 9.0) overnight and centrifuge 1 hr at 100,000 g prior to the coupling step.

B. Determine the appropriate ratio of IgG and colloidal gold (Section 3.2).

C. Adjust colloidal gold sol to pH 9.0 with 0.2 N K_2CO_3.

D. Mix IgG and sols; after 2–3 min add 1% PEG 20M (freshly prepared and filtered).

E. Centrifuge the preparation at 14,000 g (15- to 20-nm particles) or 60,000 g (5-nm particles) for 1 hr.

F. Suspend in an equal amount of Tris-buffered saline (pH 7.5) containing PEG 20M at 0.5 mg/ml (TBS–PEG). Centrifuge at low speed (250 g for 15- to 20-nm particles, 4800 g for 5-nm particles) for 20 min to remove aggregates of gold particles.

G. Repeat steps E and F.
H. Centrifuge at high speed again and suspend in a small volume of TBS–PEG, comparable to that specified in the standard procedures in Section 2.

3.3.4. Coupling Other Peptides

It will be apparent that the first efforts in complexing an untested protein should follow a simple set of rules and these are reiterated in a generalized form below. Individual species of monoclonal antibodies must also be treated as untested proteins and if direct couples are necessary, best results will be obtained if the preliminary procedures are followed.

A. Determine the pI of the untested protein.
B. Titer the binding capacity of the desired size of colloidal gold sols for the untested protein at a pH several tenths of a unit higher than the pI value.
C. Prepare the protein in as low an ionic strength as possible. It is advisable to microfilter or centrifuge the protein solution at high speed before coupling.
D. Couple the protein and colloidal gold at a predetermined pH using 10–20% more protein than the titer procedure indicates and freshly filtered PEG 20M.
E. After high-speed centrifugation to concentrate the complex and remove unbound probe, determine if there is a portion of the pellet that does not suspend readily. Remove any aggregates by low-speed centrifugation. Ultrafiltration may minimize the formation of aggregates and can be substituted for centrifugation in some cases.

A modification of this general method that minimizes aggregation problems was used to prepare avidin–colloidal gold particles (Section 3.3.2). In generalized form, it involves adjusting the pH to well above the pI, so that the protein will bind only poorly. Then the pH is rapidly adjusted below the pI of the protein, promoting strong attachment of the protein before aggregation occurs. The method may be suitable for other proteins that are prone to aggregation during the coupling process.

The question of how much salt or ionic strength can be tolerated during the coupling step is not entirely resolved. Most reports emphasize that the ionic strength of the protein solution (generally of lower volume than the colloidal gold solution) should be kept below 5 mM. However, Geoghegan and Ackerman (1977) performed couplings with colloidal gold sols that were dialyzed against 7 mM buffers (see also Goodman et al., 1979). The addition of salt to coupling mixtures reduces the strength of binding proteins. Some experimentation will be necessary to determine acceptable salt concentrations for stability of some proteins while maintaining effective coupling. For example, it may be possible to formulate the probe as a concentrated solution in a necessarily high-salt buffer, add a correspondingly small volume of the protein probe solution to the colloidal gold with rapid stirring, and following secondary stabilization with PEG 20M, raise the ionic strength, even before the centrifugation step. There are a number of important probes that are unstable at low ionic strength. Concanavalin A, for which there are mixed reports of successful coupling (Geohegan and Ackerman, 1977; Horisberger and Rosset, 1977), and IgM are two examples. Experience with IgM may be particularly valuable as IgM monoclonal antibodies become more widely used in EM ICC studies.

3.3.5. Alternative Approach for Producing Colloidal Gold Labels with Recalcitrant Probes

The production of indirect labels with reagents such as protein A or avidin is the easiest way to avoid experimentation of the type discussed above, but there are times when a direct one-step label is needed. An approach taken by Robinson *et al.* (1984) was to prepare protein A–colloidal gold particles and couple specific antisera to the particles. The same approach could be taken for preparing IgG and IgM monoclonal antibody labels, using secondary antibody-treated particles that are available from commercial sources, making the production of the monoclonal antibody–colloidal gold probe relatively simple. The only significant drawback relative to direct coupling is an expected loss of resolution, due to a larger shell of protein surrounding the gold particles. The effect would be insignificant in many experiments. A similar approach can be taken with Con A (Geoghegan and Ackerman, 1977). Horseradish peroxidase (HRP) is readily coupled to colloidal gold particles, and in turn, Con A will bind to HRP via the carbohydrate moieties of the enzyme. An active label results because of the multiple valency of Con A.

3.4. Release of Coupled Proteins

Since the exact mechanism or strength of force that holds protein and colloidal gold together is not well characterized, there is some apprehension about the stability or reproducibility of the label. Most workers have noted that the labels can be used for extended periods of time but there have been only a few attempts to quantitate stability (Goodman *et al.*, 1980, 1981). Some evidence suggests that some shedding of bound proteins occurs, especially during the early weeks after prepration if a large excess of protein is used in the coupling step.

The effects of shedding are minimal. The worst possibility is that the efficiency of colloidal gold labeling will be reduced because some untagged label will compete for binding sites on the tissue. Presumably, the colloidal gold particles freed of protein will precipitate when stored in saline solutions and will not be available to bind nonspecifically to tissue. However, in experiments for which the highest sensitivity and reproducibility are required, the label should either be freshly prepared or recentrifuged just before use.

3.5. Standardization and Storage of Preparations

Some level of standardization of the colloidal gold labels is necessary. However, in most experiments, the label is supplied in large excess relative to the sites to be labeled, to ensure maximum labeling and reduce the time necessary for achieving maximum label. Thus, exact quantitation is not required; often, adopting a standard method of label preparation provides an adequate level of standardization. However, there are instances when a more quantitative measure is needed. One obvious case is when there are large or variable losses in colloidal gold marker due to aggregation after the centrifugation step. Also, use of an absorbance value may be a simpler method of reporting procedure in published results. Colloidal gold sols have strong absorbance maxima in the range 510 to 550 nm; the variation is dependent on particle size (the larger sizes have absorbance maxima at longer wavelengths) and whether protein is bound (which causes a shift to longer wavelengths) (Geoghegan et al., 1980). Generally, 520 to 525 nm is chosen for 5- to 20-nm particles.

An important consideration in standardizing colloidal gold labels is the relation between absorbance value and number of labeling particles as one relates one size of gold particle to another. This is especially important in double-labeling experiments to ensure that both labels are present in excess particle numbers relative to the number of binding sites on the tissue. If one assumes that colloidal gold sols derived from the same amount of tetrachloroauric acid have approximately similar absorbance values, simple calculations show that when comparing 5-nm particles to 20-nm particles there are eightfold more particles in solutions of the same optical absorbance.

Colloidal gold labels should be stored at 4°C where they are generally stable for long periods of time. Few precise studies have been done, but most workers point out that their labels retained activity throughout the duration of their experiments. Many probes will retain activity for at least a year. One exception noted earlier was the coupling of protein A to 3-nm gold particles. Solutions must not be frozen since some irreversible aggregation will result (Horisberger, 1981b). However, it has been reported that addition of glycerol to 15% will prevent freezing-induced aggregation if very long storage is desired.

Sterility should be maintained either by microfiltering (0.45 μm) or addition of sodium azide to 3 mM. Sodium azide must not be used with colloidal gold–HRP complexes; the enzyme is irreversibly inactivated by azide.

4. TYPES OF LABELING MADE POSSIBLE USING
COLLOIDAL GOLD COMPLEXES

In addition to the postembedding labeling method discussed in Section 6, there are several types of EM ICC labeling that are either made possible or significantly enhanced by the use of colloidal gold.

4.1. Double or Multiple Labeling

It is often desirable to tag two or more components and simultaneously visualize their relative abundance or position. Such an approach can eliminate ambiguity in interpreting results or provide precise internal controls. The ability to use colloidal gold particles of different sizes for multiple-labeling experiments has helped overcome a shortage of distinguishable labels in EM ICC (see Figure 5).

In contemplating a double-label experiment with colloidal gold, several factors should be considered. First, the particles should be as small as possible. The microscope magnifications necessary to visualize bacterial structures are sufficiently high so that the 5-nm particles are readily seen. Also, larger particles do not label as densely as smaller ones (Tolson *et al.*, 1981). The explanation for this observation may be related to either steric hindrance, i.e., some antigens are insufficiently exposed to a large-diameter probe, or shear forces, i.e., the bonding strength of the antigen–antibody couple may be readily overcome by side shear force on larger particles. Therefore, for many double-label experiments, one should choose 5-nm particles and consider making the second label either by the reverse citrate method, or by size-fractionating ascorbate-reduced particles (Section 2.6).

A second factor in double labeling is one of logistics. The most straightforward approach to multiple labeling is to make direct colloidal gold couples to the two or more antibodies used. However, this limits flexibility by not being able to use stock of colloidal gold labels for a variety of experiments. Alternatively, exclusive indirect labeling systems can be used. For example, use the primary antibody–protein A/colloidal gold pair for one antigen and biotinylated primary antibody–avidin/colloidal gold for the second. Another possibility is to use primary antibodies raised in distinct animal species and species-specific antibody coupled to colloidal gold (Tapia *et al.*, 1983).

It would seem problematical to utilize double labeling by indirect methods using the protein A coupled to two sizes of colloidal gold. However, Roth (1982a) reports that it can be done by sequential treatments with primary antibody No. 1, protein A–colloidal gold couple No. 1; primary antibody No. 2, protein A–gold couple No. 2. Roth emphasizes that it must be done with the smallest size particle (produced by phosphorus reduction) used first. If the larger colloidal gold label is used first, there is often a spurious clustering of small

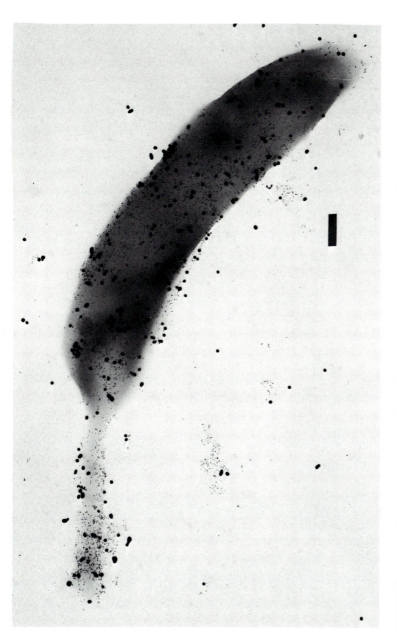

FIGURE 5. Label of a *Caulobacter crescentus* stalked cell with two sizes of colloidal gold particles. For demonstration purposes, cells were treated with antibody to the crystalline surface structure and then with protein A coupled to both 5- to 6-nm and 15- to 20-nm colloidal gold particles. The cell is viewed unfixed and unstained. The *Caulobacter* is a typical-sized bacterial cell and the two sizes of colloidal gold particles are the most readily available for double-label studies. Bar = 0.2 μm.

particles around larger particles, which may be due to weak binding of the protein A in the small particles to the Fab portion of the second IgG that coats the large particles.

Another approach to using only protein A–colloidal gold couples for double labeling is to precoat each of the labels with the specific antibody of interest, as discussed in Section 3.3. A final method is applicable only to the postembedding thin-section labeling method (see Section 6.3), where it is possible to label each side of a thin section with different labels by carefully floating sections on droplets of reagents (Bendayan, 1982b). Thus, the two indirect labels are physically separated from one another.

While two and possibly three distinct labels can be produced from colloidal gold, ferritin and Imposil are also clearly distinguishable from colloidal gold particles, and can be used along with colloidal gold labeling in many techniques. The possibilities for multiple labeling surely exceed the requirements for even the most ambitious experiments.

4.2. The GLAD Method

Larsson (1979) originally described the so-called gold-labeled antigen detection (GLAD) technique. It provides a simple but potentially useful variation of the standard indirect label method. In this method the colloidal gold particles are coated with antigen instead of antibody. The tissue is labeled with primary antibody, followed by treatment with antigen-labeled colloidal gold, which binds to unused binding sites on the bound antibody molecule. The method is useful in instances where monospecific antibody is not available, but pure antigen is. Colloidal gold renders the method practical since coupling a wide variety of molecules is relatively simple and by functional criteria, molecular conformation is only rarely altered.

4.3. Colloidal Gold Labeling and the Freeze-Fracture Method

Specific ICC labeling with the freeze-fracture/freeze-etch method is an important adjunct to whole-mount and thin-sectioning techniques, especially for work with microbes. The method allows examination of cell surfaces with much higher resolution than scanning electron microscopy (SEM). Given the small size of bacteria and the relatively large labels necessary for SEM ICC (see Section 7.1), freeze-etch or related surface replica techniques can be well worth the extra difficulty and equipment needed to prepare samples.

Since freeze-etch or surface replica methods involve viewing a platinum–carbon replica of the surface rather than the tissue itself, past efforts have required labels that have a distinctive shape rather than electron density. These labels show up in relief during the platinum shadow process. Keyhole limpet hemocyanin and ferritin are among the labels used.

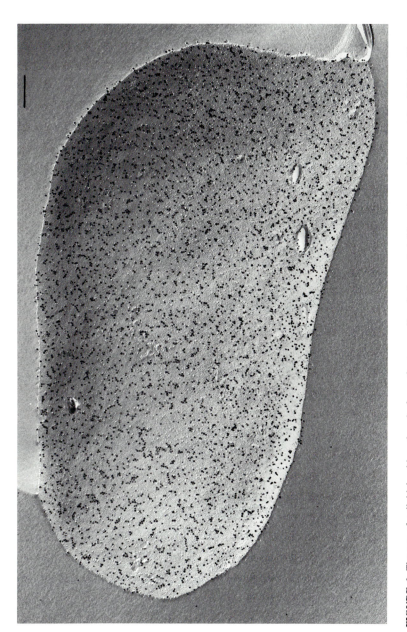

FIGURE 6. The use of colloidal gold markers in freeze-fracture replicas is demonstrated by using peroxidase-coated colloidal gold as an indirect label to detect concanavalin A-binding sites on human erythrocytes. This example of the fracture label technique is reproduced with permission from the *Journal of Cell Biology* (Pinto da Silva and Kan, 1984). Bar = 0.2 μm.

Recently, colloidal gold labels have been applied to the freeze-fracture method. Pinto da Silva and Kan (1984) introduced a "fracture label" method. Cell surfaces are specifically labeled with colloidal gold-coupled probes and the samples are prepared for freeze fracture in the usual way. Replicas are washed with water, instead of vigorous cleaning with hypochlorite or acid to remove cell material from the replicas. The result is that when viewing the outer leaflet of the plasma membrane from the inside, one can also see the colloidal gold particles attached on the membrane, which remains attached to the replica (Figure 6). It is possible to correlate labeling sites with particles or pits seen on the interior of the membrane's outer leaflet. The high density of the colloidal gold label and small size of the probe allow for high-resolution analysis.

The above replica method does not permit direct examination of labeled cell surfaces. It is also possible to look at colloidal gold-labeled surfaces via the replica technique and retain the advantages (high electron density, small size) of the particles. There are several reports of the use of colloidal gold labels with surface replica and freeze-etching techniques as applied to cultured eukaryotic cells (Mannweiler *et al.*, 1982; Robenek and Hesz, 1983). The technology should be readily transferable to the study of bacterial cell surfaces.

5. APPLICATION—ANALYSIS OF MICROBIAL SURFACES AND SURFACE STRUCTURE

5.1. General Considerations

In both bacteria and eukaryotes the most common and easiest application of EM ICC labeling is to address questions about cell surface structure and composition. Examination of surfaces by transmission electron microscopy (TEM) can be approached by thin sectioning, surface replica, freeze-etching, or simple whole-mount examination. The last method is by far the simplest.

Because bacteria are relatively small and sturdy, the whole-mount process can be effectively used to provide data with a minimum of preparative effort. The disadvantage of whole mounts is one of adequate contrast of the labels. Beam penetration of an entire bacterial cell is often poor, making it difficult to distinguish the label from underlying cell material. One can deal with the problem by examining the margins of unstained cells, where the beam penetration is better. Unfixed cells often flatten out to a greater extent than fixed cells, improving beam penetration. Searching for lysed cells can often furnish an improved image of the labeling, without affecting the experiment. Avoiding negative stain can greatly decrease electron density of the specimen, albeit at the expense of not seeing many morphological features revealed by the stain.

It is the contrast problem in whole-mount examination that colloidal gold labels have helped significantly. The labels are commonly visible without the manipulations described above, especially if 15- to 20-nm particles are used. Another advantage of colloidal gold labels for whole-mount labeling is a general absence of nonspecific binding in virtually all cell systems.

Ultrathin-section methodology has also been used for EM ICC of surface structures. Cells are surface-labeled as in the whole-mount approach, but are then fixed, embedded in epoxy resins, and cut into thin sections (see Figure 4 and Haberer and Frosch, 1982). A disadvantage is that only small portions of the surface can be analyzed since, in most cases, one only views a 100- to 150-nm-wide portion of bounding membrane suspended approximately vertically in the plastic section. A powerful advantage of this approach is the ability to correlate surface labeling with internal structures. Although there have been no published accounts, this capability will be especially pertinent when surface labeling patterns are compared to internal labeling obtained by the postembedding method discussed in Section 6. The combined use of these two methods may prove to be a very important approach to the study of synthesis and transport of membrane or secreted peptides.

5.2. Two Applications in Bacteria—Surface Structures and External Appendages

5.2.1. Surface Dynamics of a Regular Surface Structure

It is not easy to study the dynamics of surface growth in bacteria by ICC methods. The difficulty lies not in the labeling process but in the ability to manipulate the production of the surface structures or molecules under study. It is usually necessary to turn on or off the production of the particular surface component to assess the pattern of its appearance or disappearance. This requires a significant knowledge of the environmental or genetic factors that control the production of surface components. In addition, the survival of the organism may depend on the presence of the component studied to maintain membrane integrity or other vital cell function. As a consequence, there are relatively few reports describing the appearance of specific structures on the surface of bacteria. For gram-negative bacteria, the studies documenting addition of lipopolysaccharide (LPS) (Muhlradt et al., 1973), porins (Smit and Nikaido, 1978), and the bacteriophage lambda receptor of E. coli (Vos-Scheperkeuter et al., 1984) are major examples; the last example involves the use of colloidal gold labels. All these membrane components appear on the cell surface by diffuse insertions, i.e., random insertion of new components in the preexisting membrane structure.

If, however, a component is uniformly present on the surface of the cell, but is inserted by means other than a diffuse intercalation mechanism, it is relatively simple to study by indirect label methods. Such was the case for the periodic crystalline surface structure of the stalked bacterium, *Caulobacter crescentus* (Smit and Agabian, 1982). By treating cells with antibody to the surface structure, allowing a period of cell growth, and then applying the protein A–colloidal gold label, it was possible to identify the regions of the surface layer that arose by completely *de novo* growth. These areas were the stalk surface and the region of a division constriction. Growth in these areas was in contrast to a diffuse intercalation mechanism of surface layer growth over the general cell body. Figure 7 demonstrates some of those results.

The indirect labeling method was a convenient approach for this study. Growth periods of up to 2 hr between antibody labeling and viewing were necessary in the experiments. With indirect methods, it was not necessary to "encrust" the bacteria with a layer of gold particles for the growth period, thus avoiding a variety of potential artifacts.

5.2.2. Colloidal Gold Label of the *C. crescentus* Pilus

C. crescentus elaborates a few long, fragile pili at one end of the cell during the swarmer cell stage of the life cycle (Smit *et al.*, 1981). In the course of an investigation on the biogenesis of the pilus organelle, antibody was produced against the filament protein and used to decorate the pilus with colloidal gold. The most significant technological problem with the label of structures of this type is a tendency to shear them from the cell, especially during the centrifugation steps when pili are heavily coated with a layer of colloidal gold particles. However, unbound colloidal gold particles must be separated from the labeled cells. This was effectively done for the *Caulobacter* pilus labeling using a gel filtration column step in place of centrifugation (see Figure 8).

Another method of minimizing shear during labeling for such external structures is to label cells directly on the EM grid (Robinson *et al.*, 1984; Levine *et al.*, 1984). The method does require that bacterial cells adhere to the plastic support film during wash steps to remove unbound label, which will not be the case for all bacteria.

6. APPLICATION—ANALYSIS OF INTERNAL STRUCTURES AND COMPONENTS

6.1. General Considerations

Many questions regarding structure, function, and biogenesis of cellular components require label of internal structures, a much more difficult problem than surface labeling. Currently, there are three main ways to label internal structures or components for visualization in the EM:

1. EM ICC labeling of fractionated cell components
2. Postembedding labeling of ultrathin sections
3. Permeabilization of cells to permit EM labels to pass in and out of cells

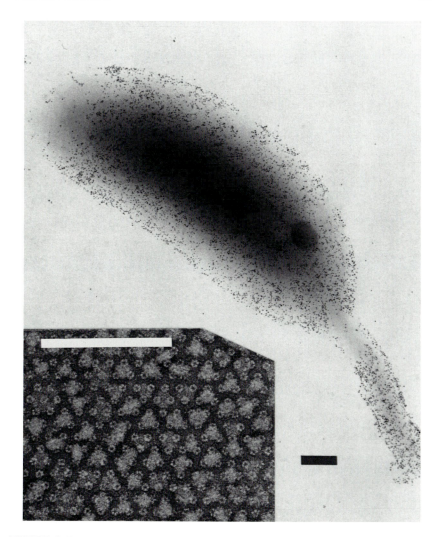

FIGURE 7. Examination of cell surface growth by colloidal gold labeling. The bacterium *C. crescentus* develops a stalk at a discrete time during the cell cycle and at a precise location in the cap. The organism also produces a complex crystalline surface structure covering the entire cell (Smit *et al.*, 1981) and whose production must be coordinated with underlying growth patterns. The inset shows a portion of an isolated fragment of intact surface layer. To produce the micrograph shown, a synchronously growing population of stalked cells was labeled with surface layer-specific antisera, allowed to grow through a division, and then treated with protein A–colloidal gold to reveal the position of bound antibody. The micrograph partially demonstrates that surface layer growth over the main cell body occurs by diffuse insertion of new material while the region surrounding a division constriction grows by completely *de novo* addition of surface layer materials, as does growth of the stalk. This is evidenced by the bare regions on the stalk and the opposite end of the cell, which is the pole generated by the preceding division. In addition, it can be seen that the stalk growth occurs at the stalk–cell junction (see Smit and Agabian, 1982). Bars = 0.2 μm.

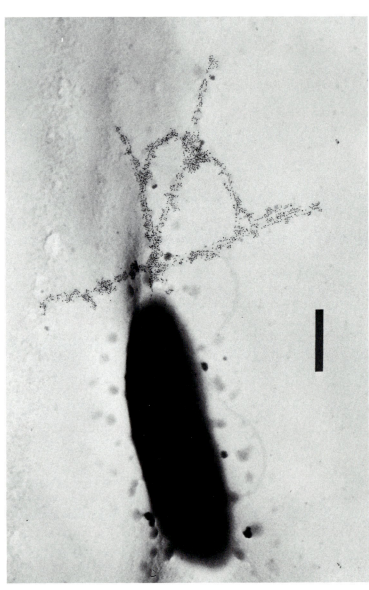

FIGURE 8. Colloidal gold label of the polar pili of *C. crescentus*. The labeling was accomplished with a direct couple of colloidal gold to pilus-specific antibody. The label was added to a growing culture of cells. Following a short incubation, the cells were directly applied to a small (0.7 × 10 cm) column and were monitored by light microscopy containing Sepharose 2B–C1 (Pharmacia) gel filtration beads. The cells fractionate with the void volume. Pooled fractions were concentrated for whole-mount electron microscopy using dialysis tubing coated with dry Sephadex G=200 beads. The gel filtration step efficiently separated cells from unbound label. The unlabeled filament seen is a flagellum, also elaborated at that pole of the cell. Bar = 0.2 μm.

The last approach is a developing technology that will likely be useful only for cells much larger and with more structural organization than typical bacteria. Bacteria, with a cell wall structure to penetrate, much smaller cell volume (i.e., it is likely that much of the cell contents can leave the cell through the same penetrations), and fewer internal organelles that can be distinguished, seem poor candidates for this method and it will not be considered in this chapter.

6.2. Labeling Subcellular Fractions

It is sometimes possible to identify internal cell structures after disrupting cells and fractionating the cell constituents. If a way can be devised to label the desired structure and then remove unbound label, localization of specific molecules within the structure can be inferred.

A good example of the use of colloidal gold labels and subcellular fractionation to precisely assess the question of subcellular organization was presented by Schenkman *et al.* (1983). This group used monoclonal antibodies directed to the LamB outer membrane protein of *E. coli* to determine the location of antigenic determinants on the surface of the protein while in its normal orientation in the outer membrane. This was accomplished using a combination of natural membrane fragments, artificial vesicles containing LamB protein, and several other pieces of biochemical and genetic data. A portion of the EM results are detailed in Figure 9. The study illustrates how a carefully designed EM ICC approach can work in concert with other types of data, to achieve a level of understanding about a problem that would be difficult to match without this methodology.

6.3. Postembedding Label of Ultrathin Sections

EM ICC labeling of ultrathin sections is a technology that has been developing for 10 to 15 years. Successful application of such a methodology provides a direct and largely unambiguous determination of the location of cell constituents. The method usually avoids the requirement for well-established biochemical fractionation procedures or genetic manipulations in order to arrive at conclusions. The application of the colloidal gold labels to the postembedding label methodology is a major factor in the recent advancement of the technology.

6.3.1. Use of Polymerized Resins

The most common method of postembedding labeling involves embedding specimens in plastic resins in an analogous, or sometimes identical manner to the standard embedding procedures developed for ultrastructure analysis. At present, for most projects, this approach should be explored first; it is technically easier to do, better ultrastructural detail usually results, and the method requires little more specialized equipment than is found in a typical EM laboratory.

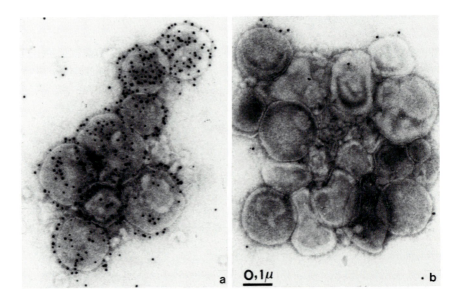

FIGURE 9. The detection of antigens in fractionated constituents of microorganisms is shown using monoclonal antibodies and protein A–colloidal gold. *E. coli* membrane vesicles were produced by sonication of spheroplasts. In (a) the LamB protein was detected by reacting the samples with an antibody directed against the exposed surface of the LamB protein followed by reaction with protein A–colloidal gold. In the control (b) the specimens were reacted with an antibody directed against an internal portion of the LamB protein prior to reaction with the protein A–colloidal gold. Reproduced with permission from the *Journal of Bacteriology* (Schenkman *et al.*, 1983). Bar = 0.1 μm.

The high-contrast colloidal gold labels are detectable even on sections of tissue treated with osmium tetroxide and stained with metal salts. (Treatment of sections with either uranyl acetate solutions at pH 4.5 or lead citrate solutions at pH 11 does not seem to significantly reduce the degree of labeling with protein A or antibody-coupled colloidal gold.) The ability to improve the tissue contrast and still see the marker molecule is a significant improvement over other labels (Figure 10). The combination of polymerized resins and high-contrast, high-resolution colloidal gold labels can result in a highly detailed localization of components within cells. As such, the approach is highly suited to the requirements of bacterial systems.

The great limitation of the resin-embedding process is whether the cell constituent under study can survive the process leading to the label. That is, are the antigenic, lectin-binding, or ligand groups of the constituent exposed to the label and in a conformation that is still recognized by the specific label? The process must be tailored to each component studied and some constituents will remain refractory to labeling.

The tailoring of the method to an individual component can be a long process since a number of variables are involved. However, there are some general guidelines that can simplify the task. Following is a brief analysis of the parts of the process that can be modified to suit a particular problem.

1. Fixation. Prior to the embedding process, the cells must be chemically cross-linked and stabilized—fixed—so that adequate ultrastructural detail is preserved. The traditional rule has been that there is an inverse relationship between the fixatives that result in the best ulstrastructure preservation and those that are the least destructive of antigenic determinants. The rule argues that a compromise between good ultrastructure and preservation of antigenicity must be made. Broadly viewed, this rule is still true; however, there appear to be numerous exceptions, such that the rule need not be applied dogmatically at the outset.

As a primary fixative, glutaraldehyde is among the best for structure preservation. However, it often greatly reduces immunoreactivity. Formaldehyde (prepared from paraformaldehyde) is frequently less harmful in this way and often provides adequate or even good structure preservation. In some cases where formaldehyde gives poor fixation, a small amount of added glutaraldehyde can greatly improve ultrastructure preservation while still allowing good label of a particular antigen (Eldred *et al.*, 1983; Tokuyasu, 1984). It has also been reported that a significant degree of restoration of antigenicity after glutaraldehyde treatment can be achieved by subsequent treatment with sodium borohydride (Eldred *et al.*, 1983). There are also reports of improved immunoreactivity of a substance by fixation with formaldehyde for a short period of time at high pH when it is chemically very reactive (Berod *et al.*, 1981).

Soluble carbodiimide compounds have also been used alone (Polak *et al.*, 1972) or in combination with aldehydes (Willingham and Yamada, 1979) as fixatives for ICC studies. Since the chemical targets of these reagents are different, carbodiimides may be a logical choice if efforts with aldehyde fixatives are unsuccessful.

Osmium tetroxide is commonly used as a secondary fixative in ultrastructure analysis since it is a potent fixative and also stains tissue well. But it is widely assumed that osmium tetroxide destroys immunoreactivity. However, numerous examples of successful ICC labeling using osmium tetroxide postfixation have been reported (Jackowski *et al.*, 1985; Bendayan and Zollinger, 1983; Bendayan, 1984; Rohde *et al.*, 1984) and it is worth an attempt to use this fixative in view of the high-quality ultrastructure preservation.

2. The Embedding Process. Three aspects of the process of taking fixed tissue from an aqueous environment to a polymerized plastic need be considered: the removal of water and replacement with unpolymerized plastic; the treatments required to polymerize the plastic; and the effects of the plastic itself on immunoreactivity.

Standard plastic resin formulations used for the thin-section technique (e.g., Epon, Araldite, Spurr's) require a transition solvent, usually ethanol, that "de-

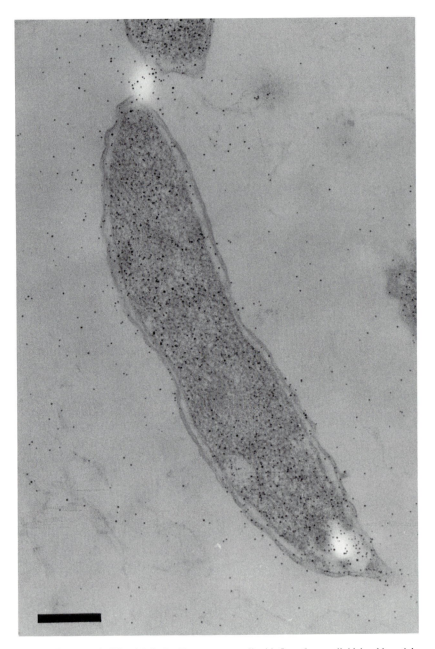

FIGURE 10. Postembedding label of a *C. crescentus* cell with 5- to 6-nm colloidal gold particles. This micrograph demonstrates the high-resolution label that is possible with small, high-contrast colloidal gold particles. Cells were fixed with paraformaldehyde and embedded in Araldite by standard methods. Sections were cut and "etched" with 10% H_2O_2 prior to incubation with affinity-

hydrates'' the specimen and allows penetration of the polymers, which are not water soluble. Many procedures also require propylene oxide as an additional transition solvent. Ethanol is infused as a graded series and the process often takes several hours. Such solvents and the time of exposure to them may have a deleterious effect on immunoreactivity. Faster dehydration can be achieved by using 2,2-dimethoxypropane (Muller and Jacks, 1975); dehydration is accomplished *in situ* in a few minutes since dimethoxypropane and water chemically react to form acetone and methanol—solvents that are miscible with embedding resins. We are not aware of a report of successful use of this dehydration method for EM ICC but the rapidity of the process may be helpful in some cases.

Other types of resins permit dehydration of samples without a transition solvent since they are water soluble before polymerization. Early formulations were based on butyl or glycol methacrylates (Figure 11); recently, the Lowicryl formulations (Altman *et al.*, 1983, 1984) offer the ability to dehydrate directly as well as additional advantages discussed below. Another approach to dehydration that has been successfully used in EM ICC is the use of low-temperature or freeze-substitution methods. Quick freezing of tissue can also replace chemical fixation in some cases. The use of low temperature can be carried all the way through the polymerization process and can serve to protect sensitive compounds (Armbruster *et al.*, 1982, 1983; Carlemalm *et al.*, 1980). A complete description of the process is the topic of another chapter in this volume.

Another aspect of the embedding process that affects immunoreactivity is the polymerization of the resin. Polymerization of Epon, Araldite, and similar resins is accomplished with tertiary amine compounds as initiators and long periods of high temperatures (55 to 70°C, 8 to 16 hr). With methacrylates and the Lowicryl formulations, polymerization can be initiated with benzoyl peroxide, a strong oxidant, or ultraviolet light. With the possible exception of ultraviolet light, all these chemicals or conditions have significant potential for reducing immunoreactivity of a cell component, via conformational or chemical alteration.

The final major consideration in the embedding process is the effect of the resin itself. Cell components in plastic are exposed to a very different environment with respect to such factors as hydrophobicity/hydrophilicity, hydrogen bonding, or ionic strength, than in their natural condition and some may not retain a semblance of their natural conformation within the resin. Hydrophilic resins such as glycol methacrylate, Lowicryl K4M, or LR Gold present an environment more similar to water than do Epon or similar resins and may offer

---

purified anti-pilus protein antibody and protein A–colloidal gold. After labeling, the sections were stained with uranyl acetate and lead citrate to improve specimen contrast. The micrograph is part of a demonstration that the subunit protein for the pilus is made and stored throughout the cytoplasm of the cell prior to assembly at one pole of the cell (see Figure 8) (Smit, unpublished). Bar = 0.2 μm.

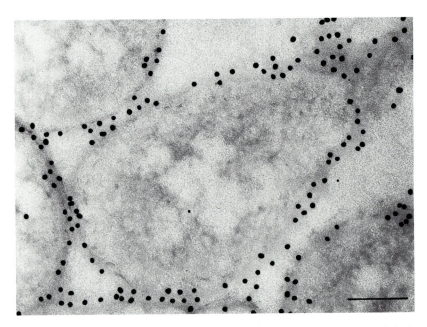

FIGURE 11. The detection of antigens in thin sections of methacrylate-embedded bacteria is demonstrated using protein A–colloidal gold. Thin sections of fixed and embedded *Pasteurella haemolytica* were reacted with antibodies raised against purified capsular antigen and protein A–colloidal gold. Reproduced with permission from the *Histochemical Journal* (Beesley *et al.*, 1984). Bar = 0.2 μm.

an improvement in the number of components that retain immunoreactivity. Major drawbacks of these hydrophilic resins include generally poorer stability in the electron beam and often the tissue has significantly less contrast.

   3. Penetration of Resins by ICC Reagents. Balanced against the advantages of plastic embedding media for EM ICC discussed above is the fact that those media are not very permeable to the labeling reagents. Typical ultrathin sections are 100 to 200 nm thick and penetration can be assumed to extend no farther than 5 to 10 nm. Sections of tissue embedded in Epon or Araldite are usually "etched" with hydrogen peroxide, sodium metaperiodate, or other solvents that are assumed to remove some of the plastic and expose more tissue to the labeling reagent. However, these chemicals have significant potential to affect antigens. Work with methacrylate resins and the PAP label indicated that no etching was needed with methacrylate resins (Childs, 1982). Presumably, the methacrylate resins [including Lowicryl (Roth, 1983b)] are at least somewhat more permeable to ICC reagents than epoxy resins.

   That plastic sections have serious limitations in the penetration of ICC reagents is an important factor in assessing the feasibility of a proposed experiment. If the constituent of interest is present in small amounts or is located diffusely in the cytoplasm, it may be difficult to get an adequate signal with this method simply because too little antigen is at the section surface. Membrane components may be the best candidates for this methodology because of a higher degree of localization. If it appears that the amount or the degree of localization of the desired components is marginal, consider the more difficult or costly alternatives discussed in the next section. The main advantage that these methods offer is greater accessibility of the ICC labeling reagents to the tissue.

6.3.2. Other Approaches to ICC Labeling of Thin Sections

At this time it is our view that alternative methods to the labeling of plastic sections should be considered only after it is apparent that immunoreactivity of the cell component studied is not maintained or that the expected location or amounts of the component argue against the use of a poorly penetrable embedding medium. The main alternatives are frozen sectioning, PEG embedding, and freeze-dry/resin embedding.

Frozen sectioning is the process by which biological material is not embedded but is directly cut into sections while frozen. The process requires an expensive modification of a microtome and considerable technical experience. The details of the process are beyond the scope of this chapter and the reader is referred to Tokuyasu (1984). Material prepared in this way may be labeled with colloidal gold (Geuze *et al.*, 1981, 1983) as well as ferritin or Imposil. As in other procedures, care must be exercised in the level of tissue staining with the ferritin or Imposil labels.

In addition to the difficulty and expense of the frozen-section method, there is often little morphological detail visible in cells prepared by current procedures. However, adequate results are possible in some cases; the work of Tinglu *et al.* (1984) is shown in Figure 12. (See also Beesley *et al.*, 1982, and Walker and Beesley, 1982.) In addition, a recently reported method, involving dehydration and plastic embedding of frozen sections *after labeling* (Keller *et al.*, 1984), allows additional staining and offers a significant improvement in morphological detail.

Presumably, ICC labeling reagents have greater access to internal components due to absence of embedding medium during labeling, but in the absence of embedding medium one must also account for the possibility that internal components may diffuse away or rearrange.

Another alternative to plastic embedding is to embed tissue in PEG (Kondo, 1984; Wolosewick, 1980; Wolosewick *et al.*, 1983). Appropriate molecular weight fractions of PEG (E.g., PEG-4000) are solid at room temperature and ultrathin sections can be prepared from PEG-infiltrated tissue. ICC reagents are able to penetrate tissue sections to an extent comparable to frozen sections. The sections may be stained followed by dehydration with ethanol and critical point-

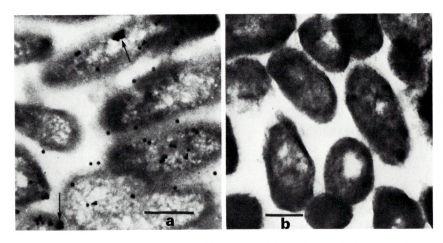

FIGURE 12. The detection of antigens in thin sections of frozen bacteria is demonstrated using colloidal gold particles coated with anti-alkaline phosphatase IgG. (a) The label tends to be clustered when associated with the cell envelopes (arrows) and as isolated particles in the cytoplasm (arrowheads). (b) Control prepared by pretreatment of the sections with free anti-alkaline phosphatase IgG prior to the addition of the colloidal gold markers. Reproduced with permission from the *Journal of Bacteriology* (Tinglu *et al.*, 1984). Bars = 0.5 μm.

drying. The method has been successfully used with colloidal gold labels. A disadvantage, relative to frozen sectioning, is that high temperature (i.e., 60°C) and alcohols, both of which must be used to embed the tissue, may affect immunoreactivity. There are only a few published reports of the PEG approach, but the ease of the procedure relative to frozen sectioning may encourage researchers to test this alternative to plastic embedding for ICC.

An alternative to standard fixation and plastic embedding procedures that may be suitable if chemical fixation must be avoided is termed freeze-dry/embedding. The tissue is quickly frozen, freeze-dried, and then infiltrated with standard plastic resins (e.g., Epon) (Dudek *et al.*, 1982). Ultrastructure preservation is reasonably good and ICC localization was successfully applied. One shortcoming is a need for a relatively uncommon rapid-freezing mechanical device.

7. APPLICATIONS IN OTHER TYPES OF MICROSCOPY

7.1. Scanning Electron Microscopy (SEM)

Because of inherent limitations in resolution, SEM is infrequently used for ICC labeling of microbial surfaces in comparison to transmission electron microscopy (TEM). For bacteria, the freeze-etch or surface replica techniques provide a much-higher-resolution examination of labeled surfaces. Yet there are instances when the capabilities of SEM are needed for labeling studies with bacteria. The best examples would be in cases where it is necessary to label and examine bacteria *in situ,* attached to mucosal surfaces or firm substrata in marine or freshwater environments.

Colloidal gold markers have been used in SEM since 1975 (Horisberger *et al.,* 1975). Unfortunately, with the standard SEM method of analysis by secondary electron imaging, colloidal gold markers can be confused with surface features of cells or can be masked by other cell structures and added specimen coatings. This is especially true when particle sizes of 15 nm or less are used as markers and the specimens are metal- or carbon-coated as is generally necessary to reduce charging and provide adequate specimen resolution. However, the problems in marker detection can be overcome with a microscope adapted for backscattered electron imaging (De Harven *et al.,* 1984). Because of the high atomic number of gold in comparison to the atomic numbers of specimen constituents, outstanding contrast between the gold markers and specimen constituents can be obtained using backscattered electron imaging. Although the gold labels are identified with certainty, the overall resolution of specimen detail is poor in the backscatter mode. To achieve both good resolution of the specimen by secondary electron imaging and identification of the gold labels by backscattered electron imaging, one can separately photograph the specimen by both modes and compare the images or compromise the recorded image by mixing both imaging modes. The former approach provides the best spatial detail and accuracy of label detection. Specimen charging can occur with uncoated specimens but can usually be reduced through use of a thin carbon coating, lower accelerating voltages, and sample preparation methods developed to improve internal conductance of the specimen (Fegia-Valasco and Arauz-Contrerces, 1981).

7.2. Colloidal Gold Labels in Light Microscopy

For light microscopy, the labeling methods using fluorescent compounds and enzymes as markers are well established and very effective. Although there is not a great need for alternative markers in light microscopy, there are some potential applications of interest for colloidal gold labels.

There are two approaches to viewing colloidal gold markers by light microscopy. The simplest but least sensitive method is direct examination by transmitted light; the labels appear red (Roth, 1982a). The sensitivity of direct examination can be increased by chemically increasing the size of the labels *in situ* by a silver precipitation procedure, which results in a black label (Holgate *et al.,* 1983). A second approach to view colloidal gold markers by light microscopy is through the use of backscattered polarized light (De Mey, 1983b). Using polarizing filters and epi-illumination, aggregates of larger gold particles (20–40 nm) can be detected as bright yellow sources of light reflected from the specimen.

The limits of this method have not as yet been determined. Possible advantages of colloidal gold markers, in comparison to fluorescent labels, are that the gold markers do not quench, and the cellular background can be easily coobserved with transmitted light. Enzyme–antibody couples avoid some of the problems of fluorescent labels, but enzymes that utilize the same substrates as the marker enzyme (e.g., peroxidase) may also be present in the specimen, resulting in background staining unless successfully blocked.

In addition to the surface labeling of whole cells, colloidal gold markers have also been applied to thin sections of paraffin-embedded specimens (Roth, 1982b, 1983a). Surprisingly, even plastic-embedded specimens can be studied by light microscopy using colloidal gold probes (Danscher and Rytter-Nörgaard, 1983).

Because of the availability of other effective labels, applications requiring colloidal gold labeling of bacteria for viewing by light microscopy are likely to be more limited. However, detection by light microscopy does provide a useful method to monitor the gold labels prior to examination by EM, and provides a way to select the labeled portions from large specimen samples. Such simultaneous monitoring in the light and electron microscope can also be accomplished by preparing colloidal gold couples with fluorescein-labeled probes (Horisberger and Vonlanthen, 1979), thereby improving the detection sensitivity for the light microscope via fluorescence microscopy.

8. CONCLUDING COMMENTS

The utility of colloidal gold particles for microscopy is now well documented and their use in ICC is rapidly becoming commonplace. In the future it will be interesting to see what other uses for colloidal gold will develop, largely spawning from microscopy applications. Already, colloidal gold particles are being used as a sensitive detection method for dot blot (Moeremans *et al.*, 1984) and Western blot immunoassays (Brada and Roth, 1984), and as a good alternative for coated red blood cells in a passive agglutination immunoassay method (Geoghegan *et al.*, 1980). It is possible to make radioactive colloidal gold particles using [198]Au or [195]Au during the reduction step (Pratten *et al.*, 1977; Winlove *et al.*, 1981; Kent and Allen, 1981), which may lead to the development of a variety of nonmicroscopic immunoassay procedures. If the degree of success achieved by microscopists is a measure, the future for colloidal gold in biological research should be bright indeed.

9. REFERENCES

Ackerman, G. A., Yang, J., and Wolken, K. W., 1983, Differential surface labeling and internalization of glucagon by peripheral leukocytes, *J. Histochem. Cytochem.* **31**:433–440.

Altman, L. G., Schneider, B. G., and Papermaster, D. S., 1983, Rapid (4 hr) method for embedding tissues in Lowicryl for immunoelectron microscopy, *J. Cell Biol.* **97**:309a.

Altman, L. G., Schneider, B. G., and Papermaster, D. S., 1984, Rapid embedding of tissues in Lowicryl K4M for immunoelectron microscopy, *J. Histochem. Cytochem.* **32**:1217–1223.

Armbruster, B. L., Carlemalm, E., Chiovetti, R., Garavito, R. M., Hobot, J. A., Kellenberger, E., and Villiger, W., 1982, Specimen preparation for electron microscopy using low temperature embedding resins, *J. Microsc. (Oxford)* **126**:77–85.

Armbruster, B. L., Garavito, R. M., and Kellenberger, E., 1983, Dehydration and embedding temperatures affect the antigenic specificity of tubulin and immunolabeling by the protein A–colloidal gold technique, *J. Histochem. Cytochem.* **31**:1380–1384.

Bayer, E. A., and Wilchek, M., 1978, The avidin–biotin complex as a tool in molecular biology, *Trends Biochem. Sci.*, **3**:257–259.

Beesley, J. E., Orpin, A., and Adlam, C., 1982, A comparison of immunoferritin, immunoenzyme and gold-labelled protein A methods for the localization of capsular antigen on frozen thin sections of the bacterium, *Pasteurella haemolytica, Histochem. J.* **14**:803–810.

Beesley, J. E., Orpin, A., and Adlam, C., 1984, An evaluation of the conditions necessary for optimal protein A–gold labelling of capsular antigen in ultrathin methacrylate sections of the bacterium *Pasteurella haemolytica, Histochem. J.* **16**:151–163.

Bendayan, M., 1982a, Protein A–gold and enzyme gold: Two novel affinity techniques for ultrastructural localization of macromolecules, *Int. Congr. Electron Microsc.* **2**:427–439.

Bendayan, M., 1982b, Double immunocytochemical labeling applying the protein A–gold technique, *J. Histochem. Cytochem.* **30**:81–85.

Bendayan, M., 1984, Concentration of amylase along its secretory pathway in the pancreatic acinar cell as revealed by high resolution immunocytochemistry, *Histochem. J.* **16**:85–108.

Bendayan, M., and Puvion, E., 1984, Ultrastructural localization of nucleic acid through several cytochemical techniques on osmium-fixed tissues, *J. Histochem. Cytochem.* **32**:1185–1191.

Bendayan, M., and Zollinger, M., 1983, Ultrastructural localization of antigenic sites on osmium-fixed tissues applying the protein A–gold technique, *J. Histochem. Cytochem.* **31**:101–109.

Berod, A., Hartman, B. K., and Pujol, D., 1981, Importance of fixation in immunohistochemistry: Use of formaldehyde solutions at variable pH for the localization of tyrosine hydroxylase, *J. Histochem. Cytochem.* **29**:844–850.

Brada, D., and Roth, J., 1984, "Golden blot"—Detection of polyclonal and monoclonal antibodies bound to antigens to nitrocellulose by protein-a-gold complexes, *Anal. Biochem.* **142**:79–83.

Breese, S. S., and Hsu, K. C., 1971, Techniques of ferritin-tagged antibodies, in: *Methods in Virology*, Vol. 5 (K. Maramorosch and H. Koprowski, eds.), pp. 399–422, Academic Press, New York.

Carlemalm, E., Garavito, R. M., and Villiger, W., 1980, Advances in low temperature embedding for electron microscopy, *Electron Microsc.* **2**:656–657.

Childs (Moriarty), G. V., 1982, Use of Immunocytochemical techniques in cellular endocrinology, in: *Electron Microscopy in Biology*, Vol. 2 (J. D. Griffith, ed.), pp. 107–174, Wiley, New York.

Cunningham, V. L., Walker, K. W., and Ackerman, G. A., 1984, Insulin complex binding to human peripheral and mitogen-stimulated lymphocytes, *J. Histochem. Cytochem.* **32**:517–525.

Danscher, G., and Rytter-Nörgaard, J. O., 1983, Light microscopic visualization of colloidal gold on resin-embedded tissue, *J. Histochem. Cytochem.* **31**:1394–1398.

de Bruijn, W. C., Emeis, J. J., and Vermeer, B. J., 1980, The application of analytical electron microscopy in the localization of individual LDL-binding sites on cell surfaces, *Artery* **8**:281–287.

De Harven, E., Leung, R., and Christensen, H., 1984, A novel approach for scanning electron microscopy of colloidal gold-labeled cell surfaces, *J. Cell Biol.* **99**:53–57.

De Mey, J., 1983a, A critical review of light and electron microscopic immunocytochemical techniques used in neurobiology, *J. Neurosci. Methods* **7**:1–18.

De Mey, J., 1983b, Colloidal gold probes in immunocytochemistry, in: *Immunocytochemistry: Practical Applications in Pathology and Biology* (J. M. Polak and S. Van Noorden, eds.), pp. 82–112, Wright, Bristol.

De Mey, J., and Moeremans, M., 1986, Preparation of colloidal gold probes and their use as markers in electron microscopy, in: *Advances in Biological Electron Microscopy,* Vol. III, Springer-Verlag, Berlin, in press.

De Mey, J., Moeremans, M., Geuens, G., Nuydens, R., and De Brabander, M., 1981, High resolution light and electron microscopic localization of tubulin with the IGS (immuno gold staining) method, *Cell Biol. Int. Rep.* **5:**889–899.

Dudek, R. W., Childs, G. V., and Boyne, A. F., 1982, Quick-freezing and freeze-drying in preparation for high quality morphology and immunocytochemistry at the ultrastructural level: Application to pancreatic beta cell, *J. Histochem. Cytochem.* **30:**129–138.

Dutton, A. H., Tokuyasu, K. T., and Singer, S. J., 1979, Iron–dextran antibody conjugates: General method for simultaneous staining of two components in high resolution immunoelectron microscopy, *Proc. Natl. Acad. Sci. USA* **76:**3392–3396.

Eldred, W. D., Zucker, C., Karten, H. J., and Yazulla, S., 1983, Comparison of fixation and penetration enhancement techniques for use in ultrastructural immunocytochemistry, *J. Histochem. Cytochem.* **31:**285–292.

Faulk, W. P., and Taylor, G. M., 1971, An immunocolloid method for the electron microscope, *Immunochemistry* **8:**1081–1083.

Fegia-Valasco, A., and Arauz-Contrerces, J., 1981, Ruthenium red-mediated osmium binding for examining uncoated biological material under the scanning electron microscope, *Stain Technol.* **56:**71–78

Fournier-Lafleche, D., Chang, A., Benichou, J. C., and Ryter, A., 1975, Immuno-labelling in frozen ultrathin sections of bacteria, *J. Microsc. Biol. Cell.* **23:**17–28.

Frens, G., 1973, Controlled nucleation for the regulation of the particle size in monodisperse gold suspensions, *Nature Phys. Sci.* **241:**20–22.

Furlan, M., Horisberger, M., Perret, B. A., and Beck, E. A., 1981, Binding of colloidal gold granules, coated with bovine factor VIII, to human platelet membranes, *Br. J. Haematol.* **48:**319–324.

Garaud, J. C., Doffoel, M., Stock, C., and Grenier, J. F., 1982, Are "G" cells the only source of gastrin in the human antrum?, *Biol. Cell* **49:**165–176.

Geoghegan, W. D., and Ackerman, G. A., 1977, Adsorption of horseradish peroxidase, ovomucoid and anti-immunoglobulin to colloidal gold for the indirect detection of concanavalin A, wheat germ agglutinin and goat anti-human immunoglobulin G on cell surfaces at the electron microscopic level: A new method, theory and application, *J. Histochem. Cytochem.* **25:**1187–1200.

Geoghagen, W. D., Ambegaonkar, S., and Calvanico, N. J., 1980, Passive gold agglutination: An alternative to passive hemagglutination, *J. Immunol. Methods* **34:**11–21.

Geuze, H. J., Slot, J. W., van der Ley, P., and Scheffer, R. C. T., 1981, Use of colloidal gold particles in double-labeling immunoelectron microscopy of ultrathin frozen tissue sections, *J. Cell Biol.* **89:**653–665.

Geuze, H. J., Slot, J. W., Strous, G. J. A. M., Lodish, H. F., and Schwartz, A. L., 1983, Intracellular site of asialoglycoprotein receptor–ligand uncoupling: Double-label immuno-electron microscopy during receptor-mediated endocytosis, *Cell* **32:**277–287.

Goding, J., 1978, Use of staphylococcal protein A as an immunological reagent, *J. Immunol. Methods* **20:**241–253.

Gonda, M. A., Gelden, R. V., and Hsu, K. C., 1979, An unlabelled antibody molecule technique using hemocyanin for the identification of type B and type C retrovirus envelope and cell surface antigens by correlative fluorescence, transmission electron, and scanning electron microscopy, *J. Histochem. Cytochem.* **27:**1445–1454.

Goodman, S. L., Hodges, G, M., Trejdosiewicz, L. K., and Livingston, D. C., 1979, Colloidal gold probes—A further evaluation, *Scanning Electron Microsc.* 619–628.

Goodman, S. L., Hodges, G. M., and Livingston, D. C., 1980, A review of the colloidal gold marker system, *Scanning Electron Microsc.* **1980 (II)**:133–146.

Goodman, S. L., Hodges, G. M., Trejdosiewicz, L. K., and Livingston, D. C., 1981, Colloidal gold markers and probes for routine application of microscopy, *J. Microsc. (Oxford)* **123**:201–213.

Green, N. M., 1975, Avidin, *Adv. Protein Chem.* **29**:85–133.

Haberer, K., and Frosch, D., 1982, Lateral mobility of membrane-bound antibodies on the surface of *Acholeplasma laidlawii:* Evidence for virus-induced cell fusion in a procaryote, *J. Bacteriol.* **152**:471–478.

Handley, D. A., and Chien, S., 1983, Colloidal gold: A pluripotent receptor probe (41697), *Proc. Soc. Exp. Biol. Med.* **174**:1–11.

Handley, D. A., Stifano, T. M., and Saunders, R. N., 1982, Cellular events of platelet receptor binding of colloidal gold–thrombin probes, *Fed. Proc.* **41**:899a.

Hicks, D., and Molday, R. S., 1984, Analysis of cell labelling for scanning and transmission electron microscopy, in: *The Science of Biological Specimen Preparation for Microscopy and Microanalysis* (J.-P. Revel, T. Barnard, and G. H. Haggis, eds.), pp. 203–219, SEM, Inc., Chicago.

Holgate, C. S., Jackson, P., Cowen, P. N., and Bird, C. C., 1983, Immunogold–silver staining: New method of immunostaining with enhanced sensitivity, *J. Histochem. Cytochem.* **31**:938–944.

Horisberger, M., 1979, Evaluation of colloidal gold as a cytochemical marker for transmission and scanning electron microscopy, *Biol. Cell.* **36**:253–258.

Horisberger, M., 1981, Colloidal gold as a cytochemical marker in electron microscopy, *Gold Bull.* **14**:90–94.

Horisberger, M., 1981b, Colloidal gold: A cytochemical marker for light and fluorescent microscopy and for transmission and scanning electron microscopy, *Scanning Electron Microsc.* **II**:9–31.

Horisberger, M., 1983, Colloidal gold as a tool in molecular biology, *Trends Biochem. Sci.* **8**:395–397.

Horisberger, M., and Rosset, J., 1977, Colloidal gold, a useful marker for transmission and scanning electron microscopy, *J. Histochem. Cytochem.* **25**:295–305.

Horisberger, M., and Vonlanthen, M., 1979, Fluorescent colloidal gold: A cytochemical marker for fluorescent and electron microscopy, *Histochemistry* **64**:115–118.

Horisberger, M., Rosset, J., and Bauer, H., 1975, Colloidal gold granules as markers for cell surface receptors in the scanning electron microscope, *Experientia* **31**:1147–1149.

Horisberger, M., Farr, D. R., and Vonlanthen, M., 1978, Ultrastructural localization of β-D-galactan in the nuclei of the myxomycete *Physarum polycephalum, Biochim. Biophys. Acta* **542**:308–314.

Hsu, K. C., 1962, Ferritin-labelled antigens and antibodies, in: *Methods in Immunology and Immunochemistry,* Vol. I (C. A. Williams, and M. W. Chase, eds.) pp. 397–404, Academic Press, New York.

Jackowski, S., Edwards, H. H., Davis, D., and Rock, C. O., 1985, Localization of acyl carrier protein in *Escherichia coli, J. Bacteriol.* **162**:5–8.

Keller, G.-A., Tokuyasu, K. T., Dutton, A. H., and Singer, S. J., 1984, An improved procedure for immunoelectron microscopy: Ultrathin plastic embedding of immunolabeled ultrathin frozen sections, *Proc. Natl. Acad. Sci. USA* **81**:5744–5747.

Kent, S. P., and Allen, F. B., 1981, Antigen-coated gold particles containing radioactive gold in the demonstration of cell surface molecules, *Histochem. J.* **72**:83–90.

Kent, S. P., and Wilson, D. V., 1975, Polysaccharides as labels for antibodies in electron microscopy, *J. Histochem. Cytochem.* **23**:169–173.

Kondo, H., 1984, Polyethylene glycol (PEG) embedding and subsequent de-embedding as a method for the structural and immuocytochemical examination of biological specimens by electron microscopy, *J. Electron Microsc. Tech.* **1**:227–241.

Korpela, J., Salonen, E. M., Kuusela, P., Sarvas, M., and Vaheri, A., 1984, Binding of avidin to bacteria and to the other membrane porin of *Escherichia coli*, *FEMS* **22**:3–10.

Larsson, L.-I., 1979, Simultaneous ultrastructural demonstration of multiple peptides in endocrine cells by a novel immunocytochemical method, *Nature (London)* **282**:743–746.

Levine, M. M., Ristaino, P., Marley, G., Smyth, C., Knutton, S., Boedeker, E., Black, R., Young, C., Clements, M. L., Cheney, C., and Patnaik, R., 1984, Coli surface antigens 1 and 3 of colonization factor antigen II-positive enterotoxigenic *Escherichia coli:* Morphology, purification, and immune responses in humans, *Infect. Immun.* **44**:409–420.

Mace, M. L., Jr., Van, N. T., and Conn, M. P., 1977, Electron microscopic localization of DNA-dependent RNA polymerase binding sites on DNA using enzyme immobilized on colloidal gold, *Cell Biol. Int. Rep.* **1**:527–534.

Mannweiler, K., Hohenberg, H., Bohn, W., and Rutter, G., 1982, Protein A gold particles as markers in replica immunocytochemistry: High resolution electron microscope investigations of plasma membrane surfaces, *J. Microsc. (Oxford)* **126**:145–149.

Moeremans, M., Daneels, G., Van Dijck, A., Langanger, G., and De Mey, J., 1984, Sensitive visualization of antigen–antibody reactions in dot and blot immune overlay assays with immunogold and immunogold/silver staining, *J. Immunol. Methods* **74**:353–360.

Montesano, R., Roth, J., Robert, A., and Orci, L., 1982, Non-coated membrane invaginations are involved in binding and internalization of cholera and tetanus toxins, *Nature (London)* **296**:651–653.

Morris, R. E., and Saelinger, C. B., 1984, Visualization of intracellular trafficking: Use of biotinylated ligands in conjunction with avidin–gold colloids, *J. Histochem. Cytochem.* **32**:124–128.

Muhlpfordt, H., 1982, The preparation of colloidal gold particles using tannic acid as an additional reducing agent, *Experientia* **38**:1127–1128.

Muhlradt, P. F., Menzel, J., Golecki, J. R., and Speth, V., 1973, Outer membrane of Salmonella: Sites of export of newly synthesized lipopolysaccharide on the bacterial surface, *Eur. J. Biochem.* **35**:471–481.

Muller, L. L., and Jacks, T. J., 1975, Rapid chemical dehydration of samples for electron microscopic examination, *J. Histochem. Cytochem.* **23**:107–110.

Peters, K. R., Rutter, G., Gschwenden, H. H., and Heller, W., 1978, Derivatized silica spheres as immunospecific markers for high resolution labelling in electron microscopy, *J. Cell Biol.* **78**:309–318.

Pinto da Silva, P., and Kan, F. W. K., 1984, Label-fracture: A method for high resolution labeling of cell surfaces, *J. Cell Biol.* **99**:1156–1161.

Polak, J. M., Kendall, P. A., Heath, C. M., and Pearse, A. G. E., 1972, Carbodiimide fixation for electron microscopy and immunoelectron cytochemistry, *Experientia* **28**:368–370.

Pratten, M. K., Williams, K. E., and Lloyd, J. B., 1977, A quantitative study of pinocytosis and intracellular proteolysis in rat peritoneal macrophages, *Biochem. J.* **168**:365–372.

Rembaun, A., and Dreyer, W. A., 1980, Immunomicrospheres: Reagents for cell labelling and separation, *Science* **208**:364–368.

Robenek, H., and Hesz, A., 1983, Dynamics of low density lipoprotein receptors in the plasma membrane of cultured human skin fibroblasts as visualized by colloidal gold in conjunction with surface replicas, *Eur. J. Cell Biol.* **31**:275–282.

Robinson, E. N., Jr., McGee, Z. A., Kaplan, J., Hammond, E. M., Larson, J. K., Buchanan, T. M., and Schoolnik, G. K., 1984, Ultrastructural localization of specific gonococcal macromolecules with antibody–gold sphere immunological probes, *Infect. Immun.* **46**:361–366.

Rohde, M., Mayer, F., and Meyer, O., 1984, Immunocytochemical localization of carbon monoxide oxidase in *Pseudomonas carboxydovorans:* The enzyme is attached to the inner aspect of the cytoplasmic membrane, *J. Biol. Chem.* **259**:14788–14792.

Romano, E. L., and Romano, M., 1977, Staphylococcal protein A bound to colloidal gold: A useful reagent to label antigen–antibody sites in electron microscopy, *Immunochemistry* **14**:711–715.

Roth, J., 1982a, The preparation of protein A–gold complexes with 3 nm and 15 nm gold particles and their use in labelling multiple antigens on ultra-thin sections, *Histochem. J.* **14**:791–801.

Roth, J., 1982b, Applications of immunocolloids in light microscopy: Preparation of protein A–silver and protein A–gold complexes and their application for localization of single and multiple antigens in paraffin sections, *J. Histochem. Cytochem.* **30**:691–696.

Roth, J., 1983a, Applications of immunocolloids in light microscopy. II. Demonstration of lectin-binding sites in paraffin sections by the use of lectin–gold or glycoprotein–gold complexes, *J. Histochem. Cytochem.* **31**:547–552.

Roth, J., 1983b, Application of lectin–gold complexes for electron microscopic localization of glycoconjugates on thin sections, *J. Histochem. Cytochem.* **31**:987–999.

Roth, J., and Binder, M., 1978, Colloidal gold, ferritin and peroxidase as markers for electron microscopic double labeling lectin techniques, *J. Histochem. Cytochem.* **26**:163–169.

Roth, J., Lucocq, J. M., and Charest, P. M., 1984, Light and electron microscopic demonstration of sialic acid residues with the lectin from *Limax flavus:* A cytochemical affinity technique with the use of fetuin–gold complexes, *J. Histochem. Cytochem.* **32**:1167–1176.

Schenkman, S., Couture, E., and Schwartz, M., 1983, Monoclonal antibodies reveal LamB antigenic determinants on both faces of the *Escherichia coli* outer membrane, *J. Bacteriol.* **155**:1382–1392.

Schwab, M. E., and Thoenen, H., 1978, Selective binding, uptake and retrograde transport of tetanus toxin by nerve terminals in the rat iris, *J. Cell Biol.* **77**:1–13.

Singer, S. J., and McLean, I. D., 1963, Ferritin-antibody conjugates as stain for electron microscopy, *Lab. Invest.* **12**:1002–1008.

Singer, S. J., and Schick, A. I., 1961, The properties of specific stains for electron microscopy prepared by conjugation of antibody with ferritin, *J. Biophys. Cytol.* **9**:519.

Slot, J. W., and Geuze, H. J., 1981, Sizing of protein A–colloidal gold probes for immunoelectron microscopy, *J. Cell Biol.* **90**:533–536.

Smit, J., and Agabian, N., 1982, Cell surface patterning and morphogenesis: Biogenesis of a periodic surface array during *Caulobacter* development, *J. Cell Biol.* **95**:41–49.

Smit, J., and Nikaido, H., 1978, Outer membrane of gram-negative bacteria. XVIII. Electron microscopic studies on porin insertion sites and growth of cell surface of *Salmonella typhimurium,* *J. Bacteriol.* **135**:687–702.

Smit, J., Hermodson, M., and Agabian, N., 1981, *Caulobacter crescentus* pilin: Purification, chemical characterization, and NH_2-terminal amino acid sequence of a structural protein regulated during development, *J. Biol. Chem.* **256**:3092–3097.

Sternberger, L. A., 1967, Electron microscope immunocytochemistry: A review, *J. Histochem. Cytochem.* **15**:139–159.

Sternberger, L. A., Hardy, P. H., Cuculis, J. S., and Meyer, H. G., 1970, The unlabelled antibody–enzyme method of immunocytochemistry: Preparation and properties of soluble antigen–antibody complex (horseradish peroxidase–antiperoxidase) and its role in identification of spirochetes, *J. Histochem. Cytochem.* **18**:315.

Takamiya, H., Batsford, S., Geldenblom, H., and Vogt, A., 1979, Immunoelectron microscopic localization of lipopolysaccharide antigens on ultrathin sections of *Salmonella typhimurium, J. Bacteriol.* **140**:261–266.

Tapia, F. J., Varndell, I. M., Probert, L., De Mey, J., and Polak, J. M., 1983, Double immunogold staining method for the simultaneous ultrastructural localization of regulatory peptides, *J. Histochem. Cytochem.* **31**:977–981.

Tinglu, G., Ghosh, A., and Ghosh, B. K., 1984, Subcellular localization of alkaline phosphatase in *Bacillus licheniformis* 749/C by immunoelectron microscopy with colloidal gold, *J. Bacteriol.* **159**:668–677.

Tokuyasu, K. T., 1984, Immuno-cryoultramicrotomy, in: *Immunolabelling for Electron Microscopy* (J. M. Polak and I. M. Varndell, eds.), pp. 71–82, Elsevier, Amsterdam.

Tolson, N. D., Boothroyd, B., and Hopkins, C. R., 1981, Cell surface labelling with gold colloid particulates: The use of avidin and staphylococcal protein A-coated gold in conjunction with biotin and Fc-bearing ligands, *J. Microsc. (Oxford)* **123:**215–226.

Turkevich, J., Stevenson, P. C., and Hillier, J., 1955, A study of the nucleation and growth processes in the synthesis of colloidal gold, *Discuss. Faraday Soc.* **11:**55–75.

Vos-Scheperkeuter, G. H., Pas, E., Brakenhoff, G. J., Nanninga, N., and Witholt, B., 1984, Topography of the insertion of LamB protein into the outer membrane of *Escherichia coli* wild-type and *lac-lamB* cells, *J. Bacteriol.* **159:**440–447.

Walker, P. D., and Beesley, J. E., 1982, Trends in the localization of bacterial antigens by immuno-electron microscopy, *Ann. N.Y. Acad. Sci.* **420:**411–421.

Willingham, M. C., and Yamada, S. S., 1979, Development of a new primary fixative for electron microscopic immunocytochemical localization of intracellular antigens in cultured cells, *J. Histochem. Cytochem.* **27:**947–960.

Winlove, C. P., Davis, J., Iacovides, A., and Chabanel, A., 1981, Radioactive gold colloid as a tracer of macromolecule transport, *Biorheology* **18:**569–578.

Wolosewick, J. J., 1980, The application of polyethylene glycol (PEG) to electron microscopy, *J. Cell Biol.* **86:**675–681.

Wolosewick, J. J., De Mey, J., and Meininger, V., 1983, Ultrastructural localization of tubulin and actin in polyethylene glycol-embedded rat seminiferous epithelium by immunogold staining, *J. Cell Biol.* **49:**219–224.

Chapter 18

X-Ray Microanalysis

H. C. Aldrich

Department of Microbiology and Cell Science
University of Florida
Gainesville, Florida 32611

1. INTRODUCTION

About 10 years ago, a new generation of computer-assisted X-ray spectrometers became available and virtually revolutionized elemental microanalysis with scanning and transmission electron microscopes. Prior to that time, considerable training and experience were required before usable results could be obtained. With the new minicomputer-based systems, biologists can obtain usable results with much less effort, and the technique is thereby more accessible to the average researcher and requires less time investment.

Energy-dispersive X-ray analysis of the elemental composition of biological specimens can be accomplished routinely on standard transmission and scanning electron microscopes. In some cases quantitation is even possible. Practical limitations have included problems with specimen preparation and the inability to detect elements lower in atomic number than sodium. Now, however, those problems largely have been overcome. In this chapter, I will briefly review the capabilities and limitations of the available instrumentation, specimen preparation methodology, and summarize the prospects for this important field.

2. PRINCIPLES

When the electron beam interacts with a specimen in a transmission or scanning microscope, several types of radiation are emitted as a result: (1) primary backscattered electrons, (2) secondary electrons, (3) Auger electrons, (4) cathodoluminescence, and (5) X-rays (Goldstein *et al.*, 1981). The first and second can be used to produce morphological images. The third and fourth are useful mainly for physical science applications. The fifth, X-rays, may be captured and counted with either a wavelength-dispersive spectrometer or an energy-dispersive spectrometer to determine the elements present in the specimen. Spatial resolution can be as good as 10 nm under ideal conditions, although 100 nm is a more realistic expectation. Wavelength-dispersive instruments can only measure one elemental energy peak with each detector during a given time period, although several spectrometers on the same microscope may be connected to a minicomputer for simultaneous counts in more than one channel. An example of the application of this technique is the work of de Chastellier and Ryter (1981), in which calcium was localized on the inner side of the plasma membrane in *Dictyostelium*. The expense and technical problems with this approach, combined with the advent of solid-state silicon–lithium detectors for energy-dispersive spectrometers, have led to the choice of the latter for most biological systems.

The unit consists of (1) a detector, usually with a beryllium window covering the solid-state silicon chip, (2) a preamplifier, (3) a pulse processor, (4) an analog-to-digital convertor, and (5) a multichannel analyzer. These are interfaced to a minicomputer, usually with a color display terminal. Images may be recorded with a camera from the cathode-ray tube of the terminal, by routing output to a plotting device, or on magnetic media such as tape, floppy disk, or hard disk drive. Data recorded on magnetic media have the advantage that they can be recalled at a later date and manipulated for optimum presentation as desired. The computer usually contains ROM (read-only memory) chips holding comparative data on known elemental spectra, background subtraction programs, mapping routines, and other useful functions. A bare-bones qualitative unit can usually be upgraded for quantitative work, digital image processing, and larger capacity data storage as the user's needs change. In fact, detectors from one company can often be interfaced to the analytical computer of another company if desired at a later date. Some users have found it advantageous to interface a single computer to two detectors, one each on a transmission microscope and a scanning microscope.

The usual operating routine would be to examine a specimen in conventional ways until a region is identified for analysis. The detector remains retracted at the column wall until needed for counting. The specimen is then tilted, if necessary, and the detector racked to within a few millimeters of the specimen. The counting system is then activated and counts are collected for one to several minutes. Numbers of pulses are displayed in histogram fashion on the terminal display. During or after counting, spectra of known elements can be superimposed over the accumulated counts, until unknown spectra match known peaks. The spectrum can be saved on a magnetic medium, photographed with Polaroid or 35-mm cameras, or output to a plotter.

3. INSTRUMENTS AND THEIR CAPABILITIES

In scanning microscopes, few problems of detector/specimen geometry are encountered, due to the large specimen chambers on these microscopes. Detectors are mounted in preexisting ports, usually to the side and slightly above the specimen stage. However, on transmission instruments, the specimen is inserted into the center of the objective polepiece, creating problems of access for an energy-dispersive detector. These problems are alleviated somewhat by mounting the specimen in a goniometer holder capable of tilting, although top entry holders can be employed. Some microscope manufacturers insert the X-ray detector from the top of the objective polepiece, but in these cases the detector cannot approach nearer the specimen than 20 or 30 mm. Other manufacturers insert the detector through the side of the objective polepiece, but this necessitates tilting the specimen as high as 45° to give the detector clear access to the grid. This type of detector can approach within 2 or 3 mm of the specimen, depending on the design of the specimen holder. This should theoretically result in greater sensitivity for this type of arrangement.

Most scanning microscopes are or can be equipped with accessories to limit the size and/or shape of the area being examined, so that X-rays are collected from a reduced area of the operator's choice. A single scan line one frame wide can also be selected, as can a single point equal to the size of the scanning spot, usually about 10 nm in diameter. These same choices are available when using a transmission instrument equipped with a scanning accessory unit (STEM). In such a modification, the minimum spot size attainable is again about 10 nm in scanning mode. Furthermore, the minimum spot size available in conventional transmission EM (CTEM) mode is smaller than that without the STEM unit installed, since the STEM installation alters the condenser lens program in CTEM mode as well. In any case, the researcher using a combined CTEM and STEM instrument can choose between the two modes when performing energy-dispersive X-ray analyses. In our experience, bulk specimens can be successfully analyzed in secondary electron imaging mode with the STEM unit, using full-frame scan, reduced area scan, or spot modes. However, with conventionally processed epoxy sections, sufficient X-ray count rate is difficult to attain in STEM mode. We have had more success reducing the condenser spot size to about 1 μm in CTEM mode and simply counting one spot for 2 to 3 min. It may well be that with freeze-dried, freeze-substituted, or freeze-hydrated specimens, scanning mode may be more profitably employed.

Low-molecular-weight elements can now be detected, using three types of technology. Ultrathin window detectors can allow elements including and above carbon to be analyzed. Second, windowless detectors are also available, but these require special precautions to minimize contamination. Third, an electron energy-loss spectrometer (EELS) detector may be placed at the base of the TEM column, under the camera chamber (Shuman *et al.*, 1981; Johnson, 1981). This detector works on an entirely different principle than that of energy-dispersive X-ray analysis. It depends on sensing loss of energy by primary electrons as they pass through a thin section during transmission microscopy, and can provide some information about chemical bonding states as well as identification of light elements (below sodium). Only a few studies have been published applying this technique to biological material (see Hutchinson and Somlyo, 1981). The reader is advised to look for additional studies of this type during the next few years. The technique has considerable promise.

4. SPECIMEN PREPARATION

Because the analytical technology has advanced so far, now the limiting factor in achieving satisfactory results is frequently specimen preparation. It is important to remember that any processing step in sample preparation can be a negative effect on the elemental analysis later. Use of any liquid, polar or nonpolar, for fixation, dehydration, or embedding, can extract diffusible and nondiffusible elements from the sample. Critical point-drying for SEM is risky. At the very least, air-dried samples and freeze-fried samples should also be analyzed for comparison. Metal specimen supports can also contribute undesirable elements to the spectrum, especially if one is analyzing for elements near aluminum, zinc, or copper. Gluing a carbon planchet on top of the aluminum SEM stub and placing the specimen atop that will usually cure or minimize this problem.

For TEM, the problems are more complex. One would often like to examine ultrathin sections to determine intracellular elemental distributions. In a few instances, such as polyphosphate inclusions in bacteria (Jensen *et al.*, 1977, 1982; Baxter and Jensen, 1980; Scherer and Bochem, 1983; Sicko-Goad *et al.*, 1975), it has been possible to accomplish this on epoxy-embedded material, providing aldehyde fixation alone is employed. This kind of inclusion is sufficiently insoluble that it resists being leached out during fixation, dehydration, and embedding. We recently found calcium and phosphorus in similar inclusions in a eukaryotic cell type (Figures 1 and 2), the zoospores of the fungi *Blastocladiella* and *Allomyces* (Aldrich *et al.*, 1984). By comparing spectra from sections to spectra from whole-mounted, air-dried cells (Figure 3), however, it was evident that some leaching had still occurred. Much stronger signal was obtained from the air-dried cells. Another problem was identified by Jensen *et al.* (1977), who found that uranyl and lead poststains can leach out all or part of polyphosphate inclusions from epoxy sections. In another study, Balkwill *et al.* (1980) used the technique to localize iron in electron-dense granules in a magnetotactic bacterium. Thus, it appears that insoluble phosphates and other compounds that resist leaching can be localized in sections of microbes with energy-

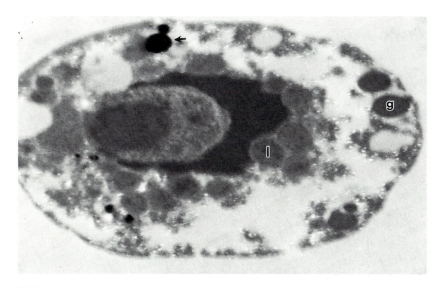

FIGURE 1. Glutaraldehyde-fixed, thin-sectioned, unstained zoospore of *Allomyces macrogynus*. Three types of globular cytoplasmic inclusions are visible: lipids (1), gamma particles (g), and a polyphosphate granule (arrow). ×20,000.

dispersive techniques. The only caveat to remember is that electron-dense stains such as osmium, uranyl, and lead should usually be avoided so as not to risk leaching other compounds during the stain procedures. Furthermore, these elements may also exhibit energy peaks near other elements of potential interest and mask the presence of low concentrations of naturally occurring elements.

Often the elements of more interest are the diffusible ions, such as calcium, potassium, phosphorus, and magnesium. More elaborate preparative protocols must be employed to retain these and many other elements *in situ* for detection. These include cryoultrathin sectioning and freeze-substitution. Recent improvements in cryosectioning attachments for the LKB, Sorvall, and Reichert ultramicrotomes have solved most of the problems inherent in the sectioning process itself. The remaining problems usually arise when freezing cells, since unfixed, uncryoprotected preparations must be used to avoid any possibilities of leaching with aqueous solutions. The reader is referred to the recent reviews of cryoultrathin sectioning (Hagler *et al.*, 1981; Hagler and Buja, 1984; Parsons *et al.*, 1984; Saubermann, 1981) for details of the techniques. Freezing protocols have been adequately covered in Chapters 8 and 9 of this volume. We have found a sequence developed by Steinbrecht and Zierold (1984) useful. This includes freezing cell suspensions in a propane jet at −178°C, consolidation of material from several such specimens into a Balzers gold specimen holder containing liquid *n*-heptane at −85°C in the cryoultramicrotome work chamber, and finally

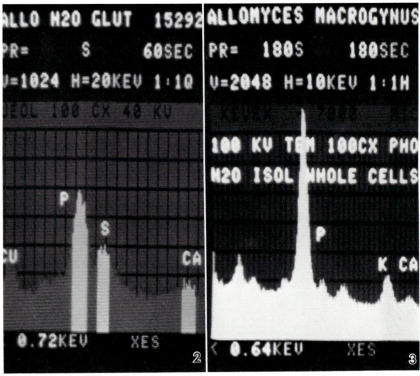

FIGURE 2. Energy-dispersive X-ray spectrum from polyphosphate granule in zoospore shown in Figure 1. Other cytoplasmic inclusions gave little or no signal.

FIGURE 3. Energy-dispersive X-ray spectrum from whole-mounted, air-dried zoospore like that shown in Figure 1. Compare with Figure 2. Note that peaks are more pronounced here, and that the potassium peak here was undetectable in fixed material, probably due to leaching during fixation.

sectioning the sample below −92°C, the freezing point of *n*-heptane, e.g., at −120°C.

Once frozen ultrathin sections are obtained, one can either freeze-dry the sections between layers of Formvar in sandwich grid assemblies (Monson and Hutchinson, 1981), or examine them in the hydrated state, using a cold transfer device and specimen holder to insert the cold sections into and maintain them within the microscope column for analysis (Dubochet *et al.*, 1983; Dubochet and McDowall, 1984). Once there, they may be etched by the microscope vacuum or examined hydrated by keeping the specimen temperature low enough to minimize sublimation of water from the section. Frozen-hydrated *Bacillus cereus* spores were analyzed very successfully by Stewart *et al.* (1980). Calcium, mag-

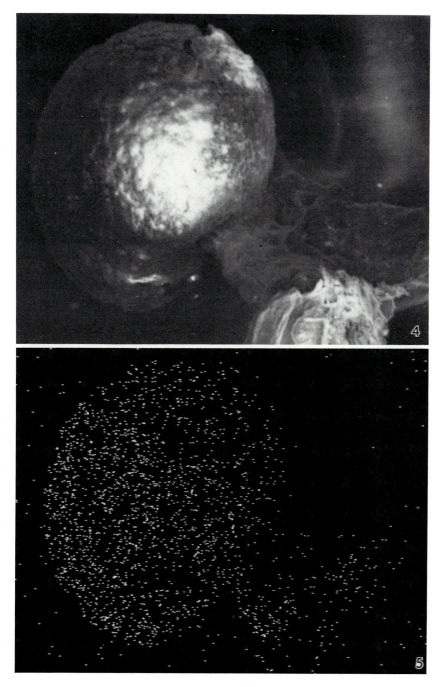

FIGURE 4. Secondary electron image of sporangium of the slime mold *Physarum*. ×100.

FIGURE 5. Calcium map of the specimen shown in Figure 4. White dots indicate locations from which calcium X-rays emanated. Entire sporangium has calcium on its surface.

nesium, manganese, and phosphorus were located in the central core of these spores. Readers interested in this particular technique are urged to examine this excellent paper in detail; it is an outstanding example of microbial applications of energy-dispersive X-ray analysis.

Freeze-substitution is somewhat less forbidding technically (see Chapter 7), and Ornberg and Reese (1981) have successfully localized calcium and other elements in mammalian tissue after this protocol.

Mapping of the locations of particular elements can be easily accomplished with most SEM and STEM instruments. One has only to define a window corresponding to the energy peaks for the element of interest, interface the mapping output from the computer to the SEM or STEM circuitry, and scan the specimen slowly while counting. White dots appear on the screen corresponding to the location of the element of interest on the specimen (Figures 4 and 5).

5. RECOMMENDATIONS AND OUTLOOK

Energy-dispersive X-ray analysis is now established as a feasible, useful technique for localization of elements in microbial cells (Holt and Beveridge, 1982). Specimen preparation still presents some technical problems, but these appear to be controllable. Some excellent papers are beginning to appear analyzing animal, plant, and microbial systems. Some experience can save time and effort, so we recommend that the neophyte collaborate with other investigators equipped to perform the techniques, at least until he or she is able to judge whether the value of the technique in his/her case justifies the additional expenditure of funds necessary to acquire the equipment. Microscope manufacturers maintain applications labs and are often willing to make their equipment available on a trial basis for investigators who need some preliminary data to justify instrument upgrading or acquisition.

6. REFERENCES

Aldrich, H. C., Barstow, W. E., and Lingle, W. L., 1984, Elemental analysis of particles in fungal zoospores, *Arch. Microbiol.* **139**:102–104.

Balkwill, D. L., Maratea, D., and Blakemore, R. P., 1980, Ultrastructure of a magnetotactic *Spirillum, J. Bacteriol.* **141**:1399–1408.

Baxter, M., and Jensen, T. E., 1980, A study of methods for in situ X-ray energy dispersive analysis of polyphosphate bodies in *Plectonema boryanum, Arch. Microbiol.* **126**:213–215.

de Chastellier, C., and Ryter, A., 1981, Calcium dependent deposits at the plasma membrane of *Dictyostelium discoideum* and their possible relation with contractile proteins, *Biol. Cell.* **40**:109–118.

Dubochet, J., and McDowall, A. W., 1984, Frozen hydrated sections, in: *The Science of Biological Specimen Preparation for Microscopy and Microanalysis* (J.-P. Revel, T. Barnard, and G. H. Haggis, eds.), pp. 147–152, SEM, Inc., Chicago.

Dubochet, J., McDowall, A. W., Menge, B., Schmid, E. N., and Lickfeld, K. G., 1983, Electron microscopy of frozen-hydrated bacteria, *J. Bacteriol.* **155**:381–390.

Goldstein, J. I., Newbury, D. E., Echlin, P., Joy, D. C., Fiori, C., and Lifshin, E., 1981, *Scanning Electron Microscopy and X-Ray Microanalysis*, Plenum Press, New York.

Hagler, H. K., and Buja, L. M., 1984, New techniques for the preparation of thin freeze dried cryosections for X-ray microanalysis, in: *The Science of Biological Specimen Preparation for Microscopy and Microanalysis* (J.-P. Revel, T. Barnard, and G. H. Haggis, eds.), pp. 161–166, SEM, Inc., Chicago.

Hagler, H. K., Burton, K., and Buja, L. M., 1981, Electron probe X-ray microanalysis of normal and injured myocardium: Methods and results, in: *Microprobe Analysis of Biological Systems* (T. E. Hutchinson and A. P. Somlyo, eds.), pp. 127–155, Academic Press, New York.

Holt, S. C., and Beveridge, T. J., 1982, Electron microscopy: Its development and application to microbiology, *Can. J. Microbiol.* **28**:1–53.

Hutchinson, T. E., and Somlyo, A. P. (eds.), 1981, *Microprobe Analysis of Biological Systems*, Academic Press, New York.

Jensen, T. E., Sicko-Goad, L., and Ayala, R. P., 1977, Phosphate metabolism in blue-green algae. III. The effect of fixation and post-staining on the morphology of polyphosphate bodies in *Plectonema boryanum, Cytologia* **42**:357–369.

Jensen, T. E., Rachlin, J. W., Jani, V., and Warkentine, B., 1982, An X-ray energy-dispersive study of cellular compartmentalization of lead and zinc in *Chlorella saccharophila* (Chlorophyta), *Navicula incerta* and *Nitschia closterium* (Bacillariophyta), *Environ. Exp. Bot.* **22**:319–328.

Johnson, D. E., 1981, Limitations to the sensitivity of energy-loss spectrometry, in: *Microprobe Analysis of Biological Systms* (T. E. Hutchinson and A. P. Somlyo, eds.), pp. 351–353, Academic Press, New York.

Monson, K. L., and Hutchinson, T. E., 1981, X-ray microanalysis of freeze-dried muscle: Techniques and problems, in: *Microprobe Analysis of Biological Systems* (T. E. Hutchinson and A. P. Somlyo, eds.), pp. 157–176, Academic Press, New York.

Ornberg, R., and Reese, T. S., 1981, Quick freezing and freeze substitution for X-ray microanalysis of calcium, in: *Microprobe Analysis of Biological Systems* (T. E. Hutchinson and A. P. Somlyo, eds.), pp. 213–228, Academic Press, New York.

Parsons, D., Bellotto, D. J., Schulz, W. W., Buja, M., and Hagler, H. K. 1984, Towards routine cryoultramicrotomy, *EMSA Bull.* **14**:49—60.

Saubermann, A., 1981, Cryosectioning of biological tissue for X-ray microanalysis of diffusible elements, in: *Microprobe Analysis of Biological Systems* (T. E. Hutchinson and A. P. Somlyo, eds.), pp. 377–396, Academic Press, New York.

Scherer, P., and Bochem, H. P., 1983, Ultrastructural investigation of Methanosarcinae and related species grown on methanol for occurrence of polyphosphate-like bodies, *Can. J. Microbiol.* **29**:1190–1199.

Shuman, H., Somlyo, A. V., and Somlyo, A. P., 1981, Electron energy-loss analysis in biology: Application to muscle and a parallel collection system, in: *Microprobe Analysis of Biological Systems* (T. E. Hutchinson and A. P. Somlyo, eds.), pp. 273–288, Academic Press, New York.

Sicko-Goad, L. M., Crang, R. E., and Jensen, T. E., 1975, Phosphate metabolism in blue-green algae. IV. In situ analysis of polyphosphate bodies by X-ray energy dispersive analysis, *Cytobiologie* **11**:430–437.

Steinbrecht, R. A., and Zierold, K., 1984, A cryoembedding method for cutting ultrathin cryosections from small frozen specimens, *J. Microsc. (Oxford)* **136**:69–75.

Stewart, M., Somlyo, A. P., Somlyo, A. V., Shuman, S., Lindsay, J. A., and Murrell, W. G., 1980, Distribution of calcium and other elements in cryosectioned *Bacillus cereus* T spores, determined by high-resolution scanning electron probe X-ray microanalysis, *J. Bacteriol.* **143**:481–491.

Index